WITHDRAWN
UTSA LIBRARIES

ADAPTIVE SIGNAL PROCESSING

Adaptive and Learning Systems for Signal Processing, Communications, and Control

Editor: Simon Haykin

Adali and Haykin / ADAPTIVE SIGNAL PROCESSING: Next Generation Solutions

Beckerman / ADAPTIVE COOPERATIVE SYSTEMS

Candy / BAYESIAN SIGNAL PROCESSING: CLASSICAL, MODERN, AND PARTICLE FILTERING METHODS

Candy / MODEL-BASED SIGNAL PROCESSING

Chen and Gu / CONTROL-ORIENTED SYSTEM IDENTIFICATION: An $\mathcal{H}_\infty$ Approach

Chen, Haykin, Eggermont, and Becker / CORRELATIVE LEARNING: A Basis for Brain and Adaptive Systems

Cherkassky and Mulier / LEARNING FROM DATA: Concepts, Theory, and Methods

Costa and Haykin / MULTIPLE-INPUT MULTIPLE-OUTPUT CHANNEL MODELS: Theory and Practice

Diamantaras and Kung / PRINCIPAL COMPONENT NEURAL NETWORKS: Theory and Applications

Farrell and Polycarpou / ADAPTIVE APPROXIMATION BASED CONTROL: Unifying Neural, Fuzzy and Traditional Adaptive Approximation Approaches

Gini and Rangaswamy / KNOWLEDGE-BASED RADAR DETECTION: Tracking and Classification

Hänsler and Schmidt / ACOUSTIC ECHO AND NOISE CONTROL: A Practical Approach

Haykin / UNSUPERVISED ADAPTIVE FILTERING: Blind Source Separation

Haykin / UNSUPERVISED ADAPTIVE FILTERING: Blind Deconvolution

Haykin and Puthussarypady / CHAOTIC DYNAMICS OF SEA CLUTTER

Haykin and Widrow / LEAST-MEAN-SQUARE ADAPTIVE FILTERS

Hrycej / NEUROCONTROL: Towards an Industrial Control Methodology

Hyvärinen, Karhunen, and Oja / INDEPENDENT COMPONENT ANALYSIS

Kristić, Kanellakopoulos, and Kokotović / NONLINEAR AND ADAPTIVE CONTROL DESIGN

Mann / INTELLIGENT IMAGE PROCESSING

Nikias and Shao / SIGNAL PROCESSING WITH ALPHA-STABLE DISTRIBUTIONS AND APPLICATIONS

Passino and Burgess / STABILITY ANALYSIS OF DISCRETE EVENT SYSTEMS

Sánchez-Peña and Sznaier / ROBUST SYSTEMS THEORY AND APPLICATIONS

Sandberg, Lo, Fancourt, Principe, Katagiri, and Haykin / NONLINEAR DYNAMICAL SYSTEMS: Feedforward Neural Network Perspectives

Sellathurai and Haykin / SPACE-TIME LAYERED INFORMATION PROCESSING FOR WIRELESS COMMUNICATIONS

Spooner, Maggiore, Ordóñez, and Passino / STABLE ADAPTIVE CONTROL AND ESTIMATION FOR NONLINEAR SYSTEMS: Neural and Fuzzy Approximator Techniques

Tao / ADAPTIVE CONTROL DESIGN AND ANALYSIS

Tao and Kokotović / ADAPTIVE CONTROL OF SYSTEMS WITH ACTUATOR AND SENSOR NONLINEARITIES

Tsoukalas and Uhrig / FUZZY AND NEURAL APPROACHES IN ENGINEERING

Van Hulle / FAITHFUL REPRESENTATIONS AND TOPOGRAPHIC MAPS: From Distortion- to Information-Based Self-Organization

Vapnik / STATISTICAL LEARNING THEORY

Werbos / THE ROOTS OF BACKPROPAGATION: From Ordered Derivatives to Neural Networks and Political Forecasting

Yee and Haykin / REGULARIZED RADIAL BIAS FUNCTION NETWORKS: Theory and Applications

ADAPTIVE SIGNAL PROCESSING

Next Generation Solutions

Edited by

Tülay Adalı

Simon Haykin

Wiley Series in Adaptive and Learning Systems for Signal Processing, Communication, and Control

The Institute of Electrical and Electronics Engineers, Inc., New York

A JOHN WILEY & SONS, INC., PUBLICATION

Copyright © 2010 by John Wiley & Sons, Inc. All rights reserved

Published by John Wiley & Sons, Inc., Hoboken, New Jersey
Published simultaneously in Canada

Cover photo: Humuhumunukunukuāpua'a, Hawaiian state fish.
Photo taken April 2007 in Honolulu, Hawaii during ICASSP 2007 where the idea for the book was first conceived. Photo copyright © 2010 by Tülay Adalı.

No part of this publication may be reproduced, stored in a retrieval system, or transmitted in any form or by any means, electronic, mechanical, photocopying, recording, scanning, or otherwise, except as permitted under Section 107 or 108 of the 1976 United States Copyright Act, without either the prior written permission of the Publisher, or authorization through payment of the appropriate per-copy fee to the Copyright Clearance Center, Inc., 222 Rosewood Drive, Danvers, MA 01923, (978) 750-8400, fax (978) 750-4470, or on the web at www.copyright.com. Requests to the Publisher for permission should be addressed to the Permissions Department, John Wiley & Sons, Inc., 111 River Street, Hoboken, NJ 07030, (201) 748-6011, fax (201) 748-6008, or online at http://www.wiley.com/go/permission.

Limit of Liability/Disclaimer of Warranty: While the publisher and author have used their best efforts in preparing this book, they make no representations or warranties with respect to the accuracy or completeness of the contents of this book and specifically disclaim any implied warranties of merchantability or fitness for a particular purpose. No warranty may be created or extended by sales representatives or written sales materials. The advice and strategies contained herein may not be suitable for your situation. You should consult with a professional where appropriate. Neither the publisher nor author shall be liable for any loss of profit or any other commercial damages, including but not limited to special, incidental, consequential, or other damages.

For general information on our other products and services or for technical support, please contact our Customer Care Department within the United States at (800) 762-2974, outside the United States at (317) 572-3993 or fax (317) 572-4002.

Wiley also publishes its books in a variety of electronic formats. Some content that appears in print may not be available in electronic formats. For more information about Wiley products, visit our web site at www.wiley.com.

Library of Congress Cataloging-in-Publication Data:

Adaptive signal processing : next generation solutions / [edited by] Tülay Adalı, Simon Haykin.
p. cm.
Includes bibliographical references.
ISBN 978-0-470-19517-8 (cloth)
1. Adaptive signal processing. I. Adalı, Tülay II. Haykin, Simon, 1931–
TK5102.9.A288 2010
621.382′2—dc22 2009031378

Printed in the United States of America

10 9 8 7 6 5 4 3 2 1

Library
University of Texas
at San Antonio

CONTENTS

Chapter 6 Nonlinear Sequential State Estimation for Solving Pattern-Classification Problems 333

Chapter 7 Bandwidth Extension of Telephony Speech 349

Index 393

PREFACE

WHY THIS NEW BOOK?

Adaptive filters play a very important role in most of today's signal processing and control applications as most real-world signals require processing under conditions that are difficult to specify *a priori.* They have been successfully applied in such diverse fields as communications, control, radar, and biomedical engineering, among others. The field of classical adaptive filtering is now well established and a number of key references—a widely used one being the book *Adaptive Filter Theory* by Simon Haykin—provide a comprehensive treatment of the theory and applications of adaptive filtering.

A number of recent developments in the field, however, have demonstrated how significant performance gains could be achieved beyond those obtained using the standard adaptive filtering approaches. To this end, those recent developments have propelled us to think in terms of a new generation of adaptive signal processing algorithms.

As data now come in a multitude of forms originating from different applications and environments, we now have to account for the characteristics of real life data:

- Non-Gaussianity;
- Noncircularity;
- Nonstationarity; and
- Nonlinearity.

Such data would typically exhibit a rich underlying structure and demand the development of new tools, hence, the writing of this new book.

ORGANIZATION OF THE BOOK

The book consists of seven chapters that are organized in five subsections as follows.

Fundamental Issues: Optimization, Efficiency, and Robustness in the Complex Domain

Chapter 1 by Adalı and Li and Chapter 2 by Ollilla and Koivunen constitute the first subsection of the book.

The first chapter of the book addresses the key problem of optimization in the complex domain, and fully develops a framework that enables taking full advantage of the power of complex-valued processing. The fundamental relationships for the derivation and analysis of adaptive algorithms in the complex domain are established based on Wirtinger calculus, and their successful application is demonstrated for two basic problems in adaptive signal processing: filtering and independent component analysis (ICA). Two important classes of filters, namely, the widely linear and nonlinear filters are studied as well as the two main approaches for performing ICA, maximum likelihood and maximization of non-Gaussianity. In the design of these solutions, the emphasis is placed on taking the full statistical information into account as well as the choice of nonlinear functions for the efficient use of information. It is shown that the framework based on Wirtinger calculus naturally addresses both of these considerations, and besides significantly simplifying the derivations and analyses, also eliminates the need for many restrictive assumptions such as circularity of signals and extends the power of many convenient tools in analysis introduced for the real-valued case to complex-valued signal processing.

The second chapter in this subsection addresses the problem of multichannel processing of complex-valued signals in cases where the underlying ideal assumptions on signal and noise models are not necessarily true or there are observations that highly deviate from the mean. Specifically, estimation techniques, which are robust both to deviations from the commonly invoked circularity and/or Gaussianity assumption and to outliers, are developed and analyzed. The methods are based on matrix-valued statistics such as M-estimators of scatter and pseudo-scatter matrices. The robustness and statistical efficiency of the methods are illustrated in several applications such as beamforming, direction-of-arrival estimation, separation of sources, detection of circularity or the number of sources. Both numerical simulations as well as analytical results are provided, employing the widely used concepts of influence function and asymptotic efficiency.

Turbo Signal Processing for Equalization

This section consists of Chapter 3 written by Regalia.

Turbo processing aims to combine receiver components via information exchange for performance enhancements, as a means of joint optimization. This chapter reviews the basic principles of turbo equalization, in which the traditionally separate techniques of channel equalization and error correction decoding are combined in an iterative loop. Various schemes are treated, including maximum *a posterior* channel estimation, decision feedback channel equalizers, blind turbo algorithms in which the channel coefficients are estimated as part of the iterative procedure, and three-level turbo schemes using differential encoding. Numerous examples are included to clarify the operation and performance features of the various constituent elements, and convergence tools and extensions to multi-user channels are outlined as well.

Tracking in the Subspace Domain

The third section of the book consists of Chapter 4 by Delmas.

Research in subspace and component-based techniques originated in statistics in the middle of the last century through the problem of linear feature extraction solved by the Karhunen–Loëve transform. It has been applied to signal processing about three decades ago and the importance of using the subspace and component-based methodology has been demonstrated by many examples in data compression, data filtering, parameter estimation, and pattern recognition. The main reason for the interest in subspace and component-based methods stems from the fact that they consist in splitting the observations into a set of desired and interfering components, and as such, they not only provide new insight into many problems, but also offer a good tradeoff between achieved performance and computational complexity. Over the past few years, new potential applications have emerged, and subspace and component methods have been adopted in new diverse areas such as smart antennas, sensor arrays, multiuser detection, speech enhancement, and radar systems to name but a few. These new applications have also underlined the importance of the development of adaptive procedures as well as ways to handle nonstationarity. In this chapter on tracking in the subspace domain, the emphasis is on the class of low-complexity decompositions for dominant and minor subspaces, and dominant and minor eigenvector tracking, while other important more classical schemes are also discussed. The algorithms are derived using different iterative procedures based on linear algebraic methods, and it is shown that the majority of these algorithms can be viewed as heuristic variations of the power method.

Nonlinear Sequential State Estimation

This subsection of the book consists of Chapter 5 by Djuric and Bugallo and Chapter 6 by Haykin and Arasaratnam.

Particle filtering belongs to a more recent generation of sequential processing methods, where the objectives are not simply the tracking of unknown states from noisy observations but also the estimation of the complete posterior distributions of the states. This is achieved by clever approximations of the posterior distributions with discrete random measures. Chapter 5 presents the essentials of particle filtering, provides details about the implementation of various types of particle filters, and demonstrates their use through various examples. It also provides some more advanced concepts including Rao–Blackwellization, smoothing, and discussion on convergence and computational issues.

The second chapter in this section addresses a novel application of the extended Kalman filter (EKF) for the training of a neural network (e.g., multilayer perceptron) to solve difficult pattern recognition problems. To be specific, the training process is viewed as a nonlinear sequential estimation problem, for which the EKF is well suited.

The chapter also includes an experimental comparison of this novel approach with the classic back-propagation algorithm and the mathematically elegant approach involving the support vector machine.

Speech—Bandwidth Extension

This last subsection of the book consists of Chapter 7 by Iser and Schmidt.

In this chapter, the authors provide an introduction to bandwidth extension of telephony speech by means of adaptive signal processing. They discuss why current telephone networks apply a limiting band pass, what kind of band pass is used, and what can be done to (re)increase the bandwidth on the receiver side without changing the transmission system. Therefore, several adaptive signal processing approaches—most of them based on the source-filter model for speech generation—are discussed. The task of bandwidth extension algorithms that make use of this model can be divided into two subtasks: an excitation signal extension part and a wideband envelope estimation part. Different methods for accomplishing these tasks, like non-linear processing, the use of signal and noise generators, or modulation approaches on the one hand, and codebook approaches, linear mapping schemes, or neural networks on the other hand, are presented. The chapter concludes with a presentation of evaluation schemes for bandwidth extension approaches.

We would like to thank our authors for their timely and valuable contributions. We would also like to express our gratitude to a number of anonymous reviewers, who provided invaluable comments and suggestions for improving the material covered in all the chapters of the book. We are hoping that you will enjoy reading the book as much as we have enjoyed putting it together.

TÜLAY ADALI
SIMON HAYKIN

CONTRIBUTORS

Tülay Adalı, University of Maryland Baltimore County, Department of Computer Science and Electrical Engineering, Baltimore, MD, USA

Ienkaran Arasaratnam, McMaster University, Hamilton, ON, Canada

Monica Bugallo, Stony Brook University, Department of Electrical and Computer Engineering, Stony Brook, NY, USA

Jean-Pierre Delmas, Institut National des Télécommunications, Département Communication, Information et Traitement de l'Information, Evry, France

Petar M. Djuric, Stony Brook University, Department of Electrical and Computer Engineering, Stony Brook, NY, USA

Simon Haykin, McMaster University, Hamilton, ON, Canada

Bernd Iser, Harman/Becker Automotive Systems, Acoustics (EDA), Ulm, Germany

Visa Koivunen, Helsinki University of Technology, Helsinki, Finland

Hualiang Li, University of Maryland Baltimore County, Department of Computer Science and Electrical Engineering, Baltimore, MD, USA

Esa Ollila, Helsinki University of Technology, Helsinki, Finland

Phillip A. Regalia, Catholic University of America, Department of Electrical Engineering and Computer Science, Washington, DC, USA

Gerhard Schmidt, Harman/Becker Automotive Systems, Acoustics (EDA), Ulm, Germany

1

COMPLEX-VALUED ADAPTIVE SIGNAL PROCESSING

Tülay Adalı and Hualiang Li
University of Maryland Baltimore County, Baltimore, MD

1.1 INTRODUCTION

Complex-valued signals arise frequently in applications as diverse as communications, radar, and biomedicine, as most practical modulation formats are of complex type and applications such as radar and magnetic resonance imaging (MRI) lead to data that are inherently complex valued. When the processing has to be done in a transform domain such as Fourier or complex wavelet, again the data are complex valued. The complex domain not only provides a convenient representation for these signals but also a natural way to preserve the physical characteristics of the signals and the transformations they go through, such as the phase and magnitude distortion a communications signal experiences. In all these cases, the processing also needs to be carried out in the complex domain in such a way that the complete information—represented by the interrelationship of the real and imaginary parts or the magnitude and phase of the signal—can be fully exploited.

In this chapter, we introduce a framework based on Wirtinger calculus that enables working completely in the complex domain for the derivation and analysis of signal processing algorithms, and in such a way that all of the computations can be performed in a straightforward manner, very similarly to the real-valued case. In the derivation of

Adaptive Signal Processing: Next Generation Solutions. Edited by Tülay Adalı and Simon Haykin
Copyright © 2010 John Wiley & Sons, Inc.

adaptive algorithms, we need to evaluate the derivative of a cost function. Since the cost functions are real valued, hence not *differentiable* in the complex domain, traditionally we evaluate derivatives separately for the real and imaginary parts of the function and then combine them to form the derivative. We show that using Wirtinger calculus, we can directly evaluate the derivatives without the need to evaluate the real and imaginary parts separately. Beyond offering simple convenience, this approach makes many signal processing tools developed for the real-valued domain readily available for complex-valued signal processing as the evaluations become very similar to the real-valued case and most results from real-valued calculus do hold and can be directly used. In addition, by keeping the expressions simple, the approach eliminates the need to make simplifying assumptions in the derivations and analyses that have become common place for many signal processing algorithms derived for the complex domain.

It is important to emphasize that the regularity condition for the applicability of Wirtinger calculus in the evaluations is quite mild, making it a very powerful tool, and also widely applicable. To reiterate the two points we have made regarding the main advantages of the approach, first, algorithm derivation and analysis become much shorter and compact compared to the traditional splitting approach. In this chapter, this advantage is demonstrated in the derivation of update rules for the multilayer perceptron and the widely linear filter, and of algorithms for independent component analysis.

However, the real advantage of the Wirtinger approach is beyond simple convenience in the derivations. Because the traditional splitting approach for the real and imaginary parts leads to long and complicated expressions, especially when working with nonlinear functions and/or second-order derivatives, one is often forced to make certain assumptions to render the evaluations more manageable. One such assumption that is commonly made is the circularity of signals, which limits the usefulness of the solutions developed since many practical signals have noncircular distributions as we discuss in Section 1.2.5. Since with Wirtinger calculus, the expressions are kept simple, we can avoid such and many other simplifying assumptions allowing one to fully exploit the power of complex processing, for example, in the derivation of independent component analysis (ICA) algorithms as discussed in Section 1.6.

Besides developing the main results for the application of Wirtinger calculus, in this chapter, we demonstrate the application of the framework to a number of powerful solutions proposed recently for the complex-valued domain, and emphasize how the Wirtinger framework enables taking full advantage of the power of complex-valued processing and of these solutions in particular. We show that the widely linear filter is to be preferred when the commonly invoked circularity assumptions on the signal do not hold, and that the fully complex nonlinear filter allows efficient use of the available information, and more importantly, show how both solutions can take full advantage of the power of Wirtinger calculus. We also show that the framework enables the development and study of a powerful set of algorithms for independent component analysis of complex-valued data.

1.1.1 Why Complex-Valued Signal Processing?

Complex domain is the natural home for the representation and processing of many signals we encounter in practice. There are four main scenarios in which complex processing is needed.

- The signal can be natively complex, where an in-phase and a quadrature component is the natural representation and enables one to fully take the relationship between the two components into account. Examples include radar and MRI signal [2] as well as many communication signals such as those using binary phase shift keying (BPSK), quadrature phase shift keying (QPSK), and quadrature amplitude modulation (QAM) as shown in Figure 1.1. The MRI signal is acquired as a quadrature signal using two orthogonal detectors as shown in Figure 1.2 [17]. Hence, the complex k-space representation is the natural one for the MRI signal, which is typically inverse Fourier-transformed into the complex image space in reconstruction resulting in complex-valued spatial domain signal.
- Harmonic analysis, in particular Fourier analysis, has been one of the most widely used tools in signal processing. More recently, complex wavelet transforms have emerged as attractive tools for signal processing as well, and in all these instances where the processing has to be performed in a transform domain, one needs to perform complex-valued signal processing.
- Analytic representation of a real-valued bandpass signal using its complex envelope is commonly used in signal processing, in particular in communications. The complex envelope representation facilitates the derivation of modulation and demodulation techniques, and the analysis of certain properties of the signal.
- There are also cases where complex domain is used to capture the relationship between the magnitude and phase or two channels of real-valued signals. Examples include wind data where a complex-valued signal is constructed using the strength and direction of wind data [37] and the magnitude of structural MRI data where the white and gray matter are combined to form a complex number to make use of their interdependence in the processing of data [116].

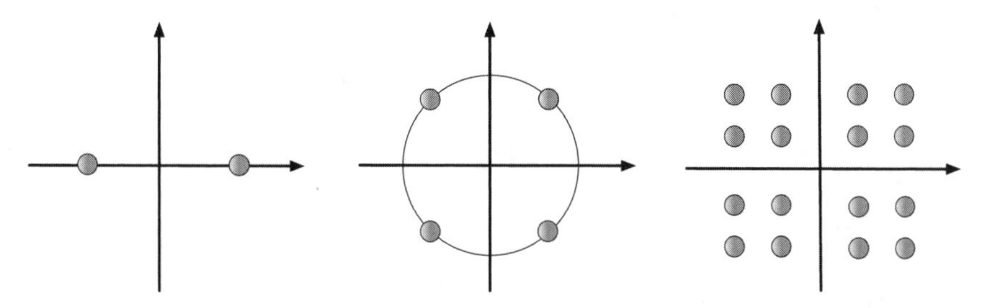

Figure 1.1 Signal constellations for BPSK, QPSK, and QAM signals.

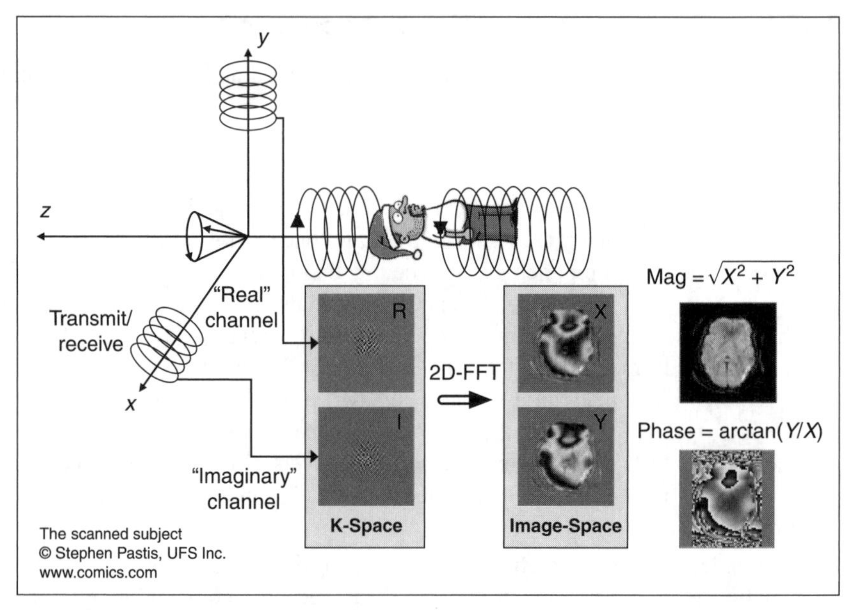

Figure 1.2 MRI signal is acquired as a quadrature signal using two orthogonal detectors, hence is inherently complex.

In all these instances, and in many similar ones, complex domain allows one to fully take advantage of the complete information in the real and imaginary channels of a given signal and thus is the natural home for the development of signal processing algorithms.

In this chapter, our focus is the description of an efficient framework such that all (or most) of the processing can be performed in the complex domain without performing transformations to and from the real domain. This point has long been a topic of debate since equivalent transformations between the two domains can be easily established, and since the real domain is the one with which we are more familiar, the question arises as to why not transform the problem into the real domain and perform all of the evaluations and analyses there. There are a number of reasons for keeping the computations and analysis in the complex domain rather than using complex-to-real transformations.

(1) Most typically, when the signal in question is complex, the cost function is also defined in the complex domain where the signal as well as the transformations the signal goes through are easily represented. It is thus desirable to keep all of the computations in the original domain rather than working with transformations to and from the real-valued domain, that is, transformations of the type: $\mathbb{C}^N \mapsto \mathbb{R}^{2N}$.

(2) Even though real-to-complex transformations are always possible using Jacobians, they are not always very straightforward to obtain, especially when the function is not invertible. In addition, when nonlinear functions are

involved, in order to transform the solution back to the complex domain, we usually have to make additional assumptions such as analyticity of the function. We give a simple example (Example 1.3) to highlight this point in Section 1.2.2.

(3) When working in the real-dimensional space with the double dimension, many quantities assume special forms. Matrices in this space usually have special block structures which can make further analysis and manipulations more complicated. In fact, these structures have been the primary motivation for invoking certain simplifying assumptions in the analysis, such as the circularity of signals. For example, this assumption is made in [13] in the derivation of an independent component analysis algorithm when computing the Hessian primarily for this reason. Circularity, which implies that the phase of the signal is uniformly distributed and hence is noninformative, is in most cases an unrealistic assumption limiting the usefulness of algorithms. The communications signals shown in Figure 1.1 as well as a number of other real-world signals can be shown not to satisfy this property, and are discussed in more detail in Section 1.2.5.

Thus, even though we can define a transformation $\mathbb{C}^N \mapsto \mathbb{R}^{2N}$, which is isomorphic, we have to remember that mathematical equivalence does not imply that the optimization, analysis, and numerical and computational properties of the algorithms will be similar in these two domains. We argue that $\mathbb{C}^N$ defines a much more desirable domain for adaptive signal processing in general and give examples to support our point. Using Wirtinger calculus, most of the processing and analysis in the complex domain can be performed in a manner very similar to the real-valued case as we describe in this chapter, thus eliminating the need to consider such transformations in the first place.

The theory and algorithms using the widely linear and the fully complex filter can be easily developed using Wirtinger calculus. Both of these filters are powerful tools for complex-valued signal processing that allow taking advantage of the full processing power of the complex domain and without having to make limiting assumptions on the nature of the signal, such as circularity.

1.1.2 Outline of the Chapter

To present the development, we first present preliminaries including a review of basic results for derivatives and Taylor series expansions, and introduce the main idea behind Wirtinger calculus that describes an effective approach for complex-valued signal processing. We define first- and second-order Taylor series expansions in the complex domain, establish the key relationships that enable efficient derivation of first- and second-order adaptive algorithms as well as performing analyses such as local stability using a quadratic approximation within a neighborhood of a local optimum. We also provide a review of complex-valued statistics, again a topic that has been, for the most part, treated in a limited form in the literature for complex signals. We carefully define circularity of a signal, the associated properties and complete statistical characterization of a complex signal, which play an important role in the subsequent discussions on widely linear filters and independent component analysis.

Next, we show how Wirtinger calculus enables derivation of effective algorithms using two filter structures that have been shown to effectively use the complete statistical information in the complex signal and discuss the properties of these filters. These are the *widely linear* and the *fully complex* nonlinear filters, two attractive solutions for the next generation signal processing systems. Even though the widely linear filter is introduced in 1995 [94], its importance in practice has not been noted until recently. Similarly, the idea of fully complex nonlinear filters is not entirely new, but the theory that justifies their use has been developed more recently [63], and both solutions hold much promise for complex-valued signal processing. In Sections 1.4 and 1.5, we present the basic theory of widely linear filters and nonlinear filters—in particular multilayer perceptrons—with fully complex activation functions using Wirtinger calculus. Finally in Section 1.6, we show how Wirtinger calculus together with fully complex nonlinear functions enables derivation of a unified framework for independent component analysis, a statistical analysis tool that has found wide application in many signal processing problems.

1.2 PRELIMINARIES

1.2.1 Notation

A complex number $z \in \mathbb{C}$ is written as $z = z_r + jz_i$ where $j = \sqrt{-1}$ and z_r and z_i refer to the real and imaginary parts. In our discussions, when concentrating on a single variable, we use the notation without subscripts as in $z = x + jy$ to keep the expressions simple. The complex conjugate is written as $z^* = z_r - jz_i$, and vectors are always assumed to be column vectors, hence $\mathbf{z} \in \mathbb{C}^N$ implies $\mathbf{z} \in \mathbb{C}^{N \times 1}$.

In Table 1.1 we show the six types of derivatives of interest that result in matrix forms along with our convention for the form of the resulting expression depending on whether the vector/matrix is in the numerator or the denominator. Our discussions in the chapter will mostly focus on the derivatives given on the top row of the table, that is, functions that are scalar valued. The extension to the other three cases given in

Table 1.1 Functions of interest and their derivatives

	Scalar Variable: $z \in \mathbb{C}$	Vector Variable: $\mathbf{z} \in \mathbb{C}^N$	Matrix Variable: $\mathbf{Z} \in \mathbb{C}^{N \times M}$
Scalar Function: $f \in \mathbb{C}$	$\dfrac{\partial f}{\partial z} \in \mathbb{C}$	$\dfrac{\partial f}{\partial \mathbf{z}} = \left[\dfrac{\partial f}{\partial z_k}\right] \in \mathbb{C}^N$	$\dfrac{\partial f}{\partial \mathbf{Z}} = \left[\dfrac{\partial f}{\partial Z_{kl}}\right] \in \mathbb{C}^{N \times M}$
Vector Function: $\boldsymbol{f} \in \mathbb{C}^L$	$\dfrac{\partial \boldsymbol{f}}{\partial z} \in \mathbb{C}^{1 \times L}$	$\dfrac{\partial \boldsymbol{f}}{\partial \mathbf{z}} = \left[\dfrac{\partial f_l}{\partial z_k}\right] \in \mathbb{C}^{N \times L}$	
Matrix Function: $\boldsymbol{F} \in \mathbb{C}^{L \times K}$	$\dfrac{\partial \boldsymbol{F}}{\partial z} \in \mathbb{C}^{K \times L}$		

the table is straightforward. The remaining three cases that are omitted from the table and that do not result in a matrix form can be either handled using the vectorization operator as in [46], or by using suitable definitions of differentials as in [7]. We introduce the vectorization operator in Section 1.2.3 and give an example of the use of the differential definition of [7] in Section 1.6.1 to demonstrate how one can alleviate the need to work with tensor representations.

The matrix notation used in Table 1.1 refers to the elements of the vectors or matrices. For the gradient vector $\nabla_{\mathbf{z}} f$, we have

$$\nabla_{\mathbf{z}} f = \frac{\partial f}{\partial \mathbf{z}} = \begin{bmatrix} \frac{\partial f}{\partial z_1} \\ \frac{\partial f}{\partial z_2} \\ \vdots \\ \frac{\partial f}{\partial z_N} \end{bmatrix}$$

and

$$\nabla_{\mathbf{Z}} f = \frac{\partial f}{\partial \mathbf{Z}} = \begin{bmatrix} \frac{\partial f}{\partial Z_{1,1}} & \frac{\partial f}{\partial Z_{1,2}} & \cdots & \frac{\partial f}{\partial Z_{1,M}} \\ \frac{\partial f}{\partial Z_{2,1}} & \frac{\partial f}{\partial Z_{2,2}} & \cdots & \frac{\partial f}{\partial Z_{2,M}} \\ \vdots & \vdots & \vdots & \vdots \\ \frac{\partial f}{\partial Z_{N,1}} & \frac{\partial f}{\partial Z_{N,2}} & \cdots & \frac{\partial f}{\partial Z_{N,M}} \end{bmatrix},$$

for the matrix gradient $\nabla_{\mathbf{Z}} f$. The $N \times L$ Jacobian matrix

$$J_{\mathbf{z}} f = \frac{\partial \mathbf{f}}{\partial \mathbf{z}}$$

is also written similarly.

In the development we present in this chapter, we emphasize the use of derivatives directly in the form given in Table 1.1 rather than splitting the derivatives into real and imaginary parts and evaluating the two separately, which is the procedure most typically used in the literature when evaluating derivatives of nonanalytic functions. Our approach keeps all expressions in the complex domain where they are typically defined, rather than transforming to and from another domain, which typically is the real domain.

As such, when evaluating complex derivatives, all conventions and formulas used in the computation of real-valued derivatives can be directly used for both analytic and nonanalytic functions. A good reference for the computation of real-valued matrix derivatives is [88]. As we show through a number of examples of interest for adaptive signal processing in Sections 1.4–1.6, these formulas can be used without much alteration for the complex case.

In the development, we use various representations for a given function $f(\cdot)$, that is, write it in terms of different arguments. When doing so, we keep the function variable, which is $f(\cdot)$ in this case, the same. It is important to note, however, that even though these representations are all equivalent, different arguments may result in quite different forms for the function. A simple example is given below.

■ EXAMPLE 1.1

For a given function $f(z) = |z|^2$, where $z = x + jy$, we can write

$$f(z, z^*) = zz^*$$

or

$$f(x, y) = x^2 + y^2.$$

It is also important to note that in some cases, explicitly writing the function in one of the two forms given above—as $f(z, z^*)$ or $f(x, y)$—is not possible. A simple example is the magnitude square of a nonlinear function, for example, $f(z) = |\tanh(z)|^2$. In such cases, the advantage of the approach we emphasize in this chapter, that is, directly working in the complex domain, becomes even more evident.

Depending on the application, one might have to work with functions defined to satisfy certain properties such as boundedness. When referring to such functions, that is, those that are defined to satisfy a given property, as well as traditional functions such as trigonometric functions, we use the terminology introduced in [61] to be able to differentiate among those as given in the next definition.

Definition 1 (Split-complex and fully-complex functions) *Functions that are defined in such a way that the real and imaginary—or the magnitude and the phase—are processed separately using real-valued functions are referred to as* split-complex *functions. An example is*

$$f(z) = \tanh x + j \tanh y.$$

Obviously, the form $f(x, y)$ follows naturally for the given example but the form $f(z, z^*)$ does not.

Complex functions that are naturally defined as $f\colon \mathbb{C} \mapsto \mathbb{C}$, on the other hand, are referred to as fully-complex *functions. Examples include trigonometric functions and their hyperbolic counterparts such as $f(z) = \tanh(z)$. These functions typically provide better approximation ability and are more efficient in the characterization of the underlying nonlinear problem structure than the split-complex functions [62].*

We define the scalar inner product between two matrices $\mathbf{W}, \mathbf{V} \in \mathcal{V}$ as

$$\langle \mathbf{W}, \mathbf{V} \rangle = \mathrm{Trace}(\mathbf{V}^H \mathbf{W})$$

such that $\langle \mathbf{W}, \mathbf{W} \rangle = \|\mathbf{W}\|^2$ and the superscript in $(\cdot)^H$ denotes the transpose of the complex conjugate. The norm we consider in this chapter is the Frobenius—also called the Euclidean—norm. For vectors, the definition simplifies to $\langle \mathbf{w}, \mathbf{v} \rangle = \mathbf{v}^H \mathbf{w}$. The definition of an inner product introduces a well-defined notion of orthogonality as well as of norm, and provides both computational and conceptual convenience. Inner product satisfies certain properties.

Properties of inner product:

positivity: $\langle \mathbf{V}, \mathbf{V} \rangle > \mathbf{0}$ for all $\mathbf{V} \in \mathcal{V}$;
definiteness: $\langle \mathbf{V}, \mathbf{V} \rangle = \mathbf{0}$ if and only if $\mathbf{V} = \mathbf{0}$;
linearity (additivity and homogeneity): $\langle \alpha(\mathbf{U} + \mathbf{W}), \mathbf{V} \rangle = \alpha \langle \mathbf{U}, \mathbf{V} \rangle + \alpha \langle \mathbf{W}, \mathbf{V} \rangle$ for all $\mathbf{W}, \mathbf{U}, \mathbf{V} \in \mathcal{V}$;
conjugate symmetry: $\langle \mathbf{W}, \mathbf{V} \rangle^* = \langle \mathbf{V}, \mathbf{W} \rangle$ for all $\mathbf{V}, \mathbf{W} \in \mathcal{V}$.

In the definition of the inner product, we assumed linearity in the first argument, which is more commonly used in engineering texts, though the alternate definition is also possible. Since our focus in this chapter is the finite-dimensional case, the inner product space also defines the Hilbert space.

A complex matrix $\mathbf{W} \in \mathbb{C}^{N \times N}$ is called symmetric if $\mathbf{W}^T = \mathbf{W}$ and Hermitian if $\mathbf{W}^H = \mathbf{W}$. Also, $\mathbf{W}$ is orthogonal if $\mathbf{W}^T\mathbf{W} = \mathbf{I}$ and unitary if $\mathbf{W}^H\mathbf{W} = \mathbf{I}$ where $\mathbf{I}$ is the identity matrix [49].

1.2.2 Efficient Computation of Derivatives in the Complex Domain

Differentiability and Analyticity Given a complex-valued function

$$f(z) = u(x, y) + jv(x, y)$$

where $z = x + jy$, the derivative of $f(z)$ at a point z_0 is written similar to the real case as

$$f'(z_0) = \lim_{\Delta z \to 0} \frac{f(z_0 + \Delta z) - f(z_0)}{\Delta z}.$$

However, different from the real case, due to additional dimensionality in the complex case, there is the added requirement that the limit should be independent of the direction of approach. Hence, if we first let $\Delta y = 0$ and evaluate $f'(z)$ by letting $\Delta x \to 0$, we have

$$f'(z) = u_x + jv_x \tag{1.1}$$

and, similarly, if we first let $\Delta x = 0$, and then $\Delta y \to 0$, we obtain

$$f'(z) = v_y - ju_y \tag{1.2}$$

where we have defined $u_x \triangleq \partial u/\partial x$, $u_y \triangleq \partial u/\partial y$, $v_x \triangleq \partial v/\partial x$, and $v_y \triangleq \partial v/\partial y$. For the existence of $f'(z)$, we thus require the equality of (1.1) and (1.2) at $z = z_0$ and in some neighborhood of z_0, which leads to the Cauchy–Riemann equations given by

$$u_x = v_y \quad \text{and} \quad v_x = -u_y. \tag{1.3}$$

A similar set of equations can be derived for other coordinate systems as well, such as polar [1]. The conditions given by (1.3) state the necessary conditions for the differentiability of $f(z)$. If, in addition, the partial derivatives of $u(x, y)$ and $v(x, y)$ exist and are continuous, then (1.3) are sufficient conditions as well.

Differentiability refers to the property of the function at a single point, and a function is called *analytic* (or holomorphic) if it is differentiable at every point in a given region. For example, the function $f(z) = z^*$ is analytic nowhere and $f(z) = 1/z^2$ is analytic for all finite $z \neq 0$. On the other hand, $f(z) = e^z$ is analytic in the entire finite z plane. Such functions are called *entire.*

In the study of analytic functions, a very fundamental result is given by Cauchy's integral theorem, which states that for a function $f(z)$ that is analytic throughout a region $\mathcal{U}$, the contour integral of $f(z)$ along any closed path lying inside $\mathcal{U}$ is zero. One of the most important consequences of Cauchy's integral theorem is a result stated by *Liouville's theorem* [95]:

> A bounded entire function must be a constant in the complex plane.

Hence, we cannot identify a function that is both bounded and analytic in the entire complex domain. Since boundedness is deemed as important for the performance—in particular stability—of nonlinear signal processing algorithms, a common practice has been to define functions that do not satisfy the analyticity requirement but are bounded (see *e.g.*, [9, 36, 45, 67, 103]). This has been the main motivation in the definition of split- and fully-complex functions given in Definition 1. The solution provides reasonable approximation ability but is an ad-hoc solution not fully exploiting the efficiency of complex representations, both in terms of parameterization (number of parameters to estimate) and in terms of learning algorithms to estimate the parameters as we cannot define true gradients when working with these functions. In Sections 1.5 and 1.6, we discuss applications of both types of functions, split nonlinear functions that are proposed to circumvent the boundedness issue, and solutions that fully exploit the efficiency of complex domain processing.

Singular Points Singularities of a complex function $f(z)$ are defined as points z_0 in the domain of the function where $f(z)$ fails to be analytic. Singular points can be at a single point, that is, *isolated*, or nonisolated as in branch cuts or boundaries. Isolated singularities can be classified as removable singularities, poles, and essential singularities [1].

- A singular point is called a *removable singular point* if we have $f(z_0) \triangleq \lim_{z \to z_0} f(z)$, that is, the limit exists even though the function is not defined at that point. In this case, the function can be written as an analytic function by simply defining the function at z_0 as $f(z_0)$.

- When we have $\lim_{z \to z_0} |f(z)| \to \infty$ for $f(z)$ analytic in a region centered at z_0, that is, in $0 < |z - z_0| < R$, we say that z_0 is a *pole* of the function $f(z)$.
- If a singularity is neither a pole nor a removable singularity, it is called an *essential singularity*, that is, the limit $\lim_{z \to z_0} f(z)$ does not exist as a complex number and is not equal to infinity either.

A simple example for a function with removable singularity is the function

$$f(z) = \frac{\sin(z - z_0)}{z - z_0}$$

which is not defined at $z = z_0$, but can be made analytic for all z by simply augmenting the definition of the function by $f(z_0) = 1$.

The function

$$f(z) = \frac{1}{(z - z_0)^N}$$

where N is an integer, is an example for a function with a pole. The pole at $z = z_0$ is called a simple pole if $N = 1$ and an Nth order pole if $N > 1$.

The essential singularity class is an interesting case and the rare example is found in functions of the form

$$f(z) = e^{1/z}.$$

This function has different limiting values for $z = 0$ depending on the direction of approach as we have $\lim_{z \to 0 \pm j} f(z) = 1$, $\lim_{z \to 0^-} f(z) = 0$, and $\lim_{z \to 0^+} f(z) = \infty$. A powerful property of essential singular points is given by Picard's theorem, which states that in any neighborhood of an essential singularity, a function, $f(z)$, assumes all values, except possibly one of them, an infinite number of times [1].

A very important class of functions that are not analytic anywhere on the complex plane are functions that are real valued, that is, $f: \mathbb{C} \mapsto \mathbb{R}$ and thus have $v(x, y) = 0$. Since the cost functions are real valued, their optimization thus poses a challenge, and is typically achieved using separate evaluations of real and imaginary parts of the function. As we discuss next, Wirtinger calculus provides a convenient framework to significantly simplify the evaluations of derivatives in the complex domain.

Wirtinger Derivatives As discussed in Section 1.2.2, differentiability, and hence analyticity are powerful concepts leading to important results such as the one summarized by Liouville's theorem. But—perhaps not surprisingly—their powerful nature also implies quite stringent conditions that need to be satisfied. When we look closely at the conditions for differentiability described by the Cauchy–Riemann equations (1.3), it is quite evident that they impose a strong structure on $u(x, y)$ and $v(x, y)$, the real and imaginary parts of the function, and consequently on

$f(z)$, as also discussed in [64]. A simple demonstration of this fact is that, *to express the derivatives of an analytic function, we only need to specify either* $u(x, y)$ *or* $v(x, y)$, *and do not need both.*

An elegant approach due to Wirtinger [115], which we explain next, relaxes this strong requirement for differentiability, and defines a less stringent form for the complex domain. More importantly, it describes how this new definition can be used for defining complex differential operators that allow computation of derivatives in a very straightforward manner in the complex domain by simply using real differentiation results and procedures.

To proceed, we first introduce the notion of *real differentiability*. In the introduction of Wirtinger calculus, the commonly used definition of differentiability that leads to the Cauchy–Riemann equations is identified as *complex differentiability*, and *real differentiability* is defined as a more flexible form.

Definition 2 *A function* $f(z) = u(x, y) + jv(x, y)$ *is called real differentiable when* $u(x, y)$ *and* $v(x, y)$ *are differentiable as functions of real-valued variables* x *and* y.

Note that this definition is quite flexible in that most nonanalytic as well as analytic functions satisfy the property as long as they have real and imaginary parts that are smooth (differentiable) functions of x and y.

To derive the form of the differential operators, we write the two real-variables as

$$x = \frac{z + z^*}{2} \quad \text{and} \quad y = \frac{z - z^*}{2j} \tag{1.4}$$

and use the chain rule to derive the form of the two derivative operators for $f(z)$ as

$$\frac{\partial f}{\partial z} = \frac{\partial f}{\partial x}\frac{\partial x}{\partial z} + \frac{\partial f}{\partial y}\frac{\partial y}{\partial z} = \frac{\partial f}{\partial x}\frac{1}{2} + \frac{\partial f}{\partial y}\frac{1}{2j}$$

and

$$\frac{\partial f}{\partial z^*} = \frac{\partial f}{\partial x}\frac{\partial x}{\partial z^*} + \frac{\partial f}{\partial y}\frac{\partial y}{\partial z^*} = \frac{\partial f}{\partial x}\frac{1}{2} - \frac{\partial f}{\partial y}\frac{1}{2j}.$$

The key point in the derivation given above is to treat the two variables z and z^* as independent from each other, which is also the main trick that allows us to make use of the elegance of Wirtinger calculus which we introduce next.

We consider a given function $f\colon \mathbb{C} \mapsto \mathbb{C}$ as a function $f\colon \mathbb{R} \times \mathbb{R} \mapsto \mathbb{C}$ by writing it as $f(z) = f(x, y)$, and make use of the underlying $\mathbb{R}^2$ structure by the following theorem [15].

Theorem 1.2.1 *Let* $f\colon \mathbb{R} \times \mathbb{R} \rightarrow \mathbb{C}$ *be a function of real variables* x *and* y *such that* $f(z, z^*) = f(x, y)$, *where* $z = x + jy$ *and that* $f(z, z^*)$ *is analytic with respect to* z^* *and* z *independently. Then,*

(1) *The partial derivatives*

$$\frac{\partial f}{\partial z} = \frac{1}{2}\left(\frac{\partial f}{\partial x} - j\frac{\partial f}{\partial y}\right) \quad \textit{and} \quad \frac{\partial f}{\partial z^*} = \frac{1}{2}\left(\frac{\partial f}{\partial x} + j\frac{\partial f}{\partial y}\right) \tag{1.5}$$

can be computed by treating z^* and z as a constant in $f(z, z^*)$ respectively; and

(2) *A necessary and sufficient condition for f to have a stationary point is that $\partial f/\partial z = 0$. Similarly, $\partial f/\partial z^* = 0$ is also a necessary and sufficient condition.*

Therefore, when evaluating the gradient, we can directly compute the derivatives with respect to the complex argument, rather than calculating individual real-valued gradients, that is, by evaluating the right side of the equations in (1.5). To do so, we write the given function $f(z)$ in the form $f(z, z^*)$ and when evaluating the derivative with respect to z, we treat z^* as a constant as done in the computation of multi-variable function derivatives, and similarly treat z as a constant when evaluating $\partial f/\partial z^*$. The requirement for the analyticity of $f(z, z^*)$ with respect to z and z^* independently is equivalent to the condition on real differentiability of $f(x, y)$ since we can move from one form of the function to the other using the simple linear transformation given in (1.4) [64, 95]. Even though the condition of real differentiability is easily satisfied, separate evaluations of real and imaginary parts has been the common practice in the literature (see *e.g.*, [34, 38, 39, 63, 67, 103]).

When $f(z)$ is analytic, that is, when the Cauchy–Riemann conditions hold in a given open set, $f(\cdot)$ becomes a function of only z, and the two derivatives, the one given in the theorem and the traditional one coincide [95]. Alternatively put, all analytic functions are independent of z^* and only depend on z. This point can be easily verified using the definitions given in (1.5) and observing that when the Cauchy–Riemann equations are satisfied, we do end up with $f'(z)$ as given in (1.1) and (1.2), and we have $f'(z^*) = 0$.

For the application of Wirtinger derivatives for scalar-valued functions, consider the following two examples.

■ EXAMPLE 1.2

Consider the real-valued function $f(z) = |z|^4 = x^4 + 2x^2y^2 + y^4$. The derivative of the function can be calculated using (1.5) as

$$f'(z) \triangleq \frac{\partial f}{\partial z} = \frac{1}{2}\left(\frac{\partial f}{\partial x} - j\frac{\partial f}{\partial y}\right) = 2x^3 + 2xy^2 - 2j(x^2y + y^3) \tag{1.6}$$

or, to make use of Wirtinger derivative, we can write the function as $f(z) = f(z, z^*) = z^2(z^*)^2$ and evaluate the derivative as

$$\frac{\partial f}{\partial z} = 2z(z^*)^2 \tag{1.7}$$

that is, by treating z^* as a constant in f when calculating the partial derivative. It can be easily shown that the two forms, (1.6) and (1.7), are equal.

We usually define functions of interest in terms of z and would like to keep the expressions in that form, hence typically, one would need to write (1.6) in terms of z. As this simple example demonstrates, depending on the function in question, this might not always be a straightforward task.

■ EXAMPLE 1.3

As another example, consider evaluation of the conjugate derivative for the real-valued function $f(z) = |g(z)|^2$ where $g(z)$ is any analytic function. Since, in general we cannot explicitly write the real and imaginary parts of such a function in terms of x and y, we write

$$g(z) = u(x, y) + jv(x, y)$$

so that we have

$$f(z) = u^2(x, y) + v^2(x, y).$$

The derivative can then be evaluated using (1.5) as

$$\begin{aligned}\frac{\partial f}{\partial z^*} &= \frac{1}{2}\left(\frac{\partial f}{\partial x} + j\frac{\partial f}{\partial y}\right)\\ &= uu_x + vv_x + j(uu_y + vv_y)\\ &= g(z)[g'(z)]^* \end{aligned} \tag{1.8}$$

where u_x, u_y, v_x, and v_y are defined in (1.1) and (1.2), and we used the Cauchy–Riemann conditions for $g(z)$ when writing the last equality.

Alternatively, we can write the function as

$$f(z) = g(z)[g(z)]^* = g(z)g(z^*)$$

where the last equality follows when we have $g^*(z)^* = g(z^*)$. Then, directly using the Wirtinger derivative we have the same form given in (1.8) for $\partial f/\partial z^*$.

The condition in Example 1.3, $g^*(z) = g(z^*)$ which also implies $[g'(z)]^* = g'(z^*)$, is satisfied for a wide class of functions. It is easy to observe that it is true for all real-valued functions, and also for all functions $g(z)$ that have a Taylor series expansion with all real coefficients in $|z| < R$. Hence, all functions that are analytic within a neighborhood of zero satisfy the equality.

Example 1.3 also underlines another important point we have made earlier in the chapter regarding the desirability of directly working in the complex domain. When using the approach that treats real and imaginary parts separately, we needed a certain relationship between the real and imaginary parts of the function to write the derivative $f'(z)$ again in terms of z. The condition in this example was satisfied by analyticity of the function as we used the Cauchy–Riemann conditions, that is, a strict relationship between the real and imaginary parts of the function.

The same approach of treating the variable and its complex conjugate as independent variables, can be used when taking derivatives of functions of matrix variables as well so that expressions given for real-valued matrix derivatives can be directly used as shown in the next example.

■ EXAMPLE 1.4

Let $g(\mathbf{Z}, \mathbf{Z}^*) = \mathrm{Trace}(\mathbf{Z}\mathbf{Z}^H)$. We can calculate the derivatives of g with respect to $\mathbf{Z}$ and $\mathbf{Z}^*$ by simply treating one variable as a constant and directly using the results from real-valued matrix differentiation as

$$\frac{\partial g}{\partial \mathbf{Z}} = \frac{\partial \mathrm{Trace}[\mathbf{Z}(\mathbf{Z}^*)^T]}{\partial \mathbf{Z}} = \mathbf{Z}^*$$

and

$$\frac{\partial g}{\partial \mathbf{Z}^*} = \mathbf{Z}$$

A good reference for real-valued matrix derivatives is [88] and a number of complex-valued matrix derivatives are discussed in detail in [46].

For computing matrix derivatives, a convenient tool is the use of differentials. In this procedure, first the matrix differential is computed and then it is written in the canonical form by identifying the term of interest. The differential of a function is defined as the part of a function $f(\mathbf{Z} + \Delta\mathbf{Z}) - f(\mathbf{Z})$ that is linear in $\mathbf{Z}$. For example when computing the differential of the function $f(\mathbf{Z}, \mathbf{Z}^*) = \mathbf{Z}\mathbf{Z}^*$, we can first write the product of the two differentials

$$(\mathbf{Z} + d\mathbf{Z})(\mathbf{Z}^* + d\mathbf{Z}^*) = \mathbf{Z}\mathbf{Z}^* + (d\mathbf{Z})\mathbf{Z}^* + \mathbf{Z}\,d\mathbf{Z}^* + d\mathbf{Z}\,d\mathbf{Z}^*$$

and take the first-order term (part of the expansion linear in $\mathbf{Z}$ and $\mathbf{Z}^*$) to evaluate the differential of the function as

$$d(\mathbf{Z}\mathbf{Z}^*) = (d\mathbf{Z})\mathbf{Z}^* + \mathbf{Z}d\mathbf{Z}^*$$

as discussed in [74, 78]. The approach can significantly simplify certain derivations. We provide an example for the application of the approach in Section 1.6.1.

Integrals of the Function $f(z, z^*)$ Though the three representations of a function we have discussed so far: $f(z)$, $f(x, y)$, and $f(z, z^*)$ are all equivalent, certain care needs to be taken when using each form, especially when using the form $f(z, z^*)$. This is the form that enables us to treat z and z^* as independent variables when taking derivatives and hence provides a very convenient representation (mapping) of a complex function in most evaluations. Obviously, the two variables are not independent as knowing z we already know its conjugate. This is an issue that needs special care in evaluations such as integrals, which is needed for example, when using $f(z, z^*)$ to denote probability density functions and calculating the probabilities with this form.

In the evaluation of integrals, when we consider $f(\cdot)$ as a function of real and imaginary parts, the definition of an integral is well understood as the integral of function $f(x, y)$ in a region $\mathcal{R}$ defined in the (x, y) space as

$$\iint\limits_{\mathcal{R}} f(x, y)\, dx\, dy.$$

However, the integral $\int\int f(z, z^*)\, dz\, dz^*$ is not meaningful as we cannot vary the two variables z and z^* independently, and cannot define the region corresponding to $\mathcal{R}$ in the complex domain. However, this integral representation serves as an intermediate step when writing the real-valued integral as a contour integral in the complex domain using Green's theorem [1] or Stokes's theorem [44, 48] as noted in [87]. We can use Green's theorem (or Stokes's theorem) along with the definitions for the complex derivative given in (1.5) to write

$$\iint\limits_{\mathcal{R}} f(x, y)\, dx\, dy = -\frac{j}{2} \oint\limits_{\mathcal{C}_{\mathcal{R}}} F(z, z^*)\, dz \tag{1.9}$$

where

$$\frac{\partial F(z, z^*)}{\partial z^*} = f(z, z^*).$$

Here, we assume that $f(x, y)$ is continuous through the simply connected region $\mathcal{R}$ and $\mathcal{C}_{\mathcal{R}}$ describes its contour. Note that by transforming the integral defined in the real domain to a contour integral when the function is written as $f(z, z^*)$, the formula takes into account the dependence of the two variables, z and z^* in a natural manner.

In [87], the application of the integral relationship in (1.9) is discussed in detail for the evaluation of probability masses when $f(x, y)$ defines a probability density function. Three cases are identified as important and a number of examples are studied as application of the formula. The three specific cases to consider for evaluation of the integral in (1.9) are when

- $F(z, z^*)$ is an analytic function inside the given contour, that is, it is a function of z only in which case the integral is zero by Cauchy's theorem;
- $F(z, z^*)$ contains poles inside the contour, which in the case of probability evaluations will correspond to probability masses inside the given region;
- $F(z, z^*)$ is not analytic inside the given contour in which case the value of the integral will relate to the size of the region $\mathcal{R}$.

We demonstrate the use of the integral formula given in (1.9) in Section 1.6.4 in the derivation of an efficient representation for the score function for complex maximum likelihood based independent component analysis.

It is also worth noting that the dependence in the variables z and z^* is different in the computation of derivatives. In [31], the author discusses polarization of an analytic identity and notes that complex-valued functions of z and z^* have linearly independent differentials dz and dz^*, and hence z and z^* are *locally* functionally independent. Still, we treat the form $f(z, z^*)$ as primarily a notational form that renders computations of derivatives simple and note the fact that special care must be taken when using the form to define quantities such as probability density functions.

Derivatives of Cost Functions The functions we typically work with in the development of signal processing algorithms are cost functions, hence these are real valued such that $f \in \mathbb{R}$. Since the class of real-valued functions is a special case of the functions considered in Theorem 1.2.1, we can employ the same procedure for this case as well and take the derivatives by treating z and z^* as independent from each other. In this chapter, we mainly consider such functions as these are the cost functions used in the derivation of adaptive signal processing algorithms. However, in the discussion, we identify the deviation, if any, from the general $f\colon \mathbb{R} \times \mathbb{R} \to \mathbb{C}$ case for completeness. Also note that when $f(z)$ is real valued, we have

$$\left(\frac{\partial f}{\partial z}\right)^* = \frac{\partial f}{\partial z^*}$$

that is, the derivative and the conjugate derivative are complex conjugates of each other.

1.2.3 Complex-to-Real and Complex-to-Complex Mappings

In this chapter, we emphasize working in the original space in which the functions are defined, even when they are not analytic. The approach is attractive for two reasons. First, it is straightforward and eliminates the need to perform transformations to and back from another space where the computations are carried out. Second, it does not increase the dimensionality of the problem. In certain cases though, in particular for the form of multidimensional transformation defined by van den Bos [110], the increase in dimensionality might offer advantages. As we discuss in this section, the $C^N \mapsto C^{2N}$ mapping given by van den Bos provides a smart way of taking advantage of Wirtinger calculus, and can lead to certain simplifications in the expression. For completeness, we discuss all major transformations that have been used in the literature for multivariate complex analysis, especially when working with non-analytic functions.

Vector-Concatenation Type Mappings The two mappings in this class, the $\overline{(\cdot)}_R$ and $\overline{(\cdot)}_C$ mappings have very different uses. The most commonly used mapping $\mathbb{C}^N \mapsto \mathbb{R}^{2N}$ takes a very simple form and is written such that

$$\mathbf{z} \in \mathbb{C}^N \mapsto \bar{\mathbf{z}}_R = \begin{bmatrix} \mathbf{z}_r \\ \mathbf{z}_i \end{bmatrix} \in \mathbb{R}^{2N} \tag{1.10}$$

and for a matrix $\mathbf{A}$ as

$$\mathbf{A} \in \mathbb{C}^{M \times N} \mapsto \bar{\mathbf{A}}_R = \begin{bmatrix} \mathbf{A}_r & -\mathbf{A}_i \\ \mathbf{A}_i & \mathbf{A}_r \end{bmatrix} \in \mathbb{R}^{2M \times 2N}. \tag{1.11}$$

It can be easily shown that $\overline{(\mathbf{Az})}_R = \bar{\mathbf{A}}_R \bar{\mathbf{z}}_R$.

The mapping provides a natural isomorphism between $\mathbb{C}^N$ and $\mathbb{R}^{2N}$, and thus is a practical approach for derivations in the complex domain. For example, in [40], the mapping is used for statistical analysis of multivariate complex Gaussian distribution and in [20] to derive the relative gradient update rule for independent component analysis.

Note that the real-vector space defined through the $\overline{(\cdot)}$ mapping is isomorphic to the standard real vector space $\mathbb{R}^{2N}$. In fact, we can define an orthogonal decomposition of the space of $2N \times 2N$ matrices such that a given matrix $\mathbf{M} \in \mathbb{R}^{2N \times 2N}$ is written in terms of four blocks of size $N \times N$ as

$$\mathbf{M} = \begin{bmatrix} \mathbf{M}_{11} & \mathbf{M}_{12} \\ \mathbf{M}_{21} & \mathbf{M}_{22} \end{bmatrix}.$$

Thus, the linear space of $2N \times 2N$ matrices can be decomposed into two orthogonal spaces: $\mathbb{R}^{2N \times 2N} = \mathcal{M}^+ \oplus \mathcal{M}^-$ where $\mathcal{M}^+$ (*resp.* $\mathcal{M}^-$) contains any matrix such that $\mathbf{M}_{11} = \mathbf{M}_{22}$ and $\mathbf{M}_{12} = -\mathbf{M}_{21}$ (*resp.* $\mathbf{M}_{11} = -\mathbf{M}_{22}$ and $\mathbf{M}_{12} = \mathbf{M}_{21}$). Hence a $2N \times 2N$ real matrix has the orthogonal decomposition $\mathbf{M} = \mathbf{M}^+ + \mathbf{M}^-$ where

$$\begin{aligned} \mathbf{M}^+ &= \frac{1}{2}\begin{bmatrix} \mathbf{M}_{11} + \mathbf{M}_{22} & \mathbf{M}_{12} - \mathbf{M}_{21} \\ \mathbf{M}_{21} - \mathbf{M}_{12} & \mathbf{M}_{11} + \mathbf{M}_{22} \end{bmatrix} \in \mathcal{M}^+ \quad \text{and} \\ \mathbf{M}^- &= \frac{1}{2}\begin{bmatrix} \mathbf{M}_{11} - \mathbf{M}_{22} & \mathbf{M}_{12} + \mathbf{M}_{21} \\ \mathbf{M}_{21} + \mathbf{M}_{12} & -\mathbf{M}_{11} + \mathbf{M}_{22} \end{bmatrix} \in \mathcal{M}^-. \end{aligned} \tag{1.12}$$

Note that the set of invertible matrices of $\mathbf{M}^+$ form a group for the usual multiplication of matrices and we have $\bar{\mathbf{A}}_R \in \mathcal{M}^+$, which is defined in (1.11).

The following are some useful properties of this complex-to-real mapping and can be verified using the isomorphism between the two spaces [20, 33, 40].

Properties of Complex-to-Real Mapping $\overline{(\cdot)}$: $\mathbb{C}^N \to \mathbb{R}^{2N}$ Let $\mathbf{A}, \mathbf{B} \in \mathbb{C}^{N \times N}$ and $\mathbf{z}, \mathbf{y} \in \mathbb{C}^N$, then for the mapping $\overline{(\cdot)}_R$, we have

(1) $\overline{(\mathbf{AB})}_R = \bar{\mathbf{A}}_R \bar{\mathbf{B}}_R$ and thus $\overline{(\mathbf{A}^{-1})}_R = (\bar{\mathbf{A}}_R)^{-1}$.
(2) $|\det(\mathbf{A})|^2 = \det(\bar{\mathbf{A}}_R)$.
(3) $\mathbf{A}$ is Hermitian if and only if $\bar{\mathbf{A}}_R$ is symmetric.
(4) $\mathbf{A}$ is nonsingular if and only if $\bar{\mathbf{A}}_R$ is nonsingular.
(5) $\mathbf{A}$ is unitary if and only if $\bar{\mathbf{A}}_R$ is orthogonal.
(6) $\mathbf{A}$ is positive definite if and only if $\bar{\mathbf{A}}_R$ is positive definite.

(7) $\mathbf{z}^H \mathbf{A} \mathbf{z} = \bar{\mathbf{z}}_R^T \bar{\mathbf{A}}_R \bar{\mathbf{z}}$.

(8) $\overline{\mathbf{z}\mathbf{y}^H}_R = 2(\bar{\mathbf{z}}_R \bar{\mathbf{y}}_R)^+$ where $(\cdot)^+$ is defined in (1.12).

In certain scenarios, for example, when working with probabilistic descriptions, or when evaluating the derivative of matrix functions, the $\mathbb{C}^N \mapsto \mathbb{R}^{2N}$ transformation can simplify the evaluations and lead to simpler forms (see *e.g.* [4, 20]).

The second mapping in this class is defined by simple concatenation of the complex vector and its complex conjugate as

$$\mathbf{z} \in \mathbb{C}^N \mapsto \bar{\mathbf{z}}_C = \begin{bmatrix} \mathbf{z} \\ \mathbf{z}^* \end{bmatrix} \in \mathbb{C}^{2N}. \tag{1.13}$$

This mapping can be useful as an intermediate step when establishing certain relationships as shown in [64] and [71]. More importantly, this vector definition provides a convenient representation for the widely linear transform, which enables incorporation of full second-order statistical information into the estimation scheme and provides significant advantages when the signal is noncircular [94]. We discuss the approach and present the main results for minimum mean square error filtering using Wirtinger calculus in Section 1.4.

Element-wise Mappings In the development that leads to the definition of Wirtinger derivatives, the key observation is the duality of the two spaces: $\mathbb{R}^2$ and $\mathbb{C}^2$ through the transformation

$$(z_r, z_i) \Longleftrightarrow (z, z^*).$$

Hence, if a function is real differentiable as a function of the two real-valued variables z_r and z_i, then it satisfies the condition for real differentiability, and the two variables, z and z^* can be treated as independent in $\mathbb{C}^2$ to take advantage of Wirtinger calculus. To extend this idea to the multidimensional case, van den Bos [110] defined the two mappings $\tilde{(\cdot)}$ given in Table 1.2 such that

$$\tilde{\mathbf{z}}_R = \begin{bmatrix} z_{r,1} \\ z_{i,1} \\ z_{r,2} \\ z_{i,2} \\ \vdots \\ z_{r,N} \\ z_{i,N} \end{bmatrix} \Longleftrightarrow \tilde{\mathbf{z}}_C = \begin{bmatrix} z_1 \\ z_1^* \\ z_2 \\ z_2^* \\ \vdots \\ z_N \\ z_N^* \end{bmatrix} \tag{1.14}$$

where $\tilde{\mathbf{z}}_R \in \mathbb{R}^{2N}$ and $\tilde{\mathbf{z}}_C \in \mathbb{C}^{2N}$. In [110], the whole development is given as an extension of Brandwood's work [15] without any reference to Wirtinger calculus in particular.

Table 1.2 Four primary mappings defined for $\mathbf{z} = \mathbf{z}_r + j\mathbf{z}_i \in \mathbb{C}^N$

	Complex-to-Real: $\mathbb{C}^N \mapsto \mathbb{R}^{2N}$	Complex-to-Complex: $\mathbb{C}^N \mapsto \mathbb{C}^{2N}$
Vector-concatenation type mappings	$\bar{\mathbf{z}}_R = \begin{bmatrix} \mathbf{z}_r \\ \mathbf{z}_i \end{bmatrix}$	$\bar{\mathbf{z}}_C = \begin{bmatrix} \mathbf{z} \\ \mathbf{z}^* \end{bmatrix}$
Element-wise mappings	$\tilde{\mathbf{z}}_R = \begin{bmatrix} z_{r,1} \\ z_{i,1} \\ \vdots \\ z_{r,N} \\ z_{i,N} \end{bmatrix}$	$\tilde{\mathbf{z}}_C = \begin{bmatrix} z_1 \\ z_1^* \\ \vdots \\ z_N \\ z_N^* \end{bmatrix}$

Since the transformation from $\mathbb{R}^2$ to $\mathbb{C}^2$ is a simple linear invertible mapping, one can work in either space, depending on the convenience offered by each. In [110], it is shown that such a transformation allows the definition of a Hessian, hence of a Taylor series expansion very similar to the one in the real-case, and the Hessian matrix $\mathbf{H}$ defined in this manner is naturally linked to the complex $\mathbb{C}^{N\times N}$ Hessian matrix. In the next section, we establish the connections of the results of [110] to $\mathbb{C}^N$ for first- and second-order derivatives such that efficient second-order optimization algorithms can be derived by directly working in the original $\mathbb{C}^N$ space where the problems are typically defined.

Relationship Among Mappings It can be easily observed that all four mappings defined in Table 1.2 are related to each other through simple linear transformations, thus making it possible to work in one domain and then transfer the solution to another. Two key transformations are given by $\bar{\mathbf{z}}_C = \mathbf{U}\bar{\mathbf{z}}_R$ and $\tilde{\mathbf{z}}_C = \tilde{\mathbf{U}}\tilde{\mathbf{z}}_R$ where

$$\mathbf{U} = \begin{bmatrix} \mathbf{I} & j\mathbf{I} \\ \mathbf{I} & -j\mathbf{I} \end{bmatrix}$$

and $\tilde{\mathbf{U}} = \mathrm{diag}\{\mathbf{U}_2, \ldots, \mathbf{U}_2\}$ where $\mathbf{U}_2 = \begin{bmatrix} 1 & j \\ 1 & -j \end{bmatrix}$. It is easy to observe that for the transformation matrices $\mathbf{U}$ defined above, we have $\mathbf{U}\mathbf{U}^H = \mathbf{U}^H\mathbf{U} = 2\mathbf{I}$ making it easy to obtain inverse transformations as we demonstrate in Section 1.3. For transformations between the two mappings, $\overline{(\cdot)}$ and $\tilde{(\cdot)}$, we can use permutation matrices that are orthogonal, thus allowing simple manipulations.

1.2.4 Series Expansions

Series expansions are a valuable tool in the study of nonlinear functions, and for analytic functions, that is, functions that are complex differentiable in a given region, the Taylor series expression assumes the same form as in the real case given by

$$f(z) = \sum_{k=0}^{\infty} \frac{f^{(k)}(z_0)}{k!}(z - z_0)^k. \tag{1.15}$$

If $f(z)$ is analytic for $|z| \leq R$, then the Taylor series given in (1.15) converges uniformly in $|z| \leq R_1 < R$. The notation $f^{(k)}(z_0)$ refers to the kth order derivative evaluated at z_0 and when the power series expansion is written for $z_0 = 0$, it is called the Maclaurin series.

In the development of signal processing algorithms (parameter update rules) and in stability analyses, the first- and second-order expansions prove to be the most useful. For an analytic function $f(\mathbf{z})$: $\mathbb{C}^N \mapsto \mathbb{C}$, we define $\Delta f = f(\mathbf{z}) - f(\mathbf{z}_0)$ and $\Delta \mathbf{z} = \mathbf{z} - \mathbf{z}_0$ to write the second-order approximation to the function in the neighborhood of $\mathbf{z}_0$ as

$$\begin{aligned} \Delta f &\approx \Delta \mathbf{z}^T \nabla_{\mathbf{z}} f + \frac{1}{2} \Delta \mathbf{z}^T \mathbf{H}(\mathbf{z}) \, \Delta \mathbf{z} \\ &= \left\langle \nabla_{\mathbf{z}} f, \ \Delta \mathbf{z}^* \right\rangle + \frac{1}{2} \left\langle \mathbf{H}(\mathbf{z}) \, \Delta \mathbf{z}, \ \Delta \mathbf{z}^* \right\rangle \end{aligned} \tag{1.16}$$

where

$$\nabla_{\mathbf{z}} f = \left. \frac{\partial f(\mathbf{z})}{\partial \mathbf{z}} \right|_{\mathbf{z}_0}$$

is the gradient evaluated at $\mathbf{z}_0$ and

$$\nabla_{\mathbf{z}}^2 f \triangleq \mathbf{H}(\mathbf{z}) = \left. \frac{\partial^2 f(\mathbf{z})}{\partial \mathbf{z} \partial \mathbf{z}^T} \right|_{\mathbf{z}_0}$$

is the Hessian matrix evaluated at $\mathbf{z}_0$. As in the real-valued case, the Hessian matrix in this case is symmetric and constant if the function is quadratic.

Second-order Taylor series expansions as given in (1.16) help summarize main results for optimization and local stability analysis. In particular, we can state the following three important observations for the real-valued case, that is, when the argument $\mathbf{z}$ and the function are real valued, by directly studying the expansion given in (1.16).

- Point $\mathbf{z}_0$ is a *local minimum* of $f(\mathbf{z})$ when $\nabla_{\mathbf{z}} f = 0$ and $\mathbf{H}(\mathbf{z})$ is positive semi-definite, that is, these are the necessary conditions for a local minimum.
- When $\mathbf{H}(\mathbf{z})$ is positive definite and $\nabla_{\mathbf{z}} f = 0$, $\mathbf{z}_0$ is *guaranteed* to be a local minimum, that is, positive-definiteness and zero gradient, together, define the sufficient condition.
- Finally, $\mathbf{z}_0$ is a *locally stable* point if, and only if, $\mathbf{H}(\mathbf{z})$ is positive definite and $\nabla_{\mathbf{z}} f = 0$, that is, in this case, the two properties define the sufficient and necessary conditions.

When deriving complex-valued signal processing algorithms, however, the functions of interest are real valued and have complex arguments $\mathbf{z}$, hence are not analytic on the complex plane. In this case, we can use Wirtinger calculus and write the expansions by treating the function $f(\mathbf{z})$ as a function of two arguments, $\mathbf{z}$ and $\mathbf{z}^*$. In this approach, the main idea is treating the two arguments as independent from each other, when they are obviously dependent on each other as we discussed. When writing the Taylor series expansion, the idea is the same. We write the series expansion for a real-differentiable function $f(\mathbf{z}) = f(\mathbf{z}, \mathbf{z}^*)$ as if $\mathbf{z}$ and $\mathbf{z}^*$ were independent variables, that is, as

$$\Delta f(\mathbf{z}, \mathbf{z}^*) \approx \langle \nabla_{\mathbf{z}} f, \Delta\mathbf{z}^* \rangle + \langle \nabla_{\mathbf{z}^*} f, \Delta\mathbf{z} \rangle + \frac{1}{2}\left\langle \frac{\partial f}{\partial \mathbf{z} \partial \mathbf{z}^T} \Delta\mathbf{z}, \Delta\mathbf{z}^* \right\rangle + \left\langle \frac{\partial f}{\partial \mathbf{z} \partial \mathbf{z}^H} \Delta\mathbf{z}^*, \Delta\mathbf{z}^* \right\rangle + \frac{1}{2}\left\langle \frac{\partial f}{\partial \mathbf{z}^* \partial \mathbf{z}^H} \Delta\mathbf{z}^*, \Delta\mathbf{z} \right\rangle. \tag{1.17}$$

In other words, the series expansion has the same form as a real-valued function of two variables which happen to be replaced by $\mathbf{z}$ and $\mathbf{z}^*$ as the two independent variables. Note that when $f(\mathbf{z}, \mathbf{z}^*)$ is real valued, we have

$$\langle \nabla_{\mathbf{z}} f, \Delta\mathbf{z}^* \rangle + \langle \nabla_{\mathbf{z}^*} f, \Delta\mathbf{z} \rangle = 2\mathrm{Re}\{\langle \nabla_{\mathbf{z}^*} f, \Delta\mathbf{z} \rangle\} \tag{1.18}$$

since in this case we have $\nabla f(\mathbf{z}^*) = [\nabla f(\mathbf{z})]^*$. Using the Cauchy–Bunyakovskii–Schwarz inequality [77], we have

$$|\Delta\mathbf{z}^H \nabla f(\mathbf{z}^*)| \leq \|\Delta\mathbf{z}\| \; \|\nabla f(\mathbf{z}^*)\|$$

which holds with equality when $\Delta\mathbf{z}$ is in the same direction as $\nabla f(\mathbf{z}^*)$. Hence, it is the gradient with respect to the complex conjugate of the variable $\nabla f(\mathbf{z}^*)$ that yields the maximum change in function $\Delta f(\mathbf{z}, \mathbf{z}^*)$.

It is also important to note that when $f(\mathbf{z}, \mathbf{z}^*) = f(\mathbf{z})$, that is, the function is analytic (complex differentiable), all derivatives with respect to $\mathbf{z}^*$ in (1.17) vanish and (1.17) coincides with (1.16). As noted earlier, the Wirtinger formulation for real-differentiable functions includes analytic functions, and when the function is analytic, all the expressions used in the formulations reduce to the traditional ones for analytic functions.

For the transformations that map the function to the real domain as those given in Table 1.2, the $\tilde{(\cdot)}_R$ and $\overline{(\cdot)}_R$ mappings, the expansion is straightforward since in this case, the expansion is written in the real domain as in

$$\Delta f(\tilde{\mathbf{z}}_R) \approx \langle \nabla_{\tilde{\mathbf{z}}_R} f(\tilde{\mathbf{z}}_R), \Delta\tilde{\mathbf{z}}_R \rangle + \frac{1}{2}\langle \mathbf{H}(\tilde{\mathbf{z}}_R)\Delta\tilde{\mathbf{z}}_R, \Delta\tilde{\mathbf{z}}_R \rangle.$$

By using the complex domain transformation defined by van den Bos (1.14), a very similar form for the expansion can be obtained in the complex domain as well, and it is given by [110]

$$\Delta f(\tilde{\mathbf{z}}_C) \approx \langle \nabla_{\tilde{\mathbf{z}}_C^*} f(\tilde{\mathbf{z}}_C), \Delta\tilde{\mathbf{z}}_C \rangle + \frac{1}{2}\langle \mathbf{H}(\tilde{\mathbf{z}}_C)\Delta\tilde{\mathbf{z}}_C, \Delta\tilde{\mathbf{z}}_C \rangle \tag{1.19}$$

where

$$\mathbf{H}(\tilde{\mathbf{z}}_C) = \left.\frac{\partial^2 f(\tilde{\mathbf{z}}_C)}{\partial \tilde{\mathbf{z}}_C^* \, \partial \tilde{\mathbf{z}}_C^T}\right|_{\tilde{\mathbf{z}}_{C_0}}.$$

When writing the expansions in the transform domain, we assume that the function $f(\cdot)$ is written in terms of the transformed arguments, for example, we have $f(\mathbf{z}) = f(\tilde{\mathbf{z}}_C)$. Hence, in the expansions given in this section, we have included the variable explicitly in all the expressions.

The two Hessian matrices, $\mathbf{H}(\tilde{\mathbf{z}}_R)$ and $\mathbf{H}(\tilde{\mathbf{z}}_C)$ are related through the mapping

$$\mathbf{H}(\tilde{\mathbf{z}}_R) = \tilde{\mathbf{U}}^H \mathbf{H}(\tilde{\mathbf{z}}_C) \tilde{\mathbf{U}}$$

where $\tilde{\mathbf{U}}$ is defined in Section 1.2.3. Since the real-valued Hessian is a symmetric matrix—we assume the existence of continuous second-order derivatives of $f(\cdot)$—and $\tilde{\mathbf{U}}\tilde{\mathbf{U}}^H = 2\mathbf{I}$, the complex Hessian matrix $\mathbf{H}(\tilde{\mathbf{z}}_C)$ is Hermitian. Hence, we can write

$$\mathbf{H}(\tilde{\mathbf{z}}_R) - \lambda \mathbf{I} = \tilde{\mathbf{U}}^H [\mathbf{H}(\tilde{\mathbf{z}}_C) - 2\lambda \mathbf{I}] \tilde{\mathbf{U}}$$

and observe that if λ is an eigenvalue of $\mathbf{H}(\tilde{\mathbf{z}}_C)$, then 2λ is an eigenvalue of $\mathbf{H}(\tilde{\mathbf{z}}_R)$. Thus, when checking whether the Hessian is a positive definite matrix—for example, for local optimality and local stability properties—one can work with either form of the Hessian. Hence, other properties of the Hessian such as its condition number, which is important in a number of scenarios for example, when deriving second-order learning algorithms, are also preserved under the transformation [110].

Even though it is generally more desirable to work in the original space where the functions are defined, which is typically $\mathbb{C}^N$, the transformations given in Section 1.2.3 can provide simplifications to the series expansions. For example, the mapping $\tilde{(\cdot)}_C$ given in (1.14) can lead to simplifications in the expressions as demonstrated in [86] in the derivation and local stability analysis of a complex independent component analysis algorithm. The use of Wirtinger calculus through the $\mathbb{R}^2 \mapsto \mathbb{C}^2$ mapping in this case leads to a simpler block structure for the final Hessian matrix $\mathbf{H}(\tilde{\mathbf{z}}_C)$ compared to $\mathbf{H}(\tilde{\mathbf{z}}_R)$, hence simplifying assumptions such as circularity of random variables as done in [13] for a similar setting can be avoided.

In this section, we concentrated on functions of vector variables. For matrix variables, a first-order expansion can be obtained in a very similar manner. For a function $f(\mathbf{Z}, \mathbf{Z}^*)$: $\mathbb{C}^{N \times M} \times \mathbb{C}^{N \times M} \to \mathbb{R}$, we have

$$\begin{aligned} \Delta f(\mathbf{Z}, \mathbf{Z}^*) &\approx \langle \nabla_{\mathbf{Z}} f, \Delta \mathbf{Z}^* \rangle + \langle \nabla_{\mathbf{Z}^*} f, \Delta \mathbf{Z} \rangle \\ &= 2\mathrm{Re}\{\langle \nabla_{\mathbf{Z}^*} f, \Delta \mathbf{Z} \rangle\} \end{aligned} \tag{1.20}$$

where $\partial f / \partial \mathbf{Z}$ is an $N \times M$ matrix whose (k, l)th entry is the partial derivative of f with respect to w_{kl} and the last equality follows only for real-valued functions. Again, it is the gradient with respect to the conjugate variable, that is, $\nabla_{\mathbf{Z}^*} f$, the quantity that defines the direction of the maximum rate of change in f with respect to $\mathbf{Z}$ not the gradient $\nabla_{\mathbf{Z}} f$.

Since the definition of a Hessian for a function of the form $f(\mathbf{Z}, \mathbf{Z}^*)$ does not result in a matrix form and cannot be written as one of the six forms given in Table 1.1, there are a number of options when working with the second-order expansions in this case. One approach is to write the expression directly in terms of each element, which is given by

$$\nabla^2_{\mathbf{Z}} f = \frac{1}{2}\sum_{m,n}\sum_{k,l} \frac{\partial^2 f}{\partial z_{mn}\partial z_{kl}} dz_{mn}dz_{kl} + \frac{1}{2}\sum_{m,n}\sum_{k,l} \frac{\partial^2 f}{\partial z^*_{mn}\partial z^*_{kl}} dz^*_{mn}dz^*_{kl}$$

$$+ \sum_{m,n}\sum_{k,l} \frac{\partial^2 f}{\partial z_{mn}\partial z^*_{kl}} dz_{mn}dz^*_{kl}.$$

Note that this form is written by evaluating the second-order term in (1.17) with respect to every entry of matrix $\mathbf{Z}$. In certain cases, second-order matrix differentials can be put into compact forms using matrix differentials introduced in Section 1.2.2 and invariant transforms as in [7]. Such a procedure allows for efficient derivations while keeping all the evaluations in the original transform domain as demonstrated in the derivation of maximum likelihood based relative gradient update rule for complex independent component analysis in [68].

Another approach for calculating differential or Hessian expressions of matrix variables is to use the vectorization operator $\text{vec}(\cdot)$ that converts the matrix to a vector form by stacking the columns of a matrix into a long column vector starting from the first column [50]. Then the analysis proceeds by using vector calculus. The approach requires working with careful definitions of functions for manipulating the variables defined as such and then their reshaping at the end. This is the approach taken in [46] for defining derivatives of functions with matrix arguments.

1.2.5 Statistics of Complex-Valued Random Variables and Random Processes

Statistical Description of Complex Random Variables and Vectors A complex-valued random variable $X = X_r + jX_i$ is defined through the joint probability density function (pdf) $f_X(x) \triangleq f_{X_r X_i}(x_r, x_i)$ provided that it exists. For a pdf $f_{X_r X_i}(x_r, x_i)$ that is differentiable with respect to x_r and x_i individually, we can write $f_{X_r X_i}(x_r, x_i) = f(x, x^*)$ where $x = x_r + jx_i$, and use the expression written in terms of x and x^* in the evaluations to take advantage of Wirtinger calculus.

Note that writing the pdf in the form $f(x, x^*)$ is mainly a representation, which in most instances, significantly simplifies the evaluations. Thus, it is primarily a computational tool. As in the case of representation of any function using the variables x and x^* rather than only x, the form is degenerate since the two variables are not independent of each other. In [87], the evaluation of probability masses using the form $f(x, x^*)$ is discussed in detail, both for continuous and mixed-distribution random variables. When evaluating expected values using a pdf written as $f(x, x^*)$, we have to thus consider the contour integrals as given in (1.9).

The joint pdf for a complex random vector $\mathbf{X} \in \mathbb{C}^N$ is extended to the form $f(\mathbf{x}, \mathbf{x}^*)$: $\mathbb{C}^N \times \mathbb{C}^N \mapsto \mathbb{R}$ similarly. In the subsequent discussion, we write the expectations with respect to the corresponding joint pdf, pdf of a scalar or vector random variable as defined here.

Second-order statistics of a complex random vector $\mathbf{X}$ are completely defined through two (auto) covariance matrices: the covariance matrix

$$\mathbf{C}_{XX} = E\{(\mathbf{X} - E\{\mathbf{X}\})(\mathbf{X} - E\{\mathbf{X}\})^H\}$$

that is commonly used, and in addition, the *pseudo-covariance* [81] matrix—also called the complementary covariance [101] or the relation matrix [92]—given by

$$\mathbf{P}_{XX} = E\{(\mathbf{X} - E\{\mathbf{X}\})(\mathbf{X} - E\{\mathbf{X}\})^T\}.$$

Expressions are written similarly for the cross-covariance matrices $\mathbf{C}_{XY}$ and $\mathbf{P}_{XY}$ of two complex random vectors $\mathbf{X}$ and $\mathbf{Y}$. The properties given in Section 1.2.3 for complex-to-real mappings can be effectively used to work with covariance matrices in either the complex- or the double-dimensioned real domain. In the sequel, we drop the indices used in matrix definitions here when the matrices in question are clear from the context, and assume that the vectors are zero mean without loss of generality.

Through their definitions, the covariance matrix is a Hermitian and the pseudo-covariance matrix is a complex symmetric matrix. As is easily shown, the covariance matrix is nonnegative definite—and in practice typically positive definite. Hence, the nonnegative eigenvalues of the covariance matrix can be identified using simple eigenvalue decomposition. For the pseudo-covariance matrix, however, we need to use Takagi's factorization [49] to obtain the spectral representation such that

$$\mathbf{P} = \mathbf{Q}\mathbf{D}\mathbf{Q}^T$$

where $\mathbf{Q}$ is a unitary matrix and $\mathbf{D} = \text{diag}\{\kappa_1, \kappa_2, \ldots, \kappa_N\}$ contains the singular values, $1 \geq \kappa_1 \geq \kappa_2 \geq \cdots \geq \kappa_N \geq 0$, on its diagonal. The values κ_n are canonical correlations of a given vector and its complex conjugate [100] and are called the *circularity coefficients* [33]—though noncircularity coefficients might be the more appropriate name—since for a second-order circular random vector, which we define next, these values are all zero.

The vector transformation $\mathbf{z} \in \mathbb{C}^N \mapsto \bar{\mathbf{z}}_C \in \mathbb{C}^{2N}$ given in (1.13) can be used to define a single matrix summarizing the second-order properties of a random vector $\mathbf{X}$, which is called the augmented correlation matrix [92, 101]

$$E\{\bar{\mathbf{X}}_C \bar{\mathbf{X}}_C^H\} = E\left\{\begin{bmatrix} \mathbf{X} \\ \mathbf{X}^* \end{bmatrix} [\mathbf{X}^H \mathbf{X}^T]\right\} = \begin{bmatrix} \mathbf{C} & \mathbf{P} \\ \mathbf{P}^* & \mathbf{C}^* \end{bmatrix}$$

and is used in the study of widely linear least mean squares filter which we discuss in Section 1.4.

Circularity Properties of a Complex Random Variable and Random Vector An important property of complex-valued random variables is related to their circular nature.

> A zero-mean complex random variable is called *second-order circular* [91] (or proper [81, 101]) when its pseudo-covariance is zero, that is,
>
> $$E\{X^2\} = 0$$
>
> which implies that $\sigma_{X_r} = \sigma_{X_i}$ and $E\{X_r X_i\} = 0$ where σ_{X_r} and σ_{X_i} are the standard deviations of the real and imaginary parts of the variable.

For a random vector $\mathbf{X}$, the condition for second-order circularity is written in terms of the pseudo-covariance matrix as $\mathbf{P} = 0$, which implies that $E\{\mathbf{X}_r\mathbf{X}_r^T\} = E\{\mathbf{X}_i\mathbf{X}_i^T\}$ and $E\{\mathbf{X}_r\mathbf{X}_i^T\} = -E\{\mathbf{X}_i\mathbf{X}_r^T\}$.

A stronger condition for circularity is based on the pdf of the random variable.

> A random variable X is called *circular in the strict-sense*, or simply *circular*, if X and $Xe^{j\theta}$ have the same pdf, that is, the pdf is rotation invariant [91].

In this case, the phase is non-informative and the pdf is a function of only the magnitude, $f_X(x) = g(|x|)$ where $g: \mathbb{R} \mapsto \mathbb{R}$, hence the pdf can be written as a function of zz^* rather than z and z^* separately. A direct consequence of this property is that $E\{X^p(X^*)^q\} = 0$ for all $p \neq q$ if X is circular. Circularity is a strong property, preserved under linear transformations, and since it implies noninformative phase, a real-valued approach and a complex-valued approach for this case are usually equivalent [109].

As one would expect, circularity implies second-order circularity, and only for a Gaussian-distributed random variable, second-order circularity implies (strict sense) circularity. Otherwise, the reverse is not true.

For random vectors, in [91], three different types of circularity are identified. A random vector $\mathbf{X} \in \mathbb{C}^N$ is called

- *marginally circular* if each component of the random vector X_n is a circular random variable;
- *weakly circular* if $\mathbf{X}$ and $\mathbf{X}e^{j\theta}$ have the same distribution for any given θ; and
- *strongly circular* if $\mathbf{X}$ and $\mathbf{X}'$ have the same distribution where $\mathbf{X}'$ is formed by rotating the corresponding entries (random variables) in $\mathbf{X}$ by θ_n, such that $X'_n = X_n e^{j\theta_n}$. This condition is satisfied when θ_k are independent and identically distributed random variables with uniform distribution in $[-\pi, \pi]$ and are independent of the amplitude of the random variables, X_n.

As the definitions suggest, strong circularity implies weak circularity, and weak circularity implies marginal circularity.

Differential Entropy of Complex Random Vectors The differential entropy of a zero mean random vector $\mathbf{X} \in \mathbb{C}^N$ is given by the joint entropy

$H(\mathbf{X}_r, \mathbf{X}_i)$, and satisfies [81]:

$$H(\mathbf{X}) \leq \log[(\pi e)^N \det(\mathbf{C})] \tag{1.21}$$

with equality if, and only if, $\mathbf{X}$ is second-order circular and Gaussian with zero mean. Thus, it is a *circular* Gaussian random variable that maximizes the entropy for the complex case. It is also worthwhile to note that orthogonality and Gaussianity, together do not imply independence for complex Gaussian random variables, unless the variable is circular.

For a noncircular Gaussian random vector, we have [33, 100]

$$H_{\text{noncirc}} = \underbrace{\log[(\pi e)^N \det(\mathbf{C})]}_{H_{\text{circ}}} + \frac{1}{2}\log \prod_{n=1}^{N} (1 - \kappa_n^2)$$

where κ_n are the singular values of $\mathbf{P}$ as defined and $\kappa_n = 0$ when the random vector is circular. Hence, the circularity coefficients provide an attractive measure for quantifying circularity and a number of those measures are studied in [100]. Since $\kappa_n \leq 1$ for all n, the second term is negative for noncircular random variables decreasing the overall differential entropy as a function of the circularity coefficients.

Complex Random Processes In [8, 27, 81, 90, 91], the statistical characterization and properties of complex random processes are discussed in detail. In particular, [91] explores the strong relationship between stationarity and circularity of a random process through definitions of circularity and stationarity with varying degrees of assumptions on the properties of the process.

In our introduction to complex random processes, we focus on discrete-time processes and primarily use the notations and terminology adopted by [81] and [91]. The covariance function for a complex discrete-time random process $X(n)$ is written as

$$c(n, m) = E\{X(n)X^*(m)\} - E\{X(n)\}E\{X^*(m)\}$$

and the correlation function as $E\{X(n)X^*(m)\}$.

To completely define the second-order statistics, as in the case of random variables, we also define the pseudo-covariance function [81]—also called the complementary covariance [101] and the relation function [91]—as

$$p(n, m) = E\{X(n)X(m)\} - E\{X(n)\}E\{X(m)\}.$$

In the sequel, to simplify the expressions, we assume zero mean random processes, and hence, the covariance and correlation functions coincide.

Stationarity and Circularity Properties of Random Processes A random signal $X(n)$ is stationary if all of its statistical properties are invariant to any given time shift (translations by the origin), or alternatively, if the family of distributions that describe the random process as a collection of random variables are all invariant to any time shift. As in the case of a random variable, the distribution for a complex random process is defined as the joint distribution of real and imaginary parts of the process.

For second-order stationarity, again we need to consider the complete characterization using the pseudo-covariance function.

A complex random process $X(n)$ is called *wide sense stationary* (WSS) if $E\{X(n)\} = m_x$, is independent of n and if

$$E\{X(n)X^*(m)\} = r(n - m)$$

and it is called *second-order stationary* (SOS) if it is WSS and in addition, its pseudo-covariance function satisfies and

$$E\{X(n)X(m)\} = p(n - m)$$

that is, it is a function of the time difference $n - m$.

Obviously, the two definitions are equivalent for real-valued signals and second-order stationarity implies WSS but the reverse is not true. In [81], second-order stationarity is identified as circular WSS and a WSS process is defined as an SOS process.

Let $X(n)$ be a second-order zero mean stationary process. Using the widely-linear transform for the scalar-valued random process $X(n)$, $\bar{\mathbf{X}}_C(n) = [X(n)\ X^*(n)]^T$ we define the spectral matrix of $\bar{\mathbf{X}}_C(n)$ as the Fourier transform of the covariance function of $\bar{\mathbf{X}}_C(n)$ [93], which is given by

$$\mathbf{C}_C(f) \triangleq \mathcal{F}\{E\{\bar{\mathbf{X}}_C(n)\bar{\mathbf{X}}_C^H(n)\}\} = \begin{bmatrix} C(f) & P(f) \\ P^*(-f) & C(-f) \end{bmatrix}$$

and where $C(f)$ and $P(f)$ denote the Fourier transforms of the covariance and pseudo-covariance functions of $X(n)$, that is, of $c(k)$ and $p(k)$ respectively.

The covariance function is nonnegative definite and the pseudo-covariance function of a SOS process is symmetric. Hence its Fourier transform also satisfies $P(f) = P(-f)$. Since, by definition, the spectral matrix $\mathbf{C}_C(f)$ has to be nonnegative definite, we obtain the condition

$$|P(f)|^2 \leq C(f)C(-f)$$

from the condition for nonnegative definiteness of $\mathbf{C}_C(f)$. The inequality also states the relationship between the power spectrum $C(f)$ and the Fourier transform of a pseudo-covariance function.

A random process is called second-order circular if its pseudo-covariance function

$$p(k) = 0, \quad \forall k$$

a condition that requires the process to be SOS.

Also, it is easy to observe that an analytic signal constructed from a WSS real signal is always second-order circular, since for an analytic signal we have $C(f) = 0$ for $f < 0$, which implies that $P(f) = 0$. An analytic signal corresponding to a nonstationary real signal is, on the other hand, in general noncircular [93].

EXAMPLES

In Figure 1.3, we show scatter plots of three random processes: (1) a circular complex autoregressive (AR) process driven by a circular Gaussian signal; (2) a 16 quadrature amplitude modulated (QAM) signal; and (3) a noncircular complex AR process driven by a circular Gaussian signal. The processes shown in the figure are circular, second-order circular, and noncircular respectively. The corresponding covariance and pseudo-covariance functions [$c(k)$ and $p(k)$] are shown in Figure 1.4, which demonstrate that for the first two processes, the pseudo-covariance function is zero since both are second-order circular.

Note that even though the 16-QAM signal is second-order circular, it is not circular as it is not invariant to phase rotations. A binary phase shift keying signal, on the other hand, is noncircular when interpreted as a complex signal, and since the signal is actually real valued, its covariance and pseudo-covariance

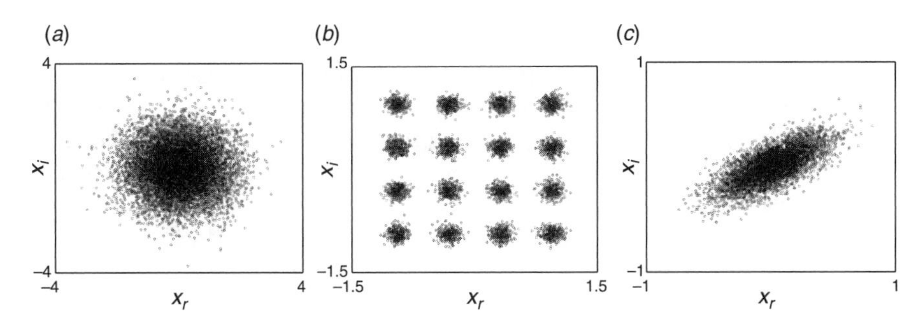

Figure 1.3 Scatter plots for a strictly (*a*) circular, (*b*) second-order circular 16-QAM, and (*c*) noncircular AR process.

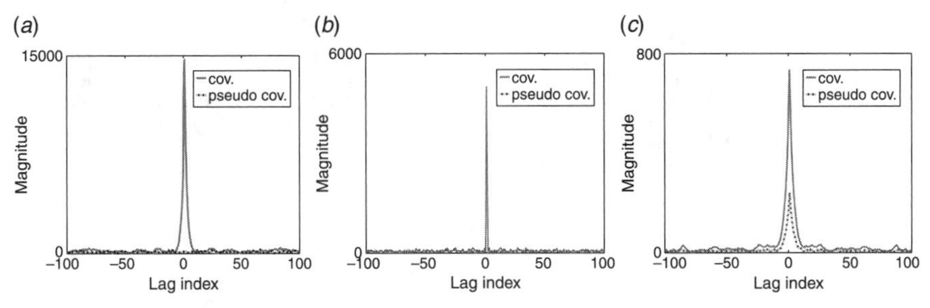

Figure 1.4 Covariance and pseudo-covariance function plots for the strictly (*a*) circular, (*b*) second-order circular 16-QAM, and (*c*) noncircular AR processes shown in Figure 1.3.

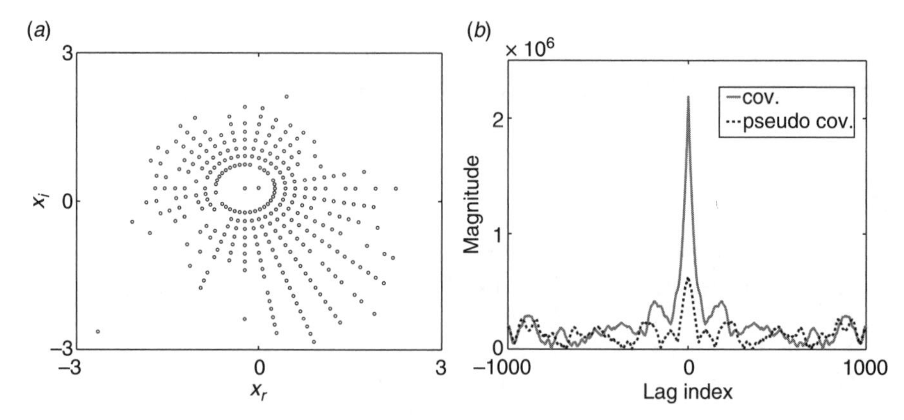

Figure 1.5 (*a*) Scatter plot and the (*b*) covariance and pseudo-covariance function plots for a sample wind data.

functions are the same. Hence, it has a non-zero pseudo-covariance function thus quantitatively verifying its noncircular nature.

In Figures 1.5 and 1.6, we show examples of real-world signals where the samples within each data set are normalized to zero mean and unit variance. The scatter plot of a sample of wind data obtained from http://mesonet.agron.iastate.edu is shown in Figure 1.5 along with its covariance and pseudo-covariance functions. The data are interpreted as complex by combining its strength as the magnitude and direction as the phase information. As observed from the scatter plot as well as its nonzero pseudo-covariance function, the signal is noncircular. Two more samples are shown in Figure 1.6. The example in Figure 1.6*a* shows

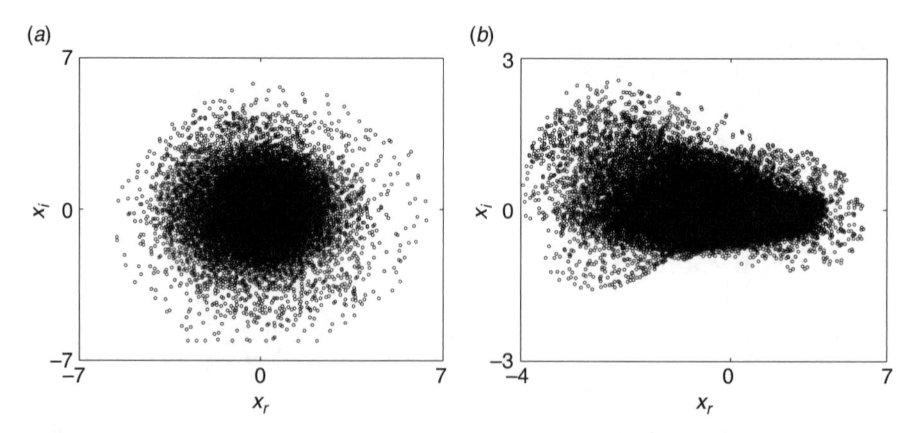

Figure 1.6 Scatter plots of (*a*) a circular (radar) data and (*b*) a noncircular (fMRI) data.

a sample Ice Multiparameter Imaging X-Band Radar (IPIX) data from the website http://soma.crl.mcmaster.ca/ipix/. As observed in the figure, the data have circular characteristics. In Figure 1.6*b*, we show the scatter plot of a functional MRI data volume. The paradigm used in the collection of the data is a simple motor task with a box-car type time-course, that is, the stimulus has periodic on and off periods. Since fMRI detects intensity changes, to evaluate the value of the fMRI signal at each voxel, we have calculated the average difference between the intensity values during the period the stimulus was "on" and "off" as a function of time. The scatter plot suggests a highly noncircular signal. The noncircular nature of fMRI data is also noted in [47] as the a large signal change in magnitude is noted as being accompanied by a corresponding change in the phase. Even though in these examples we have based the classifications on circular nature on simple visual observations, such a classification can be statistically justified by using a proper measure of noncircularity and a statistical test such as the generalized likelihood ratio test [100, 102].

As demonstrated by these examples, noncircular signals commonly arise in practice even though circularity has been a common assumption for many signal processing problems. Thus, we emphasize the importance of designing algorithms for the general case where signals may be noncircular and not to make assumptions such as circularity.

1.3 OPTIMIZATION IN THE COMPLEX DOMAIN

Most problems in signal processing involve the optimization of a real-valued cost function, which, as we noted, is not differentiable in the complex domain. Using Wirtinger calculus, however, we can relax the stringent requirement for differentiability (complex differentiability) and when the more relaxed condition of real differentiability is satisfied, can perform optimization in the complex domain in a way quite similar to the real domain. In this section, we provide the basic relationships that enable the transformation between the real and the complex domains and demonstrate how they can be used to extend basic update rules to the complex domain. We first provide a basic review of first- and second-order learning rules in the real domain and then discuss the development of appropriate tools in $\mathbb{C}^N$.

1.3.1 Basic Optimization Approaches in $\mathbb{R}^N$

Most signal processing applications use an iterative optimization procedure to determine the parameter vector $\mathbf{w}$ for a given nonlinear function $f(\mathbf{w})$: $\mathbb{R}^N \mapsto \mathbb{R}$ that cannot be directly solved for $\mathbf{w}$. We start with an initial guess for the parameter vector (weights) $\mathbf{w}(0) \in \mathbb{R}^N$ and generate a sequence of iterations for the weights as $\mathbf{w}(1), \mathbf{w}(2), \ldots, \mathbf{w}(n)$ such that the cost function $f(\mathbf{w})$ decreases (increases) until it reaches a local minimum (maximum). At each iteration n (or typically time index for most signal processing applications), the weights are updated such that

$$\mathbf{w}(n+1) = \mathbf{w}(n) + \mu\,\mathbf{d}(n)$$

where μ is the stepsize and $\mathbf{d}(n)$ is the line search direction, that is, the update vector. Without loss of generality, if we consider a minimization problem, both μ and $\mathbf{d}(n)$ should be chosen such that $f[\mathbf{w}(n+1)] < f[\mathbf{w}(n)]$. In the derivation of the form of the update vector $\mathbf{d}(n)$, Taylor series expansions discussed in Section 1.2.4 play a key role.

To derive the gradient descent (also called the steepest descent) updates for the minimization of $f(\mathbf{w})$, we write the first-order Taylor series expansion of $f(\mathbf{w})$ at $\mathbf{w}(n+1)$ as

$$f[\mathbf{w}(n+1)] = f[\mathbf{w}(n)] + \langle \mu\mathbf{d}(n), \nabla_{\mathbf{w}(n)} f \rangle$$

where $\nabla_{\mathbf{w}(n)} f$ is the gradient vector of $f(\cdot)$ at $\mathbf{w}(n)$. The inner product between the gradient and the update vector is written as

$$\langle \mathbf{d}(n), \nabla_{\mathbf{w}(n)} f \rangle = \mathbf{d}^T(n)\nabla_{\mathbf{w}(n)} f = \|\mathbf{d}(n)\|\,\|\nabla_{\mathbf{w}(n)} f\| \cos\theta$$

where θ is the angle between the two vectors. Thus, for a fixed stepsize μ and magnitude of $\mathbf{d}(n)$, maximum decrease in $f[\mathbf{w}(n)]$ is achieved when $\mathbf{d}(n)$ and $\nabla_{\mathbf{w}(n)} f$ are in reverse directions yielding the gradient descent update rule

$$\mathbf{w}(n+1) = \mathbf{w}(n) - \mu\nabla_{\mathbf{w}(n)} f.$$

Newton method, on the other hand, assumes that the function can be locally approximated as a quadratic function in the region around the optimum. Thus, to derive the Newton update, we write the Taylor series expansion of $f[\mathbf{w}(n+1)]$ up to the second order as

$$\begin{aligned} f[\mathbf{w}(n+1)] &= f[\mathbf{w}(n)] + \mathbf{d}^T(n)\nabla_{\mathbf{w}(n)} f + \frac{1}{2}\mathbf{d}^T(n)\mathbf{H}[\mathbf{w}(n)]\mathbf{d}(n) \\ &= f[\mathbf{w}(n)] + \langle \nabla_{\mathbf{w}(n)} f, \mathbf{d}(n) \rangle + \frac{1}{2}\langle \mathbf{H}[\mathbf{w}(n)]\mathbf{d}(n), \mathbf{d}(n) \rangle \end{aligned}$$

where $\mathbf{H}[\mathbf{w}(n)] \triangleq \nabla^2_{\mathbf{w}(n)} f$ is the Hessian matrix of $f(\mathbf{w})$ at $\mathbf{w}(n)$ and the stepsize μ is set to 1. Setting the derivative of this expansion [with respect to $\mathbf{d}(n)$] to zero, we obtain

$$\nabla_{\mathbf{w}(n)} f + \mathbf{H}[\mathbf{w}(n)]\mathbf{d}(n) = 0 \tag{1.22}$$

as the necessary condition for the optimum function change. The optimum direction

$$\mathbf{d}(n) = -(\mathbf{H}[\mathbf{w}(n)])^{-1}\nabla_{\mathbf{w}(n)} f \tag{1.23}$$

is called the Newton direction if $\mathbf{H}[\mathbf{w}(n)]$ is nonsingular. Newton method converges quadratically to a local optimum if $\mathbf{w}(0)$ is sufficiently close to this point and if the Hessian is positive definite. However, the method faces difficulties when the quadratic approximation is not a reasonable one at the current weight update and/or the Hessian is not positive definite. Thus a number of modifications have been proposed to the Newton method, such as performing a line search along the Newton direction, rather than using the stepsize that minimizes the quadratic model assumption. More importantly, a number of procedures are introduced that use an approximate Hessian rather than the actual Hessian that allow better numerical properties. These include the Davidon–Fletcher–Powell (DFP) method and the Broyden–Fletcher–Goldfarb–Shanno (BFGS) method [82].

Another approach is to solve (1.22) iteratively, which is desirable also when the dimensionality of the problem is high and/or the numerical properties of the Hessian are known to be poor. For the task, we can employ the well known conjugate gradient algorithm, which generates a sequence $\mathbf{d}(1), \mathbf{d}(2), \ldots, \mathbf{d}(k)$ such that $\mathbf{d}(k)$ converges to the optimal direction $-(\mathbf{H}[\mathbf{w}(n)])^{-1}\nabla_{\mathbf{w}(n)}f$.

A set of nonzero vectors $[\mathbf{c}(0), \mathbf{c}(1), \ldots, \mathbf{c}(n)]$ are said to be conjugate with respect to a symmetric positive definite matrix $\mathbf{A}$ if

$$\mathbf{c}^T(k)\mathbf{A}\mathbf{c}(l) = 0, \quad \text{for all } k \neq l$$

where, in this case $\mathbf{A} = \mathbf{H}[\mathbf{w}(n)]$.

It can be shown that for any $\mathbf{d}(0) \in \mathbb{R}^N$, the sequence $\mathbf{d}(k)$ generated by the conjugate direction algorithm as

$$\begin{aligned}
\mathbf{d}(k+1) &= \mathbf{d}(k) + \alpha_k \mathbf{c}(k) \\
\alpha(k) &= -\frac{\mathbf{q}^T(k)\mathbf{c}(k)}{\mathbf{c}^T(k)\mathbf{H}[\mathbf{w}(n)]\mathbf{c}(k)} \\
\mathbf{q}(k) &= \nabla_{\mathbf{w}(n)}f + \mathbf{H}[\mathbf{w}(n)]\mathbf{d}(k)
\end{aligned}$$

converges to the optimal solution at most N steps. The question that remains is how to construct the set of conjugate directions. Generally $\mathbf{c}(k)$ is selected to be a linear combination of $\mathbf{q}(k)$ and the previous direction $\mathbf{c}(k-1)$ as

$$\mathbf{c}(k) = -\mathbf{q}(k) + \beta(k)\mathbf{c}(k-1)$$

where

$$\beta(k) = \frac{\mathbf{q}^T(k)\mathbf{H}[\mathbf{w}(n)]\mathbf{c}(k-1)}{\mathbf{c}^T(k-1)\mathbf{H}[\mathbf{w}(n)]\mathbf{c}(k-1)}$$

is determined by the constraint that $\mathbf{c}(k)$ and $\mathbf{c}(k-1)$ must be conjugate to the Hessian matrix.

1.3.2 Vector Optimization in $\mathbb{C}^N$

Given a real-differentiable cost function $f(\mathbf{w})$: $\mathbb{C}^N \mapsto \mathbb{R}$, we can write $f(\mathbf{w}) = f(\mathbf{w}, \mathbf{w}^*)$ and take advantage of Wirtinger calculus as discussed in Section 1.2.2. The first-order Taylor series expansion of $f(\mathbf{w}, \mathbf{w}^*)$ is given by (1.18), and as discussed in Section 1.2.4, it is the gradient with respect to the *conjugate* of the variable that results in the maximum change for the complex case. Hence, the updates for gradient optimization of f is written as

$$\Delta\mathbf{w} = \mathbf{w}(n+1) - \mathbf{w}(n) = -\mu\nabla_{\mathbf{w}^*(n)} f. \tag{1.24}$$

The update given in (1.24) leads to a nonpositive increment, $\Delta f = -2\mu\|\nabla_{\mathbf{w}(n)} f\|^2$, while the update that uses $\Delta\mathbf{w} = -\mu\nabla_{\mathbf{w}(n)} f$, leads to changes of the form $\Delta f = -2\mu\mathrm{Re}\{\langle\nabla_{\mathbf{w}^*(n)} f, \nabla_{\mathbf{w}(n)} f\rangle\}$, which are not guaranteed to be nonpositive. Here, we consider only first-order corrections since μ is typically very small.

The complex gradient update rule given in (1.24) can be also derived through the relationship given in the following proposition, which provides the connection between the real-valued and the complex-valued gradients. Using the mappings defined in Table 1.2 (Section 1.2.3) and the linear transformations among them, we can extend Wirtinger derivatives to the vector case both for the first- and second-order derivatives as stated in the following proposition.

Proposition 1 *Given a function* $f(\mathbf{w}, \mathbf{w}^*)$: $\mathbb{C}^N \times \mathbb{C}^N \mapsto \mathbb{R}$ *that is real differentiable up to the second-order. If we write the function as* $f(\bar{\mathbf{w}}_R)$: $\mathbb{R}^{2N} \mapsto \mathbb{R}$ *using the definitions for* $\bar{\mathbf{w}}_C$ *and* $\bar{\mathbf{w}}_R$ *given in Table 1.2 we have*

$$\frac{\partial f}{\partial \bar{\mathbf{w}}_R} = \mathbf{U}^H \frac{\partial f}{\partial \bar{\mathbf{w}}_C^*} \tag{1.25}$$

$$\frac{\partial^2 f}{\partial \bar{\mathbf{w}}_R \partial \bar{\mathbf{w}}_R^T} = \mathbf{U}^H \frac{\partial^2 f}{\partial \bar{\mathbf{w}}_C^* \partial \bar{\mathbf{w}}_C^T} \mathbf{U} \tag{1.26}$$

where $\mathbf{U} = \begin{bmatrix} \mathbf{I} & j\mathbf{I} \\ \mathbf{I} & -j\mathbf{I} \end{bmatrix}$.

Proof 1 *Since we have* $\mathbf{U}\mathbf{U}^H = 2\mathbf{I}$, $\bar{\mathbf{w}}_C = \mathbf{U}\bar{\mathbf{w}}_R$ *and* $\bar{\mathbf{w}}_R = \frac{1}{2}\mathbf{U}^H\bar{\mathbf{w}}_C$. *We can thus write the two Wirtinger derivatives given in (1.5) in vector form as*

$$\frac{\partial f}{\partial \bar{\mathbf{w}}_C} = \frac{1}{2}\mathbf{U}^* \frac{\partial f}{\partial \bar{\mathbf{w}}_R}$$

in a single equation. Rewriting the above equality as

$$\frac{\partial f}{\partial \bar{\mathbf{w}}_R} = \mathbf{U}^T \frac{\partial f}{\partial \bar{\mathbf{w}}_C} = \mathbf{U}^H \frac{\partial f}{\partial \bar{\mathbf{w}}_C^*} \tag{1.27}$$

we obtain the first-order connection between the real and the complex gradient.

Taking the transpose of the first equality in (1.27), we have

$$\frac{\partial f}{\partial \bar{\mathbf{w}}_R^T} = \frac{\partial f}{\partial \bar{\mathbf{w}}_C^T}\mathbf{U}. \tag{1.28}$$

We regard the kth element of the two row vectors in (1.28) as two equal scalar-valued functions defined on $\bar{\mathbf{w}}_R$ *and* $\bar{\mathbf{w}}_C$, *and take their derivatives to obtain*

$$\frac{\partial\left(\frac{\partial f}{\partial \bar{\mathbf{w}}_R^T}\right)_k}{\partial \bar{\mathbf{w}}_R} = \mathbf{U}^T \frac{\partial\left(\frac{\partial f}{\partial \bar{\mathbf{w}}_C^T}\mathbf{U}\right)_k}{\partial \bar{\mathbf{w}}_C}.$$

We can then take the conjugate on each side and write the equality in vector form as

$$\frac{\partial^2 f}{\partial \bar{\mathbf{w}}_R \partial \bar{\mathbf{w}}_R^T} = \mathbf{U}^H \frac{\partial^2 f}{\partial \bar{\mathbf{w}}_C^* \partial \bar{\mathbf{w}}_C^T}\mathbf{U} = \mathbf{U}^T \frac{\partial^2 f}{\partial \bar{\mathbf{w}}_C \partial \bar{\mathbf{w}}_C^T}\mathbf{U}$$

to obtain the second-order relationship given in (1.26).

The second-order differential relationship for vector parameters given in (1.26) is first reported in [111] but is defined with respect to variables $\tilde{\mathbf{w}}_R$ and $\tilde{\mathbf{w}}_C$ using element-wise transforms given in Table 1.2. Using the mapping $\bar{\mathbf{w}}_C$ as we have shown here rather than the element-wise transform enables one to easily reduce the dimension of problem from C^{2N} to C^N. The second-order Taylor series expansion using the two forms ($\tilde{\mathbf{w}}_C$ and $\bar{\mathbf{w}}_C$) are the same, as expected, and we can write using either $\tilde{\mathbf{w}}_C$ or $\bar{\mathbf{w}}_C$

$$\Delta f \approx \Delta \bar{\mathbf{w}}_C^T \frac{\partial f}{\partial \bar{\mathbf{w}}_C} + \frac{1}{2}\Delta\, \bar{\mathbf{w}}_C^H \frac{\partial^2 f}{\partial \bar{\mathbf{w}}_C^*\, \partial \bar{\mathbf{w}}_C^T}\Delta \bar{\mathbf{w}}_C \tag{1.29}$$

as in (1.19), a form that demonstrates the fact that the $C^{2N\times 2N}$ Hessian in (1.29) can be decomposed into three $C^{N\times N}$ Hessians which are given in (1.17).

The two complex-to-real relationships given in (1.25) and (1.26) are particularly useful for the derivation of update rules in the complex domain. Next, we show their application in the derivation of the complex gradient and the complex Newton updates, and note the connection to the corresponding update rules in the real domain.

Complex Gradient Updates Given a real-differentiable function f as defined in Proposition 1, the well-known gradient update rule for $f(\bar{\mathbf{w}}_R)$ is

$$\Delta \bar{\mathbf{w}}_R = -\mu \frac{\partial f}{\partial \bar{\mathbf{w}}_R}$$

which can be mapped to the complex domain using (1.25) as

$$\Delta \bar{\mathbf{w}}_C = \mathbf{U}\Delta \bar{\mathbf{w}}_R = -\mu \mathbf{U}\frac{\partial f}{\partial \bar{\mathbf{w}}_R} = -2\mu \frac{\partial f}{\partial \bar{\mathbf{w}}_C^*}$$

The dimension of the update equation can be further decreased as

$$\begin{bmatrix} \Delta \mathbf{w} \\ \Delta \mathbf{w}^* \end{bmatrix} = -2\mu \begin{bmatrix} \dfrac{\partial f}{\partial \mathbf{w}^*} \\ \dfrac{\partial f}{\partial \mathbf{w}} \end{bmatrix} \Longrightarrow \Delta \mathbf{w} = -2\mu \frac{\partial f}{\partial \mathbf{w}^*}.$$

Complex Newton Updates

Proposition 2 *Given function $f(\cdot)$ defined in Proposition 1, Newton update in $\mathbb{R}^{2N}$ given by*

$$\frac{\partial^2 f}{\partial \bar{\mathbf{w}}_R \partial \bar{\mathbf{w}}_R^T} \Delta \bar{\mathbf{w}}_R = -\frac{\partial f}{\partial \bar{\mathbf{w}}_R} \tag{1.30}$$

is equivalent to

$$\Delta \mathbf{w} = -(\mathbf{H}_2^* - \mathbf{H}_1^* \mathbf{H}_2^{-1} \mathbf{H}_1)^{-1} \left(\frac{\partial f}{\partial \mathbf{w}^*} - \mathbf{H}_1^* \mathbf{H}_2^{-1} \frac{\partial f}{\partial \mathbf{w}} \right) \tag{1.31}$$

in $\mathbb{C}^N$, where

$$\mathbf{H}_1 \triangleq \frac{\partial^2 f}{\partial \mathbf{w} \partial \mathbf{w}^T} \quad \text{and} \quad \mathbf{H}_2 \triangleq \frac{\partial^2 f}{\partial \mathbf{w} \partial \mathbf{w}^H}. \tag{1.32}$$

Proof 2 *By using (1.25) and (1.26), the real domain Newton updates given in (1.30) can be written as*

$$\frac{\partial^2 f}{\partial \bar{\mathbf{w}}_C^* \partial \bar{\mathbf{w}}_C^T} \Delta \bar{\mathbf{w}}_C = -\frac{\partial f}{\partial \bar{\mathbf{w}}_C^*}$$

which can then put into the form

$$\begin{bmatrix} \mathbf{H}_2^* & \mathbf{H}_1^* \\ \mathbf{H}_1 & \mathbf{H}_2 \end{bmatrix} \begin{bmatrix} \Delta \mathbf{w} \\ \Delta \mathbf{w}^* \end{bmatrix} = -\begin{bmatrix} \dfrac{\partial f}{\partial \mathbf{w}^*} \\ \dfrac{\partial f}{\partial \mathbf{w}} \end{bmatrix}$$

where $\mathbf{H}_1$ and $\mathbf{H}_2$ are defined in (1.32).

We can use the formula for the inverse of a partitioned positive definite matrix ([49], p. 472) when the nonnegative definite matrix $\dfrac{\partial^2 f}{\partial \bar{\mathbf{w}}_C^* \partial \bar{\mathbf{w}}_C^T}$ is positive definite, to write

$$\begin{bmatrix} \Delta \mathbf{w} \\ \Delta \mathbf{w}^* \end{bmatrix} = -\begin{bmatrix} \mathbf{T}^{-1} & -\mathbf{H}_2^{-*} \mathbf{H}_1^* \mathbf{T}^{-*} \\ -\mathbf{T}^{-*} \mathbf{H}_1 \mathbf{H}_2^{-*} & \mathbf{T}^{-*} \end{bmatrix} \begin{bmatrix} \dfrac{\partial f}{\partial \mathbf{w}^*} \\ \dfrac{\partial f}{\partial \mathbf{w}} \end{bmatrix}$$

where $\mathbf{T} \triangleq \mathbf{H}_2^* - \mathbf{H}_1^*\mathbf{H}_2^{-1}\mathbf{H}_1$ *and* $(\cdot)^{-*}$ *denotes* $[(\cdot)^*]^{-1}$. *Since* $\dfrac{\partial^2 f}{\partial \bar{\mathbf{w}}_C^* \partial \bar{\mathbf{w}}_C^T}$ *is Hermitian, we finally obtain the complex Newton's method given in (1.31). The expression for* $\Delta\mathbf{w}^*$ *is the conjugate of (1.31).*

In [80], it has been shown that the Newton algorithm for N complex variables cannot be written in a form similar to the real-valued case. However, as we have shown, by including the conjugate of N variables, it can be written as shown in (1.31), a form that is equivalent to the Newton method in $\mathbb{R}^{2n}$. This form is also given in [110] using the variables $\tilde{\mathbf{w}}_R$ and $\tilde{\mathbf{w}}_C$, which is shown to lead to the form given in (1.31) using the same notation in [64]. Also, a quasi-Newton update is given in [117] by setting the matrix $\mathbf{H}_1$ to a zero matrix, which might not define a descent direction for every case, as also noted in [64].

1.3.3 Matrix Optimization in $\mathbb{C}^N$

Complex Matrix Gradient Gradient of a matrix-valued variable can also be written similarly using Wirtinger calculus. For a real-differentiable $f(\mathbf{W}, \mathbf{W}^*)$: $\mathbb{C}^{N\times N} \times \mathbb{C}^{N\times N} \mapsto \mathbb{R}$, we recall the first-order Taylor series expansion given in (1.20)

$$\begin{aligned}\Delta f &\approx \left\langle \Delta\mathbf{W}, \frac{\partial f}{\partial \mathbf{W}^*}\right\rangle + \left\langle \Delta\mathbf{W}^*, \frac{\partial f}{\partial \mathbf{W}}\right\rangle \\ &= 2\mathrm{Re}\left\{\left\langle \Delta\mathbf{W}, \frac{\partial f}{\partial \mathbf{W}^*}\right\rangle\right\} \qquad (1.33)\end{aligned}$$

where $\dfrac{\partial f}{\partial \mathbf{W}}$ is an $N \times N$ matrix whose (m, n)th entry is the partial derivative of f with respect to w_{mn}. As in the vector case, the matrix gradient with respect to the conjugate $\dfrac{\partial f}{\partial \mathbf{W}^*}$ defines the direction of the maximum rate of change in f with respect to the variable $\mathbf{W}$.

Complex Relative Gradient Updates We can use the first-order Taylor series expansion to derive the relative gradient update rule [21] for complex matrix variables, which is usually directly extended to the complex case without a derivation [9, 18, 34]. To write the relative gradient rule, we consider an update of the parameter matrix $\mathbf{W}$ in the invariant form $G(\mathbf{W})\mathbf{W}$ [21]. We then write the first-order Taylor series expansion for the change of the form $G(\mathbf{W})\mathbf{W}$ as

$$\begin{aligned}\Delta f &\approx \left\langle G(\mathbf{W})\mathbf{W}, \frac{\partial f}{\partial \mathbf{W}^*}\right\rangle + \left\langle G(\mathbf{W}^*)\mathbf{W}^*, \frac{\partial f}{\partial \mathbf{W}}\right\rangle \\ &= 2\mathrm{Re}\left\{\left\langle G(\mathbf{W}), \frac{\partial f}{\partial \mathbf{W}^*}\mathbf{W}^H\right\rangle\right\}\end{aligned}$$

to determine the quantity that maximizes the rate of change in the function. Using the Cauchy–Bunyakovskii–Schwarz inequality, it is clear that $G(\mathbf{W})$ has to be in the same direction as $\frac{\partial f}{\partial \mathbf{W}^*}\mathbf{W}^H$ to maximize the change. Therefore we define the complex relative gradient of $f(\cdot)$ at $\mathbf{W}$ as $\frac{\partial f}{\partial \mathbf{W}^*}\mathbf{W}^H$ to write the relative gradient update term as

$$\Delta\mathbf{W} = -\mu G(\mathbf{W})\mathbf{W} = -\mu \frac{\partial f}{\partial \mathbf{W}^*}\mathbf{W}^H\mathbf{W}. \tag{1.34}$$

Upon substitution of $\Delta\mathbf{W}$ into (1.33), we observe that $\Delta f = -2\mu\|(\partial f/\partial \mathbf{W}^*)\mathbf{W}^H\|^2_{\text{Fro}}$, that is, it is a nonpositive quantity, thus a proper update term.

Complex Matrix Newton Update To derive the matrix Newton update rule, we need to write the Taylor series expansion up to the second order with respect to matrix variables. However, since the variables are matrix quantities, the resulting Hessian in this case is a tensor with four indices.

The Taylor series expansion up to the second order can be written as

$$\begin{aligned}\Delta f \approx &\sum_{m,n} \frac{\partial f}{\partial w_{mn}} dw_{mn} + \sum_{m,n} \frac{\partial f}{\partial w^*_{mn}} dw^*_{mn} + \sum_{m,n}\sum_{k,l} \frac{\partial^2 f}{\partial w_{mn}\partial w^*_{kl}} dw_{mn}\, dw^*_{kl} \\ &+ \frac{1}{2}\sum_{m,n}\sum_{k,l} \frac{\partial^2 f}{\partial w_{mn}\partial w_{kl}} dw_{mn}\, dw_{kl} + \frac{1}{2}\sum_{m,n}\sum_{k,l} \frac{\partial^2 f}{\partial w^*_{mn}\partial w^*_{kl}} dw^*_{mn}\, dw^*_{kl}.\end{aligned}$$

For the update of a single element w_{mn}, the Newton update rule is derived by taking the partial derivatives of the Taylor series expansion with respect to the differential dw_{mn} and setting it to zero

$$\frac{\partial(\Delta f)}{\partial(dw_{mn})} = \frac{\partial f}{\partial w_{mn}} + \sum_{k,l}\left(\frac{\partial^2 f}{\partial w_{mn}\partial w_{kl}} dw_{kl} + \frac{\partial^2 f}{\partial w_{mn}\partial w^*_{kl}} dw^*_{kl}\right) = 0 \tag{1.35}$$

where we have given the expression in element-wise form in order to keep the notation simple.

The solution of Newton equation in (1.35) thus yields the element-wise matrix Newton update rule for w_{mn}. In certain applications, such as independent component analysis, the Newton equation given in (1.35) can be written in a compact matrix form instead of the element-wise form given here. This point will be illustrated in Section 1.6.1 in the derivation of complex Newton updates for maximum likelihood independent component analysis.

1.3.4 Newton-Variant Updates

As we have shown in Section 1.3.2, equations (1.25) and (1.26) given in Proposition 1 play a key role in the derivation of the complex gradient and Newton update rules. Also, they can be used to extend the real-valued Newton variations that are proposed in the literature to the complex domain such that the limitations of the Newton method can be mitigated.

Linear Conjugate Gradient (CG) Updates For the Newton's method given in (1.3.1), in order to achieve convergence, we require the search direction $\Delta\bar{\mathbf{w}}_R$ to be a descent direction when minimizing a given cost function. This is the case when the Hessian $\dfrac{\partial^2 f}{\partial\bar{\mathbf{w}}_R \partial\bar{\mathbf{w}}_R^T}$ is positive definite. However, when the Hessian is not positive definite, $\Delta\bar{\mathbf{w}}_R$ may be an ascent direction. The line search Newton-CG method is one of the strategies for ensuring that the update is of good quality. In this strategy, we solve (1.30) using the CG method, terminating the updates if $\Delta\bar{\mathbf{w}}_R^T\left(\dfrac{\partial^2 f}{\partial\bar{\mathbf{w}}_R \partial\bar{\mathbf{w}}_R^T}\right)\Delta\bar{\mathbf{w}}_R \leq 0$.

In general, a complex-valued function is defined in $\mathbb{C}^N$. Hence, writing it in the form $f(\mathbf{w}, \mathbf{w}^*)$ is much more straightforward than converting it to a function of the $2N$ dimensional real variable as in $f(\bar{\mathbf{w}}_R)$. Using a procedure similar to the derivation of complex gradient and Newton updates, the complex-valued CG updates can be derived using the real-valued version given in Section 1.3.1. Using (1.25) and (1.26), and defining $\mathbf{s} \triangleq \partial f/\partial\mathbf{w}^*$, the complex CG method can be derived as:

Complex Conjugate Gradient Updates

Given an initial gradient $\mathbf{s}(0)$;
Set $\mathbf{x}(0) = \mathbf{0}$, $\mathbf{c}(0) = -\mathbf{s}(0)$, $k = 0$;
while $|s(k)| \neq 0$

$$\alpha(k) = \frac{\mathbf{s}^H(k)\mathbf{s}(k)}{\text{Re}\{\mathbf{c}^T(k)\mathbf{H}_2\mathbf{c}^*(k) + \mathbf{c}^T(k)\mathbf{H}_1\mathbf{c}(k)\}};$$

$\mathbf{x}(k+1) = \mathbf{x}(k) + \alpha(k)\mathbf{c}(k);$
$\mathbf{s}(k+1) = \mathbf{s}(k) + \alpha(k)(\mathbf{H}_2^*\mathbf{c}(k) + \mathbf{H}_1^*\mathbf{c}^*(k));$

$$\beta(k+1) = \frac{\mathbf{s}^H(k+1)\mathbf{s}(k+1)}{\mathbf{s}^H(k)\mathbf{s}(k)};$$

$\mathbf{c}(k+1) = -\mathbf{s}(k+1) + \beta(k+1)\mathbf{c}(k);$
$k = k + 1;$
end(while)

where $\mathbf{H}_1$ and $\mathbf{H}_2$ is defined in (1.32).
The complex line search Newton-CG algorithm is given as:

for $k = 0, 1, 2, \ldots$
  Compute a search direction $\Delta\mathbf{w}$ by applying the complex CG update rule, starting at $\mathbf{x}(0) = 0$.
  Terminate when $\text{Re}\{\mathbf{c}^T(k)\mathbf{H}_2\mathbf{c}^*(k) + \mathbf{c}^T(k)\mathbf{H}_1\mathbf{c}(k)\} \leq 0$;
  Set $\mathbf{w}(k+1) = \mathbf{w}(k) + \mu\Delta\mathbf{w}$, where μ satisfies a complex Wolfe condition.
end

The complex Wolfe condition [82] can be easily obtained from the real Wolfe condition using (1.25). It should be noted that the complex CG algorithm is a linear version. It is straightforward to obtain a nonlinear version based on the linear version as shown in [82] for the real case.

Other Newton Variant Updates As shown for the derivation of complex gradient and Newton update rules, we can easily obtain complex versions of other real Newton variant methods using (1.25) and (1.26). In [70], this is demonstrated for the real-valued scaled conjugate gradient (SCG) method [79]. SCG belongs to the class of CG methods and shows superlinear convergence in many optimization problems.

When the cost function takes a least-squares form, a complex version of the Gauss–Newton algorithm can be developed as in [64]. In the Gauss–Newton algorithm, the original Hessian matrix in the Newton update is replaced with a Gauss–Newton Hessian matrix, which has better numerical properties hence providing better performance. For more general cost functions, BFGS is a popular and efficient Newton variant method [82] and can be extended to the complex domain similarly.

1.4 WIDELY LINEAR ADAPTIVE FILTERING

As discussed in Section 1.2.5, in order to completely characterize the second-order statistics of a complex random process, we need to specify both the covariance and the pseudo-covariance functions. Only when the process is circular, the covariance function is sufficient since the pseudo-covariance in this case is zero. A fundamental result in this context, introduced in [94], states that a *widely linear* filter rather than the typically used linear one provides significant advantages in minimizing the mean-square error when the traditional circularity assumptions on the data do not hold. A widely linear filter augments the data vector with the conjugate of the data, thus providing both the covariance and pseudo-covariance information for a filter designed using a second-order error criterion.

The assumption of circularity is a limiting assumption as, in practice, the real and imaginary parts of a signal typically will have correlations and/or different variances. One of the reasons for the prevalence of the circularity assumption in signal processing has been due to the inherent assumption of stationarity of signals. Since the complex envelope of a stationary signal is second-order circular [91], circularity is directly implied in this case. However many signals are not stationary, and a good number of complex-valued signals such as fMRI and wind data as shown in Section 1.2.5, do not necessarily have circular distributions. Thus, the importance of widely linear filters started to be noted and widely linear filters have been proposed for applications such as interference cancelation, demodulation, and equalization for direct sequence code-division-multiple-access systems and array receivers [23, 56, 99] implemented either in direct form, or computed adaptively using the least-mean-square (LMS) [99] or recursive least squares (RLS) algorithms [55]. Next, we present the widely linear mean-square error filter and discuss its properties, in particular when computed using LMS updates as discussed in [5]. We use the vector notation introduced in

Section 1.2.3 which allows direct extension of most main results of a linear filter to the widely linear one.

1.4.1 Linear and Widely Linear Mean-Square Error Filter

A linear filter approximates the desired sequence $d(n)$ through a linear combination of a window of input samples $x(n)$ such that the estimate of the desired sequence is

$$y(n) = \mathbf{w}^H \mathbf{x}(n)$$

where the input vector at time n is written as $\mathbf{x}(n) = [x(n)\ x(n-1)\ \cdots\ x(n-N+1)]^T$ and the filter weights as $\mathbf{w} = [w_0\ w_1\ \cdots\ w_{N-1}]^T$. The minimum mean-square error (MSE) filter is designed such that the error

$$J_L(\mathbf{w}) = E\{|e(n)|^2\} = E\{|d(n) - y(n)|^2\}$$

is minimized. To evaluate the weights $\mathbf{w}_{\text{opt}}$ given by

$$\mathbf{w}_{\text{opt}} = \arg\min_{\mathbf{w}} J_L(\mathbf{w})$$

we can directly take the derivative of the MSE with respect to $\mathbf{w}^*$ (by treating the variable $\mathbf{w}$ as a constant)

$$\begin{aligned} \frac{\partial E\{e(n)e^*(n)\}}{\partial \mathbf{w}^*} &= \frac{\partial E\{[d(n) - \mathbf{w}^H \mathbf{x}(n)][d^*(n) - \mathbf{w}^T \mathbf{x}^*(n)]\}}{\partial \mathbf{w}^*} \\ &= -E\{\mathbf{x}(n)[d^*(n) - \mathbf{w}^T \mathbf{x}^*(n)]\} \end{aligned} \tag{1.36}$$

and obtain the *complex Wiener–Hopf equation*

$$E\{\mathbf{x}(n)\mathbf{x}^H(n)\}\mathbf{w}_{\text{opt}} = E\{d^*(n)\mathbf{x}(n)\}$$

by setting (1.36) to zero. For simplicity, we assume that the input is zero mean so that the covariance and correlation functions coincide. We define the input covariance matrix $\mathbf{C} = E\{\mathbf{x}\mathbf{x}^H\}$ and the cross-covariance vector $\mathbf{p} = E\{d^*(n)\mathbf{x}\}$, to write

$$\mathbf{w}_{\text{opt}} = \mathbf{C}^{-1}\mathbf{p}$$

when the input is persistenly exciting, that is, the covariance matrix is nonsingular, which is typically the case and our assumption for the rest of the discussion in this section.

We can also compute the weight vector **w** adaptively using gradient descent updates as discussed in Section 1.3.2

$$\begin{aligned}\mathbf{w}(n+1) &= \mathbf{w}(n) - \mu \frac{\partial J_L(\mathbf{w})}{\partial \mathbf{w}^*(n)} \\ &= \mathbf{w}(n) + \mu E\{e^*(n)\mathbf{x}(n)\}\end{aligned}$$

or using stochastic gradient updates as in

$$\mathbf{w}(n+1) = \mathbf{w}(n) + \mu e^*(n)\mathbf{x}(n)$$

which leads to the popular least-mean-square (LMS) algorithm [113]. For both updates, $\mu > 0$ is the stepsize that determines the trade-off between the rate of convergence and the minimum error $J_L(\mathbf{w}_{\text{opt}})$.

Widely Linear MSE Filter A widely linear filter forms the estimate of $d(n)$ through the inner product

$$y_{WL}(n) = \mathbf{v}^H \bar{\mathbf{x}}(n) \tag{1.37}$$

where the weight vector $\mathbf{v} = [v_0 \; v_1 \; \cdots \; v_{2N-1}]^T$, that is, it has double dimension compared to the linear filter and

$$\bar{\mathbf{x}}(n) = \begin{bmatrix} \mathbf{x}(n) \\ \mathbf{x}^*(n) \end{bmatrix}$$

as defined in Table 1.2 and the MSE cost in this case is written as

$$J_{WL}(\mathbf{w}) = E\{|d(n) - y_{WL}(n)|^2\}.$$

As in the case for the linear filter, the minimum MSE optimal weight vector is the solution of

$$\frac{\partial J_{WL}(\mathbf{v})}{\partial \mathbf{v}^*} = 0$$

and results in the *widely linear* complex Wiener–Hopf equation given by

$$E\{\bar{\mathbf{x}}(n)\bar{\mathbf{x}}^H(n)\}\mathbf{v}_{\text{opt}} = E\{d^*(n)\bar{\mathbf{x}}(n)\}.$$

We can solve for the optimal weight vector as

$$\mathbf{v}_{\text{opt}} = \bar{\mathbf{C}}^{-1}\bar{\mathbf{p}}$$

where

$$\bar{\mathbf{C}} = E\{\bar{\mathbf{x}}(n)\bar{\mathbf{x}}^H(n)\} = \begin{bmatrix} \mathbf{C} & \mathbf{P} \\ \mathbf{P}^* & \mathbf{C}^* \end{bmatrix}$$

and

$$\bar{\mathbf{p}} = E\{d^*(n)\bar{\mathbf{x}}(n)\} = \begin{bmatrix} \mathbf{p} \\ \mathbf{q}^* \end{bmatrix}$$

with the definition of the pseudo-covariance matrix $\mathbf{P} = E\{\mathbf{x}(n)\mathbf{x}^T(n)\}$ and the pseudo cross covariance vector $\mathbf{q} = E\{d(n)\mathbf{x}(n)\}$ in addition to the definitions for $\mathbf{C}$ and $\mathbf{p}$ given earlier for the linear MSE filter. Matrix $\bar{\mathbf{C}}$ provides the complete second-order statistical characterization for a zero-mean complex random process and is called the augmented covariance matrix.

The minimum MSE value for the two linear models can be calculated as

$$\begin{aligned} J_{L,\min} &\triangleq J_L(\mathbf{w}_{\text{opt}}) = E\{|d(n)|^2\} - \mathbf{p}^H\mathbf{C}^{-1}\mathbf{p} \\ J_{WL,\min} &\triangleq J_{WL}(\mathbf{v}_{\text{opt}}) = E\{|d(n)|^2\} - \bar{\mathbf{p}}^H\bar{\mathbf{C}}^{-1}\bar{\mathbf{p}} \end{aligned} \tag{1.38}$$

and the difference between the two is given by [94]

$$\begin{aligned} J_{\text{diff}} &= J_{L,\min} - J_{WL,\min} \\ &= (\mathbf{q}^* - \mathbf{P}^*\mathbf{C}^{-1}\mathbf{p})^H(\mathbf{C}^* - \mathbf{P}^*\mathbf{C}^{-1}\mathbf{P})^{-1}(\mathbf{q}^* - \mathbf{P}^*\mathbf{C}^{-1}\mathbf{p}). \end{aligned} \tag{1.39}$$

Since the covariance matrix $\mathbf{C}$ is assumed to be nonsingular and thus is positive definite, the error difference J_{diff} is always nonnegative. When the joint-circularity condition is satisfied, that is, when $\mathbf{P} = \mathbf{0}$ and $\mathbf{q} = \mathbf{0}$, the performance of the two filters, the linear and the widely linear filter, coincide, and there is no gain in using a widely linear filter. It can be shown that the performance of the two filters can be equal even for cases where the input is highly noncircular (see Problems 1.6 and 1.7). However, when certain circularity properties do not hold, widely linear filters provide important advantages in terms of performance [23, 94, 101] by including the complete statistical information.

Widely Linear LMS Algorithm The widely linear LMS algorithm is written similar to the linear case as

$$\mathbf{v}(n+1) = \mathbf{v}(n) + \mu e^*(n)\bar{\mathbf{x}}(n) \tag{1.40}$$

where μ is the stepsize and $e(n) = d(n) - \mathbf{v}^H(n)\bar{\mathbf{x}}(n)$.

The study of the properties of the LMS filter, which was introduced in 1960 [114], has been an active research topic and a thorough account of these is given in [43] based on the different types of assumptions that can be invoked to simplify the analysis. With the augmented vector notation, most of the results for the behavior of the linear LMS filter can be readily extended to the widely linear one.

The convergence of the LMS algorithm depends on the eigenvalues of the input covariance matrix, which in the case of a widely linear LMS filter, is replaced by the eigenvalues of the augmented covariance matrix. A main result in this context can be described through the natural modes of the LMS algorithm [16, 43] as follows.

Define $\boldsymbol{\varepsilon}(n)$ as the weight vector error difference $\boldsymbol{\varepsilon}(n) = \mathbf{v}(n) - \mathbf{v}_{\text{opt}}$ and let the desired response be written as

$$d(n) = \mathbf{v}_{\text{opt}}^H \bar{\mathbf{x}}(n) + e_0(n).$$

When the noise term $e_0\,(n)$ is strongly uncorrelated with the input, that is, uncorrelated with $x(n)$ and its conjugate, we have

$$E\{\boldsymbol{\varepsilon}(n+1)\} = (\mathbf{I} - \mu\bar{C})E\{\boldsymbol{\varepsilon}(n)\}$$

We introduce the rotated version of the weight vector error difference $\boldsymbol{\varepsilon}'(n) = \mathbf{Q}^H\boldsymbol{\varepsilon}(n)$ where $\mathbf{Q}$ is the unitary matrix composed of the eigenvectors associated with the eigenvalues of $\bar{\mathbf{C}}$, that is, we assume that the augmented covariance matrix is written through the unitary similarity transformation $\bar{\mathbf{C}} = \mathbf{Q}\bar{\Lambda}\mathbf{Q}^H$. The mean value of the natural mode $\varepsilon_k(n)$, that is, the kth element of vector $\boldsymbol{\varepsilon}'(n)$ can then be written as

$$E\{\varepsilon_k'(n)\} = \varepsilon_k'(0)(1 - \mu\bar{\lambda}_k)^n \tag{1.41}$$

where $\bar{\lambda}_k$ is the kth eigenvalue of $\bar{\mathbf{C}}$.

Thus for the convergence of LMS updates to the true solution in the mean, the step-size has to be chosen such that

$$0 < \mu < \frac{2}{\bar{\lambda}_{\max}}$$

where $\bar{\lambda}_{\max}$ is the maximum eigenvalue of the augmented covariance matrix $\bar{\mathbf{C}}$. Also, as is evident from the expression given in (1.41), small eigenvalues significantly slow down the convergence in the mean. These conclusions hold for the linear LMS filter by simply replacing the eigenvalues of $\bar{\mathbf{C}}$ by the the eigenvalues of $\mathbf{C}$, λ_ks.

A measure typically used for measuring the eigenvalue disparity of a given matrix is the condition number (or the eigenvalue spread), which is written as

$$\kappa(\mathbf{C}) = \frac{\lambda_{\max}}{\lambda_{\min}}$$

for a Hermitian matrix $\mathbf{C}$, a property satisfied by the covariance and augmented covariance matrices.

When the signal is circular, the augmented covariance matrix assumes the block diagonal form

$$\bar{\mathbf{C}}_{\text{circ}} = \begin{bmatrix} \mathbf{C} & 0 \\ 0 & \mathbf{C}^* \end{bmatrix}$$

and has eigenvalues that occur with even multiplicity. In this case, the conditioning of the augmented covariance matrix $\bar{\mathbf{C}}$ and $\mathbf{C}$ are the same. As the noncircularity of the signal increases, the values of the entries of the pseudo covariance matrix moves away from zero increasing the condition number of the augmented covariance matrix $\bar{\mathbf{C}}$, thus the advantage of using a widely linear filter for noncircular signals comes at a cost when the LMS algorithm is used when estimating the widely linear MSE solution. An update scheme such as recursive least squares algorithm [43] which is less sensitive to the eigenvalue spread can be more desirable in such cases. In the next example, we demonstrate the impact of noncircularity on the convergence of LMS algorithm using a simple input model.

■ EXAMPLE 1.5

Define a random process

$$X(n) = \sqrt{1-\rho^2}X_r(n) + j\rho X_i(n) \tag{1.42}$$

where $X_r(n)$ and $X_i(n)$ are two uncorrelated real-valued random processes, both Gaussian distributed with zero mean and unit variance. By changing the value of $\rho \in [0, 1]$, we can change the degree of noncircularity of $X(n)$ and for $\rho = 1/\sqrt{2}$, the random process $X(n)$ becomes circular. Note that since second-order circularity implies strict-sense circularity for Gaussian signals, this model lets us to generate a circular signal as well.

If we define the random vector $\mathbf{X}(n) = [X(n)X(n-1) \cdots X(n-N+1)]^T$, we can show that the covariance matrix of $\mathbf{X}(n)$ is given by $\mathbf{C} = \mathbf{I}$, and the pseudo covariance matrix as $\mathbf{P} = (1-2\rho^2)\mathbf{I}$. The eigenvalues of the augmented covariance matrix $\bar{\mathbf{C}}$ can be shown to be $2\rho^2$ and $2(1-\rho^2)$, each with multiplicity N. Hence, the condition number is given by

$$\kappa(\bar{\mathbf{C}}) = \frac{1}{\rho^2} - 1$$

if $\rho \in [0, 1/\sqrt{2}]$ and by its inverse if $\rho \in [1/\sqrt{2}, 1]$.

In Figure 1.7, we show the convergence behavior of a linear and a widely linear LMS filter with input generated using the model in (1.42) for identification of a system with coefficients $w_{\text{opt},n} = \alpha[1 + \cos(2\pi(n-3)/5] - j[1 + \cos(2\pi(n-3)/10)])$, $n = 1, \ldots, 5$, and α is chosen so that the weight norm is unity (in this case, $\alpha = 0.432$). The input signal to noise ratio is 20 dB and the step size is fixed at

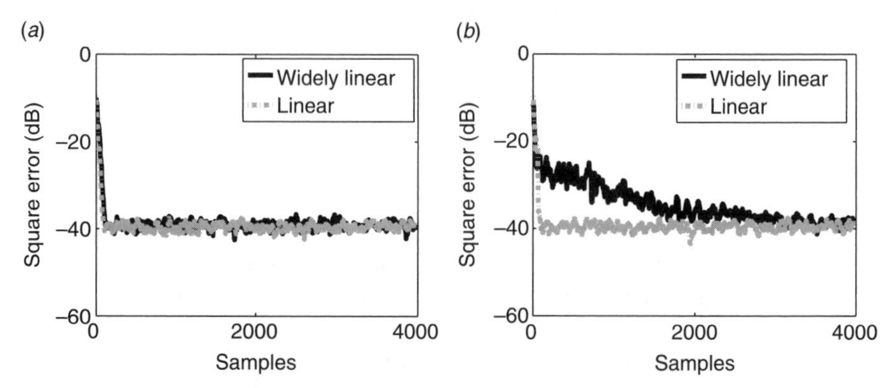

Figure 1.7 Convergence of the linear and widely linear filter for a circular input $\rho = 1\sqrt{2}$) (*a*) and a noncircular input ($\rho = 0.1$) (*b*) for a linear finite impulse response system identification problem.

$\mu = 0.04$ for all runs. In Figure 1.7*a*, we show the learning curve for a circular input, that is, $\rho = 1/\sqrt{2}$, and in Figure 1.7*b*, with a noncircular input where $\rho = 0.1$. For the first case, the condition numbers for both $\mathbf{C}$ and $\bar{\mathbf{C}}$ are approximately unity whereas for the second case, $\kappa(\mathbf{C}) \approx 1$ but $\kappa(\bar{\mathbf{C}}) \approx 100$. As expected, when the input is noncircular, the convergence rate of the widely linear LMS filter decreases. Since the lengths for the linear and widely linear filter are selected to match that of the unknown system (as 5 and 10 respectively), as discussed in Problem 6, both filters yield similar steady-state mean square error values.

In this example, even though the input is noncircular, the use of a widely linear filter does not provide an additional advantage in terms of MSE, and in addition, the convergence rate of the LMS algorithm decreases when the input is noncircular. Another observation to note for Example 1.5 is that the steady-state error variance for the widely linear filter is slightly higher compared to the linear filter. The steady-state MSE for the widely linear LMS filter can be approximated as

$$J_{WL}(\infty) = J_{WL,\min} + \frac{\mu J_{WL,\min}}{2} \sum_{k=1}^{2N} \bar{\lambda}_k$$

when the stepsize is assumed to be small. The steady-state error expression for the linear LMS filter has the same form except the very last term, which is replaced by $\sum_{k=1}^{N} \lambda_k$ where λ_k denotes the eigenvalues of $\mathbf{C}$ [43]. Since we have $\sum_{k=1}^{2N} \bar{\lambda}_k = \text{Trace}(\bar{\mathbf{C}}) = 2N\sigma^2$ and $\sum_{k=1}^{N} \lambda_k = N\sigma^2$ where $\sigma^2 = E\{|\mathbf{X}(n)|^2\}$, compared to the linear LMS filter, doubling the dimension for the widely linear filter increases the residual mean-square error compared to the linear LMS filter as expected. The difference can be eliminated by using an annealing procedure such that the step size is also adjusted such that $\mu(n) \rightarrow 0$ as $n \rightarrow \infty$.

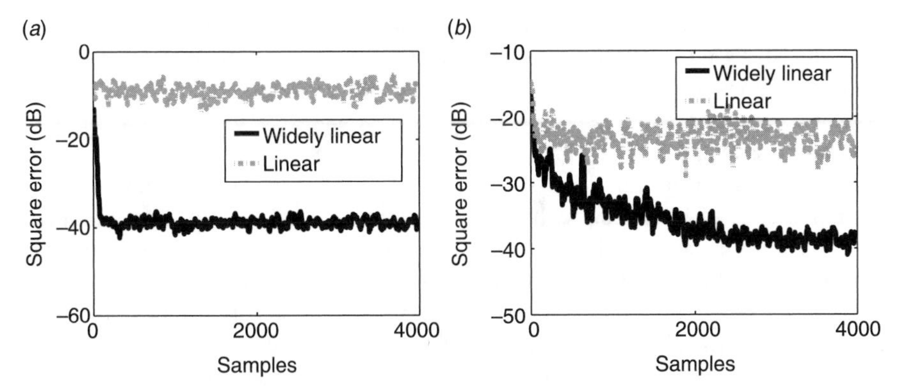

Figure 1.8 Convergence of the linear and widely linear filter for a circular input ($\rho = 1/\sqrt{2}$) (*a*) and a noncircular input ($\rho = 0.1$) (*b*) for the identification of a widely linear system.

■ EXAMPLE 1.6

In Figure 1.8, we show the learning curves for the linear and widely linear LMS filters for a widely linear channel. All the settings for the simulation are the same as those in Example 1.5 except that the unknown system output is given by

$$d(n) = \mathrm{Re}\{\mathbf{w}_{\mathrm{opt}}^H \mathbf{x}(n)\}$$

and the filter coefficients $w_{\mathrm{opt},n}$ are selected as before.

As observed in the figures, for both the circular and noncircular cases, the widely linear filter provides smaller MSEs, though its convergence is again slower for the noncircular input due to the increased eigenvalue spread.

An interesting point to note in Example 1.6 is that the advantage of using a widely linear filter—in terms of the minimum MSE that is achieved—is more pronounced in this case for circular input, even though the advantages of widely linear filters are, in general, emphasized for noncircular statistics.

For a circular input, the MSE gain by using a widely linear filter given in (1.39) reduces to

$$J_{\mathrm{diff}} = \|\mathbf{q}\|^2 = \|E\{d(n)\,\mathbf{x}(n)\}\|^2$$

and is clearly nonzero for the widely linear system chosen in this example, as observed in Figure 1.8 resulting in significant performance gain with the widely linear filter.

1.5 NONLINEAR ADAPTIVE FILTERING WITH MULTILAYER PERCEPTRONS

Neural network structures such as multilayer perceptron (MLP) and the radial basis function (RBF) filters have been successfully used for adaptive signal processing in the real domain for problems that require nonlinear signal processing capability [42]. Both the MLP and the RBF filters are shown to be universal approximators of any smooth nonlinear mapping [30, 35, 51] and their use has been extended to the complex domain, see for example [12, 14, 67, 108].

A main issue in the implementation of nonlinear filters in the complex domain has been the choice of the activation function. Primarily due to stability considerations, the importance of boundedness has been emphasized, and identified as a property an activation function should satisfy for use in a complex MLP [36, 119]. Thus, the typical practice has been the use of split-type activation functions, which are defined in Section 1.2.1. Fully-complex activation functions, as we discuss next, are more efficient in approximating nonlinear functions, and can be shown to be universal approximators as well. In addition, when a fully-complex nonlinear function is used as the activation function, it enables the use of Wirtinger calculus so that derivations for the learning rules for the MLP filter can be carried out in a manner very similar to the real-valued case, making many efficient learning procedures developed for the real-valued case readily accessible in the complex domain. These results can be extended to RBF filters in a similar manner.

1.5.1 Choice of Activation Function for the MLP Filter

As noted in Section 1.2.2, Liouville's theorem states the conflict between the boundedness and differentiability of functions in the complex domain. For example, the tanh nonlinearity that has been the most typically used activation function for real-valued MLPs, has periodic singular points as shown in Figure 1.13.

Since boundedness is deemed as important for the stability of algorithms, a practical solution when designing MLP filters for the complex domain has been to define nonlinear functions that process the real and imaginary parts separately through bounded real-valued nonlinearities as defined in Section 1.2.1 and given for the typically employed function tanh as

$$f(z) \triangleq \tanh(x) + j\tanh(y) \tag{1.43}$$

for a complex variable $z = x + jy$ where tanh: $\mathbb{R} \mapsto \mathbb{R}$. The activation function can also be defined through real-valued functions defined for the magnitude and phase of z as introduced in [45]

$$f(z) = f(re^{j\theta}) \triangleq \tanh\left(\frac{r}{m}\right)e^{j\theta} \tag{1.44}$$

where m is any number different than 0. Another such activation function is proposed in [36]

$$f(z) \triangleq \frac{z}{c + |z|/d} \tag{1.45}$$

where again c and d are arbitrary constants with $d \neq 0$. The characteristics of the activation functions given in (1.43)–(1.45) are shown in Figures 1.9–1.11. As observed

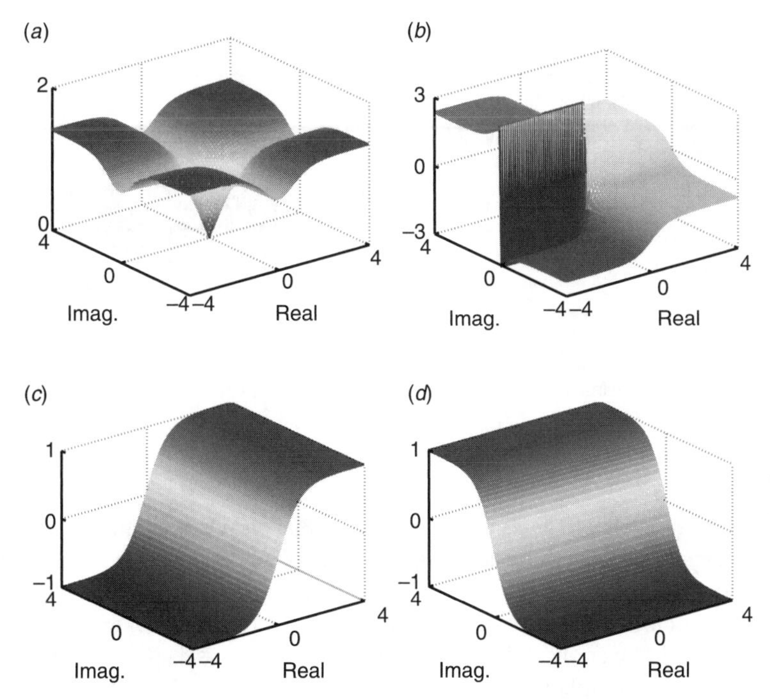

Figure 1.9 (*a*) Magnitude, (*b*) phase, (*c*) real and (*d*) imaginary responses of the split tanh function given in (1.43).

in the figures, though bounded, none of these functions can provide sufficient discrimination ability, the split function in (1.43) shown in Figure 1.9 provides a decoupled real and imaginary response while those shown in Figures 1.10 and 1.11, provide smooth radially symmetric magnitude responses and a phase response that is simply linear. The responses of the real and imaginary parts for these two functions are the same as in the case for the split function. In [61–63] examples in system identification and channel equation are provided to show that these functions cannot use the

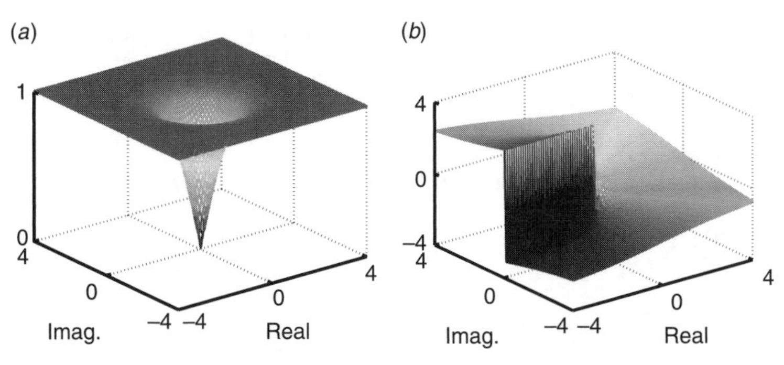

Figure 1.10 (*a*) Magnitude and (*b*) phase responses of the activation function given in (1.44) for $m = 1$.

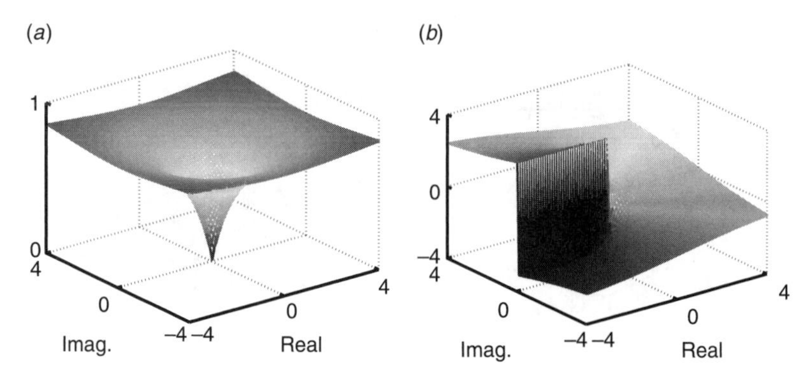

Figure 1.11 (*a*) Magnitude and (*b*) phase responses of the activation function given in (1.45) for $c = d = 1$.

phase information effectively, and in applications that introduce significant phase distortion such as equalization of saturating type channels, are not effective as complex domain nonlinear filters. Fully complex activation functions, or more simply, complex analytic functions, on the other hand provide a much more powerful modeling ability compared to split functions. It is worth noting that the universal approximation ability of MLPs that employ split activation functions as given in (1.43) can be easily shown by simply extending the universal approximation theorem from the real domain to the complex one [10]. However, as we demonstrate in this section, they cannot make efficient use of the available information.

In [63], a number of fully-complex—or simply analytic functions are proposed as activation functions and it is shown by a number of recent examples that MLPs using these activation functions provide a more powerful modeling ability compared to split functions [38, 39, 41, 61–63]. These functions all have well-defined first-order derivatives and squashing type characteristics that are generally required for nonlinear filters to be used as global approximators, such as the MLPs. These functions can be divided into four classes as

- Circular functions:

$$\tan z = \frac{e^{jz} - e^{-jz}}{j(e^{jz} - e^{-jz})} \qquad \frac{d}{dz} \tan z = \sec^2 z$$

$$\sin z = \frac{e^{jz} - e^{-jz}}{2j} \qquad \frac{d}{dz} \sin z = \cos z.$$

- Inverse circular functions:

$$\operatorname{atan} z = \int_0^z \frac{dt}{1 + t^2} \qquad \frac{d}{dz} \operatorname{atan} z = \frac{1}{1 + z^2}$$

$$\text{asin } z = \int_0^z \frac{dt}{(1-t)^{1/2}} \quad \frac{d}{dz}\text{asin } z = (1-z^2)^{-1/2}$$

$$\text{acos } z = \int_0^z \frac{dt}{(1-t^2)^{1/2}} \quad \frac{d}{dz}\text{acos } z = -(1-z^2)^{-1/2}.$$

- Hyperbolic functions:

$$\tanh z = \frac{\sinh z}{\cosh z} = \frac{e^z - e^{-z}}{e^z + e^{-z}} \quad \frac{d}{dz}\tanh z = \text{sech}^2 z$$

$$\sinh z = \frac{e^z - e^{-z}}{2} \quad \frac{d}{dz}\sinh z = \cosh z.$$

- Inverse hyperbolic functions:

$$\text{a}\tanh z = \int_0^z \frac{dt}{1-t^2} \quad \frac{d}{dz}\text{a}\tanh z = \frac{1}{1-z^2}$$

$$\text{a}\sinh z = \int_0^z \frac{dt}{(1+t^2)^{1/2}} \quad \frac{d}{dz}\text{a}\sinh z = \frac{1}{1+z^2}.$$

The magnitude and phase characteristics of these functions are shown in Figures 1.12–1.16, and for the case of the atanh function, also the responses of real and imaginary parts are shown to emphasize the variability of the responses of these functions for real and imaginary parts when compared to the split type functions shown in Figures 1.9–1.11. Note that hyperbolic functions and their trigonometric counterparts (*e.g.*, asinh and asin) have very similar responses except that they are $\pi/2$ rotated versions of each other.

In [63], three types of approximation theorems are given for MLP networks that use complex activation functions as those listed above from the trigonometric and hyperbolic family. The theorems are based on type of singularity a function possesses, as discussed in Section 1.2.2. The approximation theorems for the first two classes of functions are very general and resemble the universal approximation theorem for the real-valued feedforward MLP whereas the third approximation theorem for the complex MLP is unique in that it is uniform only in the analytic domain of convergence. As in the real case, the structure of the MLP network is a single hidden layer network as shown in Figure 1.17 with nonlinear activation functions in the hidden layer and a linear output layer. The three approximation theorems are given as [63]:

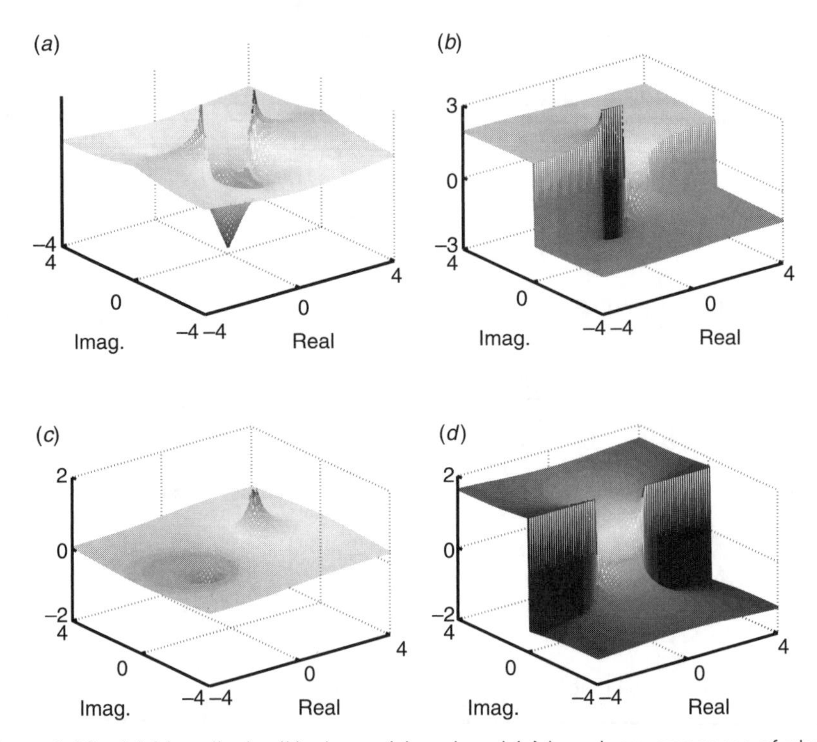

Figure 1.12 (*a*) Magnitude, (*b*) phase, (*c*) real and (*d*) imaginary responses of atanh.

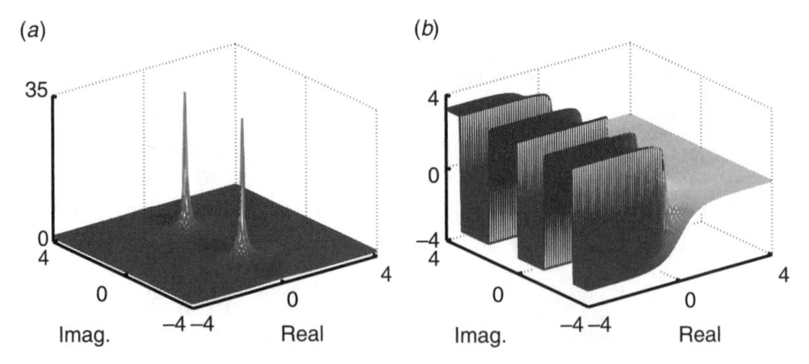

Figure 1.13 (*a*) Magnitude and (*b*) phase responses of tanh.

- MLPs that use continuous nonlinear activation functions without any singular points can achieve universal approximation of any continuous nonlinear mapping over a compact set in $\mathbb{C}^N$. Note that these functions are not bounded, as shown for the sinh function in Figure 1.16, but by bounding the region of interest using scaling, for example, for range around the unit circle for the sinh function, they can be used as activation functions and can provide good approximation as demonstrated in [63].

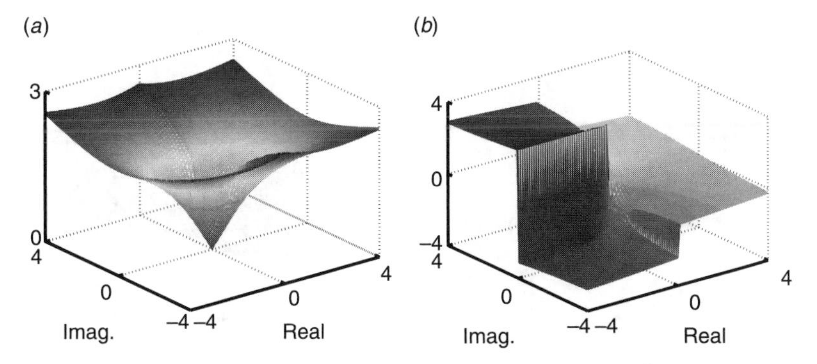

Figure 1.14 (*a*) Magnitude and (*b*) phase responses of asinh.

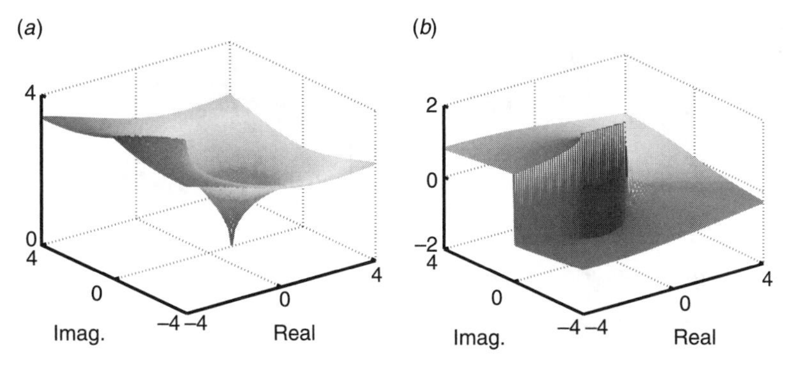

Figure 1.15 (*a*) Magnitude and (*b*) phase responses of acosh.

- The second group of functions considered are those with bounded singularities, such as branch cuts over a bounded domain and removable singularities. Examples include the asinh and acosh functions—and their trigonometric counterparts—which are all bounded complex measurable functions. It is shown that MLP using activation functions with bounded singularities provides universal approximation *almost everywhere* over a compact set in $\mathbb{C}^N$.
- Finally, the third theorem considers unbounded measurable activation functions, that is, those with poles, such as tanh and atanh and their trigonometric counterparts, as well as non-measurable nonlinear activation functions, those with essential singularity. For MLPs that use these activation functions, the approximation of any nonlinear mapping is uniform over the deleted annulus of singularity nearest to the origin. If there are multiple singularities, the radius of convergence is the shortest distance to a singularity from the origin.

Hence, complex functions such as trigonometric and hyperbolic functions can be effectively used as activation functions, and when the MLP structure involves

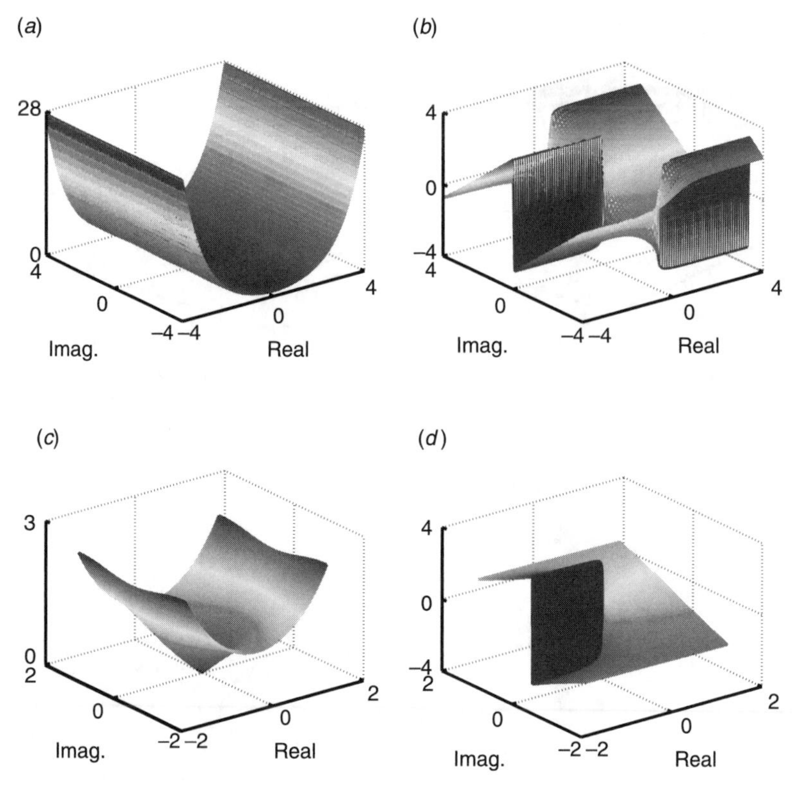

Figure 1.16 (*a, c*) Magnitude and (*b, d*) phase responses of sinh in two distinct ranges.

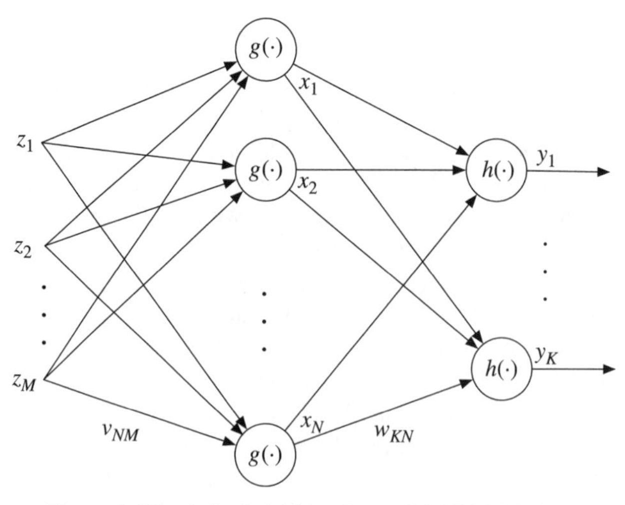

Figure 1.17 A single hidden layer (*M-N-K*) MLP filter.

such nonlinearities rather than the split type functions given in (1.43)–(1.45), the update rules for the MLP can be derived in a manner very similar to the real case as we demonstrate next for the derivation of the back-propagation algorithm.

1.5.2 Derivation of Back-Propagation Updates

For the MLP filter shown in Figure 1.17, we write the square error cost function as

$$J(\mathbf{V}, \mathbf{W}) = \sum_{k=1}^{K} (d_k - y_k)(d_k^* - y_k^*)$$

where

$$y_k = h\left(\sum_{n=1}^{N} w_{kn} x_n\right)$$

and

$$x_n = g\left(\sum_{m=1}^{M} v_{nm} z_m\right).$$

When both activation functions $h(\cdot)$ and $g(\cdot)$ satisfy the property $[f(z)]^* = f(z^*)$, then the cost function can be written as $J(\mathbf{V}, \mathbf{W}) = G(z)G(z^*)$ making it very practical to evaluate the gradients using Wirtinger calculus by treating the two variables z and z^* as independent in the computation of the derivatives. Any function $f(z)$ that is analytic for $|z| < R$ with a Taylor series expansion with all real coefficients in $|z| < R$ satisfies the property $[f(z)]^* = f(z^*)$ as noted in [6] and [71].

Examples of such functions include polynomials and most trigonometric functions and their hyperbolic counterparts (all of the functions whose characteristics are shown in Figs. 1.12–1.16), which also provide universal approximation ability as discussed in Section 1.5.1. In addition, the activation functions given in (1.43)–(1.45) that process the real and imaginary or the magnitude and phase of the signals separately also satisfy this property. Hence, there is no real reason to evaluate the gradients through separate real and imaginary part computations as traditionally done. Indeed, this approach can easily get quite cumbersome as evidenced by [12, 14, 39, 41, 62, 67, 107, 108, 118] as well as a recent book [75] where the development using Wirtinger calculus is presented as an afterthought, with the result in [6] and [71] that enables the use of Wirtinger calculus given without proper citation.

When the fully-complex functions introduced in Section 1.5.1 are used as activation functions as opposed to those given in (1.43)–(1.45), the MLP filter can achieve significantly better performance in challenging signal processing problems such as equalization of highly nonlinear channels [61, 62] both in terms of superior convergence characteristics and better generalization abilities through the efficient representation of the underlying problem structure. The nonsingularities do not pose any practical problems in the implementation, except that some care is required in the selection of their parameters when training these networks.

For the MLP filter shown in Figure 1.17, where y_k is the output and z_m the input, when the activations functions $g(\cdot)$ and $h(\cdot)$ are chosen as functions that are $\mathbb{C} \mapsto \mathbb{C}$, we can directly write the back-propagation update equations using Wirtinger derivatives as shown next.

For the output units, we have $\partial y_k / \partial w_{kn}^* = 0$, therefore

$$\begin{aligned}\frac{\partial J}{\partial w_{kn}^*} &= \frac{\partial J}{\partial y_k^*}\frac{\partial y_k^*}{\partial w_{kn}^*} \\ &= \frac{\partial[(d_k - y_k)(d_k^* - y_k^*)]}{\partial y_k^*}\frac{\partial h\left(\sum_n w_{kn}^* x_n^*\right)}{\partial w_{kn}^*} \\ &= -(d_k - y_k)h'\left(\sum_n w_{kn}^* x_n^*\right)x_n^*. \end{aligned} \tag{1.46}$$

We define $\delta_k = -(d_k - y_k)h'\left(\sum_n w_{kn}^* x_n^*\right)$ so that we can write $\partial J / \partial w_{kn}^* = \delta_k x_n^*$.

For the hidden layer or input layer, first we observe the fact that v_{nm} is connected to x_n for all m. Again, we have $\partial y_k / \partial v_{nm}^* = 0$, $\partial x_n / \partial v_{nm}^* = 0$. Using the chain rule once again, we obtain

$$\begin{aligned}\frac{\partial J}{\partial v_{nm}^*} &= \sum_k \frac{\partial J}{\partial y_k^*}\frac{\partial y_k^*}{\partial x_n^*}\frac{\partial x_n^*}{\partial v_{nm}^*} \\ &= \frac{\partial x_n^*}{\partial v_{nm}^*}\sum_k \frac{\partial J}{\partial y_k^*}\frac{\partial y_k^*}{\partial x_n^*} \\ &= g'\left(\sum_m v_{nm}^* z_m^*\right)z_m^* \sum_k \frac{\partial J}{\partial y_k^*}\frac{\partial y_k^*}{\partial x_n^*} \\ &= g'\left(\sum_m v_{nm}^* z_m^*\right)z_m^* \left(\sum_k -(d_k - y_k)h'\left(\sum_l w_{kl}^* x_l^*\right)w_{kn}^*\right) \\ &= z_m^* g'\left(\sum_m v_{nm}^* z_m^*\right)\left(\sum_k \delta_k w_{kn}^*\right). \end{aligned} \tag{1.47}$$

Thus, (1.46) and (1.47) define the gradient updates for computing the hidden and the output layer coefficients, w_{kn} and v_{nm}, through back-propagation. Note that the derivations in this case are very similar to the real-valued case as opposed to the derivations given in [12, 62, 67], where separate evaluations with respect to the real and imaginary parts are carried out, and hence the steps in the derivations assume more complicated forms. Also, the derivations for new learning rules such as those given in [38, 39, 41] can be considerably simplified by using Wirtinger derivatives. As we demonstrate in the next example, the split activation functions are not efficient in their use of the information when learning nonlinear mappings, and hence are not desirable for use as activation functions.

■ EXAMPLE 1.7

In Figure 1.18, we show the convergence characteristics of two MLP filters, one using split tanh and a second one that uses the complex tanh as the activation function. The input is generated using the same model as in Example 1.5 with $\rho = 1/\sqrt{2}$ and the nonlinear output of the system is given as $d(n) + 0.2d^2(n)$ where $d(n) = \mathbf{w}_{\text{opt}}^H \mathbf{x}(n)$ with the coefficients w_{opt} selected as in Example 1.5. The size of the input layer is chosen as 5 to match the memory of the finite impulse response component of the system, and the filter has a single output. The stepsize is chosen as 0.01 for both the split and the fully complex MLP filters and the convergence behavior is shown for two different filter sizes, one with a filter using 15 hidden nodes and a second one with 40 hidden nodes. As observed in the figures, the MLP filter using a fully complex activation function produces lower squared error value, however the performance advantage of the fully complex filter decreases when the number of hidden nodes increases as observed in Figure 1.18*b*.

The example demonstrates that the fully complex MLP filter is able to use information more efficiently, however, the performance of the filter that uses a split-type activation function can be improved by increasing the filter complexity. Note that the universal approximation of MLP filters can be demonstrated for both filter types, and the approximation result guarantees that the MLP structure can come

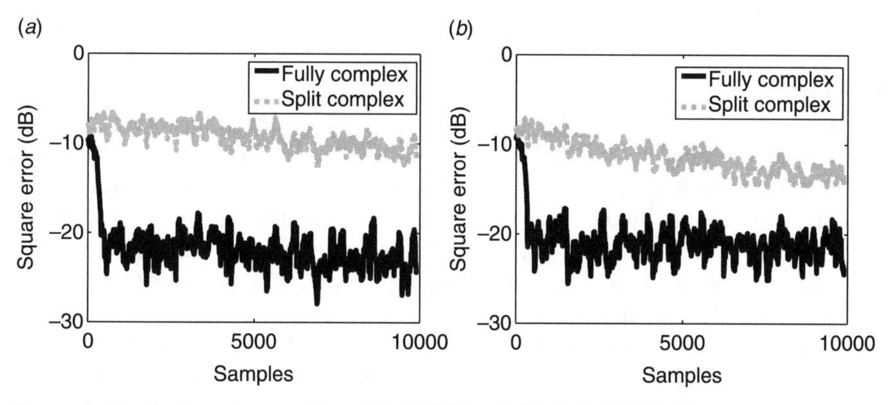

Figure 1.18 Performance of two (*a*) 5-15-1 and (*b*) 5-40-1 MLP filters for a nonlinear system identification problem using split and fully-complex tanh activation functions.

arbitrarily close to approximating any given mapping (subject to regularity conditions) *if* the number of hidden nodes chosen is sufficiently large.

The results recently given in the literature using fully complex activation functions suggest that they are promising solutions for challenging nonlinear signal processing problems, and derivation of new learning rules as well as design or selection of such activation functions is thus a research direction that deserves attention.

1.6 COMPLEX INDEPENDENT COMPONENT ANALYSIS

Independent component analysis (ICA) has emerged as an attractive analysis tool for discovering hidden factors in observed data and has been successfully applied to numerous signal processing problems in areas as diverse as biomedicine, communications, finance, and remote sensing [54]. In order to perform ICA of complex-valued data, there are a number of options. Algorithms such as joint approximate diagonalization of eigenmatrices (JADE) [22] or those using second order statistics [32, 65] achieve ICA without the need to use nonlinear functions in the algorithm. The second-order complex blind source separation algorithm, strongly uncorrelating transform (SUT) [32], though efficient, requires the sources to be noncircular and have distinct spectral coefficients. Thus a second ICA algorithm should be utilized after its application as a preprocessing step when the sources happen to be circular [34]. JADE is based on the joint diagonalization of cumulant matrices and is quite robust, however, its performance suffers as the number of sources increases, and the cost of computing and diagonalizing cumulant matrices becomes prohibitive for separating a large number of sources (see *e.g.*, [69]). On the other hand, ICA approaches that use nonlinear functions, such as maximum likelihood [89], information-maximization (Infomax) [11], nonlinear decorrelations [25, 58], and maximization of non-Gaussianity (*e.g.*, the FastICA algorithm) [53], are all intimately related to each other and present an attractive alternative for performing ICA. A number of comparison studies have demonstrated their desirable performance over other ICA algorithms such as JADE and second-order algorithms. For example, in [29], this efficiency is observed for the ICA of fMRI and fMRI-like data.

In the development of complex ICA algorithms with nonlinear functions, traditionally the same approach discussed for MLPs in Section 1.5 has been followed and a number of limitations have been imposed on the nature of complex sources either directly, or indirectly through the selection of the nonlinear function. A number of algorithms have used complex split nonlinear functions such that the real and imaginary parts (or the magnitude and phase) of the argument are processed separately through real-valued nonlinear functions [96, 103]. Another approach processes the magnitude of the argument by a real-valued function [9, 13], thus limiting the algorithm to circular sources. These approaches, while yielding satisfactory performance for a class of problems, are not effective in generating the higher order statistics required to establish independence for all possible source distributions.

In this section based on the work in [6], we concentrate on the complex ICA approaches that use nonlinear functions without imposing any limitations on the type of source distribution and demonstrate how Wirtinger calculus can be used for efficient derivation of algorithms and for working with probabilistic characterizations, which is important in the development of density matching mechanisms that play a key role in this class of ICA algorithms. We present the two main approaches for performing ICA: maximum likelihood (ML) and maximization of non-Gaussianity (MN). We discuss their relationship to each other and to other closely related ICA approaches, and in particular note the importance of source density matching for both approaches. We present extensions of source density matching mechanisms for the complex case, and note a few key points for special classes of sources, such as Gaussian sources, and those that are strictly second-order circular. We present examples that clearly demonstrate the performance equivalence of ML- and MN-based ICA algorithms when exact source matching is used for both cases.

In the development, we consider the traditional ICA problem such that

$$\mathbf{x} = \mathbf{A}\mathbf{s}$$

where $\mathbf{x}, \mathbf{s} \in \mathbb{C}^N$ and $\mathbf{A} \in \mathbb{C}^{N \times N}$, that is, the number of sources and observations are equal and all variables are complex valued.

The sources s_i where $\mathbf{s} = [s_1, \ldots, s_N]^T$ are assumed to be statistically independent and the source estimates u_i where $\mathbf{u} = [u_1, \ldots, u_N]^T$, are given by $\mathbf{u} = \mathbf{W}\mathbf{x}$. If the mixtures are whitened and sources are assumed to have unit variance, $\mathbf{WA}$ approximates a permutation matrix when the ICA problem is solved, where we assume that the mixing matrix is full rank. For the complex case, an additional component of the scaling ambiguity is the phase of the sources since all variables are assumed to be complex valued. In the case of perfect separation, the permutation matrix will have one nonzero element. Separability in the complex case is guaranteed as long as the mixing matrix $\mathbf{A}$ is of full column rank and there are no two complex Gaussian sources with the same circularity coefficient [33], where the circularity coefficients are defined as the singular values of the pseudo-covariance matrix of the source random vector. This is similar to the real-valued case where second-order algorithms that exploit the correlation structure in the mixtures use joint diagonalization of two covariance matrices [106].

1.6.1 Complex Maximum Likelihood

As in the case of numerous estimation problems, maximum likelihood theory provides a natural formulation for the ICA problem. For T independent samples $\mathbf{x}(t) \in \mathbb{C}^N$, we can write the log-likelihood function as

$$\mathcal{L}(\mathbf{W}) = \sum_{t=1}^{T} \ell_t(\mathbf{W}),$$

where

$$\ell_t(\mathbf{W}) = \log p(\mathbf{x}(t)|\mathbf{W}) = \log p_S(\mathbf{W}\mathbf{x}) + \log|\det \overline{\mathbf{W}}|$$

and the density of the transformed random variables is written through the computation of the Jacobian as

$$p(\mathbf{x}) = |\det \overline{\mathbf{W}}| p_S(\mathbf{W}\mathbf{x}) \tag{1.48}$$

where $\overline{\mathbf{W}}$ is defined in (1.11).

We use the notation that $p_S(\mathbf{W}\mathbf{x}) \triangleq \prod_{n=1}^{N} p_{S_n}(\mathbf{w}_n^H \mathbf{x})$, where $\mathbf{w}_n$ is the nth row of $\mathbf{W}$, $p_{S_n}(u_n) \triangleq p_{S_n}(u_{n_r}, u_{n_i})$ is the joint pdf of source n, $n = 1, \ldots, N$, with $u_n = u_{n_r} + ju_{n_i}$, and defined $\mathbf{W} = \mathbf{A}^{-1}$, that is, we express the likelihood in terms of the inverse mixing matrix, which provides a convenient change of parameter. Note that the time index in $\mathbf{x}(t)$ has been omitted in the expressions for simplicity.

To take advantage of Wirtinger calculus, we write each pdf as $p_{S_n}(u_r, u_i) = g_n(u, u^*)$ to define $g(\mathbf{u}, \mathbf{u}^*)$: $\mathbb{C}^N \times \mathbb{C}^N \mapsto \mathbb{R}^N$ so that we can directly evaluate

$$\frac{\partial \log g(\mathbf{u}, \mathbf{u}^*)}{\partial \mathbf{W}^*} = \frac{\partial \log g(\mathbf{u}, \mathbf{u}^*)}{\partial \mathbf{u}^*} \mathbf{x}^H \triangleq -\psi(\mathbf{u}, \mathbf{u}^*)\mathbf{x}^H \tag{1.49}$$

where $\mathbf{u} = \mathbf{W}\mathbf{x}$ and we have defined the score function $\psi(\mathbf{u}, \mathbf{u}^*)$ that is written directly by using the result in Brandwood's theorem given by (1.5)

$$\psi(\mathbf{u}, \mathbf{u}^*) = -\frac{1}{2}\left(\frac{\partial \log p_S(\mathbf{u}_r, \mathbf{u}_i)}{\partial \mathbf{u}_r} + j\frac{\partial \log p_S(\mathbf{u}_r, \mathbf{u}_i)}{\partial \mathbf{u}_i}\right). \tag{1.50}$$

When writing (1.49) and (1.50), we used a compact vector notation where each element of the score function is given by

$$\psi_n(u, u^*) = -\frac{\partial \log g_n(u_n, u_n^*)}{\partial u_n^*} = -\frac{1}{2}\left(\frac{\partial \log p_{S_n}(u_{r,n}, u_{i,n})}{\partial u_{r,n}} + j\frac{\partial \log p_{S_n}(u_{r,n}, u_{i,n})}{\partial u_{i,n}}\right). \tag{1.51}$$

To compute $\partial \log |\det \overline{\mathbf{W}}|/\partial \mathbf{W}$, we first observe that $\partial \log |\det \overline{\mathbf{W}}| = \text{Trace}(\overline{\mathbf{W}}^{-1} \partial\overline{\mathbf{W}}) = \text{Trace}(\partial\overline{\mathbf{W}}\mathbf{P}\mathbf{P}^{-1}\overline{\mathbf{W}}^{-1})$, and then choose

$$\mathbf{P} = \frac{1}{2}\begin{bmatrix} \mathbf{I} & j\mathbf{I} \\ j\mathbf{I} & \mathbf{I} \end{bmatrix}$$

to write

$$\begin{aligned} \partial \log |\det \overline{\mathbf{W}}| &= \text{Trace}(\mathbf{W}^{-1}\partial\mathbf{W}) + \text{Trace}((\mathbf{W}^*)^{-1}\partial\mathbf{W}^*) \\ &= \langle \mathbf{W}^{-H}, \partial\mathbf{W}\rangle + \langle \mathbf{W}^{-T}, \partial\mathbf{W}^*\rangle. \end{aligned} \tag{1.52}$$

Here, we have used

$$\mathbf{P}^{-1}\overline{\mathbf{W}}^{-1} = \frac{1}{2}\begin{bmatrix} \mathbf{W}^* & j\mathbf{W} \\ j\mathbf{W}^* & \mathbf{W} \end{bmatrix}^{-1} = \begin{bmatrix} (\mathbf{W}^*)^{-1} & -j(\mathbf{W}^*)^{-1} \\ -j\mathbf{W}^{-1} & \mathbf{W}^{-1} \end{bmatrix}.$$

We define $\Delta g(\mathbf{W}, \mathbf{W}^*) \partial \log |\det \overline{\mathbf{W}}|$ and write the first-order Taylor series expansion given in (1.20) as

$$\Delta g(\mathbf{W}, \mathbf{W}^*) = \langle \Delta \mathbf{W}, \nabla_{\mathbf{W}^*} \log|\det \overline{\mathbf{W}}|\rangle + \langle \Delta \mathbf{W}^*, \nabla_{\mathbf{W}} \log|\det \overline{\mathbf{W}}|\rangle$$

which, upon comparison with (1.52) gives us the required result for the matrix gradient

$$\frac{\partial \log |\det \overline{\mathbf{W}}|}{\partial \mathbf{W}^*} = \mathbf{W}^{-H}. \tag{1.53}$$

We can then write the relative (natural) gradient updates to maximize the likelihood function using Eqs. (1.34), (1.49) and (1.53) as

$$\Delta \mathbf{W} = (\mathbf{W}^{-H} - \psi(\mathbf{u})\mathbf{x}^H)\mathbf{W}^H\mathbf{W} = (\mathbf{I} - \psi(\mathbf{u})\mathbf{u}^H)\mathbf{W}. \tag{1.54}$$

The update given above and the score function $\psi(\mathbf{u})$ defined in (1.50) coincide with the one derived in [20] using a $\mathbb{C} \mapsto \mathbb{R}^{2n}$ isomorphic mapping in a relative gradient update framework and the one given in [34] considering separate derivatives.

The update equation given in (1.54) can be also derived without explicit use of the relative gradient update rule given in (1.34). We can use (1.49), (1.53), and $\partial \mathbf{u} = (\partial \mathbf{W})\mathbf{x}$, to write the first-order differential of the likelihood term $\ell_t(\mathbf{W})$ as

$$\partial \ell_t = \text{Trace}(\partial \mathbf{W} \mathbf{W}^{-1}) + \text{Trace}(\partial \mathbf{W}^* \mathbf{W}^{-*}) - \psi^H(\mathbf{u})\partial \mathbf{u} - \psi^T(\mathbf{u})\partial \mathbf{u}^*. \tag{1.55}$$

Defining $\partial \mathbf{Z} \triangleq (\partial \mathbf{W})\mathbf{W}^{-1}$, we obtain $\partial \mathbf{u} = (\partial \mathbf{W})\mathbf{x} = \partial \mathbf{W}(\mathbf{W}^{-1})\mathbf{u} = (\partial \mathbf{Z})\mathbf{u}$, $\partial \mathbf{u}^* = (\partial \mathbf{Z}^*)\mathbf{u}^*$. By treating $\mathbf{W}$ as a constant matrix, the differential matrix $\partial \mathbf{Z}$ has components ∂z_{ij} that are linear combinations of ∂w_{ij} and is a non-integrable differential form. However, this transformation allows us to easily write (1.55) as

$$\partial \ell_t = \text{Trace}(\partial \mathbf{Z}) + \text{Trace}(\partial \mathbf{Z}^*) - \psi^H(\mathbf{u})(\partial \mathbf{Z})\mathbf{u} - \psi^T(\mathbf{u})(\partial \mathbf{Z}^*)\mathbf{u}^* \tag{1.56}$$

where we have treated $\mathbf{Z}$ and $\mathbf{Z}^*$ as two independent variables using Wirtinger calculus. Therefore, the gradient update rule for $\mathbf{Z}$ is given by

$$\Delta \mathbf{Z} = \frac{\partial \ell_t}{\partial \mathbf{Z}^*} = (\mathbf{I} - \mathbf{u}^* \psi^T(\mathbf{u}))^T = \mathbf{I} - \psi(\mathbf{u})\mathbf{u}^H \tag{1.57}$$

which is equivalent to (1.54) since $\partial \mathbf{Z} = (\partial \mathbf{W})\mathbf{W}^{-1}$.

The two derivations we have given here for the score function represent a very straightforward and simple evaluation compared to those in [20, 34], and more importantly, show how to bypass a major limitation in the development of ML theory for complex valued signal processing, that is working with probabilistic descriptions using complex algebra. In the second derivation, the introduction of the differential form $\partial \mathbf{Z}$, which is not a true differential as it is not integrable, provides a convenient form and is especially attractive in evaluation of higher-order differential expressions as demonstrated in [72].

Newton Updates for ML ICA The same definition, $\partial \mathbf{Z} \triangleq (\partial \mathbf{W})\mathbf{W}^{-1}$, can be used also to derive a Newton update rule in a compact form, as opposed to the element-wise form given in (1.35). To simplify the notation, we first define $l \triangleq -\ell_t$, and consider Newton updates to minimize the negative likelihood l, and then evaluate the second-order differential of the likelihood term l.

To write the differential of the term $\partial \ell = -\partial\, \ell_t$ given in (1.56) which is a function of $\{\mathbf{Z}, \mathbf{Z}^*, \mathbf{u}, \mathbf{u}^*\}$, we use Wirtinger calculus to write $\partial(\text{Trace}(\partial \mathbf{Z}))/\partial \mathbf{Z} = \mathbf{0}$ and $\partial(\text{Trace}(\partial \mathbf{Z}^*))/\partial \mathbf{Z}^* = \mathbf{0}$. Then, the second-order differential can be written as

$$\begin{aligned}\partial^2 l &= \partial[\psi^H(\mathbf{u})\partial \mathbf{Z}\mathbf{u} + \psi^T(\mathbf{u})\partial \mathbf{Z}^*\mathbf{u}^*] \\ &= 2\text{Re}\{\mathbf{u}^T \partial \mathbf{Z}^T \eta(\mathbf{u}, \mathbf{u}^*)\partial \mathbf{Z}\mathbf{u} + \mathbf{u}^T \partial \mathbf{Z}^T \theta(\mathbf{u}, \mathbf{u}^*)\partial \mathbf{Z}^*\mathbf{y}^* + \psi^H(\mathbf{u})\partial \mathbf{Z}\partial \mathbf{Z}\mathbf{u}\}\end{aligned}$$

where $\eta(\mathbf{u}, \mathbf{u}^*)$ is a diagonal matrix with ith diagonal element

$$-\frac{\partial \log p_i(u_i, u_i^*)}{\partial u_i \partial u_i}$$

and $\theta(\mathbf{u}, \mathbf{u}^*)$ is another diagonal matrix with ith diagonal element

$$-\frac{\partial \log p_i(u_i, u_i^*)}{\partial u_i \partial u_i^*}.$$

Using some simple algebra, we can write the expected value of the second differential term as

$$E\{\partial^2 l\} = \sum_{i \neq j} [\partial z_{ij}\ \partial z_{ji}\ \partial z_{ij}^*\ \partial z_{ji}^*] \begin{bmatrix} \mathbf{H}_1 & \mathbf{H}_2 \\ \mathbf{H}_2^* & \mathbf{H}_1 \end{bmatrix} \begin{bmatrix} \partial z_{ij}^* \\ \partial z_{ji}^* \\ \partial z_{ij} \\ \partial z_{ji} \end{bmatrix} + \sum_i [\partial z_{ii}\ \partial z_{ii}^*]\, \mathbf{H}_3 \begin{bmatrix} \partial z_{ii}^* \\ \partial z_{ii} \end{bmatrix}$$

where $\mathbf{H}_1 = \begin{bmatrix} \beta_j \delta_i & 0 \\ 0 & \beta_i \delta_j \end{bmatrix}$, $\mathbf{H}_2 = \begin{bmatrix} \alpha_j \gamma_i & 1 \\ 1 & \alpha_i \gamma_j \end{bmatrix}$, $\mathbf{H}_3 = \begin{bmatrix} v_i & q_i + 1 \\ q_i^* + 1 & v_i \end{bmatrix}$, $\alpha_i = E\{u_i^2\}$, $\beta_i = E\{|u_i|^2\}$, $\gamma_i = E\{\eta_i(u_i, u_i^*)\}$, $\delta_i = E\{\theta_i(u_i, u_i^*)\}$, $q_i = E\{u_i^2 \eta_i(u_i, u_i^*)\}$, and $v_i = E\{|u_i|^2 \theta_i(u_i, u_i^*)\}$.

As given in (1.57), we have

$$\frac{\partial E\{l\}}{\partial \mathbf{Z}^*} = E\{\psi(\mathbf{u})\mathbf{u}^H\} - \mathbf{I}.$$

To derive the Newton update, we consider the diagonal and the off-diagonal elements of $E\{\partial^2 l\}$ separately. We define $\partial\tilde{\mathbf{z}}_{ii} \triangleq \begin{bmatrix} \partial z_{ii} \\ \partial z_{ii}^* \end{bmatrix}$, and can write

$$\frac{\partial E\{l\}}{\partial \tilde{\mathbf{z}}_{ii}} = \begin{bmatrix} (E\{\psi(\mathbf{u})\mathbf{u}^H\} - \mathbf{I})_{ii}^* \\ (E\{\psi(\mathbf{u})\mathbf{u}^H\} - \mathbf{I})_{ii} \end{bmatrix} \quad \text{and}$$

$$\frac{\partial^2 E\{l\}}{\partial \tilde{\mathbf{z}}_{ii}} = \mathbf{H}_3^* \partial \tilde{\mathbf{z}}_{ii}.$$

Therefore the Newton rule for updating $\partial\tilde{\mathbf{z}}_{ii}$ can be written by solving

$$\frac{\partial^2 E\{l\}}{\partial \tilde{\mathbf{z}}_{ii}} = -\frac{\partial E\{l\}}{\partial \tilde{\mathbf{z}}_{ii}}$$

as in (1.35) to obtain

$$\partial\tilde{\mathbf{z}}_{ii} = -\mathbf{H}_3^{-*} \begin{bmatrix} (E\{\psi(\mathbf{u})\mathbf{u}^H\} - \mathbf{I})_{ii}^* \\ (E\{\psi(\mathbf{u})\mathbf{u}^H\} - \mathbf{I})_{ii} \end{bmatrix} \tag{1.58}$$

and the update for ∂z_{ii}^* is simply the conjugate of ∂z_{ii}.

For each off-diagonal element pair ∂z_{ij}, we write $\partial\tilde{\mathbf{z}}_{ij} \triangleq \begin{bmatrix} \partial z_{ij} \\ \partial z_{ji} \\ \partial z_{ij}^* \\ \partial z_{ji}^* \end{bmatrix}$. As in the updates of the diagonal elements, we obtain

$$\frac{\partial E\{l\}}{\partial \tilde{\mathbf{z}}_{ij}} = \begin{bmatrix} (E\{\psi(\mathbf{u})\mathbf{u}^H\} - \mathbf{I})_{ij}^* \\ (E\{\psi(\mathbf{u})\mathbf{u}^H\} - \mathbf{I})_{ji}^* \\ (E\{\psi(\mathbf{u})\mathbf{u}^H\} - \mathbf{I})_{ij} \\ (E\{\psi(\mathbf{u})\mathbf{u}^H\} - \mathbf{I})_{ji} \end{bmatrix}$$

$$\frac{\partial^2 E\{l\}}{\partial \tilde{\mathbf{z}}_{ij}} = \begin{bmatrix} \mathbf{H}_1 & \mathbf{H}_2^* \\ \mathbf{H}_2 & \mathbf{H}_1 \end{bmatrix} \partial\tilde{\mathbf{z}}_{ij}$$

and obtain the Newton update rule for the parameters $\partial\tilde{\mathbf{z}}_{ij}$ as in the previous case

$$\partial\tilde{\mathbf{z}}_{ij} = -\begin{bmatrix} \mathbf{H}_1 & \mathbf{H}_2^* \\ \mathbf{H}_2 & \mathbf{H}_1 \end{bmatrix}^{-1} \begin{bmatrix} (E\{\psi(\mathbf{u})\mathbf{u}^H\} - \mathbf{I})_{ij}^* \\ (E\{\psi(\mathbf{u})\mathbf{u}^H\} - \mathbf{I})_{ji}^* \\ (E\{\psi(\mathbf{u})\mathbf{u}^H\} - \mathbf{I})_{ij} \\ (E\{\psi(\mathbf{u})\mathbf{u}^H\} - \mathbf{I})_{ji} \end{bmatrix} \tag{1.59}$$

where only the upper half elements of $\mathbf{Z}$ need to be updated as the lower half is given by the conjugates of the upper half elements.

Thus, the two sets of updates, (1.58) and (1.59) give the complete Newton update rule for $\partial \mathbf{Z}$. The final update rule for $\mathbf{W}$ is simply given by $\partial \mathbf{W} = \partial \mathbf{Z} \mathbf{W}$, which implies that the given Newton update can be called a *relative* Newton algorithm as its structure is similar to the relative gradient update given in (1.34). Also, note that if both Hessian terms in (1.58) and (1.59) are nonsingular, that is, positive definite, then the resulting Hessian in the updates will be equal to the identity matrix in the solution point $\mathbf{W} = \mathbf{A}^{-1}$ as discussed in [7] for the real-valued case.

1.6.2 Complex Maximization of Non-Gaussianity

Another natural cost function for performing ICA is the maximization of non-Gaussianity [28, 53]. Independence is achieved by moving the transformed mixture, that is, the independent source estimates $\mathbf{w}^H \mathbf{x}$ away from a Gaussian distribution. The natural cost in this case is negentropy that measures the entropic distance of a distribution from that of a Gaussian and can be written for the complex source as

$$\mathcal{J}(\mathbf{w}) = H(v_r, v_i) - H(u_r, u_i) \tag{1.60}$$

where $H(\cdot, \cdot)$ is the differential entropy of the given bivariate distribution and $v = v_r + jv_i$ denotes the Gaussian-distributed complex variable. Gaussian density yields the largest entropy when the covariances of the two variables v and u are fixed and attains its maximum for the circular case [81]. Hence, the first term in (1.60) is constant for a given covariance matrix, and the maximization of $\mathcal{J}(\mathbf{w})$ can be achieved by minimizing the differential entropy $H\ (u_r, u_i) = -E\ \{\log\ p_S(u_r, u_i)\}$ under a variance constraint. Hence, we can define the ICA cost function to minimize as

$$J_G(\mathbf{w}) = E\{|G(u)|^2\} = E\{|G(\mathbf{w}^H \mathbf{x})|^2\} \tag{1.61}$$

subject to a variance constraint, and choose the nonlinear function $G\colon \mathbb{C} \mapsto \mathbb{C}$ to match the source pdf, that is, as

$$p_s(u) = p_s(u_r, u_i) = K\ \exp(-|G(u)|^2)$$

where K is a constant, so that the minimization of (1.61) is equivalent to the maximization of (1.60). While writing the form of the pdf in terms of the nonlinearity $G(\cdot)$, we assumed that the expectations in (1.60) and (1.61) are written using ensemble averages over T samples using ergodic theorems. Unit variance is a typical and convenient constraint and has been a practical choice in this class of algorithms [54].

Note that for maximization of negentropy, we proceed by estimating a single source at a time, that is, an individual direction that is maximally non-Gaussian while in the case of ML estimation, the formulation leads to the estimation of all independent sources through the computation of a single demixing matrix $\mathbf{W}$. Hence, we need a mechanism to avoid convergence to the same solution when estimating multiple sources, and, in addition, to impose a bound on the variance of the estimates. When we assume that the source signals have unit variance, that is, $E\{\mathbf{s}\mathbf{s}^H\} = \mathbf{I}$, then whitening the mixtures $\mathbf{v}$ prior to ICA such that $\mathbf{x} = \mathbf{M}\mathbf{v}$ and $E\{\mathbf{x}\mathbf{x}^H\} = \mathbf{I}$ implies that the demixing matrix $\mathbf{W}$ is unitary. Therefore in this case, we can perform ICA by first computing $\max_{\|\mathbf{w}_i\|^2=1} E\{|G(u_i)|^2\}$, and after the computation of each $\mathbf{w}_i$, by performing an orthogonalization procedure such as the Gram–Schmidt procedure [77] as in [53] such that $\mathbf{w}_i$ is orthogonal to $\{\mathbf{w}_j\}$, $1 \le j < i$. The estimated sources are given by $u_i = \mathbf{w}_i^H\mathbf{x}$, $i = 1, \ldots, N$.

The cost function given in (1.61) provides a case where the $\mathbb{R}^2 \mapsto \mathbb{C}^2$ mapping used by Wirtinger calculus follows naturally. Note that the cost function can be written as

$$J_G(\mathbf{w}) = E\{G(u)(G(u))^*\} = E\{G(u)G(u^*)\} \tag{1.62}$$

where the last equality follows when $G(u)$ is analytic for $|u| < R$ with a Taylor series expansion with all real coefficients in $|u| < R$. Polynomial and most trigonometric functions and their hyperbolic counterparts satisfy this condition.

When written in the form $E\{G(u)G(u^*)\}$ as shown in (1.62), it is easy to see that the function $J_G(\mathbf{w})$ becomes complex-differentiable when considered separately with respect to the two arguments u and u^* (and consequently $\mathbf{w}$ and $\mathbf{w}^*$) if the function is chosen as an analytic function G: $\mathbb{C} \mapsto \mathbb{C}$ thus making it even easier to take advantage of Wirtinger calculus in the gradient evaluation.

For the cost function given in (1.62), the gradient is directly written as

$$\frac{\partial J_G(\mathbf{w})}{\partial \mathbf{w}^*} = E\{\mathbf{x}G(\mathbf{w}^T\mathbf{x}^*)G'(\mathbf{w}^H\mathbf{x})\} = E\{\mathbf{x}G^*(u)G'(u)\} \tag{1.63}$$

instead of evaluating the derivatives with respect to the real and imaginary parts as given in [83]. Here, we have $G'(\cdot) = dG(u)/du$. Similar to the real-valued algorithm for maximization of non-Gaussianity using gradient updates, for a general function $G(\cdot)$—a function not necessarily matched to the source pdf—we need to determine whether the cost function is being maximized or minimized by evaluating a factor γ during the updates such that $\gamma = E\{|G(u)|^2\} - E\{|G(v)|^2\}$. Since γ is a real-valued quantity and does not change the stationary points of the solution, we can simply include its sign estimate in the online updates and use $\Delta\mathbf{w} = \text{sign}(\gamma)\mu\mathbf{x}G^*(\mathbf{w}^H\mathbf{x})$ $G'(\mathbf{w}^H\mathbf{x})$ where $\mu > 0$ is the learning rate, and ensure the satisfaction of the unit norm constraints through a practical update scheme $\mathbf{w} \leftarrow \mathbf{w}/\|\mathbf{w}\|$ after each iteration of the weight vector. A more efficient update algorithm for performing ICA using the cost function in (1.61) is given in [84] using a constrained optimization formulation to ensure $\|\mathbf{w}\| = 1$ and using a modified Newton approach. The updates for this case are given by

$$\mathbf{w} \leftarrow E\{G'(u)(G')^*(u)\}\mathbf{w} - E\{G^*(u)G'(u)\mathbf{x}\} + E\{\mathbf{x}\mathbf{x}^T\}E\{G^*(u)G''(u)\}\mathbf{w}^* \tag{1.64}$$

where a following normalization step is used to ensure $\|\mathbf{w}\| = 1$ as in the gradient updates.

1.6.3 Mutual Information Minimization: Connections to ML and MN

As discussed in Sections 1.6.1 and 1.6.2, we can solve the complex ICA problem by maximizing the log likelihood function given by

$$\mathcal{L}(\mathbf{W}) = \sum_{t=1}^{T} \sum_{n=1}^{N} \log p_{S_n}(\mathbf{w}_n^H \mathbf{x}) + T \log|\det \overline{\mathbf{W}}|. \tag{1.65}$$

The weight matrix $\mathbf{W}$ to maximize the log likelihood can be computed using relative gradient update equation given in (1.54).

When using negentropy maximization as the objective, all sources can be estimated by maximizing the cost function

$$\begin{aligned} \mathcal{J}(\mathbf{W}) &= \sum_{n=1}^{N} E\{\log p_{S_n}(\mathbf{w}_n^H \mathbf{x})\} \\ &\approx \frac{1}{T} \sum_{t=1}^{T} \sum_{n=1}^{N} \log p_{S_n}(\mathbf{w}_n^H \mathbf{x}) \end{aligned} \tag{1.66}$$

under the unitary constraint for $\mathbf{W}$. The mean ergodic theorem is used to write (1.66) and when compared to the ML formulation given in (1.65), it is clear that the two objective functions are equivalent if we constrain the weight matrix $\mathbf{W}$ to be unitary for complex ML. Since $\det(\overline{\mathbf{W}}) = |\det(\mathbf{W})|^2$ [40], when $\mathbf{W}$ is unitary, the second term in (1.65) vanishes.

Similar to the real case given in [21], for the complex case, we can satisfy the unitary constraint for the weight matrix by projecting $\Delta\mathbf{W}$ to the space of skew-hermitian matrices. The resulting update equation is then given by

$$\Delta\mathbf{W} = (\mathbf{I} - \mathbf{u}\mathbf{u}^H - \psi(\mathbf{u})\mathbf{u}^H + \mathbf{u}\psi^H(\mathbf{u}))\mathbf{W}. \tag{1.67}$$

On the other hand, for the MN criterion, the weight matrix can be estimated in symmetric mode, or the individual rows of the weight matrix $\mathbf{W}$ can be estimated sequentially in a deflationary mode as in [52]. The latter procedure provides a more flexible formulation for individual source density matching than ML where each element of the score function $\psi(\mathbf{u})$ given in (1.51) needs to be matched individually.

As in the real case, the two criteria are intimately linked to mutual information. Written as the Kullback–Leibler distance between the joint and factored marginal source densities, the mutual information is given by

$$\mathcal{I}(\mathbf{W}) = D\left(\|p(\mathbf{u})\| \prod_{n=1}^{N} p_{S_n}(u_n) \right) = \sum_{n=1}^{N} H(u_n) - H(\mathbf{u})$$
$$= \sum_{n=1}^{N} H(u_n) - H(\mathbf{x}) - \log|\det\overline{\mathbf{W}}| \quad (1.68)$$

where in the last line, we have again used the complex-to-real transformation for the source density given in (1.48). Since $H(\mathbf{x})$ is constant, using the mean ergodic theorem for the estimation of entropy, it is easy to see that minimization of mutual information is equivalent to ML, and when the weight matrix is constrained to be unitary, to the MN criterion.

1.6.4 Density Matching

For all three approaches for achieving ICA, the ML, MN, and mutual information minimization discussed in Sections 1.6.1–1.6.3, the nonlinearity used in the algorithm is expected to be matched as much as possible to the density for each estimated source. Also, the desirable large sample properties of the ML estimator assume their optimal values when the score function is matched to the source pdf, for example, the asymptotic covariance matrix of the ML estimator is minimum when the score function is chosen to match the source pdfs [89]. A similar result is given for the maximization of negentropy in [52]. A number of source density adaptation schemes have been proposed for performing ICA in the real-valued case, in particular for ML-based ICA (see *e.g.*, [24, 59, 66, 112, 120]) and more recently for the complex case [84, 85] for maximization of negentropy.

The most common approach for density adaptation has been the use of a flexible parametric model and to estimate the parameters—or a number of key parameters—of the model along with the estimation of the demixing matrix. In [89], a true ML ICA scheme has been differentiated as one that estimates both the source pdfs and the demixing matrix $\mathbf{W}$, and the common form of ML ICA where the nonlinearity is fixed and only the demixing matrix is estimated is referred to as quasi-maximum likelihood. Given the richer structure of possible distributions in the two-dimensional space compared to the real-valued, that is, single dimensional case, the pdf estimation problem becomes more challenging for complex-valued ICA. In the real-valued case, a robust nonlinearity such as the sigmoid nonlinearity provides satisfactory performance for most applications [29, 54] and the performance can be improved by matching the nonlinearity to the sub- or super-Gaussian nature of the sources [66]. In the complex case, the circular/noncircular nature of the sources is another important factor affecting the performance [3, 84]. Also, obviously the unimodal versus multimodal structure of the density requires special care in both the real and the complex case. Hence, in general, it is important to take *a priori* information into account when performing source matching.

If a given source has a circular distribution, that is, $p_{S_n}(u) = g(|u|)$, the corresponding entry of the score function vector can be easily evaluated as

$$\psi_n(u) = -\frac{\partial \log g(\sqrt{uu^*})}{\partial u^*} = -\frac{u}{2|u|}\left(\frac{g'(|u|)}{g(|u|)}\right).$$

Thus, the score function always has the same phase as its argument. This is the form of the score function proposed in [9] where all sources are assumed to be circular.

If the real and imaginary parts of a given source are mutually independent, the score function takes the form

$$\psi_n(u, u^*) = -\frac{1}{2}\left(\frac{\partial \log p_{S_r}(u_r)}{\partial u_r} + j\frac{\partial \log p_{S_i}(u_i)}{\partial u_i}\right)$$

and suggests the need to use separate real-valued functions for processing the real and imaginary arguments. For example, the score function proposed in [103] for complex Infomax, $\psi(u) = \tanh(u_r) + j\ \tanh(u_i)$, is shown to provide good performance for independent and circular sources [3].

For density matching, approaches such as the Gram–Charlier and Edgeworth expansions are proposed for the real case [19], and for the complex case, bivariate expansions such as those given in [76] can be adopted. However, such expansions usually perform well for unimodal distributions that are close to the Gaussian and their estimators are very sensitive to outliers thus usually requiring large number of samples. With the added dimensionality of the problem for the complex case, in comparison to the real (univariate) case, such expansions become even less desirable for complex density matching. Limitations of such expansions are discussed in detail in [104] where an efficient procedure for least-mean-square estimation of the score function is proposed for the real case.

Next, we discuss a number of possible density models and nonlinearity choices for performing complex ICA and discuss their properties. Simple substitution of $u_r = (u + u^*)/2$ and $u_i = (u - u^*)/2j$ allows us to write a given pdf that is $p(u_r, u_i)$: $\mathbb{R} \times \mathbb{R} \mapsto \mathbb{R}$ in terms of a function $f(u, u^*)$: $\mathbb{C} \times \mathbb{C} \mapsto \mathbb{R}$. Since all smooth functions that define a pdf can be shown to satisfy the real differentiability condition, they can be used in the development of ICA algorithms and in their analyses using Wirtinger calculus.

Generalized Gaussian Density Model A generalized Gaussian density of order c of the form given in [26] can be written as a function $\mathbb{C} \times \mathbb{C} \mapsto \mathbb{R}$ as

$$f_{\mathrm{GG}}(u, u^*; \sigma_r, \sigma_i, \rho, c) = \beta \exp(-[\gamma\alpha(u, u^*)]^c) \tag{1.69}$$

where

$$\alpha(u, u^*) = \frac{(u+u^*)^2}{4\sigma_r^2} + j\frac{\rho(u^2 - u^{*2})}{2\sigma_r\sigma_i} - \frac{(u-u^*)^2}{4\sigma_i^2},$$

$$\beta = \frac{c\gamma}{\pi\Gamma(1/c)\sigma_r\sigma_i\sqrt{1-\rho^2}}, \quad \text{and} \quad \gamma = \frac{\Gamma(2/c)}{2(1-\rho^2)\Gamma(1/c)}.$$

In the above expression, σ_r and σ_i are the standard deviations of the real and imaginary parts, $\rho = \sigma_{r,i}/\sigma_r\,\sigma_i$ is the correlation coefficient between the two variables, and the distribution is assumed to be zero mean. When the shape parameter $c = 1$, the pdf takes the form of the standard bivariate Gaussian and is super-Gaussian for $0 < c < 1$ and sub-Gaussian for $c > 1$.

The score function for the pdf given in (1.69) can be evaluated by using (1.50) as

$$\psi(u, u^*) = c\gamma^c[\alpha(u, u^*)]^{c-1}\frac{\partial\alpha(u, u^*)}{\partial u^*}.$$

When the sources are circular, that is, $\sigma_r = \sigma_i = \sigma$ and $\rho = 0$, we have $\alpha(u, u^*) = uu^*/\sigma^2$, and for circular Gaussian sources ($c = 1$), the score function is linear $\psi(u, u^*) = u/2\sigma^2$ as expected, since circular Gaussian sources cannot be separated using ICA. However, noncircular Gaussians can be separated, since in this case, the score function is given by $\psi(u, u^*) = (u + u^*)/4(1-\rho^2)\sigma_r^2 - j\rho u^*/2(1-\rho^2)\sigma_r\sigma_i + (u - u^*)/4(1-\rho^2)\sigma_i^2$, and thus is nonlinear with respect to u. A simple procedure for estimating noncircular Gaussian sources using ML is given in [20]. However, the second-order approach, strongly uncorrelating transform [32, 65] provides a more efficient procedure for estimating noncircular Gaussian sources as long as the sources have unique spectral coefficients.

For the Gaussian case, we can also write the score function as in [20]

$$\psi(u, u^*) = \frac{uE\{|u|^2\} - u^*E\{u^2\}}{2(E\{|u|^2\}^2 - |E\{u^2\}|^2)}$$

to note the *widely linear* nature of the score function for Gaussian sources.

In [84], the univariate form of the generalized Gaussian density is used to model circular source densities for deriving ICA algorithms through negentropy maximization and significant performance gain is noted when the shape parameter c is updated during the estimation. Such a scheme can be adopted for ICA through ML as well and would also require the estimation of the variances of the real and imaginary parts of the sources when used for noncircular source distributions.

Mixture Model Generalized Gaussian mixture model provides a flexible alternative to source density matching, especially in cases where the sources are not unimodal. The mixture model using the generalized Gaussian kernels given in (1.69) can be written as

$$f_{\text{GM}}(u, u^*) = \sum_{k=1}^{K} \pi_k f_{\text{GG}}(u, u^*; \sigma_r, \sigma_i, \rho, c)$$

where π_k denotes the mixing proportions of the generalized Gaussian kernels. An example application of the model would be quadrature amplitude modulated (QAM) sources where the model simplifies to

$$f_{\text{QAM}}(u, u^*) = \frac{1}{K 2\pi\sigma^2} \sum_{k=1}^{K} f_{\text{G}}(u, u^*; \sigma, \mu_k) \tag{1.70}$$

where

$$f_{\text{G}}(u, u^*; \sigma, \mu_k) = \exp\left[-\frac{1}{2\sigma^2}(u - \mu_k)(u - \mu_k)^*\right]$$

since the π_ks are taken as equal and the Gaussian kernels ($c = 2$) are circular ($\sigma_r = \sigma_i = \sigma$). The parameters, μ_k are determined by the QAM scheme, which is a *prior* information, for example, are given by $\{\pm 1\}$ for 4-QAM sources, and the value of σ can be determined by the level of noise in the system, which is assumed to be Gaussian. The score function can be easily evaluated as

$$\psi_{\text{QAM}}(u, u^*) = \frac{\sum_{k=1}^{K}(u - \mu_k) f_{\text{G}}(u, u^*; \sigma, \mu_k)}{2\sigma^2 \sum_{k=1}^{K} f_{\text{G}}(u, u^*; \sigma, \mu_k)}.$$

Linear Combinations of Basis Functions In [20], the adaptive score functions of Pham and Garat [89] are extended to the complex case through $\mathbb{C}^N \mapsto \mathbb{R}^{2N}$ mappings. We can directly evaluate and write the adaptive scores in the complex domain as follows: Approximate the "true" score function $\psi_o(u, u^*)$ as a linear combination of M basis functions $\phi_m(u, u^*)$, $m = 1, \ldots, M$ such that $\psi(u, u^*) = \sum_{m=1}^{M} \gamma_m^* \phi_m(u, u^*) = \boldsymbol{\gamma}^H \boldsymbol{\phi}$ where $\boldsymbol{\gamma} = [\gamma_1, \ldots, \gamma_M]^T$ and $\boldsymbol{\phi} = [\phi_1(u, u^*), \ldots, \phi_M(u, u^*)]^T$. Then, the problem is to determine the coefficient vector γ for each source such that $E\{|\psi_o(u, u^*) - \boldsymbol{\gamma}^H \boldsymbol{\phi}|^2\}$ is minimized. The solution is given by $\boldsymbol{\gamma} = (E\{\boldsymbol{\phi}\boldsymbol{\phi}^H\})^{-1} E\{\boldsymbol{\phi}\psi_o^*(u, u^*)\}$. The term $E\{\boldsymbol{\phi}\psi_o^*(u, u^*)\}$ requires that we know the true score function, which typically is not available. The clever trick introduced in [89] allows one to bypass this limitation, and can be extended to the complex case using Wirtinger calculus as follows. We substitute the expression for $\psi_o(u, u^*)$ given in (1.51) to the integral evaluation for the expectation $E\{\boldsymbol{\phi}\psi_o^*(u, u^*)\}$ to obtain

$$E\{\boldsymbol{\phi}\psi_o^*(u, u^*)\} = -\int_{-\infty}^{\infty}\int_{-\infty}^{\infty} \boldsymbol{\alpha}(u_r, u_i) du_r du_i \tag{1.71}$$

where $\boldsymbol{\alpha}(u_r, u_i) \triangleq \boldsymbol{\phi}(\partial f_o(u, u^*)/\partial u)$, f_o denotes the true (and unknown) source pdf and, we have used $(\partial \log f_o(u, u^*)/\partial u^*)^* = \partial \log f_o(u, u^*)/\partial u$ since $f_o(u, u^*)$ is a pdf and hence real valued. Wirtinger calculus enables us to directly write

$$\phi_m(u, u^*)\frac{\partial f_o(u, u^*)}{\partial u} = \frac{\partial}{\partial u}(\phi_m(u, u^*)f_o(u, u^*)) - f_o(u, u^*)\frac{\partial \phi_m(u, u^*)}{\partial u} \tag{1.72}$$

by using the chain rule. When (1.72) is substituted into (1.71), we obtain the important equality that shows how to evaluate the coefficients for adaptive scores using expectations without knowledge of the true source distributions

$$E\{\boldsymbol{\phi}\psi_o^*(u, u^*)\} = E\left\{\frac{\partial \boldsymbol{\phi}}{\partial u}\right\} \tag{1.73}$$

which holds when the product $f_o(u, u^*)\phi_m^*(u, u^*)$ vanishes at infinity for u_r and u_i. In the evaluation for this term, we used the integral formula given in (1.9) to write the symbolic integral given in terms of u and u^* as a contour integral of a single complex variable.

In the real case, it is shown that if the set of basis functions contains at least the identity function plus some other nonlinear function, then the stability of the separation is guaranteed [89]. For the real-valued generalized Gaussian density, a combination of three basis functions $\phi_{(1)}$, $\phi_{(0.75)}$, and $\phi_{(2)}$ that correspond to the score functions with shape parameters $c = 1$, 0.75, and 2, that is, an identity (linear Gaussian score), and one corresponding to a typical super- and one to a sub-Gaussian density have been used. In the complex case, to account for the additional dimensionality, we propose to use $\phi_1 = u$, $\phi_2 = u\,\alpha_{(0.75)}(u, u^*)$, $\phi_3 = u^*\alpha_{(0.75)}(u, u^*)$, $\phi_4 = u\,\alpha_{(2)}(u, u^*)$, $\phi_5 = u^*\alpha_{(2)}(u, u^*)$ where $\alpha(u, u^*)$ is defined in (1.69). An expansion that includes these basis functions accounts for all the terms present in the evaluation of the score function $\psi(u, u^*)$ for the generalized Gaussian density given in (1.69) along with a choice similar to those to in [89] for the shape parameters. It is worth noting that it is also possible to estimate coefficients of any nonlinear approximation to the score function such as those using splines or MLPs using a criterion such as least squares. However, the approach proposed here as in [89] has the advantage of leading to a unique solution that can be easily computed.

1.6.5 Numerical Examples

Since our focus in this section has primarily been on establishing a complete framework for complex ICA, and not on algorithm implementation and density matching mechanisms, in this section, we select examples to demonstrate the relationship between the two main classes of ICA approaches, complex ML (CML) and complex MN (CMN), to each other and to other main complex ICA approaches.

We test the performance of complex maximum likelihood using the relative gradient update in (1.54), which we refer to as the CML algorithm, and the version that constrains the demixing matrix to be unitary using the update in (1.67), the CML-unitary algorithm. For maximization of non-Gaussianity, we use the modified Newton update shown in (1.64) as its performance matches that of the gradient update (1.63) when the stepsize is chosen correctly for the gradient approach [83]. We demonstrate the performance of the algorithms for three sets of sources, a set of 4-QAM and binary phase-shift keying (BPSK) sources using (1.70), and a circular set from a generalized Gaussian distribution (GGD) with different values for the shape parameter c as in (1.69). Hence, we have sources with all three types of circularity: GGDs that are strictly circular, QAM sources that are second-order circular, and noncircular BPSK sources. For the CML and CMN updates, the nonlinearity is matched to the form of the source distribution for each run, and for the 4-QAM and BPSK simulations, parameter σ in (1.70) is chosen as 0.25, which corresponds to 12 dB signal-to-noise ratio. The 4-QAM and BPSK sources are sub-Gaussian with a normalized kurtosis value of $-.885$ and $-.77$ respectively for the given σ. The GGD sources are super-Gaussian approaching to Gaussian when the shape parameter c approaches 1.

We include the performances of complex nonlinear decorrelations (C-ND) [3] using the $-\text{asinh}(u) + u$ nonlinearity for the sub-Gaussian sources, and the performances of complex FastICA [13] using the log nonlinearity, the kurtosis maximization (KM) algorithm [69], JADE with the version that uses simultaneous diagonalization of N cumulant matrices [22], and for the circular generalized Gaussian sources, complex Infomax using the nonlinear function that assumes circularity given in [9]. Since we have not considered density adaptation, all sources in a given run are generated from the same distribution, and as a result comparisons with SUT are not included since for SUT, all sources have to have distinct spectral coefficients. For CMN, we implemented symmetric orthogonalization such that all sources are estimated in parallel and the demixing matrix is orthogonalized using $\mathbf{W} \leftarrow (\mathbf{W}\mathbf{W}^H)^{1/2}\mathbf{W}$, which is noted to provide slightly better performance when all the source densities are the same [85].

As the performance index, we use the inter-symbol-interference (ISI)—or the positive separation index [73]—given by

$$\text{ISI} = \frac{1}{2N(N-1)}\left[\sum_{i=1}^{N}\left(\sum_{j=1}^{N}\frac{|p_{ij}|}{\max_k |p_{ik}|} - 1\right) + \sum_{j=1}^{N}\left(\sum_{i=1}^{N}\frac{|p_{ij}|}{\max_k |p_{kj}|} - 1\right)\right]$$

where p_{ik} are the elements of the matrix $\mathbf{P} = \mathbf{WA}$, N is the number of sources, and the lower the ISI value the better the separation performance.

Figures 1.19 and 1.20 show the ISI values for six 4-QAM and six BPSK sources with increasing number of samples and Figure 1.21, the ISI values for six circular GGD sources as the shape parameter c varies from 0.125 to 0.75, from highly super-Gaussian to closely Gaussian for 5000 samples. For both cases, the results are the average of 10 independent runs with the least ISI out of 25, as we wanted to compare the approximate best performance of all algorithms. With this selection, the standard deviation for all the algorithms were in the range 10^{-8} for small number of samples and 10^{-11} when the number of samples are increased.

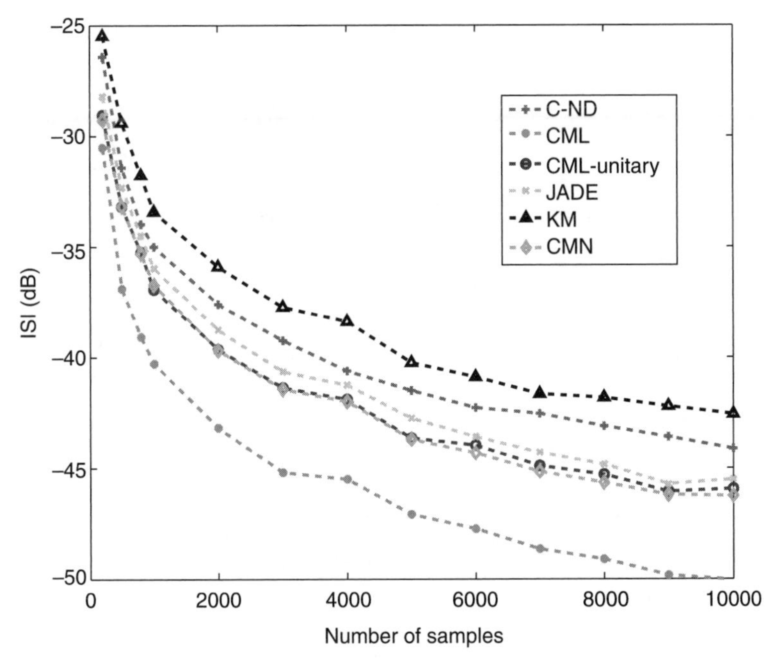

Figure 1.19 ISI as a function of number of samples for six 4-QAM sources.

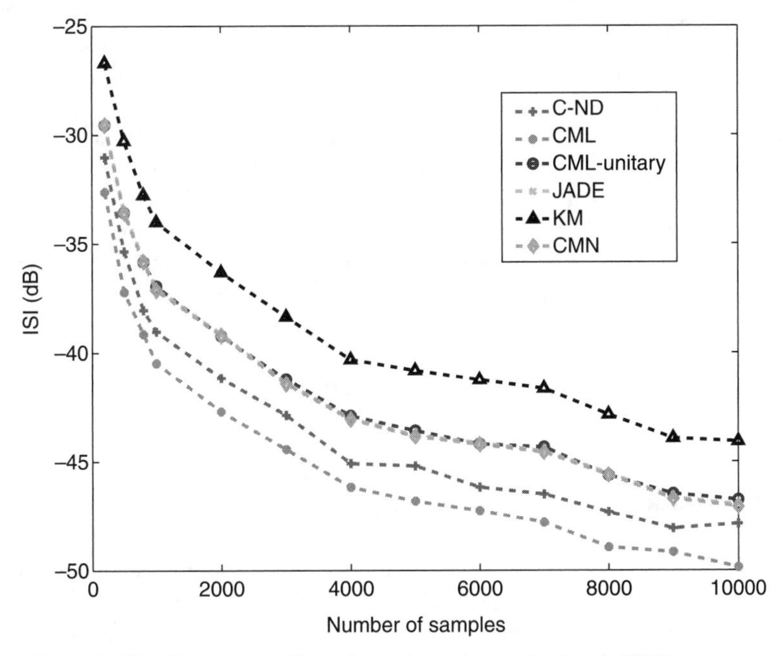

Figure 1.20 ISI as a function of number of samples for six BPSK sources.

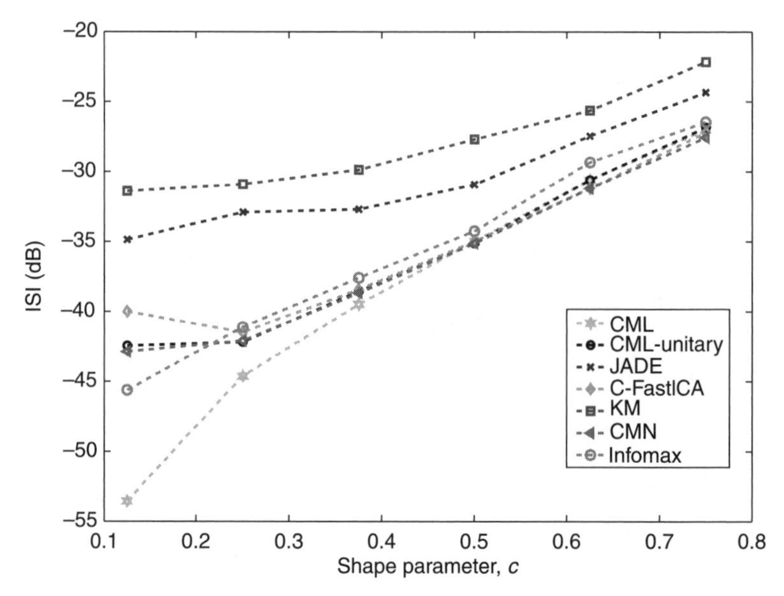

Figure 1.21 ISI as a function of the shape parameter c six GGD sources.

For all cases, we note the best performance by the CML algorithm, and the almost identical performance of the CML-unitary and CMN updates that have equivalent cost functions as discussed in Section 1.6.3. Other complex ICA approaches provide considerably satisfactory performance at a lower cost, in particular, JADE for the sub-Gaussian 4-QAM and BPSK sources, complex nonlinear decorrelations with $-\text{asinh}(u) + u$ nonlinearity for BPSK sources, and C-FastICA and complex Infomax that assume circularity for the circular GGD sources. The performance advantage of density matching comes at a computational cost as expected. The ML class of algorithms are computationally most costly when employed with density matching followed by the CMN algorithm. For example, the computational cost measured in terms of time for a single run of CML (and similarly for CML-unitary), without any optimization for implementation speed, is approximately 15 times that of KM, C-FastICA, and JADE, and three times that of CMN for six 4-QAM sources for 5000 samples. For the GGD sources, it is approximately 12 times that of KM, C-FastICA, JADE, CMN, and six times that of the ML/Infomax or nonlinear decorrelation approaches with a fixed nonlinearity.

1.7 SUMMARY

In this chapter, we provide the necessary tools for the development and analysis of algorithms in the complex domain and introduce their application to two important signal processing problems—filtering and independent component analysis. Complex-valued signal processing, we note, is not a simple extension of the real-valued case.

The definition of analyticity and commonly invoked assumptions such as circularity have been primarily motivated by this desire, such that the computations in the complex domain parallel those in the real domain. Wirtinger calculus, on the other hand, generalizes the definition of analyticity and enables development of a very convenient framework for complex-valued signal processing, which again, as desired, allows computations to be performed similar to the real case. Another important fact to note is that the framework based on Wirtinger calculus is a complete one, in the sense that the analytic case is included as a special case. Another attractive feature of the framework is that promising nonlinear structures such as fully complex (analytic) functions can be easily incorporated in algorithm development both for use within nonlinear filter structures such as MLPs and for the development of effective algorithms for performing independent component analysis. Commonly invoked assumptions such as circularity can be also easily avoided in the process, making the resulting algorithms applicable to a general class of signals, thus not limiting their usefulness. The only other reference besides this chapter—to the best of our knowledge—that fully develops the optimization framework including second-order relationships is [64], where the term $\mathbb{CR}$ calculus is used instead of Wirtinger calculus.

Though very limited in number, various textbooks have acknowledged the importance of complex-valued signals. In the most widely used book on adaptive filtering, [43], the complete development is given for complex signals starting with the 1991 edition of the book. In [43, 60, 97, 105], special sections are dedicated to complex signals, and optimization in the complex domain is introduced using the forms of the derivatives given in (1.5), also defined by Brandwood [15]. The simple trick that allows regarding the complex function as a function of two variables, $\mathbf{z}$ and $\mathbf{z}^*$, which significantly simplifies all computations, however, has not been noted in general. Even in the specific instance where it has been noted—a recent book [75] following [4, 6, 70, 71]—Wirtinger calculus is relegated to an afterthought as derivations are still given using the unnecessarily long and tedious split approach as in the previous work by the authors, for example, as in [38, 39, 41]. The important point to note is that besides simplifying derivations, Wirtinger calculus eliminates the need for many restrictive assumptions and extends the power of many convenient tools in analysis introduced for the real-valued case to the complex one. A simple example is the work in [72] where the second-order analysis of maximum likelihood independent component analysis is performed using a transformation introduced in [7] for the real-valued case while bypassing the need for any circularity assumption.

It is also interesting to note that the two forms for the derivatives given in [15], which are the correct forms and include the analytic case as well, have not been widely adopted. In a recent literature search, we noted that a significant portion of the papers published in the IEEE Transactions on Signal Processing and IEEE Transactions on Neural Networks within the past five years define the complex derivative differently than the one given in [15], which was published in 1983. The situation regarding contradictory statements and conflicting definitions in the complex domain unfortunately becomes more discouraging when we look at second-order expansions and algorithms. Even though the algorithms developed with derivative definitions other than those in (1.5) still provide reasonable—and in certain cases—equivalent processing capability, these are ad-hoc solutions and typically do not fully take advantage of the complete information and processing capability offered by the complex domain. The development we present in this chapter, on the other hand, is a complete one. When conditions such as analyticity are satisfied, or when certain assumptions such as circularity are invoked, all the results we have derived simply reduce to the versions reported earlier in the literature.

Our hope in putting together this chapter has been to describe an effective framework for the complex domain, to present all the tools under a complete and consistent umbrella, and also to attract attention to two filtering solutions for complex-valued signal processing. Widely linear and fully complex nonlinearities promise to provide effective solutions for the challenging signal processing problems of next generation systems. Both of them also open new avenues for further research and deserve much attention.

1.8 ACKNOWLEDGMENT

The work of the authors on complex-valued adaptive filtering has been supported by the National Science Foundation (Awards NSF-CCF 0635129 and NSF-IIS 0612076).

1.9 PROBLEMS

1.1 Green's theorem can be stated as [1]:

For a function $f(z) = f(x, y) = u(x, y) + jv(x, y)$, let the real-valued functions $u(x, y)$ and $v(x, y)$ along with their partial derivatives u_x, u_y, v_x, and v_y, be continuous throughout a simply connected region $\mathcal{R}$ consisting of points interior to and on a simple closed contour (described in the counter-clockwise direction) $C_{\mathcal{R}}$ in the x-y plane. We then have

$$\oint_{C_{\mathcal{R}}} (udx + vdy) = \iint_{\mathcal{R}} \left(\frac{\partial v}{\partial x} - \frac{\partial u}{\partial y} \right) dx\, dy.$$

Derive the integral formula given in (1.9) using Green's formula and the Wirtinger derivatives given in (1.5).

1.2 Verify the properties of complex-to-real mapping $\overline{(\cdot)}$: $\mathbb{C}^N \rightarrow \mathbb{R}^{2N}$ given in Section 1.2.3.

1.3 A simple way to generate samples from a circular distribution is to first generate real-valued nonnegative samples r from a selected pdf $p(r)$, and then to form the circular complex samples as

$$x + jy = rc^{j2\pi\theta} \tag{1.74}$$

where θ are samples from a uniform distribution in the range [0, 1].

We would like to generate samples from a circular generalized Gaussian distribution (GGD)—also called the exponential power distribution. We can use the procedure given in [57] to generate GGD samples with shape parameter c and scaling σ using the expression $[\texttt{gamrnd}(1/2c, \sigma)]^{1/2c}$ where the MATLAB (www.mathworks.com) function `gamrnd` generates samples from a gamma distribution with shape parameter $1/2c$ and scale parameter σ.

Explain why using this procedure directly to generate samples for the magnitude, r, will not produce samples with the same shape parameter as the bivariate case. How can you modify the expression $[\texttt{gamrnd}(1/2c, \sigma)]^{1/2c}$ so that the resulting samples will be circular-distributed GGDs with the shape parameter c when the expression given in (1.74) is used.

Hint: A simple way to check for the form of the resulting probability density function is to consider the case $c = 1$, that is, to consider the Gaussian special case.

1.4 Using the two mappings given in Proposition 1, Eqs. (1.25) and (1.26), and real-valued conjugate gradient algorithm given in Section 1.3.1, derive the complex conjugate gradient algorithm which is stated in Section 1.3.4.

1.5 Write the widely linear estimate given in (1.37) using the $\mathbb{C}^N$ notation by defining

$$\mathbf{v} = \begin{bmatrix} \mathbf{v}_1 \\ \mathbf{v}_2 \end{bmatrix}$$

and show that the optimum widely linear vector estimates can be written as

$$\mathbf{v}_{1,\mathrm{opt}} = [\mathbf{C} - \mathbf{P}\mathbf{C}^{-*}\mathbf{P}^*]^{-1}[\mathbf{p} - \mathbf{P}\mathbf{C}^{-*}\mathbf{q}^*]$$

and

$$\mathbf{v}_{2,\mathrm{opt}} = [\mathbf{C}^* - \mathbf{P}^*\mathbf{C}^{-1}\mathbf{P}]^{-1}[\mathbf{q}^* - \mathbf{P}^*\mathbf{C}^{-1}\mathbf{p}]$$

in $\mathbb{C}^N$, where $(\cdot)^{-*}$ is the complex conjugate of the inverse.

Use the forms given above for $\mathbf{v}_{1,\,\mathrm{opt}}$ and $\mathbf{v}_{2,\,\mathrm{opt}}$ to show that the mean-square error between a widely linear and linear filter J_{diff} is given by the expression in (1.38).

1.6 Given a finite impulse response system with the impulse response vector $\mathbf{w}_{\mathrm{opt}}$ with coefficients $w_{\mathrm{opt},n}$ for $n = 1, \ldots, N$.

Show that, if the desired response is written as

$$d(n) = \mathbf{w}_{\mathrm{opt}}^H \mathbf{x}(n) + v(n)$$

where $\mathbf{x}(n) = [x(n)x(n-1) \cdots x(N-1)]^T$ and both the input $x(n)$ and the noise term $v(n)$ are zero mean and $x(n)$ is uncorrelated with both $v(n)$ and $v^*(n)$, then the mean-square weight estimator is given by

$$\mathbf{w} = \mathbf{C}^{-1}\mathbf{p}$$

or by

$$\mathbf{w}^* = \mathbf{P}^{-1}\mathbf{q}$$

where the covariance and pseudo covariance matrices **C** and **P** as well as the cross covariance vectors **p** and **q** are defined in Section 1.4. Consequently, show that the mean-square error difference between a linear and a widely linear MSE filter ($J_{\text{diff}} = J_{L,\min} - J_{WL,\min}$) for this case is exactly zero, that is, using a widely linear filter does not provide any additional advantage even when the signal is noncircular.

1.7 The conclusion in Problem 1.6 can be extended to prediction of an autoregressive process given by

$$X(n) + \sum_{k=1}^{N} a_k X(n-k-1) = V(n)$$

where $V(n)$ is the white Gaussian noise. For simplicity, assume one-step ahead predictor and show that $J_{\text{diff}} = 0$ as long as $V(n)$ is a doubly white random process, that is, the covariance and the pseudo covariance functions of $V(n)$ satisfy $c(k) = c(0)\delta(k)$ and $p(k) = p(0)\delta(k)$ respectively.

1.8 For the widely linear weight vector error difference $\varepsilon(n) = \mathbf{v}(n) - \mathbf{v}_{\text{opt}}$, show that we can write the expression for the modes of the widely linear LMS algorithm given in (1.41) as

$$E\{\varepsilon'_k(n)\} = \varepsilon'_k(0)(1 - \mu\bar{\lambda}_k)^n$$

and

$$E\{|\varepsilon'_k(n)|^2\} = \frac{\mu J_{WL,\min}}{2 - \mu\bar{\lambda}_k} + (1 - \mu\bar{\lambda}_k)^{2n}\left(|\varepsilon'_k(0)|^2 - \frac{\mu J_{WL,\min}}{2 - \mu\bar{\lambda}_k}\right)$$

as shown in [16, 43] for the linear LMS algorithm. Make sure you clearly identify all assumptions that lead to the expressions given above.

1.9 Explain the importance of the correlation matrix eigenvalues on the performance of the linear and widely linear LMS filter (λ and $\bar{\lambda}$). Let input $x(n)$ be a first order autoregressive process ($N = 1$ for the AR process given in Problem 1.7) but let the white Gaussian noise $v(n)$ be noncircular such that the pseudo-covariance $E\{v^2(n)\} \neq 0$. Show that when the pseudo-covariance matrix is nonzero, the eigenvalue spread of the augmented covariance matrix $\bar{\mathbf{C}}$ will always be greater than or equal to that of the original covariance matrix $\mathbf{C}$ using the majorization theorem [49].

1.10 In real-valued independent component analysis, separation is possible as long as only one of the sources is Gaussian. In the complex case, however, as discussed in Section 1.6.4, Gaussian sources can be separated as long as they are noncircular with unique spectral coefficients.

Show that the score function for Gaussian sources can be reduced to

$$\psi_n(u) = \frac{u_r}{4\sigma_r^2} + j\frac{u_i}{4\sigma_i^2}$$

when we consider the scaling ambiguity for the complex case. Then, devise a procedure for density (score function) matching for the estimation of complex Gaussian sources.

REFERENCES

1. M. J. Ablowitz and A. S. Fokas, *Complex Variables*. Cambridge University Press, Cambridge, UK, (2003).
2. T. Adalı and V. D. Calhoun, Complex ICA of medical imaging data. *IEEE Signal Proc. Mag.*, 24(5):136–139, (2007).
3. T. Adalı, T. Kim, and V. D. Calhoun, Independent component analysis by complex nonlinearities. In *Proc. IEEE Int. Conf. Acoust., Speech, Signal Processing (ICASSP)*, V, pp. 525–528, Montreal, QC, Canada, May (2004).
4. T. Adalı and H. Li, A practical formulation for computation of complex gradients and its application to maximum likelihood. In *Proc. IEEE Int. Conf. Acoust., Speech, Signal Processing (ICASSP)*, II, pp. 633–636, Honolulu, HI, April (2007).
5. T. Adalı and H. Li, On properties of the widely linear MSE filter and its LMS implementation. In *Proc. Conf. Information Sciences and Systems*, Baltimore, MD, March (2009).
6. T. Adalı, H. Li, M. Novey, and J.-F. Cardoso, Complex ICA using nonlinear functions. *IEEE Trans. Signal Processing*, 56(9):4356–4544, (2008).
7. S.-I. Amari, T.-P. Chen, and A. Cichocki, Stability analysis of learning algorithms for blind source separation. *Neural Networks*, 10(8):1345–1351, (1997).
8. P. O. Amblard, M. Gaeta, and J. L. Lacoume, Statistics for complex variables and signals—Part 2: Signals. *Signal Processing*, 53(1):15–25, (1996).
9. J. Anemüller, T. J. Sejnowski, and S. Makeig, Complex independent component analysis of frequency-domain electroencephalographic data. *Neural Networks*, 16:1311–1323, (2003).
10. P. Arena, L. Fortuna, R. Re, and M. G. Xibilia, Multilayer perceptrons to approximate complex valued functions. *International Journal of Neural Systems*, 6:435–446, (1995).
11. A. Bell and T. Sejnowski, An information maximization approach to blind separation and blind deconvolution. *Neural Computation*, 7:1129–1159, (1995).

12. N. Benvenuto and F. Piazza, On the complex backpropagation algorithm. *IEEE Trans. Signal Processing*, 40(4):967–969, (1992).
13. E. Bingham and A. Hyvärinen, A fast fixed-point algorithm for independent component analysis of complex valued signals. *Int. J. Neural Systems*, 10:1–8, (2000).
14. D. L. Birx and S. J. Pipenberg, A complex mapping network for phase sensitive classification. *IEEE Trans. Neural Networks*, 4(1):127–135, (1993).
15. D. H. Brandwood, A complex gradient operator and its application in adaptive array theory. *Proc. Inst. Elect. Eng.*, 130(1):11–16, (1983).
16. H. J. Butterweck, A steady-state analysis of the LMS adaptive algorithm without the use of independence assumption. In *Proc. IEEE Int. Conf. Acoust., Speech, Signal Processing (ICASSP)*, pp. 1404–1407, Detroit, (1995).
17. V. D. Calhoun, *Independent Component Analysis for Functional Magnetic Resonance Imaging*. Ph.D thesis, University of Maryland Baltimore County, Baltimore, MD, 2002.
18. V. D. Calhoun and T. Adalı, Complex ICA for fMRI analysis: Performance of several approaches. In *Proc. IEEE Int. Conf. Acoust., Speech, Signal Processing (ICASSP)*, II, pp. 717–720, Hong Kong, China, April (2003).
19. J.-F. Cardoso, Blind signal separation: Statistical principles. *Proc. IEEE*, 86(10):2009–2025, (1998).
20. J.-F. Cardoso and T. Adalı, The maximum likelihood approach to complex ICA. In *Proc. IEEE Int. Conf. Acoust., Speech, Signal Processing (ICASSP)*, V, pp. 673–676, Toulouse, France, May (2006).
21. J.-F. Cardoso and B. Laheld, Equivariant adaptive source separation. *IEEE Trans. Signal Processing*, 44:3017–3030, (1996).
22. J.-F. Cardoso and A. Souloumiac, Blind beamforming for non-Gaussian signals. *IEE Proc. Radar Signal Processing*, 140:362–370, (1993).
23. P. Chevalier and F. Pipon, New insights into optimal widely linear array receivers for the demodulation of BPSK, MSK, and GMSK signals corrupted by noncircular interferences—application to SAIC. *IEEE Trans. Signal Processing*, 54(3):870–883, (2006).
24. S. Choi, A. Cichocki, and S.-I. Amari, Flexible independent component analysis. *J. VLSI Signal Processing Systems for Signal, Image, and Video Technology*, 26(1/2):25–38, (2000).
25. A. Cichocki and R. Unbehauen, Robust neural networks with online learning for blind identification and blind separation of sources. *IEEE Trans. Circuits Syst. I: Fund. Theory Apps.*, 43:894–906, (1996).
26. M. Z. Coban and R. M. Mersereau, Adaptive subband video coding using bivariate generalized gaussian distribution model. In *Proc. IEEE Int. Conf. Acoust., Speech, Signal Processing (ICASSP)*, IV, pp. 1990–1993, Atlanta, GA, May (1996).
27. P. Comon, Circularité et signaux aléatoires à temps discret. *Traitement du Signal*, 11(5):417–420, (1994).
28. P. Comon, Independent component analysis—a new concept? *Signal Processing*, 36:287–314, (1994).
29. N. Correa, T. Adalı, and V. D. Calhoun, Performance of blind source separation algorithms for fMRI analysis using a group ICA method. *Magnetic Resonance Imaging*, 25(5):684–694, (2007).

30. G. Cybenko, Approximation by superpositions of a sigmoidal function. *Mathematics of Control, Signals, and Systems*, 2:303–314, (1989).
31. J. P. D'Angelo, *Inequalities from Complex Analysis*, volume 28 of *Carus Mathematical Monographs*. Mathematical Association of America, (2002).
32. J. Eriksson and V. Koivunen, Complex-valued ICA using second order statistics. In *Proc. IEEE Int. Workshop on Machine Learning for Signal Processing (MLSP)*, pp. 183–192, Saõ Luis, Brazil, Sept. (2004).
33. J. Eriksson and V. Koivunen, Complex random vectors and ICA models: Identifiability, uniqueness and separability. *IEEE Trans. Info. Theory*, 52(3):1017–1029, (2006).
34. J. Eriksson, A. Seppola, and V. Koivunen, Complex ICA for circular and non-circular sources. In *Proc. European Signal Process. Conf. (EUSIPCO)*, Antalya, Turkey, (2005).
35. K. Funahashi, On the approximate realization of continuous mappings by neural networks. *Neural Networks*, 2:183–192, (1989).
36. G. Georgiou and C. Koutsougeras, Complex back-propagation. *IEEE Trans. Circuits Syst. II*, 39(5):330–334, (1992).
37. S. L. Goh, M. Chen, D. H. Popovic, K. Aihara, D. Obradovic, and D. P. Mandic, Complex-valued forecasting of wind profile. *Renewable Energy*, 31:1733–1750, (2006).
38. S. L. Goh and D. P. Mandic, Nonlinear adaptive prediction of complex-valued signals by complex-valued PRNN. *IEEE Trans. Signal Processing*, 53(5):1827–1836, (2005).
39. S. L. Goh and D. P. Mandic, A general fully adaptive normalised gradient descent learning algorithm for complex-valued nonlinear adaptive filters. *IEEE Trans. Neural Networks*, 18(5):1511–1516, (2007).
40. N. R. Goodman, Statistical analysis based on a certain multivariate complex Gaussian distribution. *Annals Math. Stats.*, 34:152–176, (1963).
41. A. I. Hanna and D. P. Mandic, A fully adaptive normalized nonlinear gradient descent algorithm for complex-valued nonlinear adaptive filters. *IEEE Trans. Signal Processing*, 51(10):2540–2549, (2003).
42. S. Haykin, *Neural Networks: A Comprehensive Foundation (2nd Edition) Neural Networks: A Comprehensive Foundation*. Prentice-Hall, Inc., Second edition, (1999).
43. S. Haykin, *Adaptive Filter Theory*. Prentice-Hall, Inc., Upper Saddle River, NJ, fourth edition, (2002).
44. P. Henrici, *Applied and Computational Complex Analysis*, III, Wiley, New York, NY, (1986).
45. A. Hirose, Continuous complex-valued back-propagation learning. *Electronics Letts.*, 28(20):1854–1855, (1992).
46. A. Hjõrungnes and D. Gesbert, Complex-valued matrix differentiation: Techniques and key results. *IEEE Trans. Signal Processing*, 55(6):2740–2746, (2007).
47. F. G. C. Hoogenraad, P. J. W. Pouwels, M. B. M. Hofman, J. R. Reichenbach, M. Sprenger, and E. M. Haacke, Quantitative differentiation between bold models in FMRI. *Magnetic Resonance in Medicine*, 45:233–246, (2001).
48. L. Hörmander, *An Introduction to Complex Analysis in Several Variables*. North-Holland, Oxford, UK, (1990).
49. R. A. Horn and C. R. Johnson, *Matrix Analysis*. Cambridge University Press, New York, NY, (1999).

50. R. A. Horn and C. R. Johnson, *Topics in Matrix Analysis*. Cambridge University Press, New York, NY, (1999).
51. K. Hornik, M. Stinchcombe, and H. White, Multilayer feedforward networks are universal approximators. *Neural Networks*, 2:359–366, (1989).
52. A. Hyvärinen, One-unit contrast functions for independent component analysis: A statistical analysis. In *Proc. IEEE Workshop on Neural Networks for Signal Processing (NNSP)*, pp. 388–397, Amelia Island, FL, Sept. (1997).
53. A. Hyvärinen, Fast and robust fixed-point algorithms for independent component analysis. *IEEE Trans. Neural Networks*, 10(3):626–634, (1999).
54. A. Hyvärinen, J. Karhunen, and E. Oja, *Independent Component Analysis*. Wiley, New York, NY, (2001).
55. J.-J. Jeon, RLS adaptation of widely linear minimum output energy algorithm for DS-CDMA systems. In *Proc. Adv. Industrial Conf. on Telecommunications*, pp. 98–102, (2005).
56. J.-J. Jeon, J. G. Andrews, and K.-M. Sung, The blind widely linear minimum output energy algorithm for DS-CDMA systems. *IEEE Trans. Signal Processing*, 54(5):1926–1931, (2006).
57. M. E. Johnson, Computer generation of the exponential power distributions. *Journal of Statistical Computation and Simulation*, 9:239–240, (1979).
58. C. Jutten and J. Hérault, Blind separation of sources, Part 1: An adaptive algorithm based on neuromimetic architecture. *Signal Processing*, 24(1):1–10, (1991).
59. J. Karvanen, J. Eriksson, and V. Koivunen, Pearson system based method for blind separation. In *Proc. Second Int. Workshop on ICA*, Helsinki, Finland, (2000).
60. S. M. Kay, *Fundamentals of Statistical Processing, Volume I: Estimation Theory*, Prentice Hall Signal Processing Series, Upper Saddl River, NJ (1993).
61. T. Kim and T. Adalı, Fully complex backpropagation for constant envelope signal processing. In *Proc. IEEE Workshop on Neural Networks for Signal Processing (NNSP)*, pp. 231–239. IEEE, Dec. (2000).
62. T. Kim and T. Adalı, Fully complex multi-layer perceptron network for nonlinear signal processing. *J. VLSI Signal Processing Systems for Signal, Image, and Video Technology*, 32:29–43, (2002).
63. T. Kim and T. Adalı, Approximation by fully complex multilayer perceptrons. *Neural Computation*, 15:1641–1666, (2003).
64. K. Kreutz-Delgado, ECE275A: Parameter Estimation I, Lecture Supplement on Complex Vector Calculus, New York, (2007).
65. L. De Lathauwer and B. De Moor, On the blind separation of non-circular sources. In *Proc. European Signal Process. Conf. (EUSIPCO)*, Toulouse, France, (2002).
66. T.-W. Lee, M. Girolami, and T. J. Sejnowski, Independent component analysis using an extended Infomax algorithm for mixed subgaussian and supergaussian sources. *Neural Computation*, 11:417–441, (1999).
67. H. Leung and S. Haykin, The complex backpropagation algorithm. *IEEE Trans. Signal Processing*, 39:2101–2104, (1991).
68. H. Li, *Complex-Valued Adaptive Signal Processing using Wirtinger Calculus and its Application to Independent Component Analysis*. Ph.D thesis, University of Maryland Baltimore County, Baltimore, MD, (2008).

69. H. Li and T. Adalı, Gradient and fixed-point complex ICA algorithms based on kurtosis maximization. In *Proc. IEEE Int. Workshop on Machine Learning for Signal Processing (MLSP)*, pp. 85–90, Maynooth, Ireland, Sept. (2006).
70. H. Li and T. Adalı, Optimization in the complex domain for nonlinear adaptive filtering. In *Proc. 33rd Asilomar Conf. on Signals, Systems and Computers*, pp. 263–267, Pacific Grove, CA, Nov. (2006).
71. H. Li and T. Adalı, Complex-valued adaptive signal processing using nonlinear functions. *J. Advances in Signal Processing*, 2008 (Article ID 765615, 9 pages), (2008).
72. H. Li and T. Adalı, Stability analysis of complex maximum likelihood ICA using Wirtinger calculus. In *Proc. IEEE Int. Conf. Acoust., Speech, Signal Processing (ICASSP)*, Las Vegas, NV, April (2008).
73. O. Macchi and E. Moreau, Self-adaptive source separation by direct or recursive networks. In *Proc. Int. Conf. Digital Signal Proc.*, pp. 122–129, Limasol, Cyprus, (1995).
74. J. R. Magnus and H. Neudecker, *Matrix Differential Calculus with Applications in Statistics and Econometrics*. Wiley, (1988).
75. D. Mandic and S. L. Goh, *Complex Valued Nonlinear Adaptive Filters*. Wiley, Chichester, (2009).
76. K.V. Mardia, *Families of Bivariate Distributions*. Griffen, London, (1970).
77. C. D. Meyer, *Matrix Analysis and Applied Linear Algebra*. SIAM, Philadephia, PA, (2000).
78. T. P. Minka, Old and new matrix algebra useful for statistics, http://research.microsoft.com/~minka/papers/matrix, (2000).
79. M. F. Moller, A scaled conjugate gradient algorithm for fast supervised learning. *Neural Networks*, 6:525–533, (1993).
80. D. R. Morgan, Adaptive algorithms for a two-channel structure employing allpass filters with applications to polarization mode dispersion compensation. *IEEE Trans. Circuits Syst. I: Fund. Theory Apps.*, 51(9):1837–1847, (2004).
81. F. D. Neeser and J. L. Massey, Proper complex random processes with applications to information theory. *IEEE Trans. Info. Theory*, 39:1293–1302, (1993).
82. J. Nocedal and S. J. Wright, *Numerical Optimization*. Springer, New York, NY, (2000).
83. M. Novey and T. Adalı, ICA by maximization of nongaussianity using complex functions. In *Proc. IEEE Int. Workshop on Machine Learning for Signal Processing (MLSP)*, pp. 21–26, Mystic, CT, Sept. (2005).
84. M. Novey and T. Adalı, Adaptable nonlinearity for complex maximization of nongaussianity and a fixed-point algorithm. In *Proc. IEEE Int. Workshop on Machine Learning for Signal Processing (MLSP)*, pages 79–84, Maynooth, Ireland, Sept. (2006).
85. M. Novey and T. Adalı, Complex fixed-point ICA algorithm for separation of QAM sources using Gaussian mixture model. In *Proc. IEEE Int. Conf. Acoust., Speech, Signal Processing (ICASSP)*, volume II, pages 445–448, Honolulu, HI, April (2007).
86. M. Novey and T. Adalı, Complex ICA by negentropy maximization. *IEEE Trans. Neural Networks*, 19(4):596–609, (2008).
87. S. Olhede, On probability density functions for complex variables. *IEEE Trans. Info. Theory*, 52:1212–1217, (2006).
88. K. B. Petersen and M. S. Pedersen, *The Matrix cookbook*, Oct. (2008), Version (20081110).

89. D. Pham and P. Garat, Blind separation of mixtures of independent sources through a quasi maximum likelihood approach. *IEEE Trans. Signal Processing*, 45(7):1712–1725, (1997).

90. B. Picinbono, *Random Signals and Systems*, Prentice Hall Signal Processing Series, Englewood Cliffs, NJ, (1993).

91. B. Picinbono, On circularity. *IEEE Trans. Signal Processing*, 42:3473–3482, (1994).

92. B. Picinbono, Second-order complex random vectors and normal distributions. *IEEE Trans. Signal Processing*, 44(10):2637–2640, (1996).

93. B. Picinbono and P. Bondon, Second-order statistics of random signals. *IEEE Trans. Signal Processing*, 45(2):411–419, (1997).

94. B. Picinbono and P. Chevalier, Widely linear estimation with complex data. *IEEE Trans. Signal Processing*, 43:2030–2033, (1995).

95. R. Remmert, *Theory of Complex Functions*. Springer-Verlag, Harrisonburg, VA, (1991).

96. H. Sawada, R. Mukai, S. Araki, and S. Makino, A polar-coordinate based activation function for frequency domain blind source separation. In *Proc. IEEE Int. Conf. Acoust., Speech, Signal Processing (ICASSP)*, pp. 1001–1004, May 2002.

97. A. Sayed, *Fundamentals of Adaptive Filtering*. Wiley-IEEE, Hoboken, NJ, (2003).

98. M. Scarpiniti, D. Vigliano, R. Parisi, and A. Uncini, Generalized splitting functions for blind separation of complex signals. *Neurocomputing*, (2008).

99. R. Schober, W. H. Gerstacker, and L. H.-J. Lampe, A widely linear LMS algorithm for MAI suppression for DS-CDMA. In *Proc. IEEE Int. Conf. on Communications*, pp. 2520–2525, (2003).

100. P. Schreier, Bounds on the degree of improperiety of complex random vectors. *IEEE Signal Proc. Letts.*, 15:190–193, (2008).

101. P. Schreier and L. Scharf, Second-order analysis of improper complex random vectors and processes. *IEEE Trans. Signal Processing*, 51(3):714–725, (2003).

102. P. Schreier, L. Scharf, and A. Hanssen, A generalized likelihood ratio test for impropriety of complex signals. *IEEE Signal Proc. Letts.*, 13(7):433–436, (2006).

103. P. Smaragdis, Blind separation of convolved mixtures in the frequency domain. *Neurocomputing*, 22:21–34, (1998).

104. A. Taleb and C. Jutten, Source separation in post-nonlinear mixtures. *IEEE Trans. Signal Processing*, 47(10):2807–2820, (1999).

105. C. W. Therrien, *Probability for Electrical and Computer Engineers*. CRC Press, Boca Raton, FL, (2004).

106. L. Tong, R.-W. Liu, V. C. Soon, and Y.-F. Huang, Indeterminacy and identifiability of blind identification. *IEEE Trans. Circuits Syst.*, 38(5):499–509, (1991).

107. A. Uncini and F. Piazza, Blind signal processing by complex domain adaptive spline neural networks. *IEEE Trans. Neural Networks*, 14(2):399–412, (2003).

108. A. Uncini, L. Vecci, P. Campolucci, and F. Piazza, Complex-valued neural networks with adaptive spline activation function for digital radio links nonlinear equalization. *IEEE Trans. Signal Processing*, 47(2):505–514, (1999).

109. N. N Vakhania and N. P. Kandelaki, Random vectors with values in complex Hilbert spaces. *Theory Probability Appl.*, 41(1):116–131, (1996).

110. A. van den Bos, Complex gradient and Hessian. *IEE Proc.: Vision, Image, and Signal Processing*, 141(6):380–382, (1994).

111. A. van den Bos, Estimation of complex parameters. In *10th IFAC Symp.*, volume 3, pp. 495–499, (1994).
112. N. Vlassis and Y. Motomura, Efficient source adaptivity in independent component analysis. *IEEE Trans. Neural Networks*, 12(3):559–566, (2001).
113. B. Widrow, J. Cool, and M. Ball, The complex LMS algorithm. *Proc. IEEE*, 63:719–720, (1975).
114. B. Widrow and Jr. M. E. Hopf, Adaptive switching circuits. In *IRE WESCON*, volume 4, pp. 96–104, (1960).
115. W. Wirtinger, Zur formalen theorie der funktionen von mehr komplexen veränderlichen. *Math. Ann.*, 97:357–375, (1927).
116. L. Xu, G. D. Pearlson, and V. D. Calhoun, Joint source based morphometry to identify sources of gray matter and white matter relative differences in schizophrenia versus healthy controls. In *Proc. ISMRM*, Toronto, ON, May (2008).
117. G. Yan and H. Fan, A newton-like algorithm for complex variables with application in blind equalization. *IEEE Trans. Signal Processing*, 48:553–556, (2000).
118. C.-C. Yang and N. K. Bose, Landmine detection and classification with complex-valued hybrid neural network using scattering parameters dataset. *IEEE Trans. Neural Networks*, 16(3):743–753, (2005).
119. C. You and D. Hong, Nonlinear blind equalization schemes using complex-valued multilayer feedforward neural networks. *IEEE Trans. Neural Networks*, 9(6):1442–1455, (1998).
120. L. Zhang, A. Cichocki, and S.-I. Amari, Self-adaptive blind source separation based on activation functions adaptation. *IEEE Trans. Neural Networks*, 15(2):233–244, (2004).

2

ROBUST ESTIMATION TECHNIQUES FOR COMPLEX-VALUED RANDOM VECTORS

Esa Ollila and Visa Koivunen
Helsinki University of Technology, Espoo, Finland

2.1 INTRODUCTION

In this chapter we address the problem of multichannel signal processing of complex-valued signals in cases where the underlying ideal assumptions on signal and noise models are not necessarily true. In signal processing applications we are typically interested in second-order statistics of the signal and noise. We will focus on departures from two key assumptions: circularity of the signal and/or noise as well as the Gaussianity of the noise distribution. Circularity imposes an additional restriction on the correlation structure of the complex random vector. We will develop signal processing algorithms that take into account the complete second-order statistics of the signals and are robust in the face of heavy-tailed, impulsive noise. Robust techniques are close to optimal when the nominal assumptions hold and produce highly reliable estimates otherwise. Maximum likelihood estimators (MLEs) derived under complex normal (Gaussian) assumptions on noise models may suffer from drastic degradation in performance in the face of heavy-tailed noise and highly deviating observations called outliers.

Adaptive Signal Processing: Next Generation Solutions. Edited by Tülay Adalı and Simon Haykin
Copyright © 2010 John Wiley & Sons, Inc.

Many man-made complex-valued signals encountered in wireless communication and array signal-processing applications possess circular symmetry properties. Moreover, additive sensor noise present in the observed data is commonly modeled to be complex, circular Gaussian distributed. There are, however, many signals of practical interest that are not circular. For example, commonly used modulation schemes such as binary phase shift keying (BPSK) and pulse-amplitude modulation (PAM) lead to noncircular observation vectors in a conventional baseband signal model. Transceiver imperfections or interference from other signal sources may also lead to noncircular observed signals. This property may be exploited in the process of recovering the desired signal and cancelling the interferences. Also by taking into account the noncircularity of the signals, the performance of the estimators may improve, the optimal estimators and theoretical performance bounds may differ from the circular case, and the algorithms and signal models used in finding the estimates may be different as well. As an example, the signal models and algorithms for subspace estimation in the case of noncircular or circular sources are significantly different. This awareness of noncircularity has attained considerable research interest during the last decade, see for example [1, 13, 18, 20, 22, 25, 38–41, 46, 47, 49, 50, 52–54, 57].

2.1.1 Signal Model

In many applications, the multichannel k-variate received signal $\mathbf{z} = (z_1, \ldots, z_k)^T$ (sensor outputs) is modeled in terms of the transmitted *source signals* $s_1, \ldots, s_d$ possibly corrupted by additive *noise vector* $\mathbf{n}$, that is

$$\begin{aligned}\mathbf{z} &= \mathbf{A}\mathbf{s} + \mathbf{n} \\ &= \mathbf{a}_1 s_1 + \cdots + \mathbf{a}_d s_d + \mathbf{n} \qquad (2.1)\end{aligned}$$

where $\mathbf{A} = (\mathbf{a}_1, \ldots, \mathbf{a}_d)$ is the $k \times d$ *system matrix* and $\mathbf{s} = (s_1, \ldots, s_d)^T$ contains the source signals. It is assumed that $d \leq k$. In practice, the system matrix is used to describe the array geometry in sensor array applications, multiple-input multiple-output (MIMO) channel in wireless multiantenna communication systems and mixing systems in the case of signal separation problems, for example. All the components above are assumed to be complex-valued, and $\mathbf{s}$ and $\mathbf{n}$ are assumed to be mutually statistically independent with zero mean. An example of a multiantenna sensing system with uniform linear array (ULA) configuration is depicted in Figure 2.1.

The model (2.1) is indeed very general, and covers, for example, the following important applications.

In *narrowband array signal processing*, each vector $\mathbf{a}_i$ represents a point in known array manifold (array transfer function, steering vector) $\mathbf{a}(\theta)$, that is $\mathbf{a}_i = \mathbf{a}(\theta_i)$, where θ_i is an unknown parameter, typically the direction-of-arrival (DOA) θ_i of the ith source, $i = 1, \ldots, d$. Identifying $\mathbf{A}$ is then equivalent with the problem of identifying $\theta_1, \ldots, \theta_d$. For example, in case of ULA with identical sensors,

$$\mathbf{a}(\theta) = \begin{pmatrix} 1 & e^{-j\omega} & \cdots & e^{-j(k-1)\omega} \end{pmatrix}^T,$$

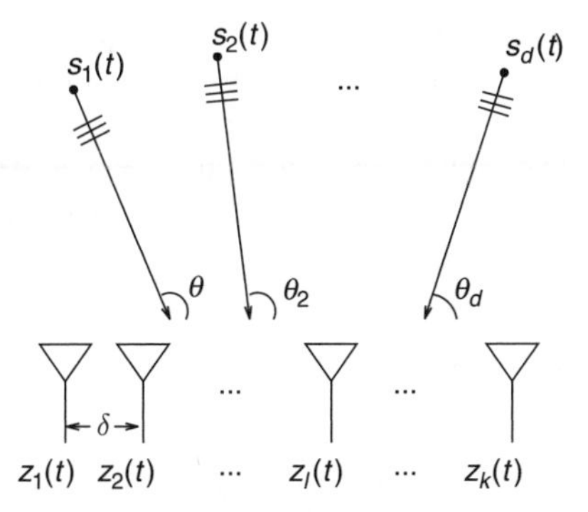

Figure 2.1 A uniform linear array (ULA) of k sensors with sensor displacement δ receiving plane waves from d far-field point sources.

where $\omega = 2\pi(\delta/\lambda)\sin(\theta)$ depends on the signal wavelength λ, the DOA θ of the signal with respect to broadside, and the sensor spacing δ. The source signal vector $\mathbf{s}$ is modeled as either deterministic or random, depending on the application.

In blind signal separation (BSS) based on independent component analysis (ICA), both the mixing system $\mathbf{A}$ and the sources $\mathbf{s}$ are unknown. The goal in ICA is to solve the mixing matrix and consequently to separate the sources from their mixtures exploiting only the assumption that sources are mutually statistically independent. In this chapter, we consider the noiseless ICA model.

Common assumptions imposed on the signal model (2.1) are as follows:

ASSUMPTION (A1) noise $\mathbf{n}$ and/or source $\mathbf{s}$ possess circularly symmetric distributions.

In addition, in the process of deriving optimal array processors, the distribution of the noise $\mathbf{n}$ is assumed to be known also, the conventional assumption being that

ASSUMPTION (A2) noise $\mathbf{n}$ possesses circular complex Gaussian distribution.

Furthermore, if $\mathbf{s}$ is modelled as stochastic, then $\mathbf{s}$ and $\mathbf{n}$ are both assumed to be independent with circular complex Gaussian distribution, and consequently, sensor output $\mathbf{z}$ also has k-variate circular complex Gaussian distribution.

In this chapter, we consider the cases where assumptions (A1) and (A2) do not hold. Hence we introduce methods for array processing and ICA that work well at circular and noncircular distributions and when the conventional assumption of normality is not valid. Signal processing examples on beamforming, subspace-based DOA estimation and source-signal separation are provided. Moreover, tests for detecting noncircularity of the data are introduced and the distributions of the test statistics are established as well. Such a test statistic can be used as a guide in choosing the appropriate array processor. For example, if the test rejects the null hypothesis of circularity, it is often wiser to choose a method that explicitly exploits the noncircularity property instead of a method that does not. For example, the generalized uncorrelating transform (GUT) method [47] that is explicitly designed for blind separation of noncircular sources has, in general, better performance in such cases than a method that does not exploit the noncircularity aspect of the sources. Uncertainties related to system matrix, for example, departures from assumed sensor array geometry and related robust estimation procedures are not considered in this chapter.

2.1.2 Outline of the Chapter

This chapter is organized as follows. First, key statistics that are used in describing properties of complex-valued random vectors are presented in Section 2.2. Essential statistics used in this chapter in the characterization of complex random vectors are the circular symmetry, covariance matrix, pseudo-covariance matrix, the strong-uncorrelating transform and the circularity coefficients. The information contained in these statistics can be exploited in designing optimal array processors. In Section 2.3, the class of complex elliptically symmetric (CES) distributions [46] are reviewed. CES distributions constitute a flexible, broad class of distributions that can model both circular/noncircular and heavy-/light-tailed complex random phenomena. It includes the commonly used circular complex normal (CN) distribution as a special case. We also introduce an adjusted generalized likelihood ratio test (GLRT) that can be used for testing circularity when sampling from CES distributions with finite fourth-order moments [40]. This test statistic is shown to be a function of circularity coefficients.

In Section 2.4, tools to compare statistical robustness and statistical efficiency of the estimators are discussed. Special emphasis is put on the concept of influence function (IF) of a statistical functional. IF describes the qualitative robustness of an estimator. Intuitively, qualitative robustness means that the impact of errors to the performance of the estimator is bounded and small changes in the data cause only small changes in the estimates. More explicitly IF measures the sensitivity of the functional to small amounts of contamination in the distribution. It can also be used to calculate the asymptotic covariance structure of the estimator. In Section 2.5, the important concepts of (spatial) scatter matrix and (spatial) pseudo-scatter matrix are defined and examples of such matrix functionals are given. These matrices will be used in developing robust array processors and blind separation techniques that work reliably for both circular/noncircular and Gaussian/non-Gaussian environments. Special emphasis is put on one particularly important class of scatter matrices, called the M-estimators of scatter, that generalize the ML-estimators of scatter matrix parameters of circular CES distributions. Then, in Section 2.6, it is demonstrated how scatter and pseudo-scatter matrices can be used in designing robust beamforming and subspace based DOA estimation methods. Also, a subspace DOA estimation method [13] designed for noncircular sources is discussed. In Section 2.7, we derive the IF of the conventional minimum variance distortionless response (MVDR) beamformer and compare it with the IF of MVDR beamformer employing a robust M-estimator of scatter in place of the conventional covariance matrix. The derived IF of a conventional MVDR beamformer reveals its vulnerability to outliers. IF is further used to compute the asymptotic variances and statistical efficiencies of the MVDR beamformers. MVDR beamformers based on robust M-estimator are shown to be robust (*i.e.*, insensitive to outliers and impulsive noise) without loosing much efficiency (accuracy) under the conventional assumption of normality. Section 2.8 considers the ICA model: we focus on Diagonalization Of Generalized covariance MAtrices (DOGMA) [49] and GUT [47] methods and illustrate how these methods are robust in face of outliers, and also fast to compute.

Notation Symbol $|\cdot|$ denotes the matrix determinant or modulus (*i.e.* $|z| = \sqrt{zz^*}$) when its argument is a complex scalar, $=_d$ reads 'has the same distribution as' and $\to_L$ means convergence in distribution or in law. Recall that every nonzero complex number has a unique (polar) representation, $z = |z|e^{j\theta}$, where $-\pi \leq \theta < \pi$ is called the *argument* of z, denoted $\theta = \arg(z)$. Complex matrix $\mathbf{G} \in \mathbb{C}^{k\times k}$ is Hermitian if $\mathbf{G}^H = \mathbf{G}$, symmetric if $\mathbf{G}^T = \mathbf{G}$ and unitary if $\mathbf{G}^H\mathbf{G} = \mathbf{I}$, where $\mathbf{I}$ denotes the identity matrix. By PDH(k) and CS(k) we denote the set of $k \times k$ positive definite hermitian and complex symmetric matrices, respectively. Recall that all the eigenvalues of PDH(k) matrix are real and positive. If $\mathbf{G}$ is a $k \times k$ diagonal matrix with diagonal elements $g_1 \ \ldots, g_k$, then we write $\mathbf{G} = \mathrm{diag}(g_i)$.

2.2 STATISTICAL CHARACTERIZATION OF COMPLEX RANDOM VECTORS

2.2.1 Complex Random Variables

A complex random variable (rva) $z = x + jy$ is comprised of a pair of real rvas $x \in \mathbb{R}$ and $y \in \mathbb{R}$. The distribution of z is identified with the joint (real bivariate) distribution of real rvas x and y

$$F_z(u + jv) \triangleq P(x \leq u, \ y \leq v).$$

In a similar manner, the probability density function (pdf) of $z = x + jy$ is identified with the joint pdf $f(x, \ y)$ of x and y, so $f(z) \equiv f(x, \ y)$. It is worth pointing out that in some applications (*e.g.*, for optimization purposes [6]) it is preferable to write the pdf $f(z)$ in the form $f(z, \ z^*)$ that separates z and its conjugate z^* as if they were independent variates. The *mean* of z is defined as $E[z] = E[x] + jE[y]$. For simplicity of presentation, we assume that $E[z] = 0$.

Characteristics of a complex rva can be described via symmetry properties of its distribution. The most commonly made symmetry assumption in the statistical signal processing literature is that of *circular symmetry*. See for example [50]. Complex rva z is said to be *circular* or, to have a *circularly symmetric* distribution, if

$$z =_d e^{j\theta}z, \qquad \forall\theta \in \mathbb{R}.$$

A circular rva z, in general, does not necessarily possess a density. However, if it does, then its pdf $f(z)$ satisfies

$$f(e^{j\theta}z) = f(z) \quad \forall\theta \in \mathbb{R}. \tag{2.2}$$

The property (2.2) can be shown to hold if, and only if, $f(z) = f(x, y)$ is a function of $|z|^2 = x^2 + y^2$, that is, $f(z) = cg(|z|^2)$ for some nonnegative function $g(\cdot)$ and normalizing constant c. Hence the regions of constant contours are circles in the complex plane, thus justifying the name for this class of distributions. rva z is said to be *symmetric*, or to have a *symmetric distribution*, if $z =_d -z$. Naturally, circular symmetry implies symmetry.

Characteristics of a complex rva z can also be described via its moments, for example, via its second-order moments. The variance $\sigma^2 = \sigma^2(z) > 0$ of z is defined as

$$\sigma^2(z) \triangleq E[|z|^2] = E[x^2] + E[y^2].$$

Note that variance does not bear any information about the correlation between the real and the imaginary part of z, but this information can be retrieved from pseudo-variance $\tau(z) \in \mathbb{C}$ of z, defined as

$$\tau(z) \triangleq E[z^2] = E[x^2] - E[y^2] + 2jE[xy].$$

Note that $E[xy] = \text{Im}[\tau(z)]/2$. The complex covariance between complex rvas z and w is defined as

$$\text{cov}(z, w) \triangleq E[zw^*].$$

Thus, $\sigma^2(z) = \text{cov}(z, z)$ and $\tau(z) = \text{cov}(z, z^*)$. If z is circular, then $\tau(z) = 0$. Hence a rva z with $\tau(z) = 0$ is called second order circular. Naturally if z or w are (or both z and w are) circular and $z \neq w$, then $\text{cov}(z, w) = 0$ as well.

Circularity quotient [41] $\varrho = \varrho(z) \in \mathbb{C}$ of a rva z (with finite variance) is defined as the quotient between the pseudo-variance and the variance

$$\varrho(z) \triangleq \frac{\text{cov}(z, z^*)}{\sqrt{\sigma^2(z)\sigma^2(z^*)}} = \frac{\tau(z)}{\sigma^2(z)}.$$

Thus we can describe $\varrho(z)$ as a measure of correlation between rva z and its conjugate z^*. The modulus

$$\lambda(z) \triangleq |\varrho(z)| \in [0, 1]$$

is referred to as the circularity coefficient [22, 41] of z. If the rva z is circular, then $\tau(z) = 0$, and consequently $\lambda(z) = 0$. Circularity coefficient measures the "amount of circularity" of zero mean rva $z = x + jy$ in that

$$\lambda(z) = \begin{cases} 0, & \text{if } x \text{ and } y \text{ are uncorrelated with equal varinaces} \\ 1, & \text{if } x \text{ or } y \text{ is zero, or } x \text{ is a linear function of } y. \end{cases} \tag{2.3}$$

Note that $\lambda(z) = 1$ if z is purely real-valued such as a BPSK modulated communication signal, or, if the signal lie on a line in the scatter plot (also called constellation or I/Q diagram) as is the case for BPSK, amplitude-shift keying (ASK), amplitude modulation (AM), or PAM-modulated communications signals. Hence a scatter plot of a rvas distrubuted as z with $\lambda = 1$ (resp. $\lambda = 0$) looks the "least circular" (resp. "most circular") in the complex plane as measured by its second-order moments. Note that λ is invariant under invertible linear transform$\{z \rightarrow s = cz \,|\, 0 \neq c \in \mathbb{C}\}$, that is, $\lambda(z) = \lambda(s)$. It is worth pointing out that circularity coefficient equals the squared eccentricity of the ellipse defined by the real covariance matrix of the composite real random vector (RV) $\mathbf{v} = (x, y)^T$ formed by stacking the real and imaginary part of $z = x + jy$, that is,

$$\lambda(z) = \frac{l_1 - l_2}{l_1 + l_2} \tag{2.4}$$

where $l_1 \geq l_2$ are the ordered eigenvalues of the 2×2 real covariance matrix $E[\mathbf{v}\mathbf{v}^T]$ of $\mathbf{v}$; see [41]. From this formula we observe that $\lambda(z) = 0$ if $l_1 = l_2$ (*i.e.*, ellipse is a sphere) and $\lambda(z) = 1$ if $l_2 = 0$ (*i.e.*, when the ellipse is elongated to a line).

Kurtosis $\kappa = \kappa(z) \in \mathbb{R}$ of z, is defined as

$$\kappa(z) \triangleq \kappa_0(z) - \lambda(z)^2$$

where

$$\kappa_0(z) \triangleq E[|z|^4]/\sigma^4(z) - 2. \tag{2.5}$$

Kurtosis $\kappa(z)$ describes "peakedness" of the density in that (a) $\kappa = 0$ if z is a rva from CN distribution, (b) $\kappa > 0$ if it has heavy-tailed ("super-Gaussian") CES distribution, and (c) $\kappa < 0$ if it has light-tailed ("sub-Gaussian") CES distribution [42]. Similar to the real case, kurtosis κ is defined via complex cumulants. If z is second-order circular, then $\kappa(z) = \kappa_0(z)$ since $\lambda(z) = 0$. Therefore we shall call $\kappa_0(z)$ as the circular kurtosis of z.

2.2.2 Complex Random Vectors

The definitions of Section 2.2.1 can be generalized for complex RVs. A complex RV

$$\mathbf{z} = \mathbf{x} + j\mathbf{y} \in \mathbb{C}^k$$

is comprised of a pair of real RVs $\mathbf{x} \in \mathbb{R}^k$ and $\mathbf{y} \in \mathbb{R}^k$. The distribution of $\mathbf{z}$ is identified with the joint real $2k$-variate distribution of real RVs $\mathbf{x}$ and $\mathbf{y}$. Hence the pdf of $\mathbf{z} = \mathbf{x} + j\mathbf{y}$ is simply the joint pdf $f(\mathbf{x}, \mathbf{y})$ of $\mathbf{x}$ and $\mathbf{y}$ (given it exists), so $f(\mathbf{z}) \equiv f(\mathbf{x}, \mathbf{y})$. The mean of $\mathbf{z}$ is defined as $E[\mathbf{z}] = E[\mathbf{x}] + jE[\mathbf{y}]$. For simplicity of presentation, we again assume that $E[\mathbf{z}] = 0$. We assume that RV $\mathbf{z}$ is non-degenerate in any subspace of $\mathbb{C}^k$.

Similar to the scalar case, random vector $\mathbf{z}$ is said to be *circular* if $\mathbf{z} =_d e^{j\theta}\mathbf{z}$ for all $\theta \in \mathbb{R}$. Naturally, the pdf $f(\mathbf{z})$ of a circular RV satisfies $f(e^{j\theta}\mathbf{z}) = f(\mathbf{z})$ for all

$\theta \in \mathbb{R}$. In the vector case, however, the term "circular" is a bit misleading since for $k \geq 2$, it does not imply that the regions of constant contours are spheres in complex Euclidean k-space.

Properties of a complex RV $\mathbf{z}$ can be described via its second-order moments. A complete second-order description of complex RV $\mathbf{z}$ is given by its covariance matrix $\mathcal{C}(\mathbf{z}) \in \mathrm{PDH}(k)$, defined as

$$\begin{aligned}\mathcal{C}(\mathbf{z}) &\triangleq E[\mathbf{z}\mathbf{z}^H] \\ &= E[\mathbf{x}\mathbf{x}^T] + E[\mathbf{y}\mathbf{y}^T] + j(E[\mathbf{y}\mathbf{x}^T] - E[\mathbf{x}\mathbf{y}^T]).\end{aligned}$$

and the pseudo-covariance matrix [38] $\mathcal{P}(\mathbf{z}) \in \mathrm{CS}(k)$, defined as

$$\begin{aligned}\mathcal{P}(\mathbf{z}) &\triangleq E[\mathbf{z}\mathbf{z}^T] \\ &= E[\mathbf{x}\mathbf{x}^T] - E[\mathbf{y}\mathbf{y}^T] + j(E[\mathbf{x}\mathbf{y}^T] + E[\mathbf{y}\mathbf{x}^T]).\end{aligned}$$

The pseudo-covariance matrix is also called relation matrix in [50] or complementary covariance matrix in [53]. Random vector $\mathbf{z}$ is said to be second-order circular [50] or proper [38] if $\mathcal{P}(\mathbf{z}) = 0$, or equivalently, if

$$E[\mathbf{x}\mathbf{x}^T] = E[\mathbf{y}\mathbf{y}^T] \text{ and } E[\mathbf{x}\mathbf{y}^T] = -E[\mathbf{y}\mathbf{x}^T] \tag{2.6}$$

The assumption (2.6) on the covariance structure of the real part $\mathbf{x}$ and imaginary part $\mathbf{y}$ of $\mathbf{z}$ is crucial in writing joint pdf $f(\mathbf{x}, \ \mathbf{y})$ of $\mathbf{x}$ and $\mathbf{y}$ with real $2k$-variate normal distribution into a complex form that is similar to the real case; see [24, 29, 61] and Section 2.3.

There can be several different ways to extend the concept of circularity quotient to the vector case. For example, since the circularity quotient can be written as $\varrho(z) = [\sigma^2(z)]^{-1}\tau(z)$, one possible extension is

$$\varrho(\mathbf{z}) \triangleq \mathcal{C}(\mathbf{z})^{-1}\mathcal{P}(\mathbf{z})$$

referred to as the circularity matrix of $\mathbf{z}$. Furthermore, since the circularity coefficient $\lambda(z)$ is the absolute value of $\varrho(z)$, that is, $\lambda(z) = \sqrt{\varrho(z)\varrho(z)^*}$, one possible way to extend this concept to the vector case, is to call the square-roots of the eigenvalues of the matrix

$$\mathcal{R}(\mathbf{z}) \triangleq \varrho(\mathbf{z})\varrho(\mathbf{z})^* \tag{2.7}$$

as the circularity coefficients of $\mathbf{z}$. The eigenvalues of $\mathcal{R}(\mathbf{z})$ are real-valued and take values on the interval [0, 1]; see Theorem 2 of [47]. Hence, also in this sense, the square-roots of the eigenvalues are valid extensions of the circularity coefficient $\lambda(z) \in [0, 1]$. Let $\lambda_i = \lambda_i(\mathbf{z}) \in [0, \ 1]$, $i = 1, \ldots, k$ denote the square-roots of the eigenvalues of the matrix $\mathcal{R}(\mathbf{z})$. In deference to [22], we shall call λ_i ($i = 1, \ldots, k$) the ith circularity coefficients of $\mathbf{z}$ and we write $\Lambda = \Lambda(\mathbf{z}) = \mathrm{diag}(\lambda_i)$ for the $k \times k$ matrix of circularity coefficients. In [54], it has been shown that circularity coefficients are the canonical correlations between $\mathbf{z}$ and its conjugate $\mathbf{z}^*$. It is easy to show that circularity coefficients are singular values of the symmetric matrix $\mathbf{K}(\mathbf{z}) \triangleq \mathbf{B}(\mathbf{z})\mathcal{P}(\mathbf{z})\mathbf{B}(\mathbf{z})^T$ (called the coherence matrix in [54]), where $\mathbf{B}(\mathbf{z})$ is any square-root matrix of $\mathcal{C}(\mathbf{z})^{-1}$ (i.e., $\mathcal{C}(\mathbf{z})^{-1} = \mathbf{B}(\mathbf{z})^H\mathbf{B}(\mathbf{z})$). This means that there exists a unitary matrix $\mathbf{U}$ such that symmetric matrix $\mathbf{K}(\mathbf{z})$ has a special form of singular value decomposition (SVD), called *Takagi factorization*, such that $\mathbf{K}(\mathbf{z}) = \mathbf{U}\Lambda\mathbf{U}^T$. Thus, if we now define matrix $\mathbf{W} \in \mathbb{C}^{k\times k}$ as $\mathbf{W} = \mathbf{B}^H\mathbf{U}$, where $\mathbf{B}$ and $\mathbf{U}$ are defined as above, then we observe that the transformed data $\mathbf{s} = \mathbf{W}^H\mathbf{z}$ satisfies

$$\mathcal{C}(\mathbf{s}) = \mathbf{U}^H\mathbf{B}\,\mathcal{C}(\mathbf{z})\mathbf{B}^H\mathbf{U} = \mathbf{I} \quad \text{and} \quad \mathcal{P}(\mathbf{s}) = \mathbf{U}^H\mathbf{B}\mathcal{P}(\mathbf{z})\mathbf{B}^T\mathbf{U}^* = \Lambda$$

that is, transformed RV $\mathbf{s}$ has (strongly-)uncorrelated components. Hence the matrix $\mathbf{W}$ is called the strong-uncorrelating transform (SUT) [21, 22].

Note that

$$\Lambda = 0 \;\Leftrightarrow\; \mathcal{P} = 0 \;\Leftrightarrow\; \mathcal{R} = 0.$$

As in the univariate case, circularity coefficients are invariant under the group of invertible linear transformations $\{\mathbf{z} \rightarrow \mathbf{s} = \mathbf{G}\mathbf{z} \,|\, \mathbf{G} \in \mathbb{C}^{k\times k} \text{ nonsingular }\}$, that is, $\lambda_i(\mathbf{z}) = \lambda_i(\mathbf{s})$. Observe that the set of circularity coefficients $\{\lambda_i(\mathbf{z}), i = 1, \ldots, k\}$ of the RV $\mathbf{z}$ does not necessarily equal the set of circularity coefficient of the variables $\{\lambda(z_i), i = 1, \ldots, k\}$ although in some cases (for example, when the components $z_1, \ldots, z_k$ of $\mathbf{z}$ are mutually statistically independent) they can coincide.

2.3 COMPLEX ELLIPTICALLY SYMMETRIC (CES) DISTRIBUTIONS

Random vector $\mathbf{z}$ of $\mathbb{C}^k$ has k-variate circular CN distribution if its real and imaginary part $\mathbf{x}$ and $\mathbf{y}$ have $2k$-variate real normal distribution and a $2k \times 2k$ real covariance matrix with a special form (2.6), that is, $\mathcal{P}(\mathbf{z}) = 0$. Since the introduction of the circular CN distribution in [24, 61], the assumption (2.6) seems to be commonly thought of as essential—although it was based on application specific reasoning—in writing the joint pdf $f(\mathbf{x}, \mathbf{y})$ of $\mathbf{x}$ and $\mathbf{y}$ into a natural complex form $f(\mathbf{z})$. In fact, the prefix "circular" is often dropped when referring to circular CN distribution, as it has due time become the commonly accepted complex normal distribution. However, rather recently, in [51, 57], an intuitive expression for the joint density of normal RVs $\mathbf{x}$ and $\mathbf{y}$ was derived without the unnecessary second-order circularity assumption (2.6) on their covariances. The pdf of $\mathbf{z}$ with CN distribution is uniquely parametrized by the covariance matrix $\mathcal{C}$ and pseudo-covariance matrix $\mathcal{P}$, the case of vanishing pseudo-covariance matrix, $\mathcal{P} = 0$, thus indicating the (sub)class of circular CN distributions.

There are many ways to represent complex random vectors and their probability distributions. The representation exploited in the seminal works of [51, 57] to derive the results is the so-called augmented signal model, where a $2k$-variate complex-valued augmented vector

$$\tilde{\mathbf{z}} \triangleq \begin{pmatrix} \mathbf{z} \\ \mathbf{z}^* \end{pmatrix}$$

is formed by stacking the complex vector and its complex conjugate $\mathbf{z}^*$. This form is also used in many different applications. The augmentation may also be performed by considering the composite real-valued vectors $(\mathbf{x}^T, \mathbf{y}^T)^T$ of $\mathbb{R}^{2k}$. These two augmented models are related via invertible linear transform

$$\begin{pmatrix} \mathbf{x} \\ \mathbf{y} \end{pmatrix} = \frac{1}{2} \begin{pmatrix} \mathbf{I} & \mathbf{I} \\ -j\mathbf{I} & j\mathbf{I} \end{pmatrix} \begin{pmatrix} \mathbf{z} \\ \mathbf{z}^* \end{pmatrix}. \tag{2.8}$$

The identity (2.8) can then be exploited as in [51] (resp. [46]) in writing the joint pdf of $\mathbf{x}$ and $\mathbf{y}$ with $2k$-variate real normal (resp. real elliptically symmetric) distribution into a complex form.

2.3.1 Definition

Definition 1 *Random vector $\mathbf{z} \in \mathbb{C}^k$ is said to have a (centered) CES distribution with parameters $\Sigma \in \mathrm{PDH}(k)$ and $\Omega \in \mathrm{CS}(k)$ if its pdf is of the form*

$$f(\mathbf{z} \mid \Sigma, \Omega) = c_{k,g} \, |\Gamma|^{-1/2} g(\Delta(\mathbf{z} \mid \Gamma)) \tag{2.9}$$

where

$$\Gamma \triangleq \begin{pmatrix} \Sigma & \Omega \\ \Omega^* & \Sigma^* \end{pmatrix} \in \mathrm{PDH}(2k)$$

and $\Delta(\mathbf{z}|\Gamma)$ is a quadratic form

$$\Delta(\mathbf{z}|\Gamma) \triangleq \frac{\tilde{\mathbf{z}}^H \Gamma^{-1} \tilde{\mathbf{z}}}{2} \tag{2.10}$$

and g:$[0, \infty) \to [0, \infty)$ is a fixed function, called the density generator, independent of Σ and Ω and $c_{k,g}$ is a normalizing constant. We shall write $\mathbf{z} \sim F_{\Sigma,\Omega} \equiv \mathrm{CE}_k(\Sigma, \Omega, g)$.

In (2.9), $c_{k,g}$ is defined as $c_{k,g} \triangleq 2(s_k \mu_{k,g})^{-1}$, where $s_k \triangleq 2\pi^k/\Gamma(k)$ is the surface area of unit complex k-sphere $\{\mathbf{z} \in \mathbb{C}^k : \|z\| = 1\}$ and

$$\mu_{k,g} \triangleq \int_0^\infty t^{k-1} g(t)\, dt$$

Naturally, $c_{k,g}$ could be absorbed into the function g, but with this notation g can be independent of the dimension k. CES distributions can also be defined more generally (without making the assuption that the probability density function exists) via their characteristic function. The functional form of the density generator $g(\cdot)$ uniquely distinguishes different CES distributions from another. In fact, any nonnegative function $g(\cdot)$ that satisfies $\mu_{k,g} < \infty$ is a valid density generator.

The covariance matrix and pseudo-covariance matrix of $\mathbf{z} \sim F_{\Sigma,\Omega}$ (if they exist) are proportional to parameters Σ and Ω, namely

$$\mathcal{C}(F_{\Sigma,\Omega}) = \sigma_{\mathcal{C}}\,\Sigma \quad \text{and} \quad \mathcal{P}(F_{\Sigma,\Omega}) = \sigma_{\mathcal{C}}\Omega \tag{2.11}$$

where the positive real-valued scalar factor σ_{C} is defined as

$$\sigma_{\mathcal{C}} \triangleq \frac{E(\delta)}{k}; \qquad \delta \triangleq \Delta(\mathbf{z}|\Gamma)$$

where the positive real rva δ has density

$$f(\delta) = \delta^{k-1} g(\delta) \mu_{k,g}^{-1}. \tag{2.12}$$

Hence, the covariance matrix of $F_{\Sigma,\Omega}$ exists if, and only if, $E(\delta) < \infty$, that is, $\mu_{k+1,g} = \int t^k g(t)\,dt < \infty$. Write

$$\mathcal{G}^{\ell} \triangleq \{g:[0, \infty) \rightarrow [0, \infty) | \mu_{k+\ell,g} < \infty\}.$$

Then $\mathrm{CE}_k(\Sigma, \Omega, g)$ with $g \in \mathcal{G}^{\ell}$ indicates the subclass of CES distributions with finite moments of order 2ℓ.

Note that the pdf $f(\mathbf{z}|\Sigma, \Omega)$ can also be parameterized via matrices [46]

$$\begin{aligned} &\mathbf{S} = \Sigma - \Omega\Sigma^{-*}\Omega^* \in \mathrm{PDH}(k) && \text{(Schur complement of } \Gamma) \\ &\mathbf{R} = \Sigma^{-1}\Omega \in \mathbb{C}^{k \times k} && \text{(``circularity matrix'')} \end{aligned}$$

in which case $\Delta(\mathbf{z}|\Gamma) = \mathbf{z}^H \mathbf{S}^{-1}\mathbf{z} - \mathrm{Re}(\mathbf{z}^H \mathbf{S}^{-1} \mathbf{R}^T \mathbf{z}^*)$ and $|\Gamma| = |\mathbf{S}|^2 |\mathbf{I} - \mathbf{R}\mathbf{R}^*|^{-1}$. If $g \in \mathcal{G}^1$ (i.e., the covariance matrix exists), then $\mathbf{R}$ is equal to circularity matrix ϱ defined in (2.7) then since the covariance matrix and pseudo-covariance matrix at $F_{\Sigma,\Omega}$ are proportional to parameters Σ and Ω by (2.11). However, $\mathbf{R}$ is a well defined parameter also in the case that the covariance matrix does not exist.

Recall that the functional form of the density generator $g(\cdot)$ uniquely distinguishes among different CES distributions. We now give examples of well-known CES distributions defined via their density generator.

■ **EXAMPLE 2.1**

The complex normal (CN) distribution, labeled Φ_k, is obtained with

$$g(\delta) = \exp(-\delta)$$

which gives $c_{k,g} = \pi^{-k}$ as the value of the normalizing constant. At Φ_k-distribution, $\sigma_C = 1$, so the parameters Σ and Ω coincide with the covariance matrix and pseudo-covariance matrix of the distribution. Thus we write $\mathbf{z} \sim \mathrm{CN}_k(\mathcal{C}, \mathcal{P})$.

■ **EXAMPLE 2.2**

The complex t-distribution with ν degrees of freedom ($0 < \nu < \infty$), labeled $T_{k,\nu}$, is obtained with

$$g_\nu(\delta) = (1 + 2\delta/\nu)^{-(2k+\nu)/2}$$

which gives $c_{k,g} = 2^k\Gamma(\frac{2k+\nu}{2})/[(\pi\nu)^k\Gamma(\frac{\nu}{2})]$ as the value of the normalizing constant. The case $\nu = 1$ is called the complex Cauchy distribution, and the limiting case $\nu \to \infty$ yields the CN distribution. We shall write $\mathbf{z} \sim \mathrm{CT}_{k,\nu}(\Sigma, \Omega)$. Note that the $T_{k,\nu}$-distribution possesses a finite covariance matrix for $\nu > 2$, in which case $\sigma_C = \nu/(\nu - 2)$.

2.3.2 Circular Case

Definition 2 *The subclass of CES distributions with* $\Omega = 0$, *labeled* $F_\Sigma = \mathrm{CE}_k(\Sigma, g)$ *for short, is called circular CES distributions.*

Observe that $\Omega = 0$ implies that $\Delta(\mathbf{z}|\Gamma) = \mathbf{z}^H\Sigma^{-1}\mathbf{z}$ and $|\Gamma| = |\Sigma|^2$. Thus the pdf of circular CES distribution takes the form familiar from the real case

$$f(\mathbf{z}|\Sigma) \equiv f(\mathbf{z}|\Sigma, 0) = c_{k,g}|\Sigma|^{-1}g(\mathbf{z}^H\Sigma^{-1}\mathbf{z}).$$

Hence the regions of constant contours are ellipsoids in complex Euclidean k-space. Clearly circular CES distributions belong to the class of circularly symmetric distributions since $f(e^{j\theta}\mathbf{z}|\Sigma) = f(\mathbf{z}|\Sigma)$ for all $\theta \in \mathbb{R}$. For example, $\mathrm{CN}_k(\mathcal{C}, 0)$, labeled $\mathrm{CN}_k(\mathcal{C})$ for short, is called the circular CN distribution (or, proper CN distribution [38]), the pdf now taking the classical [24, 61] form

$$f(\mathbf{z}|\mathcal{C}) = \pi^{-k}|\mathcal{C}|^{-1}\exp(-\mathbf{z}^H\mathcal{C}^{-1}\mathbf{z}).$$

See [33] for a detailed study of circular CES distributions.

2.3.3 Testing the Circularity Assumption

In case the signals or noise are noncircular, we need to take the full second-order statistics into account when deriving or applying signal processing algorithms. Hence, there needs to be a way to detect noncircularity. This may be achieved via hypothesis testing; see [46, 54]. In the following, we will develop a generalized likelihood ratio test (GLRT) for detecting noncircularity and establish some asymptotic properties of the test statistics.

Assume that $\mathbf{z}_1, \ldots, \mathbf{z}_n$ is an independent identically distributed (i.i.d.) random sample from a random vector $\mathbf{z} \in \mathbb{C}^k$. Sample covariance matrix (SCM)

$$\widehat{\mathcal{C}} \triangleq \frac{1}{n} \sum_{i=1}^{n} \mathbf{z}_i \mathbf{z}_i^H$$

is then the natural plug-in estimator of the covariance matrix, that is, $\hat{\mathcal{C}} \equiv \mathcal{C}(F_n)$ is the value of the covariance matrix at the empirical distribution function F_n of the sample. Similarly, sample pseudo-covariance matrix

$$\widehat{\mathcal{P}} \triangleq \frac{1}{n} \sum_{i=1}^{n} \mathbf{z}_i \mathbf{z}_i^T$$

is the plug-in estimator of the pseudo-covariance matrix. In addition, $\widehat{\mathcal{C}}$ and $\widehat{\mathcal{P}}$ are also the ML-estimators when the data is a random sample from $\text{CN}_k(\mathcal{C}, \mathcal{P})$ distribution.

In [46] and [54], a GLRT statistic was derived for the the hypothesis

$$H_0\colon \mathcal{P} = 0 \text{ when } \mathbf{z} \sim \text{CN}_k(\mathcal{C}, \mathcal{P})$$

against the general alternative $H_1\colon \mathcal{P} \neq 0$. So the purpose is to test the validity of the circularity assumption when sampling from CN distribution. The GRLT decision statistic is

$$q_n \triangleq \frac{\max\limits_{\mathcal{C},\mathcal{P}} L_n(\mathcal{C}, \mathcal{P})}{\max\limits_{\mathcal{C}} L_n(\mathcal{C}, 0)},$$

where $L_n(\mathcal{C}, \mathcal{P}) = \prod_{i=1}^{n} f(\mathbf{z}_i \,|\, \mathcal{C}, \mathcal{P})$ is the likelihood function of the sample and $f(\,\cdot \mid \cdot\,)$ the pdf of CN distribution. In [46], it was shown that

$$l_n \triangleq q_n^{-2/n} = |\mathbf{I} - \hat{\varrho}\hat{\varrho}^*| = |\mathbf{I} - \hat{\Lambda}^2| \tag{2.13}$$

where $\hat{\varrho} \triangleq \widehat{\mathcal{C}}^{-1}\widehat{\mathcal{P}}$ is the sample version of the circularity matrix ϱ and $\hat{\Lambda} = \text{diag}(\hat{\lambda}_i)$, where $\hat{\lambda}_i$ is the sample circularity coefficients, that is the square-roots of the eigenvalues of $\hat{\varrho}\hat{\varrho}^*$. This test statistic is invariant (since $\hat{\Lambda}$ is invariant) under the group of invertible linear transformations. In [40], based on general asymptotic theory of GLR-tests, the following result was shown:

Theorem 1 *Under* H_0, $-n \ln l_n \to \chi_p^2$ *in distribution, where* $p \triangleq k(k+1)$.

The test that rejects H_0 whenever $-n \ln l_n$ exceeds the corresponding chi-square $(1-\alpha)$th quantile is thus GLRT with asymptotic level α. This test statistic is, however, highly sensitive to violations of the assumption of complex normality. Therefore, in [40], a more general hypothesis was considered also

$$H_0' : \mathcal{P} = 0 \text{ when } \mathbf{z} \sim \mathrm{CE}_k(\Sigma, \Omega, g) \text{ with } g \in \mathcal{G}^2$$

Hence the purpose is to test the validity of the circularity assumption when sampling from unspecified (not necessarily normal) CES distributions with finite fourth-order moments. Denotes by $\kappa_i = \kappa(z_i)$ the marginal kurtosis of the ith variable z_i. Under H_0', the marginal kurtosis coincide, so $\kappa = \kappa_1 = \cdots = \kappa_k$. In addition, under H_0', the circularity coefficient of the marginals vanishes, that is, $\lambda(z_i) = 0$ for $i = 1, \ldots, k$, so $\kappa = E[|z_i|^4]/\sigma_i^4 - 2$, where $\sigma_i^2 = \sigma^2(z_i)$. Let $\hat{\kappa}$ be *any* consistent estimate of κ. Clearly, a natural estimate of the marginal kurtosis is the average of the sample marginal kurtosis $\hat{\kappa}_i = (\frac{1}{n}\sum_{j=1}^{n} |z_{ij}|^4)/\hat{\sigma}_i^4 - 2$, that is, $\hat{\kappa} = \frac{1}{k}\sum_{i=1}^{k} \hat{\kappa}_i$. Then, in [40], an adjusted GLRT-test statistic was shown to be asymptotically robust over the class of CES distributions with finite fourth-order moments.

Theorem 2 *Under* H_0', $\ell_n \triangleq -n \ln l_n/(1 + \hat{\kappa}/2) \to \chi_p^2$ *in distribution.*

This means that by a slight adjustment, that is, by dividing the GLRT statistic $-n \log l_n$ by $(1 + \hat{\kappa}/2)$, we obtain an adjusted test statistic ℓ_n of circularity that is valid—not just at the CN distribution, but—over the whole class of CES distributions with finite fourth-order moments. Based on the asymptotic distribution, we reject the null hypothesis at (asymptotic) α-level if $P = 1 - F_{\chi_p^2}(\ell_n) < \alpha$.

We now investigate the validity of the χ_p^2 approximation to the finite sample distribution of the adjusted GLRT-test statistic ℓ_n at small sample lengths graphically via "chi-square plots". For this purpose, let $\ell_{n,1}, \ldots, \ell_{n,N}$ denote the computed values of the adjusted GLRT-test statistic from N simulated samples of length n and let $\ell_{n,[1]} \leq \cdots \leq \ell_{n,[N]}$ denote the ordered sample, that is, the sample quantiles. Then

$$q_{[j]} = F_{\chi_p^2}^{-1}((j - 0.5)/N), \quad j = 1, \ldots, N$$

are the corresponding theoretical quantiles (where 0.5 in $(j - 0.5)/N$) is a commonly used continuity correction). Then a plot of the points $(q_{[j]}, \ell_{n,[j]})$ should resemble a straight line through the origin having slope 1. Particularly, the theoretical $(1-\alpha)$th quantile should be close to the corresponding sample quantile (e.g. $\alpha = 0.05$). Figure 2.2 depicts such chi-square plots when sampling from circular $T_{k,\nu}$ distribution

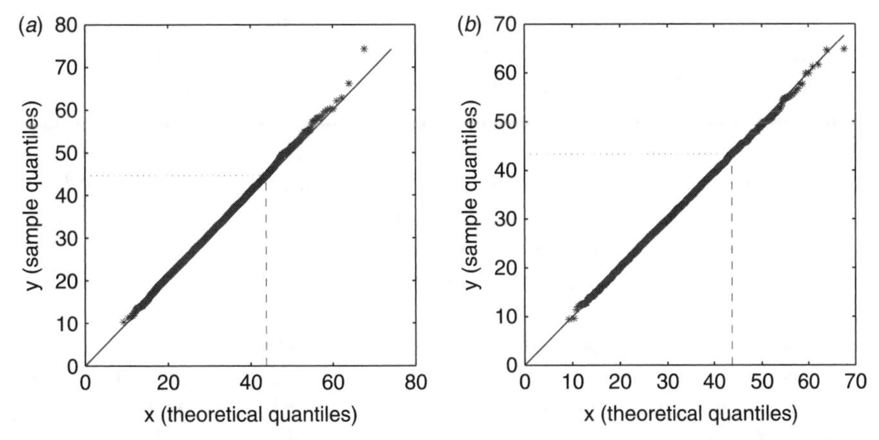

Figure 2.2 Chi-square plot when sampling from circular $T_{k,\nu}$ distribution (i.e. $\mathcal{P} = 0$) using sample length $n = 100$ (*a*) and $n = 500$ (b) The number of samples was $N = 5000$, degrees of freedom (d.f.) parameter was $\nu = 6$ and dimension was $k = 5$. The dashed vertical (resp. dotted horizontal) line indicate the value for the theoretical (resp. sample) 0.05-upper quantile.

(with $k = 5$, $\nu = 6$) using sample lengths $n = 100$ and $n = 500$ (i.e., H_0' holds). The number of samples was $N = 5000$. A very good fit with the straight line is obtained. The dashed vertical (resp. dotted horizontal) line indicates the value for the theoretical (resp. sample) 0.05-upper quantile. The quantiles are almost identical since the lines are crossing approximately on the diagonal. In generating a simulated random sample from circular $T_{k,\nu}$ distribution, we used the property that for independent RV $\mathbf{z}_0 \sim \mathrm{CN}_k(\mathbf{I})$ and rva $s \sim \chi^2_\nu$, the distribution of the composite RV

$$\mathbf{z} = \sqrt{\nu}\mathbf{z}_0/\sqrt{s} \tag{2.14}$$

follows circular $T_{k,\nu}$ distribution with $\Sigma = \mathbf{I}$, and, $\mathbf{z}' = \mathbf{G}\mathbf{z}$ has circular $T_{k,\nu}$ distribution with $\Sigma = \mathbf{G}\mathbf{G}^H$ for any nonsingular $\mathbf{G} \in \mathbb{C}^{k\times k}$.

If it is not known *a priori* whether the source signals are circular or noncircular, the decision (accept/reject) of GLRT can be used to guide the selection of the optimal array processor for further processing of the data since optimal array processors are often different for circular and noncircular cases.

We now investigate the performance of the test with a communications example.

■ EXAMPLE 2.3

Three independent random circular signals—one quadrature phase-shift keying (QPSK) signal, one 16-phase-shift keying (PSK) and one 32-quadrature amplitude modulation (QAM) signal—of equal power σ_s^2 are impinging on an $k = 8$ element ULA with $\lambda/2$ spacing from DOAs $-10°$, $15°$ and $10°$. The noise $\mathbf{n}$ has circular CN distribution with $\mathcal{C}(\mathbf{n}) = \sigma_n^2\mathbf{I}$. The signal to noise ratio (SNR) is

0.05 dB and the number of snapshots is $n = 300$. Since the noise and the sources are circular, also the marginals z_i of the array output $\mathbf{z}$ are circular as well, so $\mathcal{P}(\mathbf{z}) = 0$. Then, based on 500 Monte-Carlo trials, the null hypothesis of (second-order) circularity was falsely rejected (type I error) by GLRT test at $\alpha = 0.05$ level in 5.6 percent of all trials. Hence we observe that the GLRT test performs very well even though the Gaussian data assumption under which the GLRT test statistic l_n was derived do not hold exactly. (Since the source RV $\mathbf{s}$ is non-Gaussian, the observed array output $\mathbf{z} = \mathbf{As} + \mathbf{n}$ is also non-Gaussian.)

We further investigated the power of the GLRT in detecting noncircularity. For this purpose, we included a fourth source, a BPSK signal, that impinges on the array from DOA 35°. Apart from this additional source signal, the simulation setting is exactly as earlier. Note that the BPSK signal (or any other purely real-valued signal) is noncircular with circularity coefficient $\lambda = 1$. Consequently, the array output $\mathbf{z}$ is no longer second-order circular. The calculated GLRT-test statistic $-n \ln l_n$ correctly rejected at the $\alpha = 0.05$ level the null hypothesis of second-order circularity for all 500 simulated Monte-Carlo trials. Hence, GLRT test was able to detect noncircularity of the snapshot data (in conventional thermal circular Gaussian sensor noise) despite the fact that source signals were non-Gaussian.

2.4 TOOLS TO COMPARE ESTIMATORS

2.4.1 Robustness and Influence Function

In general, robustness in signal processing means insensitivity to departures from underlying assumptions. Robust methods are needed when precise characterization of signal and noise conditions is unrealistic. Typically the deviations from the assumptions occur in the form of outliers, that is, observed data that do not follow the pattern of the majority of the data. Other causes of departure include noise model class selection errors and incorrect assumptions on noise environment. The errors in sensor array and signal models and possible uncertainty in physical signal environment (e.g. propagation) and noise model emphasize the importance of validating all of the assumptions by physical measurements. Commonly many assumptions in multichannel signal processing are made just to make the algorithm derivation easy. For example, by assuming circular complex Gaussian pdfs, the derivation of the algorithms often leads to linear structures because linear transformations of Gaussians are Gaussians.

Robustness can be characterized both quantitatively and qualitatively. Intuitively, quantitative robustness describes how large a proportion of the observed data can be contaminated without causing significant errors (large bias) in the estimates. It is commonly described using the concept of breakdown point. Qualitative robustness on the other hand characterizes whether the influence of highly deviated observations is bounded. Moreover, it describes the smoothness of the estimator in a sense that small changes in the data should cause only small changes in the resulting estimates. We will focus on the qualitative robustness of the estimators using a very powerful tool called the influence function (IF).

Influence function is a versatile tool for studying qualitative robustness (local stability) and large sample properties of estimators, see [26, 27]. Consider the ε-point-mass contamination of the reference distribution F, defined as

$$F_{\varepsilon,\mathbf{t}}(\mathbf{z}) \triangleq (1-\varepsilon)F(\mathbf{z}) + \varepsilon\Delta_{\mathbf{t}}(\mathbf{z})$$

where $\Delta_{\mathbf{t}}(\mathbf{z})$ is a point-mass probability measure that assigns mass 1 to the point $\mathbf{t}$. Then the IF of a statistical functional $\mathbf{T}$ at a fixed point $\mathbf{t}$ and a given distribution F is defined as

$$\mathrm{IF}(\mathbf{t};\mathbf{T},F) \triangleq \lim_{\varepsilon\downarrow 0} \frac{\mathbf{T}(F_{\varepsilon,\mathbf{t}}) - \mathbf{T}(F)}{\varepsilon} = \frac{\partial}{\partial\varepsilon}\mathbf{T}(F_{\varepsilon,\mathbf{t}})|_{\varepsilon=0}. \tag{2.15}$$

One may interpret the IF as describing the effect (influence) of an infinitesimal point-mass contamination at a point $\mathbf{t}$ on the estimator, standardized by the mass of the contamination. Hence, the IF gives asymptotic bias caused by the contamination. Clearly, the effect on $\mathbf{T}$ is desired to be small or at least bounded. See [26] for a more detailed explanation of the influence function.

Let F_n denote the empirical distribution function associated with the data set $Z_n = \{\mathbf{z}_1, \ldots, \mathbf{z}_n\}$. Then a natural plug-in estimator of $\mathbf{T}(\cdot)$ is $\hat{\theta} \triangleq \mathbf{T}(F_n)$. If the estimator $\hat{\theta}$ is robust, its theoretical functional $\mathbf{T}(\cdot)$ has a bounded and continuous IF. Loosely speaking, the boundedness implies that a small amount of contamination at any point $\mathbf{t}$ does not have an arbitrarily large influence on the estimator whereas the continuity implies that the small changes in the data set cause only small changes in the estimator.

As the definition of the IF is rather technical, it is intructive to illuminate this concept with the simplest example possible. Let F denote the cumulative distribution function (c.d.f) of a real-valued random variable x symmetric about μ, so $F(\mu) = 1/2$. Then, to estimate the unknown symmetry center μ of F, two commonly used estimates are the sample mean $\bar{x} = \frac{1}{n}(x_1 + \cdots + x_n)$ and the sample median $\hat{\mu} = \mathrm{Med}(x_1, \ldots, x_n)$. The expected value and the population median

$$T_{ave}(F) = E[x] = \int x dF(x) \quad \text{and} \quad T_{med}(F) = F^{-1}\left(\frac{1}{2}\right)$$

(where $F^{-1}(q) = \inf\{x : F(x) \geq q\}$) are the statistical functionals corresponding to the sample mean and the median, respectively. Indeed observe for example, that

$$T_{ave}(F_n) = \int x dF_n(X) = \sum_x x P_{F_n}(x) = \frac{1}{n}\sum_{i=1}^{n} x_i = \bar{x}$$

since $P_{F_n}(x) = 1/n \; \forall x = x_i, \; i = 1, \ldots, n$. The value of T_{ave} at $F_{\varepsilon,t}$ is

$$\begin{aligned} T_{ave}(F_{\varepsilon,t}) &= \int x dF_{\varepsilon,t}(x) = \int x d[(1-\varepsilon)F + \varepsilon\Delta_t](x) \\ &= (1-\varepsilon)\int x dF(x) + \varepsilon \int x d\Delta_t(x) \\ &= (1-\varepsilon)T(F) + \varepsilon t \end{aligned}$$

Hence

$$IF(t; T_{ave}, F) = \frac{\partial}{\partial\varepsilon} T_{ave}(F_{\varepsilon,t})|_{\varepsilon=0} = t - \mu$$

since $T_{ave}(F) = \mu$ (as the expected value of the symmetric c.d.f F is equal to the symmetry center μ of F). The IF for the median $T_{med}(\cdot)$ is well-known to be

$$\mathrm{IF}(t; T_{med}, F) = \begin{cases} -\dfrac{1}{2f(\mu)}, & t < \mu \\ 0, & t = \mu \\ \dfrac{1}{2f(\mu)}, & t > \mu, \end{cases} \tag{2.16}$$

If the c.d.f. F is the c.d.f. of the standard normal distribution Φ (i.e., $\mu = 0$), then the above IF expressions can be written as

$$\begin{aligned} \mathrm{IF}(t; T_{ave}, \Phi) &= t, \\ \mathrm{IF}(t; T_{med}, \Phi) &= \sqrt{\frac{\pi}{2}}\,\mathrm{sign}(t). \end{aligned}$$

These are depicted in Figure 2.3. The median has bounded IF for all possible values of the contamination t, where as large outlier t can have a large effect on the mean.

■ EXAMPLE 2.4

IF of the covariance matrix. Let $\mathbf{T}(F) = \mathcal{C}(F)$ be our statistical functional of interest. The value of $\mathcal{C}(F) = \int \mathbf{z}\mathbf{z}^H dF(\mathbf{z})$ at the ε-point-mass distribution is

$$\begin{aligned} \mathcal{C}(F_{\varepsilon,\mathbf{t}}) &= \int \mathbf{z}\mathbf{z}^H d[(1-\varepsilon)F + \varepsilon\Delta_{\mathbf{t}}](\mathbf{z}) \\ &= (1-\varepsilon)\int \mathbf{z}\mathbf{z}^H dF(\mathbf{z}) + \varepsilon \int \mathbf{z}\mathbf{z}^H d\Delta_{\mathbf{t}}(\mathbf{z}) \\ &= (1-\varepsilon)\mathcal{C}(F) + \varepsilon \mathbf{t}\mathbf{t}^H \end{aligned}$$

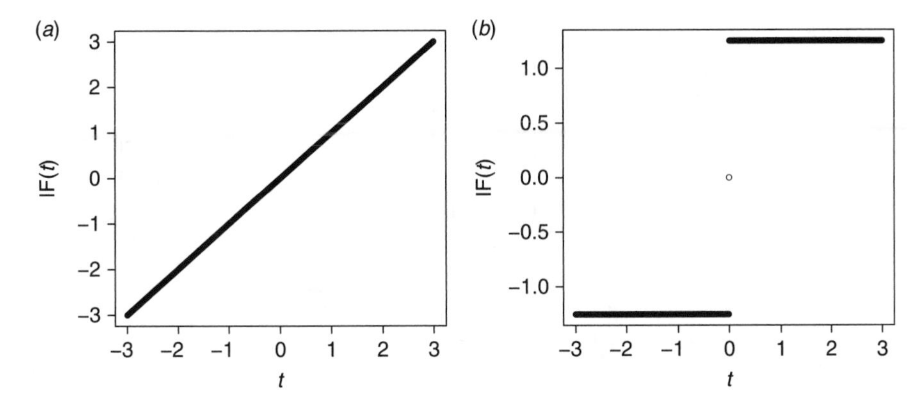

Figure 2.3 The influence functions of the mean $T_{ave}(F) = E_F(x)$ (*a*) and the median $T_{med}(F) = F^{-1}(1/2)$ (*b*) when F is the c.d.f Φ of the standard distribution. The median has bounded influence function for all possible values of the contamination t, where as large outlier t can have a large effect on the mean.

This shows that

$$\mathrm{IF}(\mathbf{t}; \mathcal{C}, F) = \frac{\partial}{\partial \varepsilon} \mathcal{C}(F_{\varepsilon,\mathbf{t}}) \mid_{\varepsilon=0} = \mathbf{t}\mathbf{t}^H - \mathcal{C} \tag{2.17}$$

where $\mathcal{C}$ denotes the value of the functional $\mathcal{C}(\cdot)$ at the reference distribution F. Thus the IF of $\mathcal{C}(\cdot)$ is unbounded with respect to standard matrix norms. This means that an infinitesimal point-mass contamination at a point $\mathbf{t}$ can have an arbitrarily large influence on the conventional covariance matrix functional, that is, it is not robust.

Note however that the IF is an asymptotic concept, characterizing stability of the estimator as n approaches infinity. Corresponding finite sample version is obtained by suppressing the limit in (2.15) and choosing $\varepsilon = 1/(n+1)$ and $F = F_n$. This yields the empirical influence function (EIF) (also called sensitivity function [26]) of the estimator $\hat{\theta} = \hat{\theta}(Z_n)$

$$\mathrm{EIF}(\mathbf{t}; \hat{\theta}, Z_n) = (n+1)[\hat{\theta}(Z_n \cup \{\mathbf{t}\}) - \hat{\theta}(Z_n)].$$

The EIF thus calculates the standardized effect of an additional observation at $\mathbf{t}$ on the estimator. In many cases, the empirical influence function $\mathrm{EIF}(\mathbf{t}; \hat{\theta}, Z_n)$ is a consistent estimator of the corresponding theoretical influence function $\mathrm{IF}(\mathbf{t}; \mathbf{T}, F)$ of the theoretical functional $\mathbf{T}(\cdot)$ of the estimator $\hat{\theta}$; cf. [17, 26].

■ EXAMPLE 2.5

EIF of the sample covariance matrix. If $\mathbf{T}(F) = \mathcal{C}(F)$ is the functional of interest, the corresponding plug-in estimator $\hat{\theta} = \mathbf{T}(F_n)$ is naturally the SCM $\widehat{\mathcal{C}}$ since

$$\mathcal{C}(F_n) = \int \mathbf{z}\mathbf{z}^H dF_n(\mathbf{z}) = \frac{1}{n} \sum_{i=1}^{n} \mathbf{z}_i \mathbf{z}_i^H = \widehat{\mathcal{C}}$$

The EIF of $\widehat{\mathcal{C}}$ is then

$$\begin{aligned}
\mathrm{EIF}(\mathbf{t}; \hat{\mathcal{C}}, Z_n) &= (n+1)[\widehat{\mathcal{C}}(Z_n \cup \{\mathbf{t}\}) - \widehat{\mathcal{C}}(Z_n)] \\
&= (n+1)\left[\frac{1}{n+1}\left\{\sum_{i=1}^{n} \mathbf{z}_i\mathbf{z}_i^H + \mathbf{t}\mathbf{t}^H\right\} - \frac{1}{n}\sum_{i=1}^{n} \mathbf{z}_i\mathbf{z}_i^H\right] \\
&= \sum_{i=1}^{n} \mathbf{z}_i\mathbf{z}_i^H + \mathbf{t}\mathbf{t}^H - \left(1+\frac{1}{n}\right)\sum_{i=1}^{n} \mathbf{z}_i\mathbf{z}_i^H \\
&= \mathbf{t}\mathbf{t}^H - \widehat{\mathcal{C}}.
\end{aligned}$$

Hence we conclude that the EIF of the SCM $\widehat{\mathcal{C}}$ is a consistent estimator of the theoretical influence function $\mathrm{IF}(\mathbf{t}; \mathcal{C}, F)$ (since $\widehat{\mathcal{C}}$ is a consistent estimator of the covariance matrix $\mathcal{C}(F)$ when Z_n is a random sample from F).

2.4.2 Asymptotic Performance of an Estimator

Earlier, we defined the CN distribution $\mathrm{CN}_k(\mathcal{C}, \Omega)$ via its density function. More generally we can define the CN distribution via its characteristic function (CF). The CF is a convenient tool for describing probability distributions since it always exists, even when the density function or moments are not well-defined. The CF of CN distribution is [51, 57]

$$\phi_{\mathcal{C},\mathcal{P}}(\mathbf{z}) = \exp\left[-\frac{1}{4}\{\mathbf{z}^H\mathcal{C}\mathbf{z} + \mathrm{Re}(\mathbf{z}^H\mathcal{P}\mathbf{z}^*)\}\right].$$

If the covariance matrix $\mathcal{C}$ is nonsingular, then CN possess the density function of Example 2.1. If $\mathcal{C}$ is singular, then the CN distribution is more commonly referred to as singular CN distribution.

For a complete second-order description of the limiting distribution of any statistic $\hat{\theta} \in \mathbb{C}^d$ we need to provide both the asymptotic covariance *and* the pseudo-covariance matrix. This may be clarified by noting that the real multivariate central limit theorem (e.g. [4], p.385) when written into a complex form reads as follows.

Complex Central Limit Theorem (CCLT) Let $\mathbf{z}_1, \ldots, \mathbf{z}_n \in \mathbb{C}^k$ be i.i.d. random vectors from F with mean $\mu \in \mathbb{C}^k$, finite covariance matrix $\mathcal{C} = \mathcal{C}(F)$ and pseudo-covariance matrix $\mathcal{P} = \mathcal{P}(F)$, then $\sqrt{n}(\bar{\mathbf{z}} - \mu) \to_L \mathrm{CN}_k(\mathcal{C}, \mathcal{P})$.

Estimator $\hat{\theta}$ of $\theta \in \mathbb{C}^d$ based on i.i.d. random sample $\mathbf{z}_1, \ldots, \mathbf{z}_n$ from F has asymptotic CN distribution with asymptotic covariance matrix $\mathrm{ASC}(\hat{\theta}; F)$ and asymptotic pseudo-covariance matrix $\mathrm{ASP}(\hat{\theta}; F)$, if

$$\sqrt{n}(\hat{\theta} - \theta) \to_L \mathrm{CN}_d(\mathrm{ASC}(\hat{\theta}; F), \mathrm{ASP}(\hat{\theta}; F))$$

If $\text{ASP}(\hat{\theta}; F) = 0$, then $\hat{\theta}$ has asymptotic circular CN distribution. By CCLT, the sample mean $\bar{\mathbf{z}} = \frac{1}{n}(\mathbf{z}_1 + \cdots + \mathbf{z}_n)$ has CN distribution with $\text{ASC}(\bar{\mathbf{z}}; F) = \mathcal{C}(F)$ and $\text{ASP}(\bar{\mathbf{z}}; F) = \mathcal{P}(F)$. Moreover, $\bar{\mathbf{z}}$ has asymptotic circular CN distribution if, and only if, F is second-order circular.

Given two competing estimators of the parameter θ, their efficiency of estimating the parameter of interest (at large sample sizes) can be established by comparing the ratio of the matrix traces of their asymptotic covariance matrices at a given reference distribution F, for example. It is very common in statistical signal processing and statistical analysis to define the asymptotic relative efficiency (ARE) of an estimator $\hat{\theta}$ as the ratio of the matrix traces of the asymptotic covariance matrices of the estimator and the optimal ML-estimator $\hat{\theta}_{\text{ML}}$. By using such a definition, the ARE of an estimator is always smaller than 1. If the ARE attains the maximum value 1, then the estimator is said to be asymptotically optimal at the reference distribution F. Later in this chapter we conduct such efficiency analysis for the MVDR beamformer based on the conventional sample covariance matrix (SCM) and the M-estimators of scatter.

Next we point out that the IF of the functional $\mathbf{T}(\cdot)$ can be used to compute the asymptotic covariance matrices of the corresponding estimator $\hat{\theta} = \mathbf{T}(F_n)$. If a functional $\mathbf{T}$ corresponding to an estimator $\hat{\theta} \in \mathbb{C}^d$ is sufficiently regular and $\mathbf{z}_1, \ldots, \mathbf{z}_n$ is an i.i.d. random sample from F, one has that [26, 27]

$$\sqrt{n}\{\hat{\theta} - \mathbf{T}(F)\} = \sqrt{n}\,[\frac{1}{n}\sum_{i=1}^{n} \text{IF}(\mathbf{z}_i; \mathbf{T}, F)\}] + o_p(1). \tag{2.18}$$

It turns out that $E[\text{IF}(\mathbf{z}; \mathbf{T}, F)] = 0$ and, hence by CCLT, $\hat{\theta}$ has asymptotic CN distribution

$$\sqrt{n}\{\hat{\theta} - \mathbf{T}(F)\} \rightarrow_L \text{CN}_d(\text{ASC}(\hat{\theta}; F), \text{ASP}(\hat{\theta}; F))$$

with

$$\text{ASC}(\hat{\theta}; F) = E[\text{IF}(\mathbf{z}; \mathbf{T}, F)IF(\mathbf{z}; \mathbf{T}, F)^H], \tag{2.19}$$

$$\text{ASP}(\hat{\theta}; F) = E[\text{IF}(\mathbf{z}; \mathbf{T}, F)\text{IF}(\mathbf{z}; \mathbf{T}, F)^T]. \tag{2.20}$$

Although (2.18) is often true, a rigorous proof may be difficult and beyond the scope of this chapter. However, given the form of the IF, equations (2.19) and (2.20) can be used to calculate an expression for the asymptotic covariance matrix and pseudo-covariance matrix of the estimator $\hat{\theta}$ in a heuristic manner.

2.5 SCATTER AND PSEUDO-SCATTER MATRICES

2.5.1 Background and Motivation

A starting point for many multiantenna transceiver and smart antenna algorithms is the array covariance matrix. For example, many direction-of-arrival (DOA) estimation algorithms such as the classical (delay-and-sum) beamformer and the Capon's MVDR beamformer require the array covariance matrix to measure the power of the beamformer output as a function of angle of arrival or departure. In addition, many high-resolution subspace-based DOA algorithms (such as MUSIC, ESPRIT, minimum norm etc.) compute the noise or signal subspaces from the eigenvectors of the array covariance matrix and exploit the fact that signal subspace eigenvectors and steering vector $\mathbf{A}$ matrix span the same subspace. See for example, [32, 55] and Section 2.6 for a overview of beamforming and subspace approaches to DOA estimation.

Since the covariance matrix is unknown, the common practice is to use the SCM $\widehat{\mathcal{C}}$ estimated from the snapshot data in place of its true unknown quantity. Although statistical optimality can often be claimed for array processors using the SCM under the normal (Gaussian) data assumption, they suffer from a lack of robustness in the face of outliers, that is, highly deviating observations and signal or noise modeling errors. Furthermore, their efficiency for heavy-tailed non-Gaussian and impulsive noise environments properties are far from optimal. It is well known that if the covariance matrix is estimated in a nonrobust manner, statistics (such as eigenvalues and eigenvectors) based on it are unreliable and far from optimal. In fact, such estimators may completely fail even in the face of only minor departures from the nominal assumptions. A simple and intuitive approach to robustify array processors is then to use robust covariance matrix estimators instead of the conventional nonrobust SCM $\widehat{\mathcal{C}}$. This objective leads to the introduction of a more general notion of covariance, called the scatter matrix.

As was explained in Section 2.2, the covariance matrix $\mathcal{C}(\mathbf{z})$ unambiguously describes relevant correlations between the variables in the case that the distribution F of $\mathbf{z}$ is circular symmetric. In the instances that F is noncircular distribution, also the information contained in the pseudo-covariance matrix $\mathcal{P}(\mathbf{z})$ can/should be exploited for example, in the blind estimation of noncircular sources or in the process of recovering the desired signal and cancelling the interferences. Therefore, in the case of noncircularity, an equally important task is robust estimation of the pseudo-covariance matrix. This objective leads to the introduction of a more general notion of pseudo-covariance, called the pseudo-scatter matrix.

2.5.2 Definition

Scatter and pseudo-scatter matrix are best described as generalizations of the covariance and pseudo-covariance matrix, respectively.

Definition 3 *Let*

$$\mathbf{s} = \mathbf{A}\mathbf{z} \quad \text{and} \quad \mathbf{v} = \mathbf{U}\mathbf{z}$$

denote the nonsingular linear and unitary transformations of $\mathbf{z} \in \mathbb{C}^k$ *for any nonsingular* $\mathbf{A} \in \mathbb{C}^{k\times k}$ *and unitary* $\mathbf{U} \in \mathbb{C}^{k\times k}$.

(a) *Matrix functional* $\mathbf{C} \in \mathrm{PDH}(k)$ *is called a scatter matrix (resp. spatial scatter matrix) if* $\mathbf{C}(\mathbf{s}) = \mathbf{A}\mathbf{C}(\mathbf{z})\mathbf{A}^H$ *(resp.* $\mathbf{P}(\mathbf{v}) = \mathbf{U}\mathbf{P}(\mathbf{z})\mathbf{U}^H$*).*

(b) *Matrix functional* $\mathbf{P} \in \text{CS}(k)$ *is called a pseudo-scatter matrix (resp. spatial pseudo-scatter matrix) if* $\mathbf{P}(\mathbf{s}) = \mathbf{A}\mathbf{P}(\mathbf{z})\mathbf{A}^T$ *(resp.* $\mathbf{P}(\mathbf{v}) = \mathbf{U}\mathbf{P}(\mathbf{z})\mathbf{U}^T$*).*

Spatial (pseudo-)scatter matrix is a broader notion than the (pseudo-)scatter since it requires equivariance only under unitary linear transformations, that is, every (pseudo-)scatter is also a spatial (pseudo-)scatter matrix. Weighted spatial covariance matrix

$$\mathcal{C}_\varphi(\mathbf{z}) \triangleq E[\varphi(\|\mathbf{z}\|^2)\mathbf{z}\mathbf{z}^H]$$

and weighted spatial pseudo-covariance matrix

$$\mathcal{P}_\varphi(\mathbf{z}) \triangleq E[\varphi(\|\mathbf{z}\|^2)\mathbf{z}\mathbf{z}^T]$$

where $\varphi(\cdot)$ denotes any real-valued weighting function on $[0, \infty)$, are examples of a spatial scatter and spatial pseudo-scatter matrix (but not of a (pseudo-)scatter matrix), respectively. Using weight $\varphi(x) = x$, we obtain matrices called the kurtosis matrix [49] and pseudo-kurtosis matrix [47]

$$\mathcal{C}_{\text{kur}}(\mathbf{z}) = E[\|\mathbf{z}\|^2\mathbf{z}\mathbf{z}^H] \quad \text{and} \quad \mathcal{P}_{\text{kur}}(\mathbf{z}) = E[\|\mathbf{z}\|^2\mathbf{z}\mathbf{z}^T]. \tag{2.21}$$

Using weight $\varphi(x) = x^{-1}$, we obtain matrices called the sign covariance matrix [49, 59] and the sign pseudo-covariance matrix [47]

$$\mathcal{C}_{\text{sgn}}(\mathbf{z}) \triangleq E[\|\mathbf{z}\|^{-2}\mathbf{z}\mathbf{z}^H] \quad \text{and} \quad \mathcal{P}_{\text{sgn}}(\mathbf{z}) \triangleq E[\|\mathbf{z}\|^{-2}\mathbf{z}\mathbf{z}^T].$$

These matrix functionals have been shown to be useful in blind separation and array signal processing problems. (cf. [8, 47, 49, 59]) and they possess very different statistical (e.g. robustness) properties. Sign covariance and sign pseudo-covariance matrices are highly robust in the face of non-Gaussian noise. The name for these matrices stems from the observation that "spatial sign" vector $\|\mathbf{z}\|^{-1}\mathbf{z}$ (a unit vector pointing towards the direction of $\mathbf{z}$) can be thought of as a generalization of the univariate sign of an observation which also provides information about the direction of the observation with respect to origin but not its magnitude. Robustness derives from the fact that they use only directional information. The use of the sign covariance matrix in high-resolution DOA estimation is briefly decribed later in this chapter.

The covariance matrix $\mathcal{C}(\cdot)$ and pseudo-covariance matrix $\mathcal{P}(\cdot)$ serve as examples of a scatter and pseudo-scatter matrix, respectively, assuming $\mathbf{z}$ has finite second-order moments. Scatter or pseudo-scatter matrices (or spatial equivalents), by their definition, do not necessarily require the assumption of finite second-order moments for its existence and are therefore capable in describing dependencies between complex random variables in more general settings than the covariance and pseudo-covariance matrix.

More general members of the family of scatter and pseudo-scatter matrices are the weighted covariance matrix and weighted pseudo-covariance matrix, defined as

$$\mathcal{C}_{\varphi,\mathcal{C}}(\mathbf{z}) \triangleq E[\varphi(\mathbf{z}^H\mathbf{C}(\mathbf{z})^{-1}\mathbf{z})\mathbf{z}\mathbf{z}^H],$$
$$\mathcal{P}_{\varphi,\mathcal{C}}(\mathbf{z}) \triangleq E[\varphi(\mathbf{z}^H\mathbf{C}(\mathbf{z})^{-1}\mathbf{z})\mathbf{z}\mathbf{z}^T]$$

respectively, where $\varphi(\cdot)$ is any real-valued weighting function on $[0, \infty)$ and $\mathbf{C}$ is any scatter matrix, for example, the covariance matrix. Note that the covariance matrix and the pseudo-covariance matrix are obtained with unit weight $\varphi \equiv 1$.

An improved idea of the weighted covariance matrices are M-estimators of scatter, reviewed in detail in the next section. M-estimators of scatter constitute a broad class which include for example MLEs of the parameter Σ of circular CES distributions F_Σ. Weighted covariance matrix can be thought of as "1-step M-estimator".

2.5.3 *M*-estimators of Scatter

One of the first proposals of robust scatter matrix estimators were M-estimators of scatter due to Maronna [37]. Extension of M-estimators for complex-valued data has been introduced and studied in [43–45, 48]. As in the real case they can be defined by generalizing the MLE.

Let $\mathbf{z}_1, \ldots, \mathbf{z}_n$ be an i.i.d. sample from a circular CES distribution $F_\Sigma = \mathrm{CE}_k(\Sigma, g)$, where $n > k$ (i.e., sample size n is larger than the number of sensors k). The MLE of Σ, is found by minimizing the negative of the log-likelihood function

$$l(\Sigma) \triangleq \prod_{i=1}^{n} \log f(\mathbf{z}_i \,|\, \Sigma)$$
$$= n \log|\Sigma| - \sum_{i=1}^{n} \log g(\mathbf{z}_i^H \Sigma^{-1} \mathbf{z}_i)$$

where we have omitted the constant term (the logarithm of the normalizing constant, $\log[c_{k,g}]$) since it does not depend on the unknown parameter Σ. By differentiating $l(\Sigma)$ with respect to Σ (using complex matrix differentiation rules [6]) shows that the MLE is a solution to estimating equation

$$\Sigma = \frac{1}{n}\sum_{i=1}^{n} \varphi_{ml}(\mathbf{z}_i^H \Sigma^{-1} \mathbf{z}_i)\mathbf{z}_i\mathbf{z}_i^H \tag{2.22}$$

where

$$\varphi_{ml}(\delta) \triangleq -\frac{g'(\delta)}{g(\delta)} \tag{2.23}$$

is a weight function that depends on the density generator $g(\cdot)$ of the underlying circular CES distribution. For the CN distribution (i.e., when $g(\delta) = \exp(-\delta)$), we have that $\varphi_{ml} \equiv 1$, which yields the SCM $\widehat{\mathcal{C}}$ as the MLE of Σ. The MLE for $T_{k,\nu}$ distribution (cf. Example 2.2), labeled MLT(ν), is obtained with

$$\varphi_{ml}(x) = \frac{2k+\nu}{\nu+2x}. \tag{2.24}$$

Note that MLT(1) is the highly robust estimator corresponding to MLE of Σ for the complex circular Cauchy distribution, and that MLT(ν) $\to \widehat{\mathcal{C}}$ (as $T_{k,\nu} \to \Phi_k$) as $\nu \to \infty$, thus the robustness of MLT(ν) estimators decrease with increasing values of ν.

We generalize (2.22), by defining the M-estimator of scatter, denoted by $\hat{\mathbf{C}}_\varphi$, as the choice of $\mathbf{C} \in \text{PDH}(k)$ that solves the estimating equation

$$\mathbf{C} = \frac{1}{n}\sum_{i=1}^{n} \varphi(\mathbf{z}_i^H \mathbf{C}^{-1}\mathbf{z}_i)\mathbf{z}_i\mathbf{z}_i^H \tag{2.25}$$

where φ is any real-valued weight function on $[0, \infty)$. Hence M-estimators constitute a wide class of scatter matrix estimators that include the MLE's for circular CES distributions as important special cases. M-estimators can be calculated by a simple iterative algorithm described later in this section.

The theoretical (population) counterpart, the M-functional of scatter, denoted by $\mathbf{C}_\varphi(\mathbf{z})$, is defined analogously as the solution of an implicit equation

$$\mathbf{C}_\varphi(\mathbf{z}) = E[\varphi(\mathbf{z}^H \mathbf{C}_\varphi^{-1}(\mathbf{z})\mathbf{z})\mathbf{z}\mathbf{z}^H]. \tag{2.26}$$

Observe that (2.26) reduces to (2.25) when F is the empirical distribution F_n, that is, the solution $\hat{\mathbf{C}}_\varphi$ of (2.25) is the natural plug-in estimator $\mathbf{C}_\varphi(F_n)$. It is easy to show that the M-functional of scatter is equivariant under invertible linear transformation of the data in the sense required from a scatter matrix. Due to equivariance, $\mathbf{C}_\varphi(F_\Sigma) = \sigma_\varphi \Sigma$, that is, the M-functional is proportional to the parameter Σ at F_Σ, where the positive real-valued scalar factor $\sigma_\varphi = \sigma_\varphi(\delta)$ may be found by solving

$$E[\varphi(\delta/\sigma_\varphi)\delta/\sigma_\varphi] = k \tag{2.27}$$

where δ has density (2.12). Often σ_φ need to be solved numerically from (2.27) but in some cases an analytic expression can be derived. Since parameter Σ is proportional to underlying covariance matrix $\mathcal{C}(F_\Sigma)$, we conclude that the M-functional of scatter is also proportional to the covariance matrix provided it exists (*i.e.*, $g \in \mathcal{G}^1$). In many applications in sensor array processing, covariance matrix is required only up to a constant scalar (see e.g. Section 2.7), and hence M-functionals can be used to define a robust class of array processors.

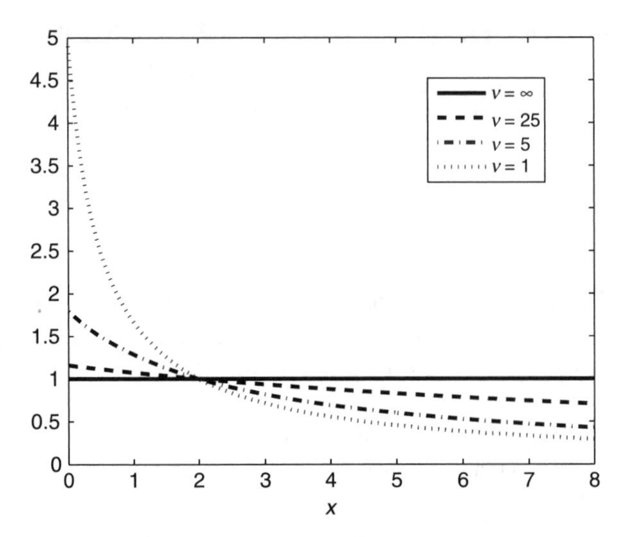

Figure 2.4 $\varphi(x)$ of MLT(ν) estimators.

By equation (2.25), $\hat{\mathbf{C}}_\varphi$ can be interpreted as a weighted covariance matrix. Hence, a robust weight function φ should descend to zero. This means that small weights are given to those observations $\mathbf{z}_i$ that are highly outlying in terms of measure $\mathbf{z}_i^H \hat{\mathbf{C}}_\varphi^{-1} \mathbf{z}_i$. It downweights highly deviating observations and consequently makes their influence in the error criterion bounded. Note that SCM $\widehat{\mathcal{C}}$ is an M-estimator that gives unit weight ($\varphi \equiv 1$) to all observations. Figure 2.4 plots the weight function (2.24) of MLT(ν) estimators for selected values of ν. Note that weight function (2.24) tends to weight function $\varphi \equiv 1$ of the SCM as expected (since $T_{k,\nu}$ tends to Φ_k distribution when $\nu \to \infty$). Thus, MLT(ν) $\approx \widehat{\mathcal{C}}$ for large values of ν.

Some examples of M-estimators are given next; See [43–45, 48] for more detailed descriptions of these estimators.

■ EXAMPLE 2.6

Huber's *M*-estimator, labeled HUB(q), is defined via weight

$$\varphi(x) = \begin{cases} 1/b, & \text{for } x \leq c^2 \\ c^2/(xb), & \text{for } x > c^2 \end{cases}$$

where c is a tuning constant defined so that $q = F_{\chi^2_{2k}}(2c^2)$ for a chosen $q (0 < q \leq 1)$ and the scaling factor $b = F_{\chi^2_{2(k+1)}}(2c^2) + c^2(1-q)/k$. The choice $q = 1$ yields $\varphi \equiv 1$, that is, HUB(1) correspond to the SCM. In general, low values of q increase robustness but decrease efficiency at the nominal circular CN model. Figure 2.5. depicts weight function of HUB(q) estimators for selected values of q.

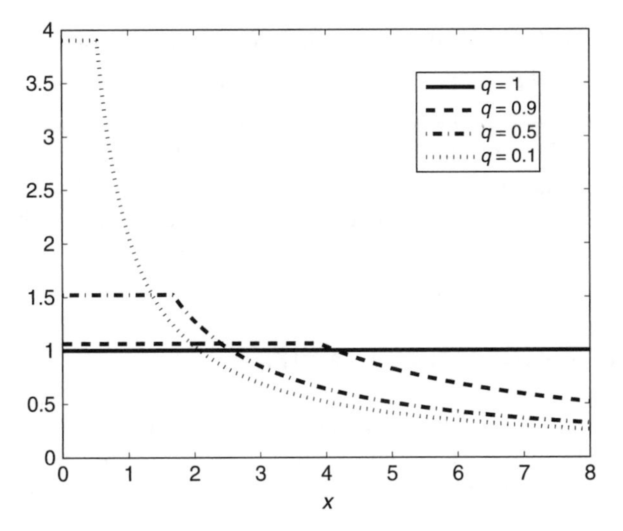

Figure 2.5 $\varphi(x)$ function of *HUB*(q) estimators.

■ EXAMPLE 2.7

Tyler's *M*-estimator of scatter [30, 44] utilizes weight function

$$\varphi(x) = k/x.$$

This *M*-estimator of scatter is also the MLE of the complex angular cental Gaussian distribution [30].

Computation of M-estimators Given any initial estimate $\widehat{\mathbf{C}}_0 \in \text{PDH}(k)$, the iterations

$$\widehat{\mathbf{C}}_{m+1} = \frac{1}{n}\sum_{i=1}^{n} \varphi(\mathbf{z}_i^H \widehat{\mathbf{C}}_m^{-1} \mathbf{z}_i)\mathbf{z}_i\mathbf{z}_i^H$$

converge to the solution $\widehat{\mathbf{C}}_\varphi$ of (2.25) under some mild regularity conditions. The authors of [27, 31, 37] consider the real case only, but the complex case follows similarly. See also discussions in [44].

As an example, let the initial estimate be the SCM, that is, $\widehat{\mathbf{C}}_0 = \widehat{\mathcal{C}}$. The first iteration, or the "1-step M-estimator", is simply a weighted sample covariance matrix

$$\widehat{\mathbf{C}}_1 = \frac{1}{n}\sum_{i=1}^{n} w_i \mathbf{z}_i\mathbf{z}_i^H, \quad w_i = \varphi(\mathbf{z}_i^H \widehat{\mathcal{C}}^{-1} \mathbf{z}_i).$$

If $\varphi(\cdot)$ is a robust weighting function, then $\widehat{\mathbf{C}}_1$ is a robustified version of $\widehat{\mathcal{C}}$. At the second iteration step, we calculate $\widehat{\mathbf{C}}_2$ as a weighted sample covariance matrix using weights $w_i = \varphi(\mathbf{z}_i^H \widehat{\mathbf{C}}_1^{-1} \mathbf{z}_i)$ and proceed analogously until the iterations $\hat{\mathbf{C}}_3, \hat{\mathbf{C}}_4, \ldots$ "converge", that is, $\|\mathbf{I} - \hat{\mathbf{C}}_{m-1}^{-1} \hat{\mathbf{C}}_m\| < \varepsilon$, where $\| \cdot \|$ is a matrix norm and ε is predetermined tolerance level, for example, $\varepsilon = 0.001$. To reduce computation time, one can always stop after m (e.g. $m = 4$) iterations and take the "m-step M-estimator" $\hat{\mathbf{C}}_m$ as an approximation for the true M-estimator $\hat{\mathbf{C}}_\varphi$. MATLAB functions to compute MLT(ν), HUB(q) and Tyler's M-estimator of scatter are available at http://wooster.hut.fi/~esollila/MVDR/.

2.6 ARRAY PROCESSING EXAMPLES

Most array processing techniques and smart antenna algorithms employ the SCM $\widehat{C}$. In the case of heavy-tailed signals or noise, it may give poor performance. Hence, robust array processors that perform reliably and are close to optimal in all scenarios are of interest.

2.6.1 Beamformers

Beamforming is among the most important tasks in sensor array processing. Consequently, there exists a vast amount of research papers on beamforming techniques, see [36, 55, 58] for overviews.

Let us first recall the beamforming principles in narrowband applications. In receive beamforming, the beamformer weight vector $\mathbf{w}$ linearly transforms the output signal $\mathbf{z}$ of array of k sensors to form the beamformer output

$$y = \mathbf{w}^H \mathbf{z}$$

with an aim of enhancing the signal-of-interest (SOI) from look direction (DOA of SOI) $\tilde{\theta}$ and attenuating undesired signals (interferers) from other directions. The (look direction dependent) beam response or gain is defined as

$$b(\theta) \triangleq \mathbf{w}^H \mathbf{a}(\theta)$$

where $\mathbf{a}(\theta)$ is the array response (steering vector) to DOA θ. The modulus squared $|b(\theta)|^2$ as a function of θ is called the beampattern or antenna pattern. Then, beamformer output power

$$P(\tilde{\theta}) \triangleq E[|y|^2] = \mathbf{w}^H C(\mathbf{z}) \mathbf{w} \tag{2.28}$$

should provide an indication of the amount of energy coming from the fixed look direction $\tilde{\theta}$. Plotting $P(\theta)$ as a function of look direction θ is called the spatial spectrum. The d highest peaks of the spatial spectrum correspond to the beamformer DOA estimates.

The beamformer weight vector $\mathbf{w}$ is chosen with an aim that it is statistically optimum in some sense. Naturally, different design objectives lead to different beamformer weight vectors. For example, the weight vector for the classic beamformer is

$$\mathbf{w}_{\mathrm{BF}} \triangleq \frac{\tilde{\mathbf{a}}}{\tilde{\mathbf{a}}^H \tilde{\mathbf{a}}}$$

where $\tilde{\mathbf{a}} = \mathbf{a}(\tilde{\theta})$ denotes the array response for fixed look direction $\tilde{\theta}$. The classic Capon's [7] MVDR beamformer chooses $\mathbf{w}$ as the minimizer of the output power while constraining the beam response along a specific look direction $\tilde{\theta}$ of the SOI to be unity

$$\min_{\mathbf{w}} \mathbf{w}^H \mathcal{C}(\mathbf{z})\mathbf{w} \quad \text{subject to} \quad \mathbf{w}^H \tilde{\mathbf{a}} = 1.$$

The well-known solution to this constrained optimization problem is

$$\mathbf{w}_{\mathcal{C}}(\mathbf{z}) \triangleq \frac{\mathcal{C}(\mathbf{z})^{-1}\tilde{\mathbf{a}}}{\tilde{\mathbf{a}}^H \mathcal{C}(\mathbf{z})^{-1}\tilde{\mathbf{a}}}. \tag{2.29}$$

Observe that Capon's beamformer weight vector is data dependent whereas the classic beamformer weight $\mathbf{w}_{\mathrm{BF}}$ is not, that is, $\mathbf{w}_{\mathcal{C}}(\cdot)$ is a statistical functional as its value depends on the distribution F of $\mathbf{z}$ via the covariance matrix $\mathcal{C}(F)$. The spectrum (2.28) for the classic and Capon's beamformers can now be written as

$$P_{\mathrm{BF}}(\theta) \triangleq \mathbf{a}(\theta)^H \mathcal{C}(\mathbf{z})\mathbf{a}(\theta) \tag{2.30}$$

$$P_{\mathrm{CAP}}(\theta) \triangleq [\mathbf{a}(\theta)^H \mathcal{C}(\mathbf{z})^{-1}\mathbf{a}(\theta)]^{-1} \tag{2.31}$$

respectively. (See Section 6 in [55]). Note that MVDR beamformers do not make any assumption on the structure of the covariance matrix (unlike the subspace-methods of the next section) and hence can be considered as a "nonparametric method" [55].

In practice, the DOA estimates for the classic and Capon's beamformer are calculated as the d highest peaks in the estimated spectrums $\hat{P}_{\mathrm{BF}}(\theta)$ and $\hat{P}_{\mathrm{CAP}}(\theta)$, where the true unknown covariance matrix $\mathcal{C}(\mathbf{z})$ is replaced by its conventional estimate, the SCM $\hat{\mathcal{C}}$. An intuitive approach in obtaining robust beamformer DOA estimates is to use robust estimators instead of the SCM in (2.30) and (2.31), for example, the M-estimators of scatter. Rigorous statistical robustness and efficiency analysis of MVDR beamformers based on M-estimators of scatter is presented in Section 2.7.

2.6.2 Subspace Methods

A standard assumption imposed by subspace methods is that the additive noise $\mathbf{n}$ is spatially white, that is $\mathcal{C}(\mathbf{n}) = \sigma^2\mathbf{I}$. We would like to stress that this assumption does not imply that $\mathbf{n}$ is second-order circular, that is, $\mathbf{n}$ can have non-vanishing pseudo-covariance matrix. Since source $\mathbf{s}$ and noise $\mathbf{n}$ are assumed to be mutually statistically independent, the array covariance matrix of the array output $\mathbf{z} = \mathbf{As} + \mathbf{n}$ can be written in the form

$$\mathcal{C}(\mathbf{z}) = \mathbf{A}\mathcal{C}(\mathbf{s})\mathbf{A}^H + \sigma^2\mathbf{I} \tag{2.32}$$

where $\mathbf{A} = \mathbf{A}(\theta)$ is the array response matrix parametrized by the vector of DOAs $\theta = (\theta_1, \ldots, \theta_d)^T$. Low rank signal model is assumed where $d < k$. Due to the structure (2.32), the $k - d$ smallest eigenvalues of $\mathcal{C}(\mathbf{z})$ are equal to σ^2 and the corresponding eigenvectors $\mathbf{e}_{d+1}, \ldots, \mathbf{e}_k$ are orthogonal to the columns of $\mathbf{A}$. These eigenvectors span the noise subspace and the eigenvectors $\mathbf{e}_1, \ldots, \mathbf{e}_d$ corresponding to d largest eigenvalues span the signal subspace (the column space of $\mathbf{A}$).

The subspace DOA estimation methods are based on different properties of the signal/noise subspaces. Some subspace methods also impose additional assumptions on the array geometry (e.g. ESPRIT). Essentially, subspace methods need to solve the following two problems.

1. Find an estimate $\widehat{\mathbf{E}}_s$ of the signal subspace $\mathbf{E}_s = \begin{pmatrix} \mathbf{e}_1 \cdots \mathbf{e}_d \end{pmatrix}$ and/or estimate $\widehat{\mathbf{E}}_n$ of the noise subspace $\mathbf{E}_n = \begin{pmatrix} \mathbf{e}_{d+1} \cdots \mathbf{e}_k \end{pmatrix}$.
2. Find estimate $\hat{\theta}$ of the DOAs which best optimizes the selected error criterion, for example, find $\hat{\theta}$ such that distance between subspace $\mathbf{A}(\hat{\theta})$ and the estimated subspace $\widehat{\mathbf{E}}_S$ is minimal in some sense.

Commonly, the subspace methods differ only in how they approach problem 2 since the estimates of signal and noise subspaces are calculated from the eigenvectors of the conventional, nonrobust SCM $\widehat{\mathcal{C}}$. Solving problem 1 reliably, however, is much more crucial since no matter how clever criterion is used or how distances between subspaces are measured in problem 2, the DOA estimates will be unreliable if the estimates of the subspaces are unreliable. In other words, accuracy and efficiency of the subspace method depends largely on the accuracy and efficiency of the estimates of the noise or signal subspaces. Again, to obtain robust subspace methods it is sensible to use estimates of noise or signal subspaces based on eigenvectors of the M-estimators of scatter for example.

The classical MUSIC method is based on the orthogonality of the signal and noise subspace and the fact that $\mathbf{A}$ and $\mathbf{E}_s$ span the same subspace. Because of the orthogonality of the signal and noise subspace, $\mathbf{E}_n^H\mathbf{a}(\theta) = 0$, or equivalently

$$\|\mathbf{E}_n^H\mathbf{a}(\theta)\|^2 = \mathbf{a}(\theta)^H\mathbf{E}_n\mathbf{E}_n^H\mathbf{a}(\theta) = 0$$

at the DOAs $\theta_1, \ldots, \theta_d$. Then, the MUSIC methods find DOA estimates as the d highest peaks of the MUSIC spectrum

$$P_M(\theta) \triangleq [\mathbf{a}(\theta)^H\widehat{\mathbf{E}}_n\widehat{\mathbf{E}}_n^H\mathbf{a}(\theta)]^{-1}$$

Clearly, if noise subspace $\widehat{\mathbf{E}}_n$ is unreliably estimated (e.g. via eigenvectors of the SCM when the noise is non-Gaussian or impulsive), then the obtained MUSIC DOA estimators are unreliable. For robust estimation of noise subspace one may use for example, eigenvectors of M-estimators of scatter, or, eigenvectors of the sample plug-in estimate

$$\widehat{\mathcal{C}}_{\text{sgn}} \triangleq \mathcal{C}_{\text{sgn}}(F_n) = \frac{1}{n}\sum_{i=1}^{n} \|\mathbf{z}\|^{-2}\mathbf{z}_i\mathbf{z}_i^H \tag{2.33}$$

of the sign covariance matrix $\mathcal{C}_{\text{sgn}}(\mathbf{z})$ as in [59].

A weighted signal subspace fitting (SSF) approach, for example, finds DOAs via criterion function

$$\hat{\theta} = \arg\min_{\theta} \text{Tr}[\Pi_{\mathbf{A}}^{\perp}\widehat{\mathbf{E}}_s\mathbf{Y}\widehat{\mathbf{E}}_s^H]$$

where $\Pi_{\mathbf{A}}^{\perp} = \mathbf{I} - \mathbf{A}(\mathbf{A}^H\mathbf{A})^{-1}\mathbf{A}^H$ is a projection matrix onto the noise subspace and $\mathbf{Y}$ is some weighting matrix. The estimated optimal weighting matrix $\widehat{\mathbf{Y}}_{opt}$ is a diagonal matrix, whose diagonal elements are certain functions of the estimated eigenvalues of the covariance matrix $\mathcal{C}(\mathbf{z})$. Hence, reliable and accurate estimation of DOAs via weighted SSF approach requires robust estimation of the signal subspace $\mathbf{E}_s$ and eigenvalues of the covariance matrix. These can be obtained, for example, using eigenvectors and eigenvalues of robust M-estimators instead of the SCM.

■ EXAMPLE 2.8

Four independent random signals, QPSK, 16-PSK, 32-QAM and BPSK signal of equal power σ_s^2, are impinging on a $k = 8$ element ULA with $\lambda/2$ spacing from DOAs $-10°$, $15°$, $10°$ and $35°$. The simulation setting is as in Example 2.3, except that now we consider two different noise environments. In the first setting, noise $\mathbf{n}$ has circular Gaussian distribution $\text{CN}_k(\sigma_n^2\mathbf{I})$, and in the second setting noise has circular Cauchy distribution $\text{CT}_{k,1}(\sigma_n^2\mathbf{I})$. Note that the Cauchy distribution does not have finite variance and σ_n^2 is the scale parameter of the distribution. In both simulation settings, the signal to noise ratio (SNR) is $10\log_{10}(\sigma_s^2/\sigma_n^2) = 20$ dB and the number of snapshots is $n = 300$. The number of signals ($d = 4$) is assumed to be known *a priori*. We then estimated the noise subspace $\widehat{\mathbf{E}}_n$ from eigenvectors of the SCM $\widehat{\mathcal{C}}$, sample sign covariance matrix (2.33) and MLT(1) estimator. Typical MUSIC spectrums associated with different estimators are shown in Figure 2.6 for both the Gaussian and Cauchy noise settings. All the estimators are able to resolve the four sources correctly in the Gaussian noise case: in fact, the differences in the spectrums are very minor, that is, they provide essentially the same DOA estimates. In the Cauchy noise case, however, MUSIC based on the classical sample estimator $\widehat{\mathcal{C}}$ is not able to resolve the sources. The robust estimators, the sign covariance matrix and the MLT(1) estimator, however, yield reliable estimates of the DOAs. Based on the sharpness of the peaks, the MLT(1) estimator is performing better than the sample sign covariance matrix $\widehat{\mathcal{C}}_{\text{sgn}}$.

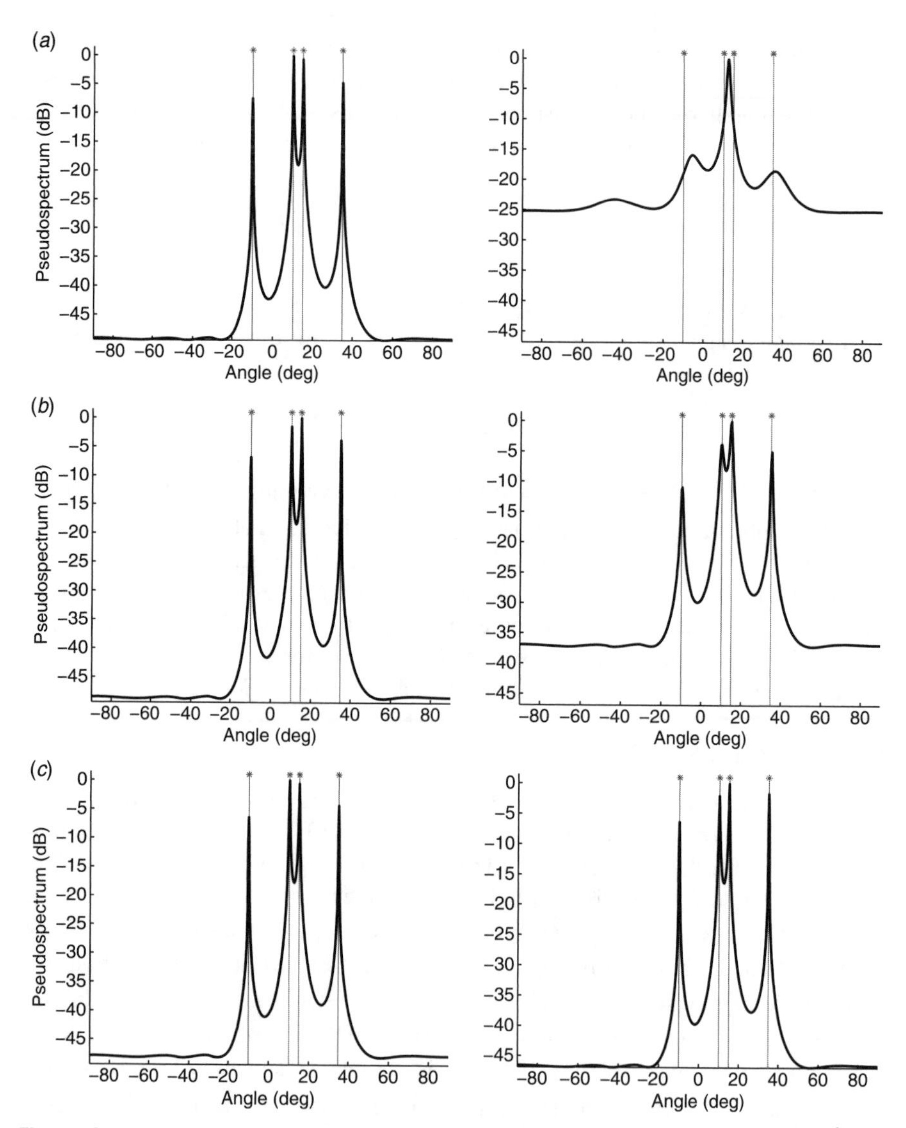

Figure 2.6 MUSIC spectrums when the noise subspace is estimated using SCM $\widehat{C}$ (*a*), sample sign covariance matrix (*b*) and MLT(1) estimator (*c*) in circular Gaussian (first column) and Cauchy (second column) noise. Sources are independent random QPSK, 16-PSK, 32-QAM and BPSK signals that arrive at 8-element ULA from DOAs $-10°$, $15°$, $10°$ and $35°$.

2.6.3 Estimating the Number of Sources

An equally important problem to DOA estimation is the estimation of the number of sources. The subspace based methods introduced in the previous section usually assume that the number of source signals is known *a priori*. In practise, the number of sources d is often not known and needs to be estimated from the data. The commonly used minimum description length (MDL)-based information theoretical criterion, obtains the estimate $\hat{d}$ for the number of signals d as an integer $p \in (0, 1, \ldots, k-1)$ which minimizes the criterion [60]

$$\text{MDL}(p) \triangleq -\log\left(\frac{(\prod_{i=p+1}^{k} l_i)^{1/(k-p)}}{\frac{1}{k-p}\sum_{i=p+1}^{k} l_i}\right)^{(k-p)n} + \frac{1}{2}p(2k-p)\log n$$

where $l_1, l_2, \ldots, l_k$ denote the (ordered) eigenvalues of the SCM $\widehat{\mathcal{C}}$ arranged in descending order. Instead of using the eigenvalues of SCM, it is desirable for purposes of reliable estimation in non-Gaussian noise to employ eigenvalues of some robust estimator of covariance, for example, M-estimator of scatter, instead of the SCM. We demonstrate this via a simulation study.

The ULA contains $k = 8$ sensors with half a wavelength interelement spacing. Two uncorrelated Gaussian signals with equal power 20 dB from DOAs $\theta_1 = -5^\circ$ and $\theta_2 = 5^\circ$ are impinging on the array. The components of the additive noise $\mathbf{n}$ are modeled as i.i.d. with complex symmetric α-stable (SαS) distribution [56] with dispersion $\gamma = 1$ and values α ranging from $\alpha = 1$ (complex Cauchy noise) to $\alpha = 2$ (complex Gaussian noise). Simulation results are based on 500 Monte Carlo runs with $n = 300$ as the sample size. Figure 2.7 depicts the relative proportion of correct estimation results using MDL criterion, when the eigenvalues are obtained from SCM $\widehat{\mathcal{C}}$ and robust MLT(1), HUB(0.9) and HUB(0.5) estimators. The performance of the classic MDL employing the SCM is poor: it is able to estimate the number of signals

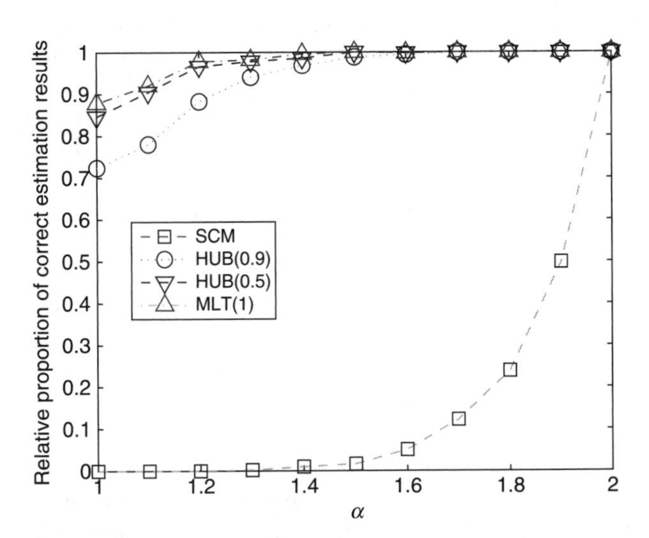

Figure 2.7 Simulation results for estimation of number of sources using the MDL criterion based on the SCM, HUB(0.9), HUB(0.5) and MLT(1)-estimators. There are $d = 2$ Gaussian source signals in SαS distributed noise for $1 \leq \alpha \leq 2$. The number of sensors is $k = 8$ and number of snapshot is $n = 300$.

reliably only for $\alpha = 2$, that is, the Gaussian case. However, the robust M-estimator is able to estimate the number of sources reliably for large range of α-values. Among the robust M-estimators, MLT(1) has the best performance.

2.6.4 Subspace DOA Estimation for Noncircular Sources

We now describe the Root-MUSIC-like method presented in [13]. As usual, assume that the signal $\mathbf{s}$ and noise $\mathbf{n}$ in the array model (2.1) are uncorrelated with zero-mean. The method further requires the following additional assumptions.

(1) The array is ULA (in order to facilitate using polynomial rooting).

(2) Noise $\mathbf{n}$ is second-order circular and spatially white, that is $\mathcal{C}(\mathbf{n}) = \sigma^2\mathbf{I}$ and $\mathcal{P}(\mathbf{n}) = 0$.

(3) Sources signals s_i, $i = 1, \ldots, d$ are uncorrelated in the sense that $\mathcal{C}(\mathbf{s}) = \text{diag}(\sigma^2(s_i))$ and $\mathcal{P}(\mathbf{s}) = \text{diag}(\tau(s_i))$.
Under these assumptions,

$$\mathcal{C}(\mathbf{z}) = \mathbf{A}\mathcal{C}(\mathbf{s})\mathbf{A}^H + \sigma^2\mathbf{I}, \quad \mathcal{P}(\mathbf{z}) = \mathbf{A}\mathcal{P}(\mathbf{s})\mathbf{A}^T$$

where as earlier $\mathbf{A} = \mathbf{A}(\theta)$ denotes the array response matrix. Further assume that

(4) $\mathcal{P}(\mathbf{s}) = \mathcal{C}(\mathbf{s})\Phi$, where $\Phi = \text{diag}(e^{j\phi_i})$.

Assumption (4) means that the circularity coefficient of the sources are equal to unity, that is, $\lambda(s_i) = 1$ for $i = 1, \ldots, d$, which by (2.3) implies that transmitted source signal s_i must be real-valued, such as AM or BPSK modulated signals, or the real part $\text{Re}(s_i)$ of the transmitted signal is a linear function of the imaginary part $\text{Im}(s_i)$. If (1)–(4) hold, then the covariance matrix for the augmented signal vector $\tilde{\mathbf{z}}$ is

$$\mathcal{C}(\tilde{\mathbf{z}}) = \begin{pmatrix} \mathbf{A} \\ \mathbf{A}^*\Phi^* \end{pmatrix} \mathcal{C}(\mathbf{s}) \begin{pmatrix} \mathbf{A} \\ \mathbf{A}^*\Phi^* \end{pmatrix}^H + \sigma^2\mathbf{I}. \tag{2.34}$$

Now by performing eigenvalue decomposition $\mathcal{C}(\tilde{\mathbf{z}})$ we may find d dimensional signal subspace and $2k - d$ dimensional orthogonal noise subspace. Thus Root-MUSIC-like direction finding algorithms can be designed; see [13] for details. By exploiting the noncircularity property we obtain extra degrees of freedom since noncircularity allows resolving more sources than sensors. Again, in the face heavy-tailed noise or outlying observations, a robust estimate of the array covariance matrix $\mathcal{C}(\mathbf{z})$ and pseudo-covariance matrix $\mathcal{P}(\mathbf{z})$ can be used instead of the conventional estimators, $\widehat{\mathcal{C}}$ and $\widehat{\mathcal{P}}$. We wish to point out, however, that the four assumptions stated above are not necessary for all subspace DOA estimation methods for noncircular sources; see for example, [1].

2.7 MVDR BEAMFORMERS BASED ON *M*-ESTIMATORS

MVDR beamformer weight functional $\mathbf{w}_{\mathcal{C}}(F)$ defined in (2.29) requires the covariance matrix $\mathcal{C}(F)$ only up to a constant scalar. Since both the covariance matrix $\mathcal{C}(F)$ and the M-functional of scatter $\mathbf{C}_{\varphi}(F)$ are proportional to parameter Σ of a CES distribution $F_{\Sigma} = \mathrm{CE}_k(\Sigma, g)$, we can define a class of MVDR-beamformers based on M-estimators $\hat{\mathbf{C}}_{\varphi}$ of scatter which all estimate the same population quantity when sampling from a CES distribution F_{Σ}.

Definition 4 *MVDR beamformer weight vector based on M-functional* $\mathbf{C}_{\varphi}(\mathbf{z})$ *of scatter, labelled* φ*-MVDR, is defined as*

$$\mathbf{w}_{\varphi}(\mathbf{z}) \triangleq \frac{\mathbf{C}_{\varphi}(\mathbf{z})^{-1}\tilde{\mathbf{a}}}{\tilde{\mathbf{a}}^H \mathbf{C}_{\varphi}(\mathbf{z})^{-1}\tilde{\mathbf{a}}}$$

where $\tilde{\mathbf{a}} = \mathbf{a}(\tilde{\theta})$ *is the nominal array response vector for fixed look direction* $\tilde{\theta}$ *(assumed to be known exactly).*

Then the φ-MVDR beamformer-based DOA estimates can be found from the spectrum

$$P_{\varphi}(\theta) = [\mathbf{a}(\theta)^H \mathbf{C}_{\varphi}(\mathbf{z})^{-1}\mathbf{a}(\theta)]^{-1}$$

If we are using weight $\varphi \equiv 1$ (in which case $\mathbf{C}_{\varphi} \equiv \mathcal{C}$), then the corresponding $\mathbf{w}_{\varphi}(\mathbf{z})$ and $P_{\varphi}(\theta)$ correspond to the conventional MVDR beamformer weight functional $\mathbf{w}_{\mathcal{C}}(\mathbf{z})$ and spectrum $P_{\mathrm{CAP}}(\theta)$ in (2.29) and (2.31), respectively.

Define

$$\mathbf{w} \triangleq \mathbf{G}\tilde{\mathbf{a}} \quad \text{where} \quad \mathbf{G} \triangleq \frac{\Sigma^{-1}}{\tilde{\mathbf{a}}^H \Sigma^{-1}\tilde{\mathbf{a}}}. \tag{2.35}$$

Since the M-functional $\mathbf{C}_{\varphi}(\mathbf{z})$ is proportional to Σ at F_{Σ}, it follows that

$$\mathbf{w}_{\varphi}(F_{\Sigma}) = \mathbf{w}$$

provided that $\mathbf{C}_{\varphi}(F_{\Sigma})$ exists. Since in practice, the true $\mathbf{C}_{\varphi}$ is unknown, we replace it by the M-estimator $\hat{\mathbf{C}}_{\varphi}$, which yields

$$\hat{\mathbf{w}}_{\varphi} \triangleq \mathbf{w}_{\varphi}(F_n) = \frac{\hat{\mathbf{C}}_{\varphi}^{-1}\tilde{\mathbf{a}}}{\tilde{\mathbf{a}}^H \hat{\mathbf{C}}_{\varphi}^{-1}\tilde{\mathbf{a}}}$$

as the plug-in estimator. The optimal weight at F_Σ employs the MLE of Σ (i.e., $\varphi = \varphi_{ml}$, where φ_{ml} is given by (2.23)) and is hereafter denoted by $\hat{\mathbf{w}}_{mle}$.

EXAMPLE 2.9

Our simulation setting is as follows. A $k = 4$ sensor ULA with $\lambda/2$ spacing received two ($d = 2$) uncorrelated circular Gaussian signals of equal variance σ_s^2 with DOAs at $-10°$ (SOI) and $15°$ (interferer). In the first setting (A), noise $\mathbf{n}$ has circular Gaussian distribution $\text{CN}_k(\sigma_n^2\mathbf{I})$, and in the second setting (B), noise has circular Cauchy distribution $\text{CT}_{k,1}(\sigma_n^2\mathbf{I})$. Note that the Cauchy distribution does not have finite variance and σ_n^2 is the scale parameter of the distribution. In both A and B, the SNR (dB) is defined using scale parameters as $10\log_{10}[\sigma_s^2/\sigma_n^2] = 15$ dB. The number of snapshots is $n = 500$.

Figure 2.8 depicts the estimated φ-MVDR beampatterns for look direction $-10°$ for settings A and B averaged over 100 realizations. Also plotted are the estimated spectrums. The employed *M*-estimators are the SCM [i.e., HUB(1)], MLT(1) and HUB(0.9). In the Gaussian noise case (setting A), the beampatterns are similar, in

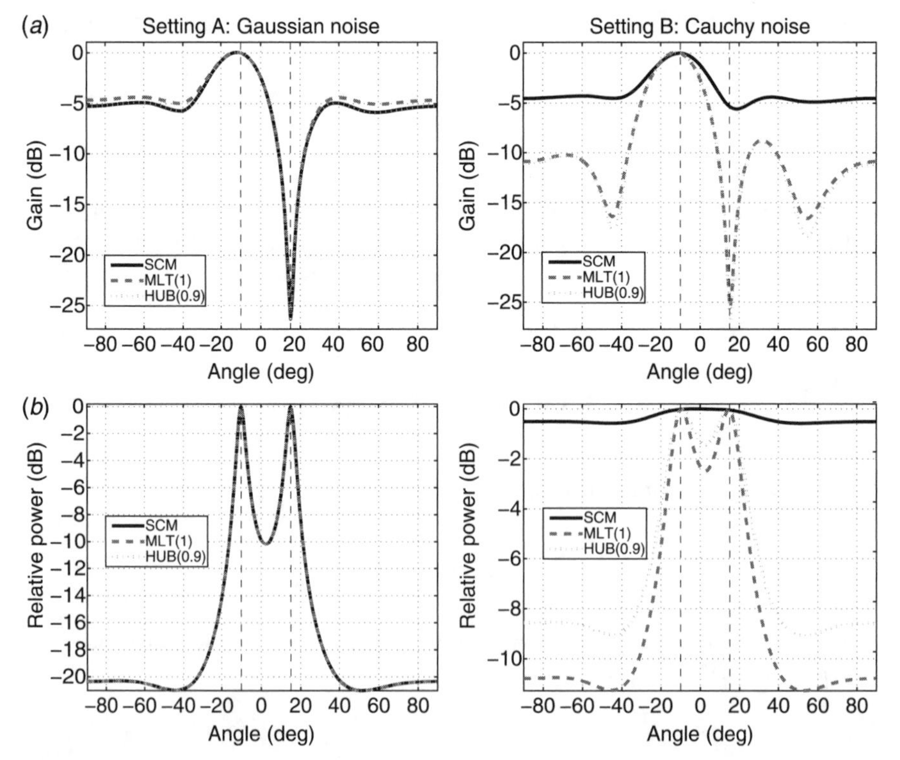

Figure 2.8 Averaged φ-MVDR beampatterns (*a*) and spectrums (*b*) for setting A and B ($n = 500$, SOI at $-10°$, interferer at $15°$). In Gaussian noise, all estimators perform comparably. In Cauchy noise, SCM fails, but robust HUB(0.9) and MLT(1) estimators perform very well.

fact, overlapping for SCM and HUB(0.9). The estimated spectrums associated with the different estimators are overlapping in the Gaussian case, so they provide essentially the same DOA estimates. In the Cauchy noise case (setting B), however, the conventional MVDR fails completely and can not resolve the two sources: the estimated beampattern and spectrum are flat and the mainlobe and the peaks cannot be well identified. Beampatterns associated with MLT(1) and HUB(0.9), however, show a narrow mainlobe centered at the look direction and a deep null at DOA of interference. Also spectrums for MLT(1) and HUB(0.9) show two sharp peaks at the DOAs of the sources. Hence the performance loss is negligible by employing MLT(1) or HUB(0.9) instead of the SCM in nominal Gaussian noise conditions. However, significant gain in performance is obtained when the noise is heavy-tailed Cauchy.

2.7.1 The Influence Function Study

First we derive the IF of the conventional MVDR functional. First note that the conventional MVDR functional can be written in the form

$$\mathbf{w}_{\mathcal{C}}(F) = \mathcal{C}(F)^{-1}\mathbf{g}(F)$$

where

$$\mathbf{g}(F) \triangleq \frac{\tilde{\mathbf{a}}}{\tilde{\mathbf{a}}^H \mathcal{C}(F)^{-1}\tilde{\mathbf{a}}}$$

is the normalized steering vector that satisfies $\|\mathcal{C}(F)^{-1/2}\mathbf{g}(F)\| = 1$.

Applying the product rule of differentiation on the identity $\mathcal{C}(F_{\varepsilon,\mathbf{t}})\mathcal{C}^{-1}(F_{\varepsilon,\mathbf{t}}) = \mathbf{I}$ shows that

$$\mathrm{IF}(\mathbf{t}; \mathcal{C}^{-1}, F) = -\mathcal{C}^{-1}\mathrm{IF}(\mathbf{t}; \mathcal{C}, F)\mathcal{C}^{-1}$$

where $\mathcal{C}$ is the value of the covariance matrix at the reference distribution F. Substituting the expression (2.17) for the IF of the covariance matrix $\mathcal{C}(\cdot)$ into the result above yields the following expression

$$\mathrm{IF}(\mathbf{t}; \mathcal{C}^{-1}, F) = -\mathcal{C}^{-1}\mathbf{t}\mathbf{t}^H\mathcal{C}^{-1} + \mathcal{C}^{-1} \tag{2.36}$$

for the IF of the inverse of the covariance matrix.

Now, using the product rule of differentiation, the IF of the conventional MVDR functional $\mathbf{w}_{\mathcal{C}}$ can be split into two parts

$$\begin{aligned}\mathrm{IF}(\mathbf{t}; \mathbf{w}_{\mathcal{C}}, F) &= \frac{\partial}{\partial\varepsilon}\mathcal{C}(F_{\varepsilon,\mathbf{t}})^{-1}\mathbf{g}(F_{\varepsilon,\mathbf{t}})|_{\varepsilon=0} \\ &= \mathrm{IF}(\mathbf{t}; \mathcal{C}^{-1}, F)\mathbf{g} + \mathcal{C}^{-1}\mathrm{IF}(\mathbf{t}; \mathbf{g}, F)\end{aligned}$$

where $\mathbf{g}$ is the value of $\mathbf{g}(\cdot)$ at F. The second part in the above IF expression can be written in the form

$$\begin{aligned} C^{-1}\mathrm{IF}(\mathbf{t};\, \mathbf{g}, F) &= C^{-1}\tilde{\mathbf{a}}\frac{\partial}{\partial\varepsilon}[\tilde{\mathbf{a}}^H C^{-1}(F_{\varepsilon,\mathbf{t}})\tilde{\mathbf{a}}]^{-1}|_{\varepsilon=0} \\ &= -C^{-1}\tilde{\mathbf{a}}[\tilde{\mathbf{a}}^H C^{-1}\tilde{\mathbf{a}}]^{-2}\,\tilde{\mathbf{a}}^H \mathrm{IF}(\mathbf{t};\, C^{-1}, F)\tilde{\mathbf{a}} \\ &= -\mathbf{w}_C\,\tilde{\mathbf{a}}^H \mathrm{IF}(\mathbf{t};\, C^{-1}, F)\mathbf{g}. \end{aligned}$$

Thus, the IF of $\mathbf{w}_C(\cdot)$ can now be written as

$$\mathrm{IF}(\mathbf{t};\, \mathbf{w}_C, F) = [\mathbf{I} - \mathbf{w}_C\tilde{\mathbf{a}}^H]\mathrm{IF}(\mathbf{t};\, C^{-1}, F)\mathbf{g}.$$

Using the IF expression (2.36) of the inverse of the covariance matrix shows that

$$\begin{aligned} \mathrm{IF}(\mathbf{t};\, C^{-1}, F)\mathbf{g} &= [-\,C^{-1}\mathbf{t}\mathbf{t}^H C^{-1} + C^{-1}]\mathbf{g} \\ &= -C^{-1}\mathbf{t}\mathbf{t}^H\mathbf{w}_C + \mathbf{w}_C. \end{aligned}$$

Thus the IF of $\mathbf{w}_C(\cdot)$ can be written

$$\mathrm{IF}(\mathbf{t};\, \mathbf{w}_C, F) = [\mathbf{I} - \mathbf{w}_C\tilde{\mathbf{a}}^H][-\,C^{-1}\mathbf{t}\mathbf{t}^H\mathbf{w}_C + \mathbf{w}_C].$$

By noting that $[\mathbf{I} - \mathbf{w}_C\tilde{\mathbf{a}}^H]\mathbf{w}_C = 0$ (due to the MVDR gain constraint $\mathbf{w}_C^H\tilde{\mathbf{a}} = 1$), shows that

$$\mathrm{IF}(\mathbf{t};\, \mathbf{w}_C, F) = [\mathbf{w}_C\tilde{\mathbf{a}}^H - \mathbf{I}]C^{-1}\mathbf{t}\mathbf{t}^H\mathbf{w}_C. \tag{2.37}$$

This is a compact expression for the IF of $\mathbf{w}_C(\cdot)$ that also neatly reveals the vulnerability of the conventional MVDR weight vector to outliers. Clearly, contamination at a point $\mathbf{t}$ with large norm $\|\mathbf{t}\|$ has an effect proportional to $\|\mathbf{t}\|^2$ the IF. We may also rewrite the IF expression (2.37) as

$$\mathrm{IF}(\mathbf{t};\, \mathbf{w}_C, F) = r^2[\mathbf{w}_C\tilde{\mathbf{a}}^H - \mathbf{I}]C^{-1/2}\mathbf{u}\mathbf{u}^H C^{1/2}\mathbf{w}_C$$

where $r^2 = \mathbf{t}^H C^{-1}\mathbf{t}$ and $\mathbf{u} = C^{-1/2}\mathbf{t}/r$ is a unit vector. This expression shows that the norm of the IF grows linearly with r (since $\mathbf{u}$ remains bounded).

Let us now consider the case that the reference distribution F is a circular CES distribution $F_\Sigma = \mathrm{CE}_k(\Sigma, g)$. In this case, since $C(F_\Sigma) = \sigma_C\Sigma$ and $\mathbf{w}_C(F_\Sigma) = \mathbf{w}$, the IF expression can be written as follows in Theorem 3.

Theorem 3 *The IF of the conventional MVDR functional* $\mathbf{w}_C(\cdot)$ *at a circular CES distribution* $F_\Sigma = \mathrm{CE}_k(\Sigma, g)$ *is given by*

$$\mathrm{IF}(\mathbf{t};\, \mathbf{w}_C, F) = \frac{r^2}{\sigma_C}[\mathbf{w}\tilde{\mathbf{a}}^H - \mathbf{I}]\Sigma^{-1/2}\mathbf{u}\mathbf{u}^H\Sigma^{1/2}\mathbf{w}$$

where $r^2 = \mathbf{t}^H\Sigma^{-1}\mathbf{t}$, $\mathbf{u} = \Sigma^{-1/2}\mathbf{t}/r$ *a unit vector and* $\mathbf{w} = \mathbf{G}\tilde{\mathbf{a}}$ *is defined in* (2.35).

It is now interesting to compare the IF of $\mathbf{w}_C(\cdot)$ to the general expression of the IF of any φ-MVDR functional $\mathbf{w}_\varphi(\cdot)$ derived in [48] and stated below.

Theorem 4 *With the notations as in Theorem 3, the influence function of φ-MVDR functional $\mathbf{w}_\varphi(\cdot)$ at a CES distribution $F_\Sigma = \mathrm{CE}_k(\Sigma, g)$ is given by*

$$\mathrm{IF}(\mathbf{t}; \mathbf{w}_\varphi; F_\Sigma) = \frac{\alpha_\varphi(r^2)}{\sigma_\varphi}[\mathbf{w}\tilde{\mathbf{a}}^H - \mathbf{I}]\Sigma^{-1/2}\mathbf{u}\mathbf{u}^H\Sigma^{1/2}\mathbf{w}$$

where σ_φ is the solution to (2.27) *and*

$$\alpha_\varphi(x) = \frac{\varphi(x/\sigma_\varphi)x}{1 + c_\varphi}; \quad c_\varphi = \frac{1}{k(k+1)}E\left[\varphi'\left(\frac{\delta}{\sigma_\varphi}\right)\frac{\delta^2}{\sigma_\varphi^2}\right] \tag{2.38}$$

and δ is a positive real rva with the pdf (2.12).

Theorem 4 shows that the IF of $\mathbf{w}_\varphi(\cdot)$ is continuous and bounded if, and only if, $\varphi(x)x$ is continuous and bounded. This follows by noting that when $\|\mathbf{t}\|$, or equivalently $r = \|\Sigma^{-1/2}\mathbf{t}\|$, grow to infinity, $\mathbf{u} = \Sigma^{-1/2}\mathbf{t}/r$ remains bounded. Hence, to validate the qualitative robustness of φ-MVDR beamformers we only need to validate that $\varphi(x)x$ is bounded. Theorem 4 also shows that $\mathrm{IF}(\tilde{\mathbf{a}}; \mathbf{w}_\varphi, F_\Sigma) = 0$, that is, if the contamination point $\mathbf{t}$ equals the array response $\tilde{\mathbf{a}}$, then it causes zero influence on the functional. We wish to point out that if $\mathbf{w}_\varphi(\cdot)$ is the conventional MVDR functional (i.e., $\varphi \equiv 1$), then the IF expression of Theorem 4 indeed gives the IF expression of Theorem 3. For example, $\mathbf{w}_\varphi$ based on HUB(0.9) or MLT(1) functionals are robust, that is, they have continuous and bounded IFs, since their $\varphi(\cdot)$ functions are downweighting observations with large magnitude as shown in Figures 2.4 and 2.5.

We wish to point out that in beamforming literature, "robust" more commonly refers to *robustness to steering errors* (imprecise knowledge of the array response $\tilde{\mathbf{a}}$ may be due to uncertainty in array element locations, steering directions and calibration errors) and *robustness in the face of insufficient sample support* that may lead to rank deficient SCM or inaccurate estimates of the array covariance matrix. Thus, the lack of robustness is caused by misspecified or uncertain system matrices or due to the fact that we do not have sufficient sample support to build up the rank of the array covariance and pseudocovariance matrices, not because of uncertainty in the probability models.

The diagonal loading of the SCM is one of the most popular techniques to overcome the problems in the modeling system matrix or rank deficiency. Then we use $(\hat{C} + \gamma\mathbf{I})$, $\gamma \in \mathbb{R}$, in place of $\hat{C}$, which may not be full rank and hence not invertible. For this type of robustness study, see for example, [12, 16, 23, 35] and references therein. Here the term "robust" refers to statistical robustness to outliers [26], commonly measured by the concept of the IF. We wish to point out that robustness (as measured by the IF) of the MVDR beamformer remains unaltered by diagonally loading the covariance matrx C, that is, using $C_\gamma(F) = C(F) + \gamma\mathbf{I}$, where γ is some constant diagonal loading term not dependent on the distribution F of $\mathbf{z}$. Although (statistical) robustness of the MVDR weight functional is not improved with diagonal loading, it provides, however, other kinds of robustness by improving the condition number of the estimated array covariance matrix. Naturally, IF is an asymptotic concept, and it is not the correct tool to analyze the performance in sample starved scenarios.

■ EXAMPLE 2.10

We now compute the EIFs of estimated φ-MVDR beamformer weights $\hat{\mathbf{w}}_\varphi$ for data sets $Z_n = \{\mathbf{z}_1, \ldots, \mathbf{z}_n\}$ simulated as in setting A of Example 2.9. In the setting A, the observed snapshots $\mathbf{z}_1, \ldots, \mathbf{z}_n$ form an i.i.d. random sample from four-variate circular complex normal distribution $\Phi_4 = \text{CN}_4(\mathcal{C})$ with covariance matrix $\mathcal{C}(\mathbf{z}) = \sigma_s^2 \mathbf{A}\mathbf{A}^H + \sigma_n^2 \mathbf{I}$, where $\mathbf{A} = \begin{pmatrix} \tilde{\mathbf{a}} & \mathbf{a}_1 \end{pmatrix} \in \mathbb{C}^{4\times 2}$ denotes the array response matrix of ULA for DOAs at 10° (SOI) and 15° (interferer). Let the k = four-variate contaminating vector $\mathbf{t}$ be such that only the first component $t_1 = u_1 + jv_1$ is allowed to vary, and the remaining components have fixed values: $t_i = \tilde{a}_i$,

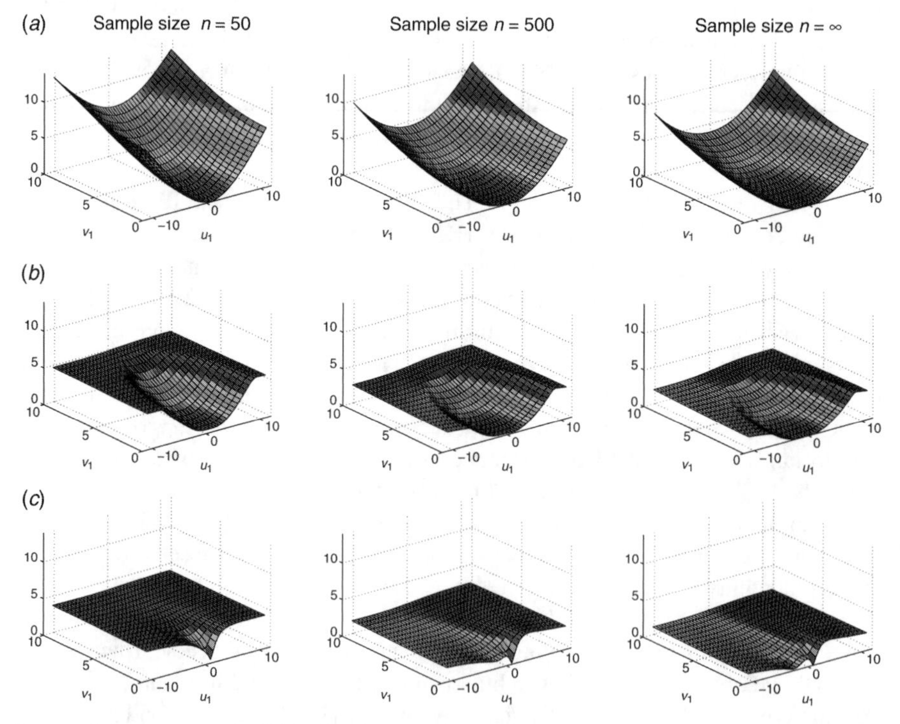

Figure 2.9 Averaged $\|\text{EIF}(\mathbf{t}; \hat{\mathbf{w}}_\varphi, Z_n)\|$ for the φ-MVDR beamformers that employ the SCM (*a*), HUB(0.9) estimator (*b*) and MLT(1) estimator (*c*). Sample Z_n ($n = 50, 500$) are generated as in setting A of Example 2.9; $n = \infty$ corresponds to the plot of $\|\text{IF}(\mathbf{t}, \mathbf{w}_\varphi, F_\Sigma)\|$. The contamination point $\mathbf{t} = (t_1, \mathbf{t}_2^T)^T$ is such that $\mathbf{t}_2$ is fixed and only $t_1 = u_1 + jv_1 \in \mathbb{C}$ is allowed to vary.

where $\tilde{a}_i$ denotes the ith component of the array response $\tilde{\mathbf{a}}$. An informative picture on the effect of contamination $t_1 = u_1 + jv_1$ on $\hat{\mathbf{w}}_\varphi$ is obtained by the surface plot of the norm of the empirical influence function $\|\text{EIF}(\mathbf{t}; \hat{\mathbf{w}}_\varphi, Z_n\|)$ with respect to v_1 and v_1. The EIFs in Figure 2.9 are averages over 100 realizations. Sample lengths are $n = 50, 500, \infty$, where the surface plots under $n = \infty$ correspond to the asymptotic value $\|\text{IF}(\mathbf{t}, \hat{\mathbf{w}}_\varphi, \Phi_4)\|$. As expected, we observe that when the sample size grows (from $n = 50$ to $n = 500$), the calculated EIF surfaces more accurately resemble the corresponding theoretical IF surface. However, at the small sample size ($n = 50$), the relative influence of an additional observation on the estimator is a bit larger than that the IF would indicate. The surface plots neatly demonstrate the nonrobustness of the conventional MVDR beamformer for both the finite and large sample cases: outlying points with large values of u_1 and/or v_1 have bounded influence in the cases of HUB(0.9) or MLT(1) but large and unbounded influence when the conventional SCM is employed.

Efficiency Study Using the IF of $\mathbf{w}_\varphi(\cdot)$ (cf. Theorem 4) and equations (2.19) and (2.20) as the definitions for the asymptotic covariance matrix and pseudo-covariance matrix of the estimator, the next theorem was proved in [48].

Theorem 5 *With the notations as in Theorem 4, the asymptotic covariance matrix of the estimated φ-MVDR weight $\hat{\mathbf{w}}_\varphi$ when sampling from $F_\Sigma = \text{CE}_k(\Sigma, g)$ is*

$$\text{ASC}(\hat{\mathbf{w}}_\varphi; F_\Sigma) = \lambda_\varphi(\mathbf{G} - \mathbf{w}\mathbf{w}^H)$$

where

$$\lambda_\varphi = \frac{E[\varphi^2(\delta/\sigma_\varphi)(\delta/\sigma_\varphi)^2]}{k(k+1)(1+c_\varphi)^2}.$$

The asymptotic pseudo-covariance matrix of $\hat{\mathbf{w}}_\varphi$ vanishes, that is, $\text{ASP}(\hat{\mathbf{w}}_\varphi; F_\Sigma) = 0$.

Note that the ASC of $\hat{\mathbf{w}}_\varphi$ depends on the selected M-estimator and on the functional form of the CES distribution F_Σ *only* via the real-valued positive multiplicative constant λ_φ. (Observe that the matrix term $\mathbf{G} - \mathbf{w}\mathbf{w}^H$ does not depend on the choice of φ and on F_Σ only via Σ.) Hence comparisons of this single scalar index is needed only. It is a surprising result that ASP vanishes, which means that $\hat{\mathbf{w}}_\varphi$ has asymptotic circular CN distribution.

Note also that $\text{ASC}(\hat{\mathbf{w}}_\varphi; F_\Sigma)$ is singular and of rank $k - 1$ (since the nullspace of $\mathbf{G} - \mathbf{w}\mathbf{w}^H$ has dimension 1 due to MVDR constraint $\mathbf{w}^H\tilde{\mathbf{a}} = 1$, so $(\mathbf{G} - \mathbf{w}\mathbf{w}^H)\tilde{\mathbf{a}} = 0$). Thus the asymptotic CN distribution of $\hat{\mathbf{w}}_\varphi$ is singular. This is an expected result since singular distributions commonly arise in constrained parameter estimation problems, where the constraint imposes a certain degree of deterministicity to the estimator.

Table 2.1 Asymptotic relative efficiencies of $\hat{\mathbf{w}}_\varphi$ based on HUB(q) and MLT(ν) estimators at Φ_k, $T_{k,5}$ and $T_{k,1}$ distributions (and dimensions $k = 2, 3, 5, 10$). Recall that HUB(1) correspond to SCM

	Normal (Φ_k)			$T_{k,5}$			Cauchy ($T_{k,1}$)		
$k =$	2	5	10	2	5	10	2	5	10
HUB(1)	1	1	1	.407	.378	.360	0	0	0
HUB(.9)	.961	.986	.994	.963	.940	.933	.770	.798	.833
HUB(.5)	.823	.923	.961	.949	.986	.988	.975	.964	.964
HUB(.1)	.698	.853	.921	.849	.960	.988	.963	.993	.997
MLT(5)	.906	.921	.941	1	1	1	.827	.927	.969
MLT(2)	.825	.882	.925	.968	.987	.995	.976	.992	.997
MLT(1)	.769	.861	.917	.923	.972	.990	1	1	1

The *asymptotic relative efficiency* (ARE) of $\hat{\mathbf{w}}_\varphi$ can thus be calculated as

$$\mathrm{ARE}(\hat{\mathbf{w}}_\varphi; F_\Sigma) = \frac{\mathrm{Tr}[\mathrm{ASC}(\hat{\mathbf{w}}_{mle}; F_\Sigma)]}{\mathrm{Tr}[\mathrm{ASC}(\hat{\mathbf{w}}_\varphi; F_\Sigma)]} = \frac{\lambda_{mle}}{\lambda_\varphi}$$

where λ_{mle} correspond to λ_φ value for the $\hat{\mathbf{w}}_{mle}$. Thus ARE $(\hat{\mathbf{w}}_\varphi; F_\Sigma) \leq 1$. Table 2.1 reports the AREs of $\hat{\mathbf{w}}_\varphi$ based on HUB(q) and MLT(ν) estimators under circular complex normal (Φ_k), circular complex Cauchy ($T_{k,1}$) and circular $T_{k,5}$ distributions for some choices of dimension k. Recall that the HUB(1) estimator corresponds to the SCM. At CN distribution, HUB(0.9) has the best performance among its robust alternatives. Furthermore, efficiencies for HUB(q) and MLT(ν) are increasing with the dimension: for example, at $k = 10$, HUB(0.9) experiences only 0.006 percent efficiency loss and MLT(1) which ranks the lowest, has a moderate 8.3 percent loss in efficiency. Hence adding more sensors to the array increases the (asymptotic) efficiency of the estimated φ-MVDR beamformers based upon the above robust M-estimators. At $T_{k,5}$ distribution, employed M-estimators are superior to the conventional MVDR beamformer based on SCM. At Cauchy distribution, all the robust M-estimators are performing very well and their efficiencies are increasing with the dimension. To conclude, these asymptotic efficiencies clearly favor φ-MVDR beamformers based upon *robust* M-estimators since they combine a high efficiency with appealing robustness properties.

2.8 ROBUST ICA

Independent component analysis (ICA) is a relatively recent technique of multivariate data analysis and signal processing (see [14, 15, 28] and their bibliographies) used for example, in several applications including wireless communications, audio and speech signal separation, biomedical signal processing, image processing, feature extraction, and data-mining.

The main bulk of research in the field so far has concentrated on the real-valued ICA model, but the complex-valued ICA model has attained increasing interest during the last decade due to its potential applications, for example, for convolutive source separation in the frequency domain and for source separation of complex-valued data arising in several application areas such as magnetic resonance imaging (MRI) or antenna array signal processing (such as radar data). Therefore, many efforts have been pursued to establish identifiability/uniqueness/separability theory for complex-valued ICA model, to generalize existing techniques to complex case, and to derive new estimators that better exploit the specific nature of the complex-valued model; see [3, 5, 8, 11, 18, 19, 21, 22, 34, 39, 47, 49] to name only a few.

In this section, we assume that sensor outputs follow the noiseless complex-valued ICA model, that is

$$\mathbf{z} = \mathbf{A}\mathbf{s},$$

where $\mathbf{s} \in \mathbb{C}^d$ has mutually *statistically independent* components $s_1, \ldots, s_d$, and without any loss of generality, assume that $E[\mathbf{s}] = 0$. As is common in ICA, we assume that the number of sources is equal to the number of sensors, so $k = d$ and that the mixing matrix $\mathbf{A} \in \mathbb{C}^{k \times k}$ is nonsingular. Due to fundamental indeterminacy [22], ICA should be understood as the determination of a matrix $\mathcal{B}$, called the separating matrix, that satisfies

$$\hat{\mathbf{s}} = \mathcal{B}\mathbf{z} = \mathbf{D}\mathbf{s}$$

where $\mathbf{D}$ is a $k \times k$ scaled permutation matrix, that is, $\hat{\mathbf{s}}$ contains permuted and scaled components of $\mathbf{s}$. For the separation to be possible (up to above ambiguities), at most one of the sources can possess circular CN distribution, but sources can have CN distribution with distinct circularity coefficient [22].

We now review two classes of ICA estimators that are based on the concepts of (spatial) scatter and pseudo-scatter matrices.

2.8.1 The Class of DOGMA Estimators

Fourth Order Blind Identification (FOBI) method Assume for a moment that the source RV $\mathbf{s}$ has finite variances and therefore $\mathcal{C}(\mathbf{z}) = \mathbf{A}\mathcal{C}(\mathbf{s})\mathbf{A}^H$ exists and is of full rank. Compute the whitening matrix $\mathbf{B} = \mathbf{B}(\mathbf{z}) \in \mathbb{C}^{k \times k}$ as the square-root matrix of $\mathcal{C}(\mathbf{z})^{-1}$, that is, $\mathbf{B}^H\mathbf{B} = \mathcal{C}^{-1}$. There exists many methods to compute $\mathbf{B}$ (e.g. Cholesky decomposition), but for our purposes, we do not need to specify any particular one. Then the whitened mixture $\mathbf{v} = \mathbf{B}\mathbf{z}$ is uncorrelated, that is, $\mathcal{C}(\mathbf{v}) = \mathbf{I}$, and also follows the ICA model

$$\mathbf{v} = \tilde{\mathbf{A}}\mathbf{s} \quad \text{with} \quad \tilde{\mathbf{A}} = \mathbf{B}\mathbf{A}.$$

The identity $\mathcal{C}(\mathbf{v}) = \tilde{\mathbf{A}}\mathcal{C}(\mathbf{s})\tilde{\mathbf{A}}^H = \mathbf{I}$ shows that the scaled mixing matrix

$$\tilde{\mathbf{A}}\mathcal{C}(\mathbf{s})^{1/2} = \begin{pmatrix} \tilde{\mathbf{a}}_1\sigma_1 & \tilde{\mathbf{a}}_2\sigma_2 & \cdots & \tilde{\mathbf{a}}_k\sigma_k \end{pmatrix}$$

is a unitary matrix, where $\sigma_i = (E[|s_i|^2])^{1/2}$. Since in any case, the scales of the columns $\tilde{\mathbf{a}}_i$ of $\tilde{\mathbf{A}}$ can not be identified due to fundamental indeterminacy of ICA, we may contend that $\tilde{\mathbf{A}}$ is a unitary matrix (without any loss of generality) . Thus the separating matrix of the whitened mixture is a unitary matrix as well, so $\mathbf{U}^H\mathbf{v} = \hat{\mathbf{s}}$ for some unitary matrix $\mathbf{U} \in \mathbb{C}^{k\times k}$, and thus $\mathbf{W} = \mathbf{U}^H\mathbf{B}$ is a separating matrix for the original mixture. Cardoso [8] has shown that, if $\mathbf{U} = \mathbf{U}(\mathbf{v})$ is the matrix of eigenvectors of kurtosis matrix $\mathcal{C}_{\text{kur}}(\mathbf{v})$ defined in (2.21), then the FOBI functional $\mathbf{W}(\mathbf{z}) = \mathbf{U}(\mathbf{v})^H\mathbf{B}(\mathbf{z})$ is a separating matrix provided that sources have finite and distinct circular kurtosis values, that is, $\kappa_0(s_i) \neq \kappa_0(s_j)$ for all $i \neq j \in \{1, \ldots, k\}$, where $\kappa_0(\cdot)$ is defined in (2.5).

FOBI is perhaps the simplest method to solve the ICA problem proposed thus far. Since the FOBI functional can be computed via standard matrix decompositions operating on matrices $\mathcal{C}(\mathbf{z})$ and $\mathcal{C}_{\text{kur}}(\mathbf{v})$, it is also computationally (among) the most efficient approaches to ICA. It has some drawbacks, however. First, the necessity of distinct circular kurtosis values restricts the applicability of the method to some extent since sources with identical distributions (and hence with identical kurtosis) may occur frequently in some applications. Second, the assumption on finite kurtosis clearly confines permissible distributions of the sources the method can separate since fourth-order moments do not exist for many heavy-tailed distributions. Third, the method is not robust, since the covariance matrix (used for whitening) and the kurtosis matrix are highly non-robust.

In order to separate sources with identical distribution and identical kurtosis, FOBI was later generalized to joint approximate diagonalization of eigen-matrices (JADE) [9, 11] which is based on joint diagonalization of several cumulant matrices. However, JADE still demands finite fourth-order moments, is not robust, and has the disadvantage that simplicity and computational efficiency of the FOBI algorithm is lost.

Generalization of FOBI Let $\mathbf{C}_1(\cdot)$ denote any scatter matrix and $\mathbf{C}_2(\cdot)$ denote any spatial scatter matrix functional with IC-property, by which we mean that if $\mathbf{s}$ has independent components, then $\mathbf{C}_1(\mathbf{s})$ and $\mathbf{C}_2(\mathbf{s})$ are diagonal matrices, that is

$$\mathbf{C}_1(\mathbf{s}) = \text{diag}([\mathbf{C}_1(\mathbf{s})]_{ii}) \quad \text{and} \quad \mathbf{C}_2(\mathbf{s}) = \text{diag}([\mathbf{C}_2(\mathbf{s})]_{ii}). \tag{2.39}$$

One can easily verify that the covariance matrix $\mathcal{C}(\cdot)$ and the kurtosis matrix $\mathcal{C}_{\text{kur}}(\cdot)$ possess IC-property. M-functionals of scatter $\mathbf{C}_\varphi(\cdot)$ do not in general satisfy IC-property. However, if the sources s_i are assumed to be symmetric, that is, $s_i =_d -s_i$ for $i = 1, \ldots, k$, then the requirement of IC-property can be dropped [47], since for symmetric independent sources, any scatter or spatial scatter matrix automatically possess the IC-property. Even if the sources are not symmetric, a symmetricized version of M-estimator (or of any scatter or spatial scatter matrix) can be easily constructed that automatically possesses IC-property, see [47, 49] for details.

DOGMA (Diagonalization Of Generalized covariance MAtrices) algorithm: DOGMA functional $\mathbf{W}(\mathbf{z}) \in C^{k\times k}$ is calculated as follows.

(a) Calculate the square-root matrix $\mathbf{B}_1(\mathbf{z})$ of $\mathbf{C}_1(\mathbf{z})^{-1}$, so $\mathbf{B}_1(\mathbf{z})^H\mathbf{B}_1(\mathbf{z}) = \mathbf{C}_1(\mathbf{z})^{-1}$, and the whitened data $\mathbf{v} = \mathbf{B}_1(\mathbf{z})\mathbf{z}$ (so $\mathbf{C}_1(\mathbf{v}) = \mathbf{I}$).

(b) Calculate the EVD of $\mathbf{C}_2(F)$ of the whitened data

$$\mathbf{C}_2(\mathbf{v}) = \mathbf{U}_2(\mathbf{v})\Lambda_2(\mathbf{v})\mathbf{U}_2(\mathbf{v})^H \tag{2.40}$$

where $\Lambda_2(\mathbf{v})$ is a diagonal matrix of eigenvalues of $\mathbf{C}_2(\mathbf{v})$ and $\mathbf{U}_2(\mathbf{v})$ is a unitary matrix with the respective eigenvectors as columns.

(c) Set $\mathbf{W}(\mathbf{z}) = \mathbf{U}_2(\mathbf{v})^H\mathbf{B}_1(\mathbf{z})$.

Note that FOBI is a DOGMA functional $\mathbf{W}(\mathbf{z})$ with choices $\mathbf{C}_1 = \mathcal{C}$ and $\mathbf{C}_2 = \mathcal{C}_{\text{kur}}$. Since the IC-property is required, the spatial sign covariance matrix $\mathcal{C}_{\text{sgn}}(F)$ for example can not be employed as the choice of $\mathbf{C}_2$ unless sources have symmetric distributions. We wish to point out that $\mathbf{W}(\mathbf{z}) = \mathbf{U}_2(\mathbf{v})^H\mathbf{B}_1(\mathbf{z})$ simultaneously diagonalizes $\mathbf{C}_1(\mathbf{z})$ and $\mathbf{C}_2(\mathbf{z})$, namely, for transformed data $\hat{\mathbf{s}} = \mathbf{W}(\mathbf{z})\mathbf{z}$ it holds that

$$\mathbf{C}_1(\hat{\mathbf{s}}) = \mathbf{I} \quad \text{and} \quad \mathbf{C}_2(\hat{\mathbf{s}}) = \Lambda_2(\mathbf{v}).$$

Hence, we call the above algorithm DOGMA (Diagonalization Of Generalized covariance MAtrices).

Note that in the step (b) of the algorithm $\Lambda_2(\mathbf{v}) = \text{diag}(\delta_1, \ldots, \delta_k)$ is a diagonal matrix with eigenvalues $\delta_1, \ldots, \delta_k$ of $\mathbf{C}_2(\mathbf{v})$ on its diagonal elements. It can be shown (cf. Theorem 1 of [49]) that

$$\Lambda_2(\mathbf{v}) = \mathbf{C}_2(\mathbf{C}_1(\mathbf{s})^{-1/2}\mathbf{s}).$$

For example, in the case of the FOBI functional (i.e., $\mathbf{C}_1 = \mathbf{C}$ and $\mathbf{C}_2 = \mathbf{C}_{\text{kur}}$), the eigenvalues are easily calculated as

$$\begin{aligned}\Lambda_2(\mathbf{v}) = \mathcal{C}_{\text{kur}}(\mathcal{C}(\mathbf{s})^{-1/2}\mathbf{s}) &= \mathcal{C}(\mathbf{s})^{-1/2}E[(\mathbf{s}^H\mathcal{C}(\mathbf{s})^{-1}\mathbf{s})\mathbf{s}\mathbf{s}^H]\mathcal{C}(\mathbf{s})^{-1/2} \\ &= \text{diag}(\kappa_{0,1}, \ \ldots, \ \kappa_{0,k}) + (k+1)\mathbf{I}\end{aligned} \tag{2.41}$$

where $\kappa_{0,i} = \kappa_0(s_i)$ denotes the circular kurtosis of the ith source rva s_i.

The following result has been proved in [49].

Theorem 6 *Under the assumptions*

D1: $\mathbf{C}_1(\mathbf{s})$ *and* $\mathbf{C}_2(\mathbf{s})$ *exists, and*

D2: *eigenvalues* $\delta_1, \ldots, \delta_k$ *of* $\mathbf{C}_2(\mathbf{v})$ *are distinct, that is,* $\delta_i \neq \delta_j$ *for all* $i \neq j \in \{1, \ldots, k\}$, *the DOGMA functional* $\mathbf{W}(\mathbf{z}) = \mathbf{U}_2(\mathbf{v})^H\mathbf{B}_1(\mathbf{z})$ *is a separating matrix for the complex-valued ICA model.*

As an example, consider the FOBI functional. Then, the assumption D1 is equivalent with the assumption that sources has finite circular kurtosis values. Assumption D2 implies (due to (2.41)) that sources have distinct values of circular kurtosis, that is, $\kappa_0(s_i) \neq \kappa_0(s_j)$ for all $i \neq j \in (1, \ldots, k)$.

For more properties of DOGMA functionals, see [49], where also alternative formulations of the method are derived along with efficient computational approach to compute the estimator.

2.8.2 The Class of GUT Estimators

Let $\mathbf{C}(\cdot)$ denote any scatter matrix functional and $\mathbf{P}(\cdot)$ denote any spatial pseudo-scatter matrix functional with IC-property (i.e., they reduce to diagonal matrices when F is a cdf of a random vector with independent components). As already mentioned, the covariance matrix is an example of a scatter matrix that possesses IC-property. Pseudo-kurtosis matrix $\mathcal{P}_{\text{kur}}(\cdot)$ defined in (2.21) is an example of a spatial pseudo-scatter matrix that possesses IC-property. Sign pseudo-covariance matrix $\mathcal{P}_{\text{sgn}}(\cdot)$ for example do not necessarily possess IC-property. However, as mentioned earlier, for symmetric independent sources, any scatter or spatial pseudo-scatter matrix automatically possesses IC-property. Again if the sources are not symmetric, a symmetricized version of any scatter or spatial pseudo-scatter matrix can be easily constructed that automatically possesses IC-property, see [47, 49] for details.

Definition 5 *Matrix functional* $\mathbf{W} = \mathbf{W}(\mathbf{z}) \in \mathbb{C}^{k \times k}$ *of* $z \in \mathbb{C}^{k}$ *is called the Generalized Uncorrelating Transform (GUT) if transformed data* $\mathbf{s} = \mathbf{W}z$ *satisfies*

$$\mathbf{C}(\mathbf{s}) = \mathbf{I} \quad \text{and} \quad \mathbf{P}(\mathbf{s}) = \Lambda \tag{2.42}$$

where $\boldsymbol{\Lambda} = \boldsymbol{\Lambda}(\mathbf{s}) = \text{diag}(\lambda_i)$ *is a real nonnegative diagonal matrix, called the circularity matrix, and* $\lambda_i = [\mathbf{P}(\mathbf{s})]_{ii} \geq 0$ *is called the ith circularity coefficient,* $i = 1, \ldots, \mathrm{k}$.

The GUT matrix with choices $\mathbf{C} = \mathcal{C}$ and $\mathbf{P} = \mathcal{P}$ corresponds to the SUT [21, 22] described in Section 2.2.2. Essentially, GUT matrix $\mathbf{W}(\cdot)$ is a data transformation that jointly diagonalizes the selected scatter and spatial pseudo-scatter matrix of the transformed data $\mathbf{s} = \mathbf{Wz}$. Note that the pseudo-covariance matrix employed by SUT is a pseudo-scatter matrix, whereas in Definition 5, we only require $\mathbf{C}(\cdot)$ to be a spatial pseudo-scatter matrix.

GUT algorithm

(a) Calculate the square-root matrix $\mathbf{B}(\mathbf{z})$ of $\mathbf{C}(\mathbf{z})^{-1}$, so $\mathbf{B}(\mathbf{z})^H \mathbf{B}(\mathbf{z}) = \mathbf{C}(\mathbf{z})^{-1}$, and the whitened data $\mathbf{v} = \mathbf{B}(\mathbf{z})\mathbf{z}$ (so $\mathbf{C}(\mathbf{v}) = \mathbf{I}$).

(b) Calculate Takagi's factorization (symmetric SVD) of $\mathbf{P}(\cdot)$ for the whitened data $\mathbf{v}$

$$\mathbf{P}(\mathbf{v}) = \mathbf{U}\Lambda\mathbf{U}^T \tag{2.43}$$

where $\mathbf{U} = \mathbf{U}(\mathbf{v}) \in \mathbb{C}^{k \times k}$ is a unitary matrix (i.e., the Takagi factor of $\mathbf{P}(\mathbf{v})$) and Λ is the circularity matrix (i.e., the singular values of $\mathbf{P}(\mathbf{v})$ are the circularity coefficients $\lambda_i = [\mathbf{P}(\mathbf{s})]_{ii}$ appearing in (2.42)).

(c) Set $\mathbf{W}(\mathbf{z}) = \mathbf{U}(\mathbf{v})^H \, \mathbf{B}(\mathbf{z})$.

In step (a), the data is *whitened* in the sense that $\mathbf{C}(\mathbf{v}) = \mathbf{I}$. Naturally, if the selected scatter matrix is the covariance matrix, then the data is whitened in the conventional sense. Since the whitening transform $\mathbf{B}$ is unique only up to left-multiplication by a unitary matrix, GUT matrix $\mathbf{W} = \mathbf{U}^H \, \mathbf{B}$ is also a whitening transform in the conventional sense but with an additional property that it diagonalizes the selected spatial pseudo-scatter matrix.

As revealed in step (b) of the algorithm, the circularity matrix $\Lambda = \mathrm{diag}(\lambda_i)$ has singular values $\lambda_1, \ldots, \lambda_k$ of $\mathbf{P}(\mathbf{v})$ as its diagonal elements. It has been shown in Theorem 3 of [47] that the circularity coefficient is

$$\lambda_i = |[\mathbf{P}(\tilde{\mathbf{s}})]_{ii}|$$

where $\tilde{\mathbf{s}} = \mathbf{C}(\mathbf{s})^{-1/2}\mathbf{s}$ denotes the whitened source whose ith component is $\tilde{s}_i = s_i/\sqrt{[\mathbf{C}(\mathbf{s})]_{ii}}$. For example, consider the SUT functional (i.e., $\mathbf{C} = \mathcal{C}, \; \mathbf{P} = \mathcal{P}$). Then

$$\lambda_i = |[\mathcal{P}(\tilde{\mathbf{s}})]_{ii}| = |\tau(\tilde{s}_i)| = |\tau(s_i)|/\sigma^2(s_i) = \lambda(s_i) \tag{2.44}$$

that is, the ith circularity coefficient is equal to the circularity coefficient of the ith source s_i. Next consider the case that the GUT functional employs $\mathbf{C} = \mathcal{C}$ and $\mathbf{P} = \mathcal{P}_{\mathrm{kur}}$. Then

$$\lambda_i = |[\mathcal{P}_{\mathrm{kur}}(\tilde{\mathbf{s}})]_{ii}| = |[E[(\tilde{\mathbf{s}}^H\tilde{\mathbf{s}})\tilde{\mathbf{s}}\tilde{\mathbf{s}}^T]]_{ii}| = |E[|\tilde{s}_i|^2\tilde{s}_i^2] + (k-1)\tau(\tilde{s}_i)\,|.$$

Hence the ith circularity coefficient λ_i is the modulus of a weighted sum of a 4th-order and 2nd-order moment of the ith whitened source $\tilde{s}_i$.

The following result has been proved in [47].

Theorem 7 *Under the assumptions*

G1: $\mathbf{C}(\mathbf{s})$ *and* $\mathbf{P}(\mathbf{s})$ *exists, and*

G2: *circularity coefficients* $\lambda_1, \ldots, \lambda_k$ *(the singular values of* $\mathbf{P}(\mathbf{v})$ *are distinct, that is,* $\lambda_i \neq \lambda_j$ *for all* $i \neq j \in \{1, \ldots, k\}$*, the GUT functional* $\mathbf{W}(\mathbf{z}) = \mathbf{U}(\mathbf{v})^H\mathbf{B}(\mathbf{z})$ *is a separating matrix for the complex-valued ICA model.*

As an example, consider the SUT functional. Then assumption G1 is equivalent with the assumption that sources has finite variances. Assumption G2 implies (due to (2.44)) that sources have distinct circularity coefficients.

It is important to observe that the GUT algorithm contains a built-in warning: since the circularity coefficients $\lambda_1, \ldots, \lambda_k$ are also necessarily extracted, then the detection of two close or almost equal circularity coefficients is an indication that the corresponding sources may not be reliably separated. Also, assumption G2 is needed to separate all the sources: GUT matrix is not able to separate the sources that have identical circularity coefficients, but the rest of the sources are separated; cf. [47].

For more properties of GUT functionals, see [47], where also alternative formulations of the method are derived along with efficient computational approaches to compute the estimator.

2.8.3 Communications Example

Let $\widehat{\mathcal{B}}$ denote an estimator of the separating matrix. The performance of the separation is often investigated via *interference matrix* $\widehat{\mathbf{G}} = \widehat{\mathcal{B}}\mathbf{A}$. Due to fundamental indeterminacy of ICA, perfect separation implies that $\hat{\mathbf{G}}$ is a scaled permutation matrix. The quality of the separation is then assessed by calculating the widely used performance index (PI) [2]

$$\mathrm{PI}(\hat{\mathbf{G}}) = \frac{1}{2k(k-1)}\left\{\sum_{i=1}^{k}\left(\sum_{j=1}^{k}\frac{|\hat{g}_{ij}|}{\max_\ell |\hat{g}_{i\ell}|} - 1\right) + \sum_{j=1}^{k}\left(\sum_{i=1}^{k}\frac{|\hat{g}_{ij}|}{\max_\ell |\hat{g}_{\ell j}|} - 1\right)\right\}.$$

where $\hat{g}_{ij} = [\hat{\mathbf{G}}]_{ij}$. Under perfect separation $\mathrm{PI}(\hat{\mathbf{G}}) = 0$. When the estimator fails to separate the sources, the value of the PI increases. The PI is scaled so that the maximum value is 1. If the separating matrix estimator $\hat{\mathcal{B}}$ is equivariant (in the sense advocated in [10], Section II-C), as is the case for GUT and DOGMA estimators, then $\hat{\mathbf{G}}$ (and thus PI) does not depend on the mixing matrix $\mathbf{A}$, and hence one could set $\mathbf{A} = \mathbf{I}$ in the simulations without any loss of generality.

In our simulation studies, GUT estimators employing the following choices of scatter matrix $\mathbf{C}$ and spatial pseudo-scatter matrix $\mathbf{P}$ are used: *gut1* employs covariance matrix and pseudo-covariance matrix (i.e., SUT), *gut2* employs covariance matrix and pseudo-kurtosis matrix, *gut3* employs covariance matrix and sign pseudo-covariance matrix (SPM), *gut4* employs HUB (0.9) estimator and SPM, *gut5* employs Tyler's *M*-estimator of scatter and SPM. We compare the results to *jade* [11], *fobi* [8], complex FastICA with deflationary approach and contrast $G_2(x) = \log(0.1 + x)$ [5] (denoted as *fica*), complex fixed point algorithm using kurtosis based contrast and symmetric orthogonalization [19] (denoted as *cfpa*), gradient based kurtosis maximization algorithm [34] (denoted as *kmg*) and DOGMA estimator employing Tyler's *M*-estimator and HUB (0.9) estimator as the choices of scatter matrices $\mathbf{C}_1(\cdot)$ and $\mathbf{C}_2(\cdot)$, respectively (denoted as *d1*).

In our simulation setting, three independent random signals—a BPSK signal, a 8-QAM signal and a circular Gaussian signal of equal power σ_s^2 are impinging on $k = 3$ element Uniform Linear Array (ULA) with half a wavelength interelement spacing from DOAs -20°, 5° and 35°. Note that BPSK and 8-QAM signals are

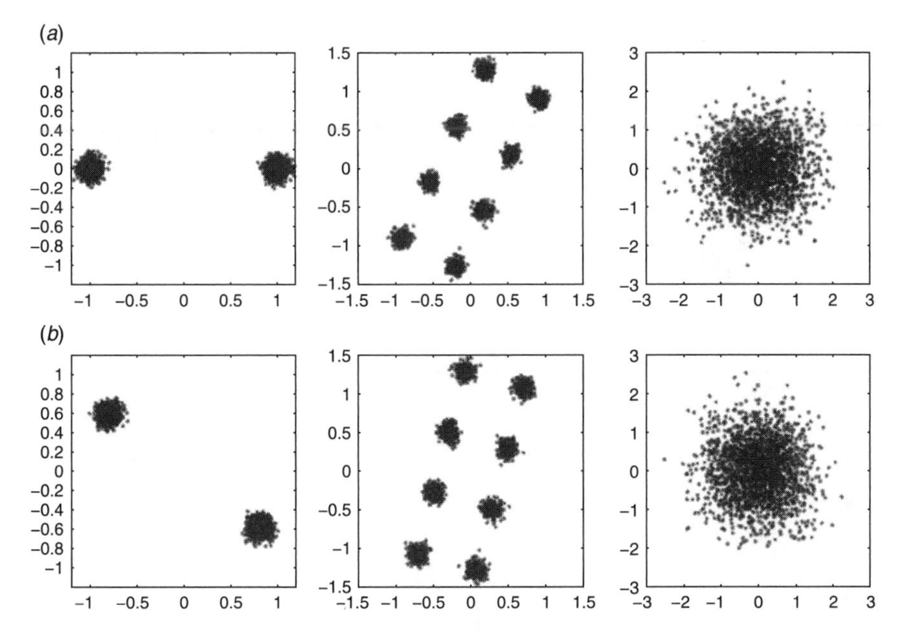

Figure 2.10 Estimated source signal constellations obtained by *jade* (*a*) and GUT method *gut5* (*b*). BPSK, 8-QAM and circular Gaussian source signals are clearly discernible on the left, middle and right plots.

second-order non-circular. The above random communications signals are symmetric, and hence any pair of scatter and spatial pseudo-scatter matrices can be employed in the definition of the GUT matrix and DOGMA matrix. The array outputs are corrupted by additive noise $\mathbf{n}$ with a circular complex Gaussian distribution with covariance matrix $\mathcal{C}(\mathbf{n}) = \sigma_n^2\mathbf{I}$. The signal to noise ratio (SNR) is $10\log_{10}(\sigma_s^2/\sigma_n^2) = 20\,\mathrm{dB}$ and the number of snapshots is $n = 2000$. The estimated source signal constellations obtained by *jade* and *gut5* are shown in Figure 2.10. Both of the methods were able to separate the sources as BPSK, 8-QAM and circular Gaussian source signals are clearly discernible. Table 2.2. shows the average performance of all the ICA estimators over 100 simulation runs. *jade* and *kmg* are performing the best but GUT methods are not far behind. *fobi* and *d1*, however, stand out as they do not quite reach the same level of performance as the others.

Table 2.2 Mean values of $-10\log_{10}(\mathrm{PI}(\widehat{\mathrm{G}}))$ computed from 100 array snapshot data sets

	gut1	gut2	gut3	gut4	gut5	Fica	Jade	Cfpa	Kmg	d1	Fobi
Without outliers	17.5	15.2	16.7	16.7	16.9	16.5	18.7	16.8	18.4	12.2	12.7
With outliers	8.5	2.7	11.4	16.6	16.8	3.6	2.5	2.6	2.4	12.2	2.6

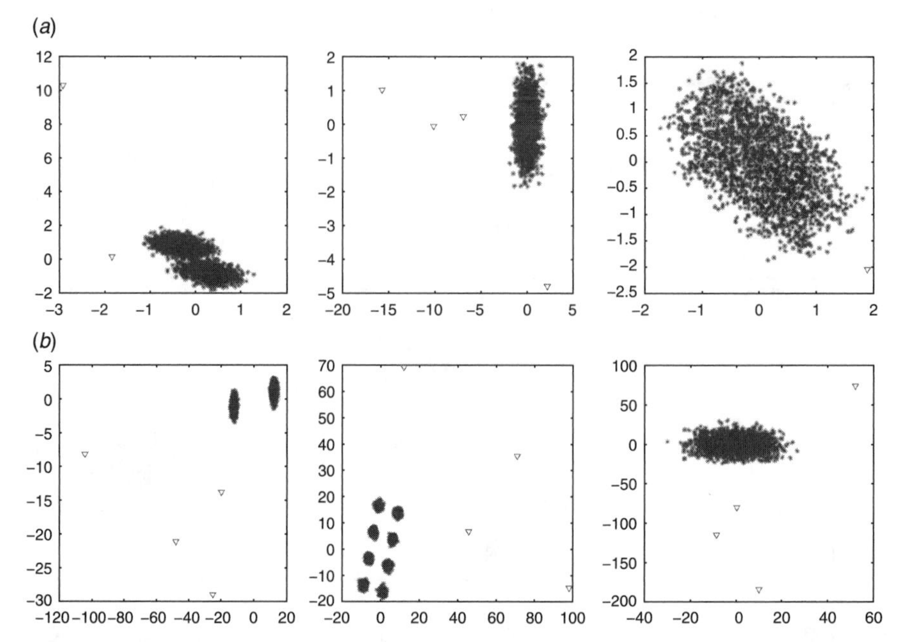

Figure 2.11 Estimated source signal constellations obtained by *jade* (*a*) and GUT method *gut5* (*b*) for the data with four outliers. Signals that correspond to outliers are marked by triangle. Robust GUT method *gut5* is unaffected by outliers, which are clearly detected in the plots whereas *jade* fails completely: outliers are able to destroy the patterns of BPSK and 8-QAM signals.

To illustrate the reliable performance of the robut GUT matrices under contamination, four observations were replaced by an outlier $\mathbf{z}_{\text{out}}$ generated as

$$\mathbf{z}_{\text{out}} = (b_1\ u_1\ z_{1,\max}, \ldots, b_4 u_4 z_{4,\max})^T$$

where $z_{i,\max}$ is the element of the ith row of the sample data matrix $Z_n = (\mathbf{z}_1 \ldots \mathbf{z}_n) \in \mathbb{C}^{k\times n}$ with largest modulus, $u_i \sim$ Unif(1,5) and b_i is a rva with a value of -1 or 1 with equal probability $\frac{1}{2}$. Note that $z_{i,\text{ out}} = b_i\ u_i\ z_{i,\max}$ points to the same or opposite direction as $z_{i,\max}$ but its magnitude is at least as big, but at most 5 times larger than that of $z_{i,\max}$. Note that only 0.2 percent (4 observations out of $n = 2000$) of the data is contaminated. Figure 2.11 depicts the estimated source signal constellations obtained by *jade* and *gut5*. As can be seen, only *gut5* is able to separate the sources and is unaffected by outliers, which are clearly detected in the plots. *jade* on the other hand fails completely: BPSK and 8-QAM sources are no longer discernible in the plots. The reliable performance of the robust ICA methods is evident in Table 2.2.: only the robust GUT methods *gut4* and *gut5* and the robust DOGMA estimator *d1* are able to separate the sources.

Table 2.3 gives average computation times. As can be seen *gut1* (i.e., the SUT) and *fobi* are the fastest to compute whereas *fica* is the slowest. To compute *fobi* we used the

Table 2.3 Average computation times (in milliseconds) over 100 runs for array snapshot data sets without outliers and with four outliers generated as $\mathbf{z}_{out}$

	gut1	gut2	gut3	gut4	gut5	Fica	Jade	Cfpa	Kmg	d1	Fobi
Without outliers	1.15	2.04	2.02	8.70	8.76	25.73	5.87	9.08	9.98	14.77	1.71
With outliers	1.19	2.16	2.12	10.13	10.49	59.52	6.36	17.30	17.61	17.82	1.77

fast algorithm of [49]. Also observe that the occurrence of outliers severely increases the computation times of the iterative fixed-point algorithms *fica*, *kmg* and *cfpa*, whereas computation times for the other methods are only slightly affected by outliers.

2.9 CONCLUSION

In this chapter we focused on multichannel signal processing of complex-valued signals in cases where the underlying ideal assumptions on signal and noise models do not necessarily hold. We considered departures from two key assumptions, that is circularity of the signal and/or noise as well as Gaussianity of the noise distribution. A comprehensive description of the full second-order statistics of complex random vectors was provided since the conventional covariance matrix alone suffices only in the case of circular signals. A detection scheme for noncircularity was developed. This facilitates using the full second-order statistics of the signals and appropriate algorithms in the presence of noncircularity. Moreover, estimators and multichannel signal processing algorithms that also take into account the noncircularity of the signals and are robust in the face of heavy-tailed noise were introduced. Their robustness and efficiency were analyzed. Example applications in beamforming, subspace-based direction finding and blind signal separation were presented.

2.10 PROBLEMS

2.1 Show that circularity coefficient $\lambda(z)$ satisfies (2.3) and (2.4).

2.2 Based on the definition q_n of the GLRT decision statistic, derive the equation (2.13) for l_n and verify using arguments based on the properties of eigenvalues that the test statistic l_n is invariant under invertible linear transformations of the data.

2.3 MATLAB assignment. Let us have a uniform linear array of eight elements. Two QPSK modulated signals are impinging on the array from directions of arrival 72° and 66°. Our $N = 200$ observations are contaminated by complex white second order circular Gaussian noise such that the signal to noise ratio (SNR) is 10 dB. Plot the MUSIC pseudo-spectrum. Study the traces of array covariance and pseudo-covariance matrices. What can you say about the circularity of the observations?

2.4 MATLAB assignment. Let us have similar array configuration as above. Now two BPSK modulated signals are impinging the array. Our $N = 200$ observations are contaminated by complex white second order circular Gaussian noise such that the signal to noise ratio (SNR) is 10 dB. Plot the MUSIC pseudo spectrum. Study the traces of array covariance and pseudo-covariance matrices. What can you say about the circularity of the observations?

2.5 MATLAB assignment. Consider again a uniform linear array of eight elements. Two QPSK modulated signals are impinging on the array from directions of arrival 72° and 66°. Now we have an ε-contaminated noise model obeying mixture of two complex second order circular Gaussian distributions $f(\mathbf{n}) = (1 - \varepsilon) f(0, \sigma^2 \mathbf{I}) + \varepsilon f(0, 50\sigma 2\ I)$. With $\varepsilon = 0.1$ we have 10 percent outliers present in the sample. The signal to noise ratio would be 10 dB in the absence of outliers. Plot the MUSIC pseudo-spectrum using sample covariance matrix based estimator and sign covariance matrix based estimator. What can you say about the robustness of the estimators?

2.6 MATLAB assignment. Write a function called `glrtcirc(Z)` for the GLRT test statistic $-n \ln l_n$ of circularity, where the argument `Z` of the function is a $k \times n$ snapshot data matrix. After doing so,

a) generate $k \times n$ data matrix $\mathbf{Z}$ consisting of n indenpendent random samples from circular $T_{k,\upsilon}$ using result (2.14). Transform the data by $\mathbf{Z} \rightarrow \mathbf{S} = \mathbf{GZ}$, where $\mathbf{G}$ can be any nonsingular $k \times k$ matrix. Then verify that computed values of test statistics `glrtcirc(Z)` and `glrtcirc(S)` coincide, meaning that GLRT test statistic is invariant under invertible linear data transformations.

b) Write a function for the adjusted GLRT test statistic $-l_n/(1 + \hat{\kappa}/2)$ of circularity and reproduce the chi-square plots of Figure 2.2.

2.7 Show that the kurtosis matrix $\mathcal{C}_{\text{kur}}(\cdot)$ and the pseudo-kurtosis matrix $\mathcal{P}_{\text{kur}}(\cdot)$ defined in (2.21) possess IC-property. If sources are symmetric, that is, $s_i = d - s_i$, for $i = 1, \ldots, d$, then show that any scatter matrix $\mathbf{C}(\cdot)$ or pseudo-scatter matrix $\mathbf{P}(\cdot)$ possess IC-property.

REFERENCES

1. H. Abeida and J.-P. Delmas, MUSIC-like estimation of direction of arrival for noncircular sources. *IEEE Trans. Signal Processing*, 54(7), (2006).
2. S. I. Amari, A. Cichocki, and H. H. Yang, "A new learning algorithm for blind source separation," in D. S. Touretzky, M. C. Mozer, and M. E. Hasselmo, Eds., *Advances in Neural Information Processing Systems 8*, pages 757–763. MIT Press, Cambridge, MA, (1996).
3. J. Anemüller, T. J. Sejnowski, and S. Makeig, Complex independent component analysis of frequency-domain electroencephalographic data. *Neural Networks*, 16: 1311–1323, (2003).
4. P. Billingsley, *Probability and Measure* (3rd ed). Wiley, New York, 1995.

5. E. Bingham and A. Hyvarinen, A fast fixed-point algorithm for independent component analysis of complex-valued signals. *Int. J. of Neural Systems*, 10(1): 1–8, (2000).
6. D. H. Brandwood, A complex gradient operator and its applications in adaptive array theory. *IEE Proc. F and H*, 1: 11–16, (1983).
7. J. Capon, High resolution frequency-wavenumber spectral analysis. *Proceedings of the IEEE*, 57(8): 1408–1418, (1969).
8. J. F. Cardoso, "Source separation using higher order moments," in *Proc. IEEE Int. Conf. on Acoustics, Speech and Signal Processing (ICASSP'89)*, pages 2109–2112, Glasgow, UK, 1989.
9. J. F. Cardoso, High-order contrasts for independent component analysis. *Neural Computation*, 11(1): 157–192, (1999).
10. J.-F. Cardoso and B. H. Laheld, Equivariant adaptive source separation. *IEEE Trans. Signal Processing*, 44(12): 3017–3030, (1996).
11. J. F, Cardoso and A. Souloumiac, Blind beamforming for non-gaussian signals. *IEE Proceedings-F*, 140(6): 362–370, (1993).
12. B. D. Carlson, Covariance matrix estimation errors and diagonal loading in adaptive arrays. *IEEE Trans. Aerosp. Electron. Syst.*, 24(4): 397–401, (1988).
13. P. Chargé, Y. Wang, and J. Saillard, A non-circular sources direction finding methods using polynomial rooting. *Signal Processing*, 81: 1765–1770, (2001).
14. A. Cichocki and S-I. Amari, *Adaptive Blind Signal and Image Processing*. John Wiley, New York, 2002.
15. P. Comon, Independent component analysis—a new concept? *Signal Processing*, 36: 287–314, (1994).
16. H. Cox, R. M. Zeskind, and M. M. Owen, Robust adaptive beamforming. *IEEE Trans. Acoust., Speech, Signal Processing*, 35(10): 1365–1376, (1987).
17. C. Croux, Limit behavior of the empirical influence function of the median. *Statistics & Probability Letters*, 37: 331–340, (1998).
18. L. De Lathauwer and B. De Moore, "On the blind separtion of non-circular source," in *Proc. 11th European Signal Processing Conference (EUSIPCO 2002)*, Toulouse, France, September 2002.
19. S. C. Douglas, Fixed-point algorithms for the blind separation of arbitrary complex-valued non-gaussian signal mixtures. *EURASIP J. Advances in Signal Processing*, 2007 (Article ID 36525, 15 pages), (2007).
20. Scott C. Douglas, "Fixed-point FastICA algorithms for the blind separation of complex-valued signal mixtures," in *Proc. 39th Asilomar Conf. on Signals, Systems and Computers (ACSSC'05)*, pages 1320–1325, 2005.
21. J. Eriksson and V. Koivunen, "Complex-valued ICA using second order statistics," in *Proc. IEEE Workshop on Machine Learning for Signal Processing (MLSP'04)*, Sao Luis, Brazil, 2004.
22. J. Eriksson and V. Koivunen, Complex random vectors and ICA models: Identifiability, uniqueness and seperability. *IEEE Trans. Inform. Theory*, 52(3): 1017–1029, (2006).
23. A. B. Gershman, "Robust adaptive beamforming: an overview of recent trends and advances in the field," in *Proc. 4th International Conference on Antenna Theory and Techniques*, pages 30–35, September 9–12, 2003.

24. N. R. Goodman, Statistical analysis based on certain multivariate complex Gaussian distribution (an introduction). *Annals Math. Statist.*, 34: 152–177, (1963).
25. M. Haardt and F. Römer, "Enhancements of unitary ESPRIT for non-circular sources," in *Proc. Int. Conf. Acoustics, Speech and Signal Processing (ICASSP'04)*, Montreal, Canada, May 2004.
26. F. R. Hampel, E. M. Ronchetti, P. J. Rousseeuw, and W. A. Stahel, *Robust Statistics: The Approach Based on Influence Functions*. Wiley, New York, 1986.
27. P. J. Huber, *Robust Statistics*. Wiley, New York, 1981.
28. A. Hyvärinen, J. Karhunen, and E. Oja, *Independent Component Analysis*. John Wiley, New York, 2001.
29. S. M. Kay, *Fundamentals of Statistical Signal Processing*. Prentice-Hall, New Jersey, 1993.
30. J. T. Kent, Data analysis for shapes and images. *J. Statist. Plann. Inference*, 57: 181–193, (1997).
31. J. T. Kent and D. E. Tyler, Redescending M-estimates of multivariate location and scatter. *Ann. Statist.*, 19(4): 2102–2119, (1991).
32. H. Krim and M. Viberg, Two decades of array signal processing: the parametric approach. *IEEE Signal Processing Mag.*, 13(4): 67–94, (1996).
33. P. R. Krishnaiah and J. Lin, Complex elliptically symmetric distributions. *Comm. Statist. - Th. and Meth.*, 15: 3693–3718, (1986).
34. H. Li and T. Adalı, A class of complex ICA algorithms based on the kurtosis cost function. *IEEE Trans. Neural Networks*, 19(3): 408–420, (2008).
35. J. Li, P. Stoica, and Z. Wang, On robust capon beamforming and diagonal loading. *IEEE Trans. Signal Processing*, 51(7): 1702–1715, (2003).
36. G. M. Manolakis, K. I. Vinay, and S. M. Kogon, *Statistical and adaptive signal processing*. McGraw-Hill, Singapore, 2000.
37. R. A. Maronna, Robust M-estimators of multivariate location and scatter. *Ann. Statist.*, 5(1): 51–67, (1976).
38. F. D. Neeser and J. L. Massey, Proper complex random processes with applications to information theory. *IEEE Trans. Inform. Theory*, 39(4): 1293–1302, (1993).
39. M. Novey and T. Adalı, On extending the complex FastICA algorithm to noncircular sources. *IEEE Trans. Signal Processing*, 56(5), (2008).
40. E. Ollila and V. Koivunen, "Adjusting the generalized likelihood ratio test of circularity robust to non-normality," in *Proc. 10th IEEE Int. Workshop on Signal Processing Advances in Wireless Comm. (SPAWC'09)*, pages 558–562, Perugia, Italy, June 21–24, 2009.
41. E. Ollila, On the circularity of a complex random variable. *IEEE Signal Processing Letters*, 15: 841–844, (2008).
42. E. Ollila, J. Eriksson, and V. Koivunen, Complex univariate distributions – part II: complex normal distribution and its extensions. Technical report, Signal Processing Labaratory, Helsinki Univerisity of Technology, 02150 Espoo, Finland, 2006.
43. E. Ollila and V. Koivunen, "Influence functions for array covariance matrix estimators," in *Proc. IEEE Workshop on Statistical Signal Processing (SSP'03)*, pages 445–448, St. Louis, USA, September 28– Oct. 1, 2003.

44. E. Ollila and V. Koivunen, "Robust antenna array processing using M-estimators of pseudo-covariance," in *Proc. 14th IEEE Int. Symp. on Personal, Indoor and Mobile Radio Comm. (PIMRC'03)*, pages 2659–2663, Beijing, China, September 7–10, 2003.
45. E. Ollila and V. Koivunen, "Robust space-time scatter matrix estimator for broadband antenna arrays," in *Proc. 58th IEEE Vehicular Technology Conference (VTC'03-Fall)*, volume 1, pages 55–59, Orlando, USA, October 6–9, 2003.
46. E. Ollila and V. Koivunen, "Generalized complex elliptical distributions," in *Proc. Third IEEE Sensor Array and Multichannel Signal Processing Workshop (SAM'04)*, Barcelona, Spain, June 18–21, 2004.
47. E. Ollila and V. Koivunen, Complex ICA using generalized uncorrelating transformation. *Signal Processing*, 89(4): 365–377, (2009).
48. E. Ollila and V. Koivunen, Influence function and asymptotic efficiency of scatter matrix based array processors: case MVDR beamformer. *IEEE Trans. Signal Processing*, 57(1): 247–259, (2009).
49. E. Ollila, H. Oja, and V. Koivunen, (in press) Complex-valued ICA based on a pair of generalized covariance matrices. *Computational Statistics and Data Analysis*, 52(7): 3789–3805, (2008).
50. B. Picinbono, On circularity. *IEEE Trans. Signal Processing*, 42(12): 3473–3482, (1994).
51. B. Picinbono, Second order complex random vectors and normal distributions. *IEEE Trans. Signal Processing*, 44(10): 2637–2640, (1996).
52. B. Picinbono and P. Chevalier, Widely linear estimation with complex data. *IEEE Trans. Signal Processing*, 43(8): 2030–2033, (1995).
53. P. J. Schreier and L. L. Scharf, Second-order analysis of improper complex random vectors and processes. *IEEE Trans. Signal Processing*, 51(3): 714–725, (2003).
54. P. J. Schreier, L. L. Scharf, and A. Hanssen, A generalized likelihood ratio test for impropriety of complex signals. *IEEE Signal Processing Letters*, 13(7): 433–436, (2006).
55. P. Stoica and R. Moses, *Introduction to spectral analysis*. Prentice-Hall, Upper Saddle River, 1997.
56. P. Tsakalides and C. L. Nikias, The robust covariation based MUSIC (roc-MUSIC) algorithm for bearing estimation in impulsive noise environments. *IEEE Trans. Signal Processing*, 44(7): 1623–1633, (1995).
57. A. van den Bos, The multivariate complex normal distribution–a generalization. *IEEE Trans. Inform. Theory*, 41(2): 537–539, (1995).
58. B. D. Van Veen and K. M. Buckley, Beamforming: a versatile approach to spatial filtering. *IEEE ASSP magazine*, April (1988).
59. S. Visuri, H. Oja, and V. Koivunen, Subspace-based direction of arrival estimation using nonparametric statistics. *IEEE Trans. Signal Processing*, 49(9): 2060–2073, (2001).
60. T. Wax and T. Kailath, Detection of signals by information theoretic criteria. *IEEE Trans. Acoust., Speech, Signal Processing*, 33(2): 387–392, (1985).
61. R. A. Wooding, The multivariate distribution of complex normal variables. *Biometrika*, 43: 212–215, (1956).

3

TURBO EQUALIZATION

Philip A. Regalia
Catholic University of America, Washington, DC

3.1 INTRODUCTION

The turbo decoding algorithm [1–7] ranks unquestionably among the major practical achievements in modern telecommunications, validating the theoretical error correction capabilities of realizable codes predicted from Shannon's work [8].

The essence of the algorithm is to harness information exchange between lower-level receiver functions, thus demonstrating the performance advantages achievable by combining traditionally compartmentalized functions into a bigger picture. In contrast to global optimization procedures which can prove unwieldily complex, the "turbo principle" aims instead to couple basic building blocks boasting computational efficiency into iterative algorithms that transform information exchange into progressive performance improvements. The standard recipe for turbo receivers, in which extrinsic information is extracted from one block to usurp the role held for *a priori* probabilities within another, can be given countless voicings depending on which elements are combined; choices include synchronization, equalization, and decoding in the traditional monouser paradigm, and can be expanded to user separation, key negotiation, and access control in multiuser settings.

The intent of this chapter is to relate measurable progress in a particular instance of iterative receiver design, namely turbo equalization which combines equalization and error correction decoding. This setting is sufficiently rich to illustrate the performance advantages induced by cooperative information exchange between traditionally separated receiver functions, yet sufficiently complicated to convince researchers that the

Adaptive Signal Processing: Next Generation Solutions. Edited by Tülay Adalı and Simon Haykin
Copyright © 2010 John Wiley & Sons, Inc.

gains from cooperative processing do not come automatically, but rather are contingent on a proper understanding of the information exchange mechanisms involved. Undoubtedly similar considerations will intervene in more detailed studies of other candidate iterative receiver configurations, some of which are mentioned in the closing sections of this chapter.

Our treatment first reviews the context of intersymbol interference and its relation to high-speed communication networks. We then develop in detail the various decoding algorithms for error correction codes and memory-laden channels, and more importantly how the information exchange between these operations fits into the turbo decoding paradigm. Traditional performance measures such as bit error rate and channel capacity are accommodated along the way.

3.2 CONTEXT

All modern communication systems employ error correction coding to combat the effects of channel noise and interference. An additional degradation arises when multiple paths exist between the transmitter and receiver, as each path will exhibit a different gain and time delay. Such paths occur naturally in wireless environments when multiple reflections of the transmitted signal impinge on the receiver, but also plague wireline systems in the presence of impedance mismatch. This is because the resulting nontrivial reflection component of the scattering parametrization of a transmission line likewise induces a multipath environment. When these multiple transmission paths combine at the receiver, the composite signal becomes a mixture of variously delayed copies of the transmitted signal. If the difference in delay times between these copies exceeds the symbol period, successive transmitted symbols overlap as seen from the receiver, giving intersymbol interference. For a fixed "delay spread" between such multiple trajectories, any increase in the symbol rate must translate into a decrease in the symbol period (defined as the inverse of the symbol rate), increasing the likelihood of interference between successive symbols. Accordingly, intersymbol interference becomes a formidable degradation in all communication systems aiming for high data rates, whether they be wireless or wireline [9, 10].

In the presence of intersymbol interference, the receiver contends with a sequence of received symbols of which each is some combination of past transmitted symbols. Whenever superposition applies, a combination of past symbols must assume the mathematical form of a convolution between the transmitted symbols and a channel impulse response. In this model, the k-th term (say) of the channel impulse response reflects the influence wielded on the most recent channel output by a symbol transmitted k symbol periods in the past.

If the channel impulse response has multiple nonzero terms, the channel frequency response will generally deviate from flat; a convolutive channel is thus a frequency-selective channel. Equalizers aim to restore a flat response in the frequency domain, or equivalently, an ideal impulse response in the time domain, thus removing intersymbol interference. Equalizers can take many forms.

- Linear equalizers are simply linear filters whose frequency responses approximate the inverse of the channel frequency response. Finite impulse response filters are typically used due to their guaranteed stability and ease of adaptation [11, 12].
- Optimal equalizers typically use some variant of the Viterbi algorithm [13] or the forward-backward algorithm [14–18], and aim to infer the most likely channel input based on knowledge of the channel impulse response, for the given observed channel output sequence.
- Nonlinear classifiers [19–29] set up decision boundaries in a multidimensional space spanned by successive channel outputs. They present useful alternatives when optimal equalizers present an otherwise prohibitive complexity, which will not be the case in this chapter as we restrict our attention to linear channels.

Equalization algorithms traditionally have worked independently of the channel coding and decoding phases. Turbo equalization is a refined approach in which the equalizer and decoder work in tandem, and is inspired by the success of the turbo decoding algorithm for concatenated codes. By this approach, a convolutive channel is perceived as a "parastic" encoder in the signal chain. The error correction encoder and the channel thus combine as a serially concatenated code, for which the iterative decoding algorithm for turbo codes [1–7] may be adapted directly [30]. When successful, the iterative algorithm generally gives a lower symbol error rate compared to systems that decouple these operations.

We first review the basics of iterative decoding, and then develop in detail the specific blocks required in turbo equalization.

3.3 COMMUNICATION CHAIN

The basic communication system we consider consists of an error correction encoder, an interleaver, and the communication channel, as depicted in Figure 3.1. In this initial set-up, which focuses on the single user case, the blocks are detailed as follows.

- The encoder has rate K/N, meaning it takes in a block of K bits $c_1, \ldots, c_K$ (each drawn from $\{0, 1\}$), and puts out a block of $N > K$ bits. The encoder is also assumed to be systematic, meaning that the first K output bits are copies of the K input bits; these are called the information bits. The remaining bits $c_{K+1}, \ldots, c_N$ are the parity check bits.

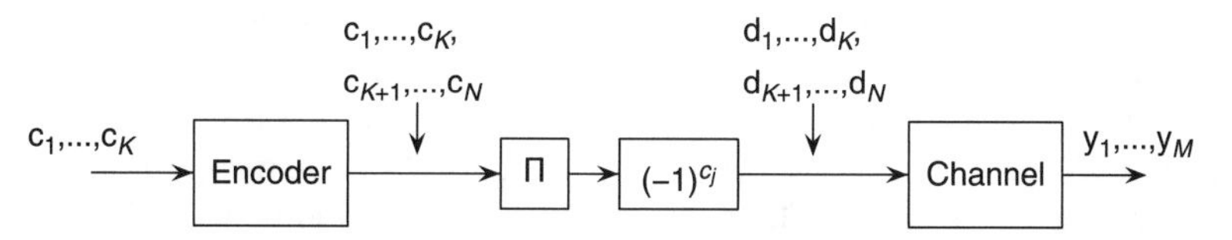

Figure 3.1 Communications chain with an error correction encoder and a noisy multipath channel. The interleaver (or permutation) is denoted Π.

- The interleaver, denoted Π, permutes the order of the N bits in the block. The index map is denoted $j = \Pi(i)$ as i runs from 1 to N. Interleavers have traditionally been used to break up bursty error patterns, but also contribute to the distance properties of concatenated codes [3].
- Conversion to antipodal form is performed as

$$d_i = (-1)^{c_j}, \quad \text{with } j = \Pi(i)$$

so that each symbol d_i is either $+1$ or -1.

- The channel induces intersymbol interference and noise according to

$$y_i = \sum_{l=0}^{L} h_l \, d_{i-l} + b_i \tag{3.1}$$

where $\{h_l\}_{l=0}^{L}$ is the channel impulse response sequence, and b_i is an additive noise term. Due to the delay spread (here, the support length $L+1$ of the channel impulse response $\{h_i\}$), the output sequence will extend over $M = N + L$ samples.

The decoding problem is to deduce estimates, denoted $\hat{c}_1, \ldots, \hat{c}_N$, of the transmitted bits $c_1, \ldots, c_N$, given the received signal values $y_1, \ldots, y_M$. Let $\mathbf{c} = (c_1, \ldots, c_N)$ be a vector collecting the N transmitted bits, and likewise for $\hat{\mathbf{c}} = (\hat{c}_1, \ldots, \hat{c}_N)$ and $\mathbf{y} = (y_1, \ldots, y_M)$. Optimal decoding relies on the *a posteriori* probability mass function $\Pr(\mathbf{c}|\mathbf{y})$, opting traditionally for either of the following two criteria.

- The wordwise optimal configuration is the code word $\hat{\mathbf{c}}$ which maximizes the *a posteriori* probability mass function

$$\hat{\mathbf{c}} = \arg\max_{\mathbf{c}} \mathbf{Pr}(\mathbf{c}|\mathbf{y}).$$

The maximization is over the 2^K candidate configurations of the code word vector $\mathbf{c}$.

- The bitwise optimal solution is obtained by first calculating the marginal probability functions

$$\Pr(c_i|\mathbf{y}) = \sum_{c_j, j \neq i} \Pr(\mathbf{c}|\mathbf{y}), \quad i = 1, 2, \ldots, N$$

and opting for $\hat{c}_i = 0$ if $\Pr(c_i = 0|\mathbf{y}) > \Pr(c_i = 1|\mathbf{y})$, and $\hat{c}_i = 1$ otherwise.

In either case, the number of evaluations of the *a posteriori* probability mass function $\Pr(\mathbf{c}|\mathbf{y})$ is 2^K, because $\mathbf{c}$ is constructed from K information bits. As such, comparing these evaluations and/or marginalizing them presents a complexity that grows exponentially with the block length. For this reason, a direct evaluation of these probabilities is usually not feasible. The complexity problem is traditionally addressed by restricting the encoder and channel to have a convolutional structure, as reviewed in Section 3.5. First, however, we review some standard manipulations of the *a posteriori* probability mass function.

Using Bayes's rule, we may develop $\Pr(\mathbf{c}|\mathbf{y})$ as[1]

$$\Pr(\mathbf{c}|\mathbf{y}) = \Pr(\mathbf{c})\,\frac{\Pr(\mathbf{y}|\mathbf{c})}{\Pr(\mathbf{y})}$$

in which the following applies.

- The denominator term $\Pr(\mathbf{y})$ is the joint probability distribution for the channel outputs $\mathbf{y} = (y_1, \ldots, y_M)$, constituting the *evidence*, evaluated for the received data. As it contributes the same factor to each evaluation of $\Pr(\mathbf{c}|\mathbf{y})$ (interpreted as a function of $\mathbf{c}$ for *fixed* $y_1, \ldots, y_M$), it may be omitted with no effect on either the wordwise or bitwise estimator.
- The numerator term $\Pr(\mathbf{y}|\mathbf{c})$ is the channel likelihood function, taking as many evaluations as there are configurations of $\mathbf{c}$ (namely, 2^K); the vector $\mathbf{y}$ is fixed at the received channel output values.
- The leading term $\Pr(\mathbf{c})$ is the *a priori* probability mass function for configuration $\mathbf{c}$. If this probability is uniform, that is, $\Pr(\mathbf{c}) = 1/2^K$ for each configuration of $\mathbf{c}$, then maximizing the *a posteriori* probability $\Pr(\mathbf{c}|\mathbf{y})$ versus $\mathbf{c}$ reduces to maximizing the likelihood function $\Pr(\mathbf{y}|\mathbf{c})$ versus $\mathbf{c}$.

3.4 TURBO DECODER: OVERVIEW

By interpreting the channel as a 'parasitic' convolutional encoder, the communication chain assumes a serially concatenated structure, allowing direct application of the turbo decoding algorithm. In the absence of an interleaver, the cascade of the encoder and channel would appear as a convolution, and the Viterbi algorithm could be applied to perform optimal decoding. An interleaver is usually included, however, in order to break up bursty error patterns. It also destroys the convolutional structure of the overall code. Iterative decoding splits the task into two local operations, comprising an equalizer which estimates the channel input given its output, and the the decoder which estimates the code word from the channel input. The 'turbo effect' is triggered by coupling the information between these two local operators.

The equalizer aims to calculate the marginal probabilities

$$\begin{aligned}\Pr(d_i|\mathbf{y}) &= \sum_{d_j, j\neq i} \Pr(\mathbf{d}|\mathbf{y}) \\ &= \sum_{d_j, j\neq i} \Pr(\mathbf{d})\,\frac{\Pr(\mathbf{y}|\mathbf{d})}{\Pr(\mathbf{y})} \\ &\propto \sum_{d_j, j\neq i} \Pr(\mathbf{d})\Pr(\mathbf{y}|\mathbf{d}), \qquad i = 1, 2, \ldots, N,\end{aligned} \tag{3.2}$$

[1]For notational simplicity, we use $\Pr(\cdot)$ to denote a probability mass function if the underlying distribution is discrete (like $\mathbf{c}$), or a probability density function if the underlying distribution is continuous (like $\mathbf{y}$).

evaluated for the received vector $\mathbf{y}$. Note that we drop the term $\Pr(\mathbf{y})$ in the final line since it does not vary with our hypothesis for $\mathbf{d}$. The computational complexity of calculating these marginals would appear to be $\mathcal{O}(2^N)$, but given that $\mathbf{y}$ is obtained from $\mathbf{d}$ via a convolution, the complexity will decrease to a number linear in n if we use trellis decoding, to be illustrated in Section 3.5. As it turns out, the complexity reduction will be successful provided the *a priori* probability mass function $\Pr(\mathbf{d})$ factors into the product of its marginals

$$\Pr(\mathbf{d}) = \prod_{i=1}^{N} \Pr(d_i).$$

This factorization is, strictly speaking, incorrect, since the variables d_i are derived by interleaving the output from an error correction code, which imparts redundancy among its elements for error control purposes. This factorization, by contrast, treats the d_i as *a priori* independent variables. Nonetheless, if we turn a blind eye to this shortcoming in the name of efficiency, the term $\Pr(\mathbf{d})$ will contribute a factor $\Pr(d_i)$ to each term of the sum in (3.2), so that we may rewrite the *a posteriori* probabilities as

$$\Pr(d_i|\mathbf{y}) \propto \Pr(d_i) \underbrace{\sum_{d_j, j\neq i} \Pr(\mathbf{y}|\mathbf{d}) \prod_{l\neq i} \Pr(d_l)}_{\text{extrinsic probability}}, \qquad i = 1, 2, \ldots, N.$$

This extrinsic probability for symbol i is so named because it is seen to depend on symbols other than d_i [although d_i still enters in via the likelihood function $\Pr(\mathbf{y}|\mathbf{d})$], in contrast to the first term $\Pr(d_i)$ which depends only on d_i. As this variable will appear frequently, we denote it as

$$T_i(d_i) \stackrel{\delta}{=} \sum_{d_j, j\neq i} \Pr(\mathbf{y}|\mathbf{d}) \prod_{l\neq i} \Pr(d_l), \qquad i = 1, 2, \ldots, N$$

where the scale factor δ is chosen so that evaluations sum to one: $T_i(-1) + T_i(1) = 1$. Observe that, due to the summing operation on the right-hand side, the function $T_i(\cdot)$ behaves as a type of marginal probability, dependent on the sole bit d_i.

Now, the outer decoder aims to infer the bits contained in $\mathbf{c}$ from the symbols contained in $\mathbf{d}$, according to

$$\begin{aligned}\Pr(c_j|\mathbf{d}) &= \sum_{c_i, i\neq j} \Pr(\mathbf{c}|\mathbf{d}) \\ &\propto \sum_{c_i, i\neq j} \Pr(\mathbf{d}|\mathbf{c}) \Pr(\mathbf{c}).\end{aligned}$$

If the information bits $c_1, \ldots, c_K$ are each equiprobable, then the *a priori* probability function $\Pr(\mathbf{c}) = \Pr(c_1, \ldots, c_K, c_{K+1}, \ldots, c_N)$ behaves as a scaled indicator function for the outer code

$$\Pr(\mathbf{c}) = \begin{cases} 1/2^K, & \text{if } \mathbf{c} = (c_1, \ldots, c_K, c_{K+1}, \ldots, c_N) \text{ is a code word;} \\ 0, & \text{otherwise.} \end{cases}$$

Since the scale factor $1/2^K$ will contribute a common factor to all terms in the calculations to follow, we may remove it, thus obtaining the indicator function $\phi(\mathbf{c})$ for the code

$$\phi(\mathbf{c}) = \begin{cases} 1, & \text{if } \mathbf{c} \text{ is a code word;} \\ 0, & \text{otherwise.} \end{cases}$$

The likelihood function $\Pr(\mathbf{d}|\mathbf{c})$ is in principle a deterministic function that factors are

$$\Pr(\mathbf{d}|\mathbf{c}) = \prod_{i=1}^{N} \Pr(d_i|c_j), \qquad \text{with } j = \Pi(i)$$

since d_i is determined from c_j alone, with $j = \Pi(i)$ the index map from the interleaver. The values in $\mathbf{d}$ are not known with certainty; however, only estimates of them are available from the inner decoder. Many techniques may be used to inject these estimates; the most common is to inject the extrinsic probabilities introduced above. Specifically, first de-interleave the values $\{T_i(d_i)\}$ and associate them with the bits (c_j) according to

$$\begin{aligned} T_j(c_j = 0) &= T_i(d_i = +1) \\ T_j(c_j = 1) &= T_i(d_i = -1) \qquad \text{with } j = \Pi(i). \end{aligned}$$

Observe that since each bit d_i is determined from a sole bit c_j, we may equally regard $T_i(d_i)$ as a function of c_j rather than d_i, and reindex the values according to the interleaver $j = \Pi(i)$. (The use of T_j and T_i can, admittedly, be confusing at first sight, although we shall always include the argument c_j or d_i to distinguish interleaved from de-interleaved values.) We may then usurp the channel likelihood values as

$$\Pr(d_i|c_j) \longleftarrow T_j(c_j).$$

This then gives, for the marginal calculations

$$\begin{aligned} \Pr(c_j|\mathbf{d}) \longleftarrow & \sum_{c_i, i \neq j} \phi(\mathbf{c}) \prod_{l=1}^{N} T_l(c_l) \\ &= T_j(c_j) \underbrace{\sum_{c_i, i \neq j} \phi(\mathbf{c}) \prod_{l \neq i} T_l(c_l)}_{\text{extrinsic probability}} \end{aligned}$$

in which we again expose an extrinsic probability, obtained by factoring out the term $T_j(c_j)$ from each addend in the j-th marginal sum. This extrinsic probability again appears frequently, so we denote it as

$$U_j(c_j) \triangleq \sum_{c_i, i \neq j} \phi(\mathbf{c}) \prod_{l \neq j} T_l(c_l), \quad j = 1, 2, \ldots, N$$

in which the scale factor δ is chosen to ensure that evaluations sum to one: $U_j(0) + U_j(1) = 1$. Note again that, due to the summation on the right-hand side, the resulting expression is a function of the sole bit c_j. Although we are ultimately interested in these marginal probabilities for the information bits $c_1, \ldots, c_K$, the above equations can be applied to the parity check bits $c_{K+1}, \ldots, c_N$ as well.

The loop is now closed by interleaving the extrinsic probabilities from the outer decoder according to

$$\begin{aligned} U_i(d_i = +1) &= U_j(c_j = 0) \\ U_i(d_i = -1) &= U_j(c_j = 1) \qquad \text{with } j = \Pi(i) \end{aligned}$$

and then replacing the *a priori* probabilities $\Pr(d_i)$ used in the inner decoder

$$\Pr(d_i) \longleftarrow U_i(d_i).$$

The inner decoder then begins its calculations anew. A basic flowgraph illustrating the information exchange between the two decoders is sketched in Figure 3.2, in which a superscript (m) denotes an iteration index. The algorithm is summarized as follows.

1. Initialize the "pseudo priors" $U_i^{(0)}(d_i) = \frac{1}{2}$ for all d_i, and collect the channel output measurements $\mathbf{y}$. Run the following steps for iterations $m = 0, 1, 2, \ldots$.
2. Calculate the "pseudo posteriors" from the inner encoder as

$$[\Pr(d_i|\mathbf{y})]^{(m)} \propto U_i^{(m)}(d_i) \underbrace{\sum_{d_j, j \neq i} \Pr(\mathbf{y}|\mathbf{d}) \prod_{l \neq i} U_l^{(m)}(d_l)}_{\propto T_i^{(m)}(d_i)}$$

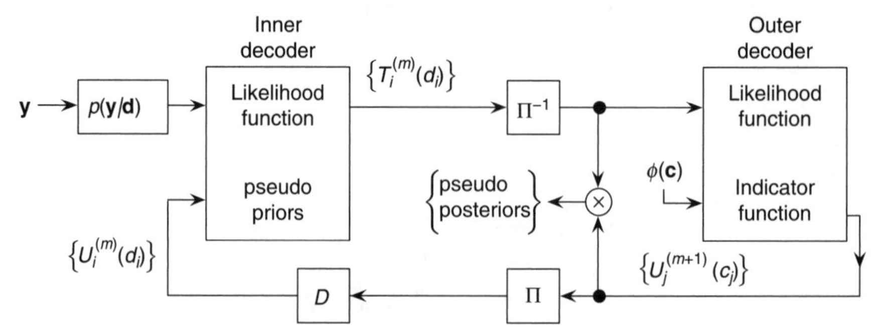

Figure 3.2 Showing the information exchange between the two decoders, and the formation of pseudo posteriors.

in which the resulting $T_i^{(m)}(d_i)$ is scaled so that evaluations sum to one: $T_i^{(m)}(-1) + T_i^{(m)}(+1) = 1$ for each i. (The term pseudo posterior is employed since the pseudo priors $U_i^{(m)}(d_i)$ are not generally the true *a priori* probabilities.)

3. De-interleave the extrinsic probabilities from the inner decoder

$$T_j^{(m)}(c_j = 0) = T_i^{(m)}(d_i = +1),$$
$$T_j^{(m)}(c_j = 1) = T_i^{(m)}(d_i = -1), \quad \text{with } j = \Pi(i)$$

for $i = 1, 2, \ldots, N$, which variables will replace the channel likelihood values in the outer decoder.

4. Calculate the pseudo-posteriors from the outer encoder as

$$[\Pr(c_j|\mathbf{d})]^{(m)} \propto T_j^{(m)}(c_j) \underbrace{\sum_{c_i, i \neq j} \phi(\mathbf{c}) \prod_{l \neq i} T_l^{(m)}(c_l)}_{\propto U_j^{(m+1)}(c_j)}$$

in which $U_j^{(m+1)}(c_j)$ is scaled so that evaluations sum to one

$$U_j^{(m+1)}(0) + U_j^{(m+1)}(1) = 1 \qquad \text{for all } j.$$

The pseudo posteriors $\Pr(c_j|\mathbf{d})$ should likewise be scaled such that

$$\Pr(c_j = 0|\mathbf{d}) + \Pr(c_j = 1|\mathbf{d}) = 1, \qquad \text{for all } j.$$

5. Interleave the extrinsic probabilities from the outer encoder as

$$U_i^{(m+1)}(d_i = +1) = U_j^{(m+1)}(c_j = 0),$$
$$U_i^{(m+1)}(d_i = -1) = U_j^{(m+1)}(c_j = 1), \quad \text{with } j = \Pi(i)$$

for $i = 1, 2, \ldots, N$, which variables will replace the priors in the inner decoder. Increment the iteration counter to $m + 1$, and return to step 2.

The algorithm iterates until convergence is observed, or a maximum number of iterations is reached. Before detailing the inner workings of steps 2 and 4 in Section 3.5, we recall some basic properties applicable to this scheme.

3.4.1 Basic Properties of Iterative Decoding

Observe first from steps 2 and 4 in the previous section, that the pseudo posterior marginals from either decoder are

$$[\Pr(d_i|\mathbf{y})]^{(m)} \propto U_i^{(m)}(d_i)\, T_i^{(m)}(d_i)$$
$$[\Pr(c_j|\mathbf{d})]^{(m)} \propto U_j^{(m+1)}(c_j)\, T_j^{(m)}(c_j), \quad \text{with } j = \Pi(i) \text{ and } d_i = (-1)^{c_j}$$

Now, a stationary point corresponds to no change in the variables from one iteration to the next, giving $U_j^{(m+1)}(c_j) = U_j^{(m)}(c_j)$ for all j. By inspection, this corresponds to the two decoders producing the same pseudo posterior values. Conversely, if the pseudo posteriors from the two decoders agree, that is, $[\Pr(d_i = 1|\mathbf{y})]^{(m)} = [\Pr(c_j = 0|\mathbf{d})]^{(m)}$ with $j = \Pi(i)$, then we must have $U_j^{(m+1)}(c_j) = U_j^{(m)}(c_j)$, giving a stationary point of the iterative procedure. We summarize as [4]:

Property 1 (Consensus property) *A stationary point of the iterative decoding algorithm occurs if and only if the two decoders reach consensus on the pseudo posteriors.*

Property 2 *The iterative decoding algorithm so described always admits a stationary point.*

A proof is developed in [31, 32], and sketched here. A classic result of nonlinear mathematics, known as the Brouwer fixed point theorem [33], asserts that any continuous function from a closed, bounded and convex set into itself admits a fixed point. To apply this result in our present context, consider the variables $U_j^{(m)}(c_j)$ for $i = 1, \ldots, N$, at a given iteration m. As these variables are pseudo probabilities, their values are restricted to the hypercube

$$0 \leq U_j^{(m)}(c_j) \leq 1, \quad j = 1, 2, \ldots, N.$$

This hypercube is clearly a closed, bounded, and convex set. As the updated values $U_j^{(m+1)}(c_j)$ at iteration $m+1$ remain in this hypercube, the iterative decoding procedure may be viewed as a function that maps a closed, bounded, and convex set into itself. Upon showing that this function is continuous [31], the conditions of the Brouwer fixed point theorem are satisfied, to establish the existence of a fixed point.

Convergence to a fixed point, conversely, is more difficult to establish, and is taken up in Section 3.9. We present first, however, the detailed workings of the forward-backward algorithm for calculating marginals.

3.5 FORWARD-BACKWARD ALGORITHM

The forward-backward algorithm is an efficient procedure to calculate marginal probabilities when the underlying observables are produced from a Markov chain. Its origins in the communications literature are usually credited to [14] in the context of channel equalization, and [17] in the context of error correction decoding, although the same basic algorithm may be recognized under various guises [15, 16, 34, 35]. For clarity we illustrate the algorithm for a specific encoder and channel, with the understanding that other choices of encoder and channel are readily accommodated once the basics are established.

Consider the "(5,7)" recursive systematic encoder sketched in Figure 3.3. (The label "(5,7)" applies since the connection branches below and above the delay line are 1 0 1 and 1 1 1, respectively, where 1 denotes a connection branch and 0 its

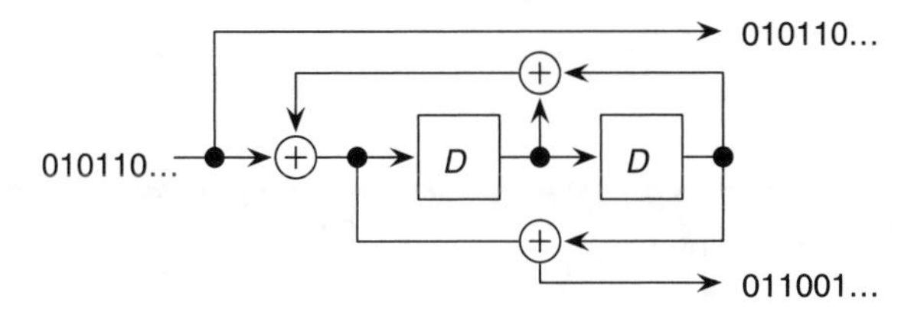

Figure 3.3 A recursive systematic "(5,7)" encoder of rate 1/2.

absence; these are the binary representations of 5 and 7, respectively.) The encoder has rate 1/2 since, at each sample instant, the system takes one bit in and produces two output bits. The encoder has the deliberate property that one of the output bits is a copy of the input bit; the encoder is thus systematic [7, 9]. The internal registers comprise the encoder's state, which, according to the two bits stored, has four configurations: 00, 01, 10, or 11. Letting x_{0j} and x_{1j} denote the binary variables stored in the state registers at time j, the system admits the state-space description

$$\begin{bmatrix} x_{0j} \\ x_{1j} \end{bmatrix} = \begin{bmatrix} 1 & 1 \\ 1 & 0 \end{bmatrix} \begin{bmatrix} x_{0,j-1} \\ x_{1,j-1} \end{bmatrix} \oplus \begin{bmatrix} 1 \\ 0 \end{bmatrix} u_j$$

$$\begin{bmatrix} c_j \\ c_{j+K} \end{bmatrix} = \begin{bmatrix} 0 & 0 \\ 1 & 0 \end{bmatrix} \begin{bmatrix} x_{0,j-1} \\ x_{1,j-1} \end{bmatrix} \oplus \begin{bmatrix} 1 \\ 1 \end{bmatrix} u_j$$

in which addition is performed modulo 2

$$\begin{aligned} 0 \oplus 0 &= 0 \\ 0 \oplus 1 &= 1 \\ 1 \oplus 0 &= 1 \\ 1 \oplus 1 &= 0. \end{aligned}$$

Observe that, since the encoder is systematic, we have $c_j = u_j$; for non-systematic encoders, this relation would not hold. Each input sequence $u_1, u_2, \ldots, u_k$ generates a state sequence and an output sequence, which is captured by a trellis diagram, as sketched in Figure 3.4. Each path through the trellis is associated with a codeword $\mathbf{c} = [c_1, \ldots, c_{2k}]$, obtained by reading off the input and output bits from each branch of the path. The set of all codewords will be denoted $\mathcal{C}$.

■ EXAMPLE 3.1

Path Tracing. Figure 3.5 shows a path forged by a particular input sequence, using $K = 6$ input bits; each branch is labeled with the corresponding two bits produced, as c_j/c_{j+K}. For the particular path forged through the trellis, the codeword

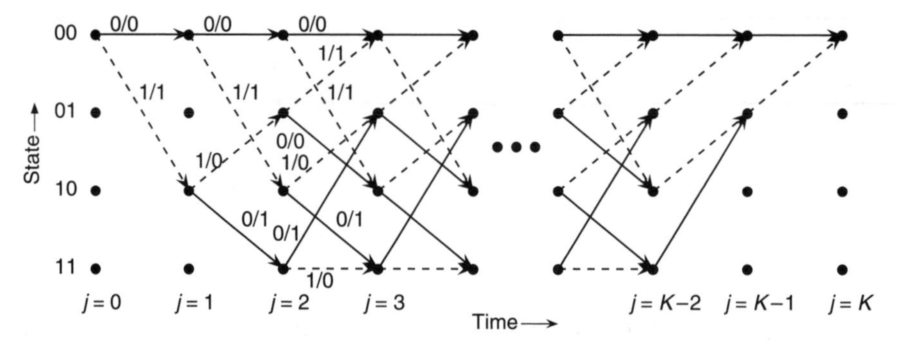

Figure 3.4 Trellis diagram for the encoder of Figure 3.3, which captures state transitions and outputs. Each path through the trellis corresponds to a code word. Transitions as solid lines are provoked by an input bit of 0, while transitions as dashed lines are provoked by an input bit of 1.

bits may be read off as

$$\begin{array}{cccccc} c_1 = 1 & c_2 = 1 & c_3 = 0 & c_4 = 0 & c_5 = 0 & c_6 = 1 \\ c_7 = 1 & c_8 = 0 & c_9 = 0 & c_{10} = 1 & c_{11} = 1 & c_{12} = 1. \end{array}$$

Observe that the final state returns to the same configuration as the initial state (here, the all-zero state). The final two input bits may always be chosen to force the final state $\mathbf{x}_K$ to the all-zero state, if desired. This can be exploited in the decoding algorithm to be reviewed presently.

We consider first the decoder operation using a channel that merely adds background noise, but is without intersymbol interference for the time being. In this simplified setting, the sequence of output bits is converted to an antipodal form and immersed in background noise, according to

$$v_j = (-1)^{c_j} + b_j, \quad j = 1, 2, \ldots, 2K \tag{3.3}$$

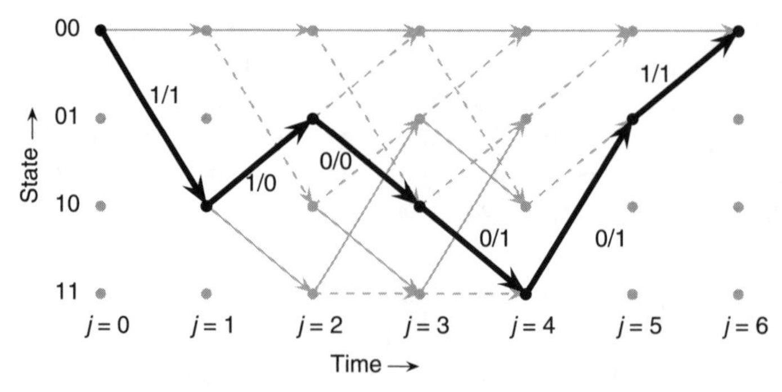

Figure 3.5 Illustrating a particular path through the trellis, using an input block of length 6.

in which v_j denotes the channel output at this stage rather than y_j, to avoid confusion with Section 3.5.1 treating the case involving intersymbol interference. Optimal decoding is accomplished with the aid of the *a posteriori* probability mass function

$$\Pr(\mathbf{c}|\mathbf{v}) = \frac{\Pr(\mathbf{c})\Pr(\mathbf{v}|\mathbf{c})}{\Pr(\mathbf{v})}$$

which is evaluated for the particular $\mathbf{v} = [v_1, \ldots, v_{2K}]$ received, as $\mathbf{c}$ varies among the set of code words $\mathcal{C}$ from the encoder. The bitwise optimal solution is obtained by thresholding the marginal evaluations from this function

$$\sum_{\substack{\mathbf{c}\in C\\ c_j=1}} \Pr(\mathbf{c}|\mathbf{v}) \gtrless \sum_{\substack{\mathbf{c}\in C\\ c_j=0}} \Pr(\mathbf{c}|\mathbf{v}), \quad j = 1, 2, \ldots, 2K$$

and choosing $\hat{c}_j = 1$ if the left-hand side is greater than the right-hand side, and $\hat{c}_j = 0$ otherwise.

Now, at each time step j, the encoder produces two bits, c_j and c_{j+K}, which we shall combine into a quaternary symbol $C_j = [c_j \; c_{j+K}]$ taking four configurations. Similarly, the channel outputs v_j and v_{j+K} [cf. (3.3)] may be combined into a composite $V_j = [v_j \; v_{j+K}]$ conveying information on the encoder output at time j.

The trellis description shows that at any time instant j, the internal state of the encoder assumes one of four configurations, *viz.*

$$\underbrace{\mathbf{x}_j = \begin{bmatrix}0\\0\end{bmatrix}}_{S_0} \quad \text{or} \quad \underbrace{\mathbf{x}_j = \begin{bmatrix}0\\1\end{bmatrix}}_{S_1} \quad \text{or} \quad \underbrace{\mathbf{x}_j = \begin{bmatrix}1\\0\end{bmatrix}}_{S_2} \quad \text{or} \quad \underbrace{\mathbf{x}_j = \begin{bmatrix}1\\1\end{bmatrix}}_{S_3}$$

in which $S_0, \ldots, S_3$ denote the state configurations so indicated. The *a posteriori* probabilities $\Pr(c_j|\mathbf{v})$ may then be related to the state transition probabilities $\Pr(\mathbf{x}_{j-1} = S_{m'}, \mathbf{x}_j = S_m|\mathbf{v})$. As an example

$$\begin{aligned}\Pr(c_j = 0|\mathbf{v}) = {} & \Pr(\mathbf{x}_{j-1} = S_0, \mathbf{x}_j = S_0|\mathbf{v}) \\ & + \Pr(\mathbf{x}_{j-1} = S_1, \mathbf{x}_j = S_2|\mathbf{v}) \\ & + \Pr(\mathbf{x}_{j-1} = S_2, \mathbf{x}_j = S_3|\mathbf{v}) \\ & + \Pr(\mathbf{x}_{j-1} = S_3, \mathbf{x}_j = S_1|\mathbf{v})\end{aligned}$$

because the four state transitions in the probabilities on the right-hand side encompass the event that $c_j = 0$.

The state transition probabilities can be expressed using three conditional probability sequences

$$\begin{aligned}\alpha_j(m) &= \Pr(\mathbf{x}_j = S_m \mid V_1, \ldots, V_j) \\ \beta_j(m) &= \Pr(V_{j+1}, \ldots, V_K \mid \mathbf{x}_j = S_m) \\ \gamma_j(m', m) &= \Pr(\mathbf{x}_j = S_m; V_j \mid \mathbf{x}_{j-1} = S_{m'})\end{aligned}$$

for $m = 0, 1, 2, 3$ and $m' = 0, 1, 2, 3$, since

$$\begin{aligned}\alpha_{j-1}(m')\,\gamma_j(m', m)\,\beta_j(m) &= \Pr(\mathbf{x}_{j-1} = S_{m'}, \mathbf{x}_j = S_m, \mathbf{v}) \\ &= \Pr(\mathbf{x}_{j-1} = S_{m'}, \mathbf{x}_j = S_m|\mathbf{v})\Pr(\mathbf{v})\end{aligned}$$

in which the scale factor $\Pr(\mathbf{v})$ does not vary with the state transition (m', m).

As shown in [17], the sequence $\alpha_j(m)$ may be computed from the recursion

$$\alpha_j(m) = \sum_{m'} \alpha_{j-1}(m')\,\gamma_j(m', m)$$

where the sum is among all transitions (m', m) that connect state configuration $S_{m'}$ at stage $j-1$ to configuration S_m at stage j. This defines the forward recursion. For the example encoder of Figure 3.4, these calculations become

$$\begin{aligned}\alpha_j(0) &= \alpha_{j-1}(0)\,\gamma_j(0, 0) + \alpha_{j-1}(1)\,\gamma_j(1, 0) \\ \alpha_j(1) &= \alpha_{j-1}(2)\,\gamma_j(2, 1) + \alpha_{j-1}(3)\,\gamma_j(3, 1) \\ \alpha_j(2) &= \alpha_{j-1}(0)\,\gamma_j(0, 2) + \alpha_{j-1}(1)\,\gamma_j(1, 2) \\ \alpha_j(3) &= \alpha_{j-1}(2)\,\gamma_j(2, 3) + \alpha_{j-1}(3)\,\gamma_j(3, 3)\end{aligned}$$

as illustrated in Figure 3.6(a). In practice, the values $\alpha_j(m)$ should be scaled to sum to one at each stage

$$\alpha_j(0) + \alpha_j(1) + \alpha_j(2) + \alpha_j(3) = 1$$

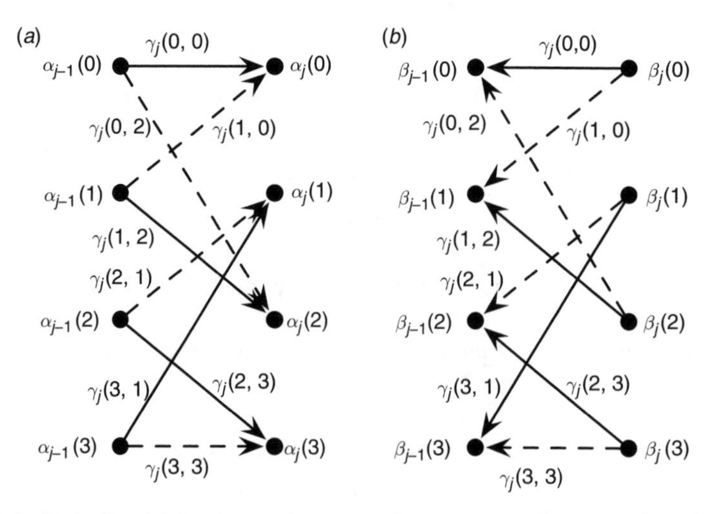

Figure 3.6 Illustrating (a) the forward recursion for one stage of the trellis from Figure 3.4; and (b) the backward recursion.

Assuming the encoder is initialized to $\mathbf{x}(0) = S_0$, the forward recursion is initialized as

$$\alpha_0(0) = 1, \quad \alpha_0(1) = \alpha_0(2) = \alpha_0(3) = 0.$$

In the same vein, a recursion for the sequence $\beta_i(m)$ may be developed as [17]

$$\beta_{j-1}(m) = \sum_{m'} \gamma_j(m, m')\, \beta_j(m')$$

where the sum again is among all transitions $\gamma_j(m, m')$ which connect state configuration $S_{m'}$ at time j to state configuration S_m at time $j-1$. These define the backward recursions. Again for the example encoder of Figure 3.4, these calculations appear as

$$\begin{aligned}
\beta_{j-1}(0) &= \gamma_j(0, 0)\, \beta_j(0) + \gamma_j(0, 2)\, \beta_j(2) \\
\beta_{j-1}(1) &= \gamma_j(1, 0)\, \beta_j(0) + \gamma_j(1, 2)\, \beta_j(2) \\
\beta_{j-1}(2) &= \gamma_j(2, 1)\, \beta_j(1) + \gamma_j(2, 3)\, \beta_j(3) \\
\beta_{j-1}(3) &= \gamma_j(3, 1)\, \beta_j(1) + \gamma_j(3, 3)\, \beta_j(3)
\end{aligned}$$

as sketched in Figure 3.6(b). Again, the values $\beta_{j-1}(m)$ should be scaled to sum to one

$$\beta_{j-1}(0) + \beta_{j-1}(1) + \beta_{j-1}(2) + \beta_{j-1}(3) = 1.$$

If the input sequence to the encoder is chosen to drive the final state $\mathbf{x}_K$ to S_0, then the backward recursion is initialized as

$$\beta_K(0) = 1, \quad \beta_K(1) = \beta_K(2) = \beta_K(3) = 0.$$

If instead no such constraint is placed on the trellis, then the backward recursion may be initialized with a uniform distribution

$$\beta_K(0) = \beta_K(1) = \beta_K(2) = \beta_K(3) = \tfrac{1}{4}.$$

To specify the terms $\gamma_j(m, m')$, we repeat the definition here

$$\begin{aligned}
\gamma_j(m', m) &= \Pr(\mathbf{x}_j = S_m;\ V_j \,|\, \mathbf{x}_{j-1} = S_{m'}) \\
&= \Pr(V_j \,|\, \mathbf{x}_j = S_m, \mathbf{x}_{j-1} = S_{m'}) \Pr(\mathbf{x}_j = S_m \,|\, \mathbf{x}_{j-1} = S_{m'}).
\end{aligned}$$

The transition probability $\Pr(\mathbf{x}_j = S_m \,|\, \mathbf{x}_{j-1} = \mathbf{S}_{m'})$ is the *a priori* probability that the encoder input at time j provoked the transition from configuration $S_{m'}$ to S_m. The likelihood term $\Pr(V_j \,|\, \mathbf{x}_j = S_m, \mathbf{x}_{j-1} = \mathbf{S}_{m'})$ reduces to the channel likelihood function for the measurement V_j, given the encoder output induced by the state transition from $S_{m'}$

to S_m. Illustrating for the trellis diagram of Figure 3.4, consider the transition from S_0 at time $j-1$ to S_0 at time j (any topmost branch). This corresponds to an encoder output of $C_j = [0\ \ 0]$, so that

$$\begin{aligned}\gamma_j(0,0) &= \Pr(\mathbf{x}_j = S_0;\ V_j \,|\, \mathbf{x}_{j-1} = S_0) \\ &= \Pr(V_j \,|\, \mathbf{x}_j = S_0, \mathbf{x}_{j-1} = S_0) \Pr(\mathbf{x}_j = S_0 \,|\, \mathbf{x}_{j-1} = S_0) \\ &= \Pr(V_j \,|\, C_j = [0\ \ 0]) \Pr(C_j = [0\ \ 0]) \\ &= \Pr(v_j | c_j = 0) \Pr(v_{j+K} | c_{j+K} = 0) \Pr(c_j = 0) \Pr(c_{j+K} = 0)\end{aligned}$$

in which we substitute $V_j = [v_j\ \ v_{j+K}]$ and $C_j = [c_j\ \ c_{j+K}] = [0\ \ 0]$, and assume a memoryless channel, so that v_j is affected only by c_j. This final form exposes the transition probabilities $\Pr(v_j|c_j)$ connecting the encoder outputs to the received symbols, as well as the *a priori* bit probabilities $\Pr(c_j)$. By repeating a similar development for each branch of the trellis of Figure 3.4, we have

$$\begin{aligned}\gamma_i(0,0) = \gamma_i(1,2) &= \Pr(v_j|c_j = 0) \Pr(v_{j+K}|c_{j+K} = 0) \Pr(c_j = 0) \Pr(c_{j+K} = 0) \\ \gamma_i(2,3) = \gamma_i(3,1) &= \Pr(v_j|c_j = 0) \Pr(v_{j+K}|c_{j+K} = 1) \Pr(c_j = 0) \Pr(c_{j+K} = 1) \\ \gamma_i(2,1) = \gamma_i(3,3) &= \Pr(v_j|c_j = 1) \Pr(v_{j+K}|c_{j+K} = 0) \Pr(c_j = 1) \Pr(c_{j+K} = 0) \\ \gamma_i(0,2) = \gamma_i(1,0) &= \Pr(v_j|c_j = 1) \Pr(v_{j+K}|c_{j+K} = 1) \Pr(c_j = 1) \Pr(c_{j+K} = 1)\end{aligned}$$

in which the evaluations for c_j/c_{j+K} follow the branch labels of the trellis. All choices (m', m) that do not have a transition branch give $\gamma_i(m', m) = 0$.

Finally, the *a posteriori* symbol probabilities at a given stage are [17]

$$\Pr(c_j = 0 \,|\, V_1, \ldots, V_k) = \delta_j \sum_{m,m':c_j=0} \alpha_{j-1}(m)\, \gamma_j(m, m')\, \beta_j(m')$$

$$\Pr(c_j = 1 \,|\, V_1, \ldots, V_k) = \delta_j \sum_{m,m':c_j=1} \alpha_{j-1}(m)\, \gamma_i(m, m')\, \beta_j(m')$$

with a similar expression for c_{j+K}, in which the scale factor δ_j is chosen to ensure that the two evaluations sum to one. For the encoder of Figure 3.4, these evaluations would appear as

$$\begin{aligned}\Pr(c_j = 0 \,|\, V_1, \ldots, V_k) \propto\ & \alpha_{j-1}(0)\, \gamma_j(0,0)\, \beta_j(0) + \alpha_{j-1}(1)\, \gamma_j(1,2)\, \beta_j(2) \\ & + \alpha_{j-1}(2)\, \gamma_j(2,3)\, \beta_j(3) + \alpha_{j-1}(3)\, \gamma_j(3,1)\, \beta_j(1) \\ \Pr(c_j = 1 \,|\, V_1, \ldots, V_k) \propto\ & \alpha_{j-1}(0)\, \gamma_j(0,2)\, \beta_j(2) + \alpha_{j-1}(1)\, \gamma_j(1,0)\, \beta_j(0) \\ & + \alpha_{j-1}(2)\, \gamma_j(2,1)\, \beta_j(1) + \alpha_{j-1}(3)\, \gamma_j(3,3)\, \beta_j(3)\end{aligned}$$

in which the two expressions account for all transitions at time j for which $c_j = 0$ and $c_j = 1$, respectively. The decoded bit is then taken as $\hat{c}_j = 0$ if the first probability is larger than the second, and $\hat{c}_j = 1$ otherwise. Similar expressions may be developed for c_{j+K} at stage j of the trellis.

■ EXAMPLE 3.2

Extrinsic Information Extraction. Closer inspection of the expression for $\Pr(c_j = 0 \mid V_1, \ldots, V_K)$ above reveals that each γ_j term in the sum—namely $\gamma_j(0, 0)$, $\gamma_j(1, 2)$, $\gamma_j(2, 3)$, and $\gamma_j(3, 1)$—contains a common factor $\Pr(v_j|c_j = 0)$ $\Pr(c_j = 0) = \Pr(v_j;\ c_j = 0)$. Similarly, each γ_j term in $\Pr(c_j = 1 \mid V_1, \ldots, V_K)$ contains a factor $\Pr(v_j;\ c_j = 1)$. The extrinsic probability $U_j(c_j)$ for bit c_j is obtained by factoring $\Pr(v_j;\ c_j)$ from $\Pr(c_j \mid V_1, \ldots, V_K)$. As such, rather than calculating $\Pr(c_j = 0 \mid V_1, \ldots, V_K)$ as written above, only to divide out $\Pr(v_j;\ c_j = 0)$, one can rewrite the expression for the extrinsic probability evaluations as

$$
\begin{aligned}
U_j(c_j = 0) \propto\ & \alpha_{j-1}(0) \Pr(v_{j+K};\ c_{j+K} = 0)\, \beta_j(0) \\
& + \alpha_{j-1}(1) \Pr(v_{j+K};\ c_{j+K} = 0)\, \beta_j(2) \\
& + \alpha_{j-1}(2) \Pr(v_{j+K};\ c_{j+K} = 1)\, \beta_j(3) \\
& + \alpha_{j-1}(3) \Pr(v_{j+K};\ c_{j+K} = 1)\, \beta_j(1) \\
U_j(c_j = 1) \propto\ & \alpha_{j-1}(0) \Pr(v_{j+K};\ c_{j+K} = 1)\, \beta_j(2) \\
& + \alpha_{j-1}(1) \Pr(v_{j+K};\ c_{j+K} = 1)\, \beta_j(0) \\
& + \alpha_{j-1}(2) \Pr(v_{j+K};\ c_{j+K} = 0)\, \beta_j(1) \\
& + \alpha_{j-1}(3) \Pr(v_{j+K};\ c_{j+K} = 0)\, \beta_j(3).
\end{aligned}
$$

This avoids divisions, and in particular avoids any division by zero which can otherwise occur if any probabilities tend to zero during the iterative procedure.

■ EXAMPLE 3.3

Likelihood Ratio Calculation. The likelihood values $\Pr(v_j|c_j)$ for an additive white Gaussian noise channel, as assumed in (3.3), become

$$
\begin{aligned}
\Pr(v_j|c_j = 0) &= \frac{1}{\sqrt{2\pi}\,\sigma} \exp\left(-\frac{(v_j - 1)^2}{2\sigma^2} \right) \\
\Pr(v_j|c_j = 1) &= \frac{1}{\sqrt{2\pi}\,\sigma} \exp\left(-\frac{(v_j + 1)^2}{2\sigma^2} \right)
\end{aligned}
$$

which are evaluated for the received channel output v_j. Here $\sigma^2 = E(b_i^2)$ denotes the background noise variance, and $(-1)^{c_j} \in \{-1, +1\}$ is the channel input. In practice, it is more convenient to first calculate the likelihood ratio

$$r = \frac{\Pr(v_j|c_j = 1)}{\Pr(v_j|c_j = 0)} = \exp\left(-\frac{2v_j}{\sigma^2}\right)$$

and then use the scaled likelihood values

$$\Pr(v_k|c_k = 0) \longleftarrow \frac{1}{1+r}$$
$$\Pr(v_k|c_k = 1) \longleftarrow \frac{r}{1+r}.$$

This avoids the squaring operation applied to v_j—with its concomitant dynamic range problems—and also ensures that the scaled likelihood values sum to one.

3.5.1 With Intersymbol Interference

We consider now the influence of a channel that induces intersymbol interference. Each output symbol from the channel depends on successive inputs to the channel, and these successive inputs define a channel state. A trellis diagram may be drawn that relates channel state transitions to channel inputs and outputs. This faciliates application of the forward-backward algorithm for estimating the input symbols to the channel. Indeed, this context of deducing input symbols to a channel with memory motivates an early derivation of the forward-backward algorithm [14].

In our setting, the outputs from the outer encoder are interleaved and converted to binary form, according to our model

$$d_i = (-1)^{c_j}, \quad \text{with } j = \Pi(i).$$

The symbols (d_i) are then sent over a channel which induces intersymbol interference and adds noise, according to

$$y_i = \underbrace{\sum_l h_l\, d_{i-l}}_{\triangleq\, a_i} + b_i$$

where a_i is the noise-free channel output and b_i is the background noise. Since the channel has binary inputs $d_i = \pm 1$, each noise-free channel output sample a_i is restricted to a finite constellation consisting of the sums and differences of the impulse response coefficients.

For illustration purposes, consider a simple 3-tap channel, for which the received channel output becomes

$$y_i = h_0\, d_i + h_1\, d_{i-1} + h_2\, d_{i-2} + b_i$$

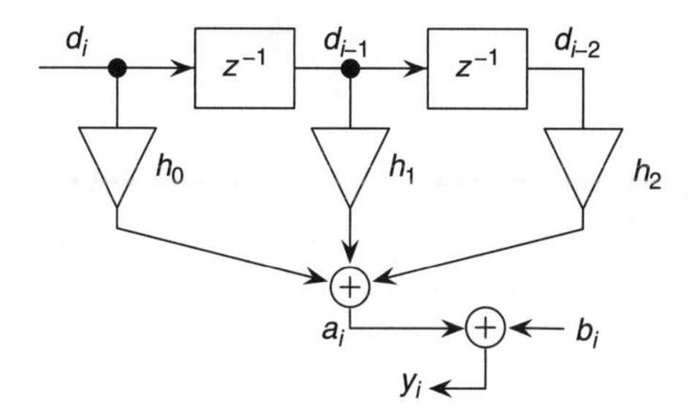

Figure 3.7 Illustrating a simple 3-tap channel.

as sketched in Figure 3.7. This system admits a state-space description as

$$\begin{bmatrix} x_{0i} \\ x_{1i} \end{bmatrix} = \begin{bmatrix} 0 & 0 \\ 1 & 0 \end{bmatrix} \begin{bmatrix} x_{0,i-1} \\ x_{1,i-1} \end{bmatrix} + \begin{bmatrix} 1 \\ 0 \end{bmatrix} d_i$$

$$a_i = [h_1 \ \ h_2] \begin{bmatrix} x_{0,i-1} \\ x_{1,i-1} \end{bmatrix} + h_0 \, d_i$$

with $y_i = a_i + b_i$, and can be written in the trellis diagram of Figure 3.8. The figure also lists the constellation values $\mathcal{H}_0, \ldots, \mathcal{H}_7$ from which each noise-free output a_i is drawn.

The forward-backward algorithm can again be run as in the previous section. The forward recursion for the α terms proceeds as

$$\begin{aligned}
\alpha_i(0) &= \alpha_{i-1}(0)\, \gamma_i(0, 0) + \alpha_{i-1}(2)\, \gamma_i(2, 0) \\
\alpha_i(1) &= \alpha_{i-1}(0)\, \gamma_i(0, 1) + \alpha_{i-1}(2)\, \gamma_i(2, 1) \\
\alpha_i(2) &= \alpha_{i-1}(1)\, \gamma_i(1, 2) + \alpha_{i-1}(3)\, \gamma_i(3, 1) \\
\alpha_i(3) &= \alpha_{i-1}(1)\, \gamma_i(1, 3) + \alpha_{i-1}(3)\, \gamma_i(3, 3)
\end{aligned}$$

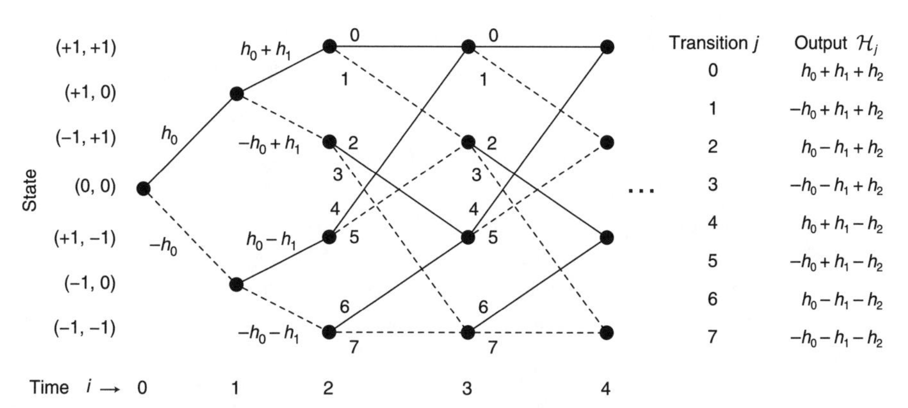

Figure 3.8 Trellis diagram for the channel of Figure 3.7. Solid lines indicate an input of +1 to the channel, while dashed lines indicate −1.

while the backward recursion for the β terms then runs as

$$\begin{aligned}
\beta_{i-1}(0) &= \gamma_i(0, 0)\,\beta_i(0) + \gamma_i(0, 1)\,\beta_i(1)\\
\beta_{i-1}(1) &= \gamma_i(1, 2)\,\beta_i(2) + \gamma_i(1, 3)\,\beta_i(3)\\
\beta_{i-1}(2) &= \gamma_i(2, 0)\,\beta_i(0) + \gamma_i(2, 1)\,\beta_i(1)\\
\beta_{i-1}(3) &= \gamma_i(3, 2)\,\beta_i(2) + \gamma_i(3, 3)\,\beta_i(3).
\end{aligned}$$

The changes here concern the channel likelihood evaluations $\gamma_i(m', m)$, of which there are eight at each trellis section. These may be tabulated as follows.

$$\begin{aligned}
\gamma_i(0, 0) &= \Pr(d_i = +1)\,\mathcal{N}_i(\mathcal{H}_0, \sigma^2), & \mathcal{H}_0 &= h_0 + h_1 + h_2\\
\gamma_i(0, 1) &= \Pr(d_i = -1)\,\mathcal{N}_i(\mathcal{H}_1, \sigma^2), & \mathcal{H}_1 &= -h_0 + h_1 + h_2\\
\gamma_i(1, 2) &= \Pr(d_i = +1)\,\mathcal{N}_i(\mathcal{H}_2, \sigma^2), & \mathcal{H}_2 &= h_0 - h_1 + h_2\\
\gamma_i(1, 3) &= \Pr(d_i = -1)\,\mathcal{N}_i(\mathcal{H}_3, \sigma^2), & \mathcal{H}_3 &= -h_0 - h_1 + h_2\\
\gamma_i(2, 0) &= \Pr(d_i = +1)\,\mathcal{N}_i(\mathcal{H}_4, \sigma^2), & \mathcal{H}_4 &= h_0 + h_1 - h_2\\
\gamma_i(2, 1) &= \Pr(d_i = -1)\,\mathcal{N}_i(\mathcal{H}_5, \sigma^2), & \mathcal{H}_5 &= -h_0 + h_1 - h_2\\
\gamma_i(3, 2) &= \Pr(d_i = +1)\,\mathcal{N}_i(\mathcal{H}_6, \sigma^2), & \mathcal{H}_6 &= -h_0 - h_1 + h_2\\
\gamma_i(3, 3) &= \Pr(d_i = -1)\,\mathcal{N}_i(\mathcal{H}_7, \sigma^2), & \mathcal{H}_7 &= -h_0 - h_1 - h_2
\end{aligned} \tag{3.4}$$

in which each Gaussian term is summarized using the notation

$$\mathcal{N}_i(\mathcal{H}_j, \sigma^2) = \frac{1}{\sqrt{2\pi}\,\sigma}\exp\left(-\frac{(y_i - \mathcal{H}_j)^2}{2\sigma^2}\right).$$

The *a posteriori* probabilities then become

$$\begin{aligned}
\Pr(d_i = +1|\mathbf{y}) \propto\ & \alpha_{i-1}(0)\,\gamma_i(0, 0)\,\beta_i(0) + \alpha_{i-1}(1)\,\gamma_i(1, 2)\,\beta_i(2)\\
&+ \alpha_{i-1}(2)\,\gamma_i(2, 0)\,\beta_i(0) + \alpha_{i-1}(3)\,\gamma_i(3, 2)\,\beta_i(2)\\
\Pr(d_i = -1|\mathbf{y}) \propto\ & \alpha_{i-1}(0)\,\gamma_i(0, 1)\,\beta_i(1) + \alpha_{i-1}(1)\,\gamma_i(1, 3)\,\beta_i(3)\\
&+ \alpha_{i-1}(2)\,\gamma_i(2, 1)\,\beta_i(1) + \alpha_{i-1}(3)\,\gamma_i(3, 3)\,\beta_i(3)
\end{aligned}$$

in which the former (respectively, latter) probability is obtained by summing terms $\alpha_{i-1}(m')\gamma_i(m', m)\beta_i(m)$ over transitions (m', m) provoked by $d_i = +1$ (resp. $d_i = -1$).

Each γ_i term in the expression for $\Pr(d_i = +1|\mathbf{y})$ [resp. $\Pr(d_i = -1|\mathbf{y})$] contains a factor $\Pr(d_i = +1)$ [resp. $\Pr(d_i = -1)$]. As such, we may rewrite these posterior probabilities as

$$\begin{aligned}
\Pr(d_i = +1|\mathbf{y}) &\propto \Pr(d_i = +1)\,T_i(d_i = +1)\\
\Pr(d_i = -1|\mathbf{y}) &\propto \Pr(d_i = -1)\,T_i(d_i = -1)
\end{aligned}$$

in which we expose the extrinsic probability values as

$$\begin{aligned} T_i(d_i = +1) \propto\ & \alpha_{i-1}(0)\mathcal{N}_i(0, \sigma^2)\,\beta_i(0) + \alpha_{i-1}(1)\mathcal{N}_i(2, \sigma^2)\,\beta_i(2) \\ & + \alpha_{i-1}(2)\mathcal{N}_i(4, \sigma^2)\,\beta_i(0) + \alpha_{i-1}(3)\mathcal{N}_i(6, \sigma^2)\,\beta_i(2) \\ T_i(d_i = -1) \propto\ & \alpha_{i-1}(0)\mathcal{N}_i(1, \sigma^2)\,\beta_i(1) + \alpha_{i-1}(1)\mathcal{N}_i(3, \sigma^2)\,\beta_i(3) \\ & + \alpha_{i-1}(2)\mathcal{N}_i(5, \sigma^2)\,\beta_i(1) + \alpha_{i-1}(3)\mathcal{N}_i(7, \sigma^2)\,\beta_i(3) \end{aligned}$$

■ EXAMPLE 3.4

Closing the Loop. To couple the channel decoder with the error correction decoder, the likelihood calculations illustrated in Example 3.3 are now replaced with the extrinsic probabilities $T_i(d_i)$ calculated above

$$\begin{aligned} \Pr(v_j|c_j = 0) &\leftarrow T_i(d_i = +1) \\ \Pr(v_j|c_j = 1) &\leftarrow T_i(d_i = -1), \qquad \text{with } j = \Pi(i) \end{aligned}$$

and the *a priori* probabilities in the outer decoder are taken to be uniform: $\Pr(c_j = 0) = \Pr(c_j = 1) = \frac{1}{2}$. Similarly, the extrinsic probabilities $U_j(c_j)$ from Example 3.2 then replace the *a priori* probabilities $\Pr(d_i)$ in the inner decoder for the next iteration

$$\begin{aligned} \Pr(d_i = +1) &\leftarrow U_j(c_j = 0) \\ \Pr(d_i = -1) &\leftarrow U_j(c_j = 1), \qquad \text{with } j = \Pi(i). \end{aligned}$$

This completes the specification of the various terms of Figure 3.2, for this basic example.

3.6 SIMPLIFIED ALGORITHM: INTERFERENCE CANCELER

The main limitation of the iterative decoding algorithm presented thus far concerns the inner decoder. Its complexity grows exponentially with the channel impulse response length. A simplified structure uses a linear equalizer in a decision feedback configuration, in order to cancel the intersymbol interference [36–38].

The basic structure is sketched in Figure 3.9, using two filters with transfer functions

$$\begin{aligned} P(z) &= z^{-L}H(z^{-1}) = h_L + h_{L-1}z^{-1} + \cdots + h_1 z^{-L+1} + h_0 z^{-L} \\ Q(z) &= z^{-L}[H(z^{-1})\,H(z) - r_0] \end{aligned} \tag{3.5}$$

in which L is the channel delay spread, z^{-1} denotes the backward shift operator (so that $z^{-1}\,d_i = d_{i-1}$), and $r_0 = \sum_k h_k^2$ is the energy of the channel impulse response. With the

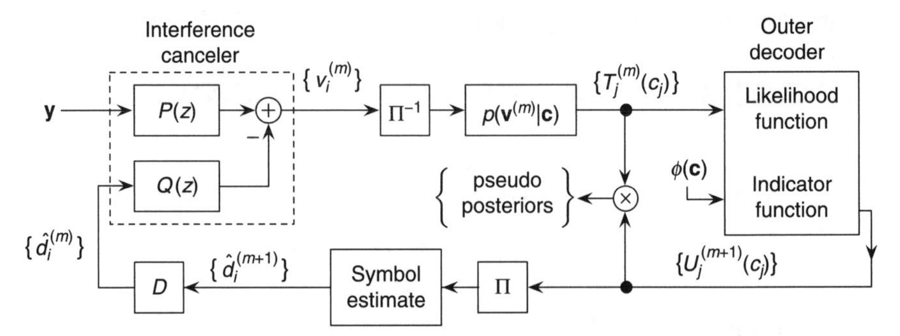

Figure 3.9 Turbo equalizer using a linear interference canceler in place of the inner decoder.

unit delay operator convention, we observe that

$$H(z)\,d_i = \sum_{l=0}^{L} h_l\, z^{-l} d_i = \sum_{l=0}^{L} h_l\, d_{i-l}$$

which correctly captures the convolution operation of the channel.

The feedforward filter $P(z)$ is recognized as the "matched filter" [9] to the channel, that is, having an impulse response which is a time-reversed version of the channel impulse response, but shifted to be causal. The feedback filter $Q(z)$ has an impulse response that behaves as the autocorrelation function of the channel impulse response minus the center term, but again shifted to be causal. The impulse response terms $\{q_k\}$ of $Q(z)$ are thus derived from the channel autocorrelation terms according to

$$q_{k+L} = r_k = \sum_i h_i\, h_{i+k}, \qquad k = \pm 1, \pm 2, \ldots, \pm L.$$

With $\hat{d}_i$ denoting the estimated channel symbols produced by the decoder, the output of the interference canceler becomes

$$\begin{aligned} v_i &= P(z)\,H(z)\,d_i - Q(z)\,\hat{d}_i + P(z)\,b_i \\ &= z^{-L} H(z^{-1})\,H(z)\,d_i - [H(z^{-1})\,H(z^{-1}) - r_0]\,\hat{d}_i + P(z)\,b_i \\ &= r_0\, d_{i-L} + \sum_{\substack{k=0 \\ k \neq L}}^{2L} q_k\,(d_{i-k} - \hat{d}_{i-k}) + \sum_{k=0}^{L} p_k\, b_{i-k}. \end{aligned}$$

The first term is the desired channel symbol delayed by L samples and scaled by r_0, the next term represents the residual intersymbol interference, and the final term is the filtered noise. If the channel decoder correctly estimates the channel symbols, then $\hat{d}_i = d_i$ and the intersymbol interference terms vanish.

■ EXAMPLE 3.5

Optimality of $P(z)$ and $Q(z)$. We show here how the choices of $P(z)$ and $Q(z)$ from (3.5) maximize the signal to noise ratio (SNR) subject to an interference cancellation constraint. To this end, let

$$P(z) = p_0 + p_1 z^{-1} + \cdots + p_L z^{-L}$$

be an arbitrary filter having degree L, where L is the channel degree. If the decoder correctly produces channel symbol estimates, giving $\hat{d}_i = d_i$, the intersymbol interference will vanish at the equalizer output provided we choose $Q(z)$ according to

$$Q(z) = P(z)\,H(z) - \hat{r}\,z^{-L}$$

in which

$$\hat{r} = \sum_{k=0}^{L} h_k\, p_{L-k}$$

is the central term of the impulse response of $P(z)\,H(z)$. The equalizer output then reduces to

$$v_i = \hat{r}\,d_{i-L} + \sum_{k=0}^{L} p_k\, b_{i-k}$$

in which the first term is a scaled desired signal (delayed by L samples), and the second term is the filtered noise. If the noise sequence (b_i) is white, then the signal to (filtered) noise ratio becomes

$$\mathrm{SNR} = \frac{E(d_{i-L}^2)}{E(b_i^2)}\,\frac{\hat{r}^2}{\sum_k p_k^2}.$$

Since $\hat{r} = \sum_k h_k\, p_{L-k}$, the Cauchy–Schwarz inequality may be invoked to show that

$$\hat{r}^2 = \left(\sum_{k=0}^{L} h_k\, p_{L-k}\right)^2 \leq \left(\sum_{k=0}^{L} h_k^2\right)\left(\sum_{k=0}^{L} p_k^2\right)$$

in which equality holds if and only if $h_k = \beta\, p_{L-k}$ for some constant β. The signal to filtered noise ratio is thus upper bounded as

$$\mathrm{SNR} \leq \frac{E(d_{i-L}^2)}{E(b_i^2)}\left(\sum_{k=1}^{L} h_k^2\right).$$

The right-hand side coincides with the SNR seen from the channel output. Thus, the choice $p_k = h_{L-k}$, corresponding to $P(z) = z^{-L}\,H(z^{-1})$, maximizes the SNR at the equalizer output, subject to an interference cancellation constraint.

The coupling elements between the interference canceler and the outer decoder are slightly modified from the iterative decoding version. The equalizer outputs (v_i), once deinterleaved (cf. the "Π^{-1}" box in Figure 3.9), generate the likelihood terms $\Pr(v_i|c_j=0)$ and $\Pr(v_i|c_j=1)$ [with $j=\Pi(i)$] as in Example 3.3. These replace the pseudo probabilities $\{T_j^{(m)}(c_j)\}$ that feed the forward-backward algorithm for the outer decoder. The feedback filter $Q(z)$ is fed with symbol estimates that may be derived either from the pseudo posteriors $\Pr(c_j=1|\mathbf{v})$ or the extrinsic probabilities $U_j(c_j=1)$ as calculated by the outer decoder; use of the latter tends to reduce the occurence of limit cycles in the resulting iterative algorithm [39]. Now, the symbol map $d_i=(-1)^{c_j}$ [with $j=\Pi(i)$] transforms $c_j=0$ to $d_i=1$, and $c_j=1$ to $d_i=-1$. Replacing c_j with its soft estimate $U_j(1)$ gives two formulas for generating soft symbol estimates $\hat{d}_i$.

- *Linear Interpolant*: One sets

$$\hat{d}_i = 1 - 2U_j(1), \quad \text{with } j=\Pi(i).$$

- *Trigonometic Interpolant*: Since $-1=e^{\iota\pi}$ (with $\iota=\sqrt{-1}$), we may recognize our symbol map as

$$d_i = (-1)^{c_j} = e^{\iota c_j \pi} = \Re(e^{\iota c_j \pi}) = \cos(c_j\pi), \quad \text{with } j=\Pi(i).$$

  Replacing c_j by its soft estimate $U_j(1)$ then gives

$$\hat{d}_i = \cos[U_j(1)\pi] \qquad \text{with } j=\Pi(i).$$

In practice, the former variant often gives a lower bit error rate after convergence, and is also simpler to compute. Other variants of soft symbol feedback are developed in [40–43].

■ EXAMPLE 3.6

Performance Comparison Using "Easy" Channel. To illustrate the performance of the turbo equalizer algorithms, we consider the (5,7) encoder from Figure 3.3, and the "Proakis A" channel from ([9], Fig. 10.2-5(a)), but scaled to unit norm

$$\begin{array}{ll}
h_0 = 0.03250 & h_6 = 0.292528 \\
h_1 = -0.040629 & h_7 = 0.000000 \\
h_2 = 0.056880 & h_8 = -0.170641 \\
h_3 = 0.170641 & h_9 = 0.024377 \\
h_4 = -0.406289 & h_{10} = 0.585057. \\
h_5 = 0.585057 &
\end{array}$$

Figure 3.10 shows the bit error rate using 15,000 information symbols per block, and averaged over 200 independent runs. For each run, a new information sequence, a new interleaver, and a new noise realization are generated. The

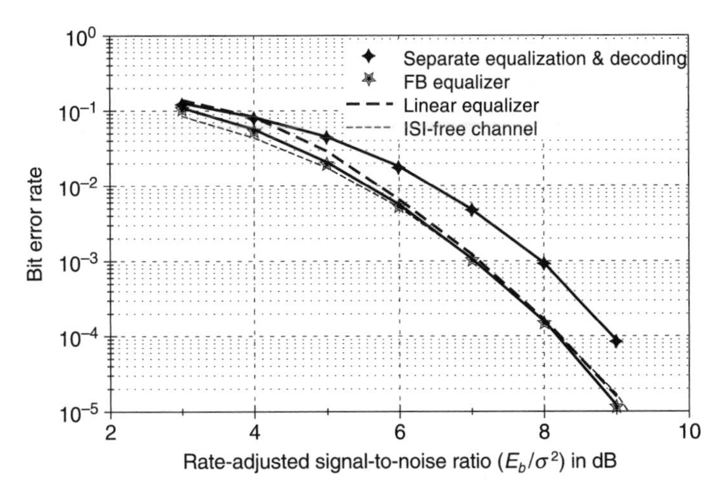

Figure 3.10 Bit error rate for the iterative decoder and interference cancellation schemes, using a (5,7) encoder and a "Proakis A" channel.

horizontal axis is the rate-adjusted SNR "E_b/σ^2". (This is the raw SNR divided by the code rate of $1/2$, or multipled by 2, since each information bit requires two transmitted bits from the encoder, effectively doubling the signal power devoted to each information bit, compared to the case in which no coding is used.)

The top-most curve is the bit error rate after the initial iteration of the turbo decoder, when the influence of the feedback loop has yet to kick in. This is precisely the performance obtained using separate receiver components for equalization and decoding. The starred line gives the bit error rate after 20 iterations of the iterative decoding algorithm, while the bold dashed line shows the bit error performance using instead a linear interference canceler. Although the bit error performance for either is comparable, the complexity is not. The forward-backward algorithm for the inner decoder requires $2^{11} = 2048$ states per trellis section, presenting a significantly higher storage and computational complexity than the linear interference canceler. The thin dashed line, finally, shows the bit error rate of the outer encoder/decoder when used over a channel having additive white Gaussian noise, but devoid of intersymbol interference. The two turbo decoder algorithms are observed to give bit error rates approaching this limit, unlike the single-iteration algorithm that does not benefit from the turbo effect.

■ EXAMPLE 3.7

Performance Comparison: "Harder" Channel. We use the same set-up as in the previous example, except for the channel which is now the Proakis B channel from [9, Fig. 10.2-5(b)]:

$$h_0 = 0.408 \qquad h_1 = 0.816 \qquad h_2 = 0.408$$

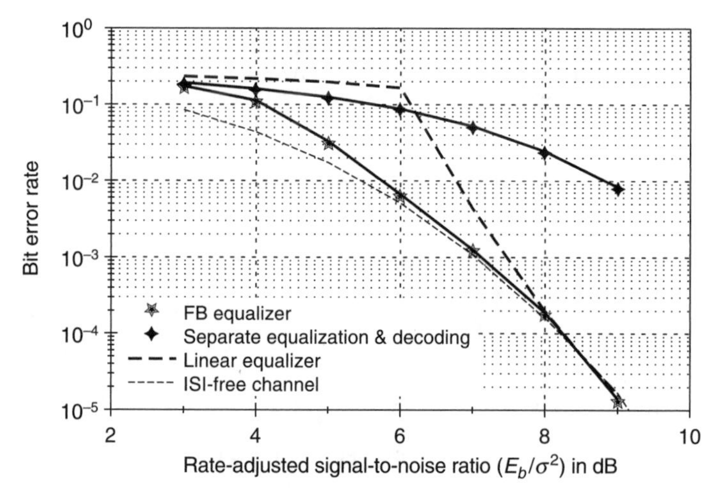

Figure 3.11 Bit error rate for the iterative decoder and interference cancellation schemes, using a (5,7) encoder and a Proakis B channel.

This channel imparts a greater amount of intersymbol interference than the previous channel. Figure 11 shows the bit error rate versus the rate-adjusted SNR again for the forward-backward channel decoder and for the linear decision feedback equalizer. The gains from the turbo effect are more prominent in view of the single iteration curve that shows the performance in the absence of the turbo effect. We observe also that poorer SNRs provoke a greater performance degradation in the linear equalizer scheme.

3.7 CAPACITY ANALYSIS

The turbo decoding algorithm attracted significant initial attention due to its ability to approach Shannon capacity on additive noise channels [1]. As the turbo equalizer setup is simply a serially concatenated coding scheme, one would naturally ask if some capacity-approaching property is inherited. In particular, by interpreting the convolutional channel as part of a concatenated code, can the capacity be increased beyond that attainable in a channel without intersymbol interference? We answer here in the negative: Subject to a constraint on the SNR at the channel output, the presence of intersymbol interference can only decrease channel capacity. As such, any gains brought by diversity from multiple transmission paths occur solely due to an increase in the SNR: Multiple transmission paths result in greater signal power at the receiver whenever the multiple paths combine constructively.

We first review the notion of channel capacity in terms of the number of messages that may be reliably distinguished at the channel output [44], and then examine a log-determinant form [45, 46] for channel capacity that lends itself readily to the context at hand.

Consider a message block composed of N successive symbols, where N is user-selected and presumably large enough but otherwise arbitrary. The N symbols are transmitted over an error-prone channel which includes, in our context, background noise and intersymbol interference. In general, the channel output will be a corrupted version of its input.

As an illustrative example, suppose we have a code book comprising two messages (each containing N symbols). The first is sent when we wish to transmit a zero, the second to transmit a one. If the two messages are initially quite similar, then the channel distortion will likely blur any distinction between them, obscuring the information they are supposed to encode from the channel output. If, conversely, the two messages are quite disparate in form, then their distorted versions at the channel output stand a better chance of remaining distinguishable, allowing the receiver to confidently read a zero or a one. This is the simplest example of code design, aiming to transmit one bit of information per message. The salient feature is not so much that the channel output match the channel input, but rather that distinct messages at the channel input remain distinguishable at the channel output.

The operational definition of channel capacity relates to the maximum number of messages, say μ, that are distinguishable (with high enough probability) at the channel output. The logarithm to base 2 of this quantity, say $K = \log_2 \mu$, is the number of bits required of a binary counter which enumerates these distinguishable messages. This is, in effect, the number of bits that each message can convey—take K information bits from a source, and interpret them as the bits of a K-bit binary counter. The value of the counter is the index of a message to send over the channel. Provided the receiver can distinguish separate messages, it need only identify the index of the message sent, whose K-bit binary representation contains the K-information bits that were to be communicated. Since each message is comprised of N symbols, the effective communication rate is K/N bits per channel use on average.

So what is this maximum number of distinguishable messages, or better still, its logarithm? The analytic solution was derived by Shannon [8] as

$$\text{Capacity} = \max_{\Pr(\mathbf{d})} I(\mathbf{d}; \mathbf{y}) \qquad \text{(in bits per message sent)}$$

in which the mutual information I($\mathbf{d}$; $\mathbf{y}$) is maximized over all probability distributions $\Pr(\mathbf{d})$ of the input, where $\mathbf{d}$ and $\mathbf{y}$ are vectors containing the input and output sequences, respectively. The mutual information expression varies according to whether the channel inputs and outputs are discrete or continuous. If both inputs and outputs are discrete, then

$$I(\mathbf{d}; \mathbf{y}) = \sum_{\mathbf{d},\mathbf{y}} \Pr(\mathbf{d}, \mathbf{y}) \log_2 \frac{\Pr(\mathbf{d}, \mathbf{y})}{\Pr(\mathbf{d}) \Pr(\mathbf{y})}$$

whereas if both are continuous, then

$$I(\mathbf{d}; \mathbf{y}) = \int \Pr(\mathbf{d}, \mathbf{y}) \log_2 \frac{\Pr(\mathbf{d}, \mathbf{y})}{\Pr(\mathbf{d}) \Pr(\mathbf{y})} \, d\mathbf{d} \, d\mathbf{y}$$

provided the integral exists. The joint probability Pr(**d**, **y**) can be factored as Pr(**y**|**d**) Pr(**d**), revealing its dependence on the channel transition probability function Pr(**y**|**d**).

A closed-form expression for the maximized mutual information is often elusive, but can be obtained in some special cases. Perhaps the most common of these is a linear model for which

$$\mathbf{y} = \mathbf{H}\,\mathbf{d} + \mathbf{b}.$$

If the input and output sequences are continuous amplitude processes and the noise is Gaussian and white, so that $E(\mathbf{b}\mathbf{b}^T) = \sigma^2\mathbf{I}$, and the input is normalized to unit power, so that $E(d_i^2) = 1$, the capacity follows the log-determinant formula [45, 46]

$$\text{Capacity} = \frac{1}{2}\log_2 \det\left(\mathbf{I} + \frac{\mathbf{H}^T\mathbf{H}}{\sigma^2}\right). \tag{3.6}$$

This capacity bound applies to our setting once we recognize that the channel model from (3.1) may be written in matrix form as

$$\underbrace{\begin{bmatrix} y_1 \\ y_2 \\ y_3 \\ \vdots \\ y_{L-1} \\ y_L \\ \vdots \\ y_{L+N-1} \end{bmatrix}}_{\mathbf{y}} = \underbrace{\begin{bmatrix} h_0 & 0 & 0 & 0 & \cdots & 0 \\ h_1 & h_0 & 0 & 0 & \cdots & 0 \\ h_2 & h_1 & h_0 & 0 & \cdots & 0 \\ \vdots & \ddots & \ddots & \ddots & \ddots & \vdots \\ h_L & h_{L-1} & \cdots & h_0 & \ddots & 0 \\ 0 & h_L & h_{L-1} & \cdots & h_0 & \vdots \\ \vdots & \ddots & \ddots & \ddots & \ddots & \vdots \\ 0 & \cdots & 0 & 0 & 0 & h_L \end{bmatrix}}_{\mathbf{H}} \underbrace{\begin{bmatrix} d_1 \\ d_2 \\ d_3 \\ \vdots \\ d_N \end{bmatrix}}_{\mathbf{d}} + \underbrace{\begin{bmatrix} b_1 \\ b_2 \\ b_3 \\ \vdots \\ b_{L-1} \\ b_L \\ \vdots \\ b_{L+N-1} \end{bmatrix}}_{\mathbf{b}}$$

in which the matrix **H**, of dimensions $(L + N - 1) \times$ N, is a convolution matrix. Its grammian $\mathbf{H}^T\mathbf{H}$ is thus a symmetric Toeplitz matrix built from the channel autocorrelation sequence

$$\mathbf{H}^T\mathbf{H} = \begin{bmatrix} r_0 & r_1 & \cdots & r_{N-1} \\ r_1 & r_0 & \ddots & \vdots \\ \vdots & \ddots & \ddots & r_1 \\ r_{N-1} & \cdots & r_1 & r_0 \end{bmatrix}, \qquad \text{with } r_k = \sum_j h_j\, h_{j+k}.$$

It follows that the matrix argument $\mathbf{I}+\mathbf{H}^T\mathbf{H}/\sigma^2$ from (3.6) is also a symmetric Toeplitz matrix

$$\left(\mathbf{I}+\frac{\mathbf{H}^T\mathbf{H}}{\sigma^2}\right)_{ij}=\begin{cases} r_{|i-j|}/\sigma^2, & i\neq j;\\ 1+r_0/\sigma^2, & i=j.\end{cases}$$

We then have the following upper bound on capacity.

Theorem 1 *The capacity is upper bounded in terms of the SNR as*

$$\textit{Capacity} \leq \frac{N}{2}\log_2\left(1+\frac{r_0}{\sigma^2}\right) \qquad \textit{(in bits per block use)}$$

with the upper bound attained if and only if $r_k=0$ *for* $k\neq 0$.

We note that, since the channel is assumed finite impulse response (FIR), the autocorrelation lags r_k will vanish for $k\neq 0$ if, and only if, the channel impulse response has a sole nonzero term; this gives a channel with no intersymbol interference. As such, subject to a power constraint on $r_0=\sum_k h_k^2$, the presence of intersymbol interference can only reduce channel capacity.

To verify the theorem, we recall that Hadamard's inequality of matrix theory ([47], p. 477), [44] asserts that the determinant of a positive definite matrix is upper bounded by the product of its diagonal elements, with equality if, and only if, the matrix is diagonal. Applying this inequality to $\mathbf{I}+\mathbf{H}^T\mathbf{H}/\sigma^2$, we have

$$\begin{aligned}\text{Capacity} &= \frac{1}{2}\log_2\det\left(\mathbf{I}+\frac{\mathbf{H}^T\mathbf{H}}{\sigma^2}\right)\\ &\leq \frac{1}{2}\log_2\left(1+\frac{r_0}{\sigma^2}\right)^N=\frac{N}{2}\log_2\left(1+\frac{r_0}{\sigma^2}\right).\end{aligned}$$

Since the off-diagonal elements of $\mathbf{I}+\mathbf{H}^T\mathbf{H}/\sigma^2$ are $r_{|i-j|}/\sigma^2$., equality holds in the bound if, and only if, $r_k=0$ for $k\neq 0$.

We remark that the bound gives the capacity for N uses of the channel; upon normalizing by the block length N, we recover Shannon's expression $\frac{1}{2}\log_2(1+r_0/\sigma^2)$ for the per-symbol capacity of an additive white Gaussian noise channel.

■ EXAMPLE 3.8

Diversity versus Signal Power. Consider an ideal channel with $h_0=1$ and $h_k=0$ for $k\geq 1$, and a modified channel with $g_0=g_1=1$, and $g_k=0$ for $k\geq 2$, having thus an additional impulse response term which induces intersymbol interference. Taking $\sigma^2=1$, the capacity for the first channel, divided by N, evaluates to 0.5 bits per channel use. In the second case, the capacity increases beyond 0.7 bits per channel use, which is sometimes attributed to increased diversity brought from the presence of the second impulse response term g_1. In fact, the seeming increase in capacity comes from a larger SNR, as $g_0^2+g_1^2>h_0^2$.

We should note that the capacity bound using the log-determinant formula above is attainable using an input that is Gaussian distributed [44]. When using a binary input, the attainable capacity is generally lower. We close this section by visiting the capacity bound for the binary input case.

In the absence of intersymbol interference, the mutual information between channel input and channel output becomes, for the binary case

$$I(d;y) = \Pr(d=+1)\int_y \Pr(y|d=+1)\log\frac{\Pr(y|d=+1)}{\Pr(y)}\,dy + \Pr(d=-1)\int_y \Pr(y|d=-1)\log\frac{\Pr(y|d=-1)}{\Pr(y)}\,dy$$

using for the output probability

$$\Pr(y) = \Pr(y|d=+1)\Pr(d=+1) + \Pr(y|d=-1)\Pr(d=-1)$$

in which the conditional probability $\Pr(y|d)$ is Gaussian

$$\Pr(y|d=\pm 1) = \frac{1}{\sqrt{2\pi}\,\sigma}\exp\left(-\frac{(y\mp 1)^2}{2\sigma^2}\right).$$

Maximizing the mutual information $I(d;y)$ versus the input probabilities $\Pr(\mathrm{d}=\pm 1)$ gives a uniform distribution on the input: $\Pr(d=+1)=\Pr(d=-1)=\frac{1}{2}$. The resulting channel capacity (per channel use) may then be plotted against the SNR $10\log_{10}(1/\sigma^2)$ in dB, as in Figure 3.12, which shows also the capacity bound

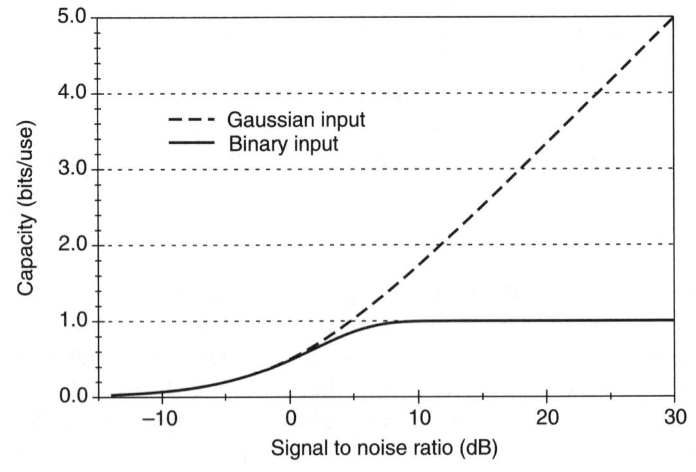

Figure 3.12 Channel capacity per channel use for Gaussian and binary inputs, in the absence of intersymbol interference.

$\frac{1}{2}\log_2(1 + 1/\sigma^2)$ applicable in the Gaussian case. We observe, as expected, that the capacity levels off at 1 bit per channel use for the binary input case.

3.8 BLIND TURBO EQUALIZATION

The schemes reviewed thus far all assume that the channel parameters, comprised of the impulse response terms (h_k) and the background noise variance $\sigma^2 = E(b_i^2)$, are known to the receiver. In practice, these quantities must be estimated at the receiver, using either some sort of channel identification procedure, or by integrating the channel estimation step into the iterative procedure.

Channel identification can be accomplished by incorporating a training sequence into the transmitted sequence, which training sequence is known to the receiver, and using adaptive filtering techniques [11, 12] to identify (an approximation to) the channel, and/or adapt the interference canceler [37, 38, 48]. Alternatively, blind channel identification methods may be employed [49–54], provided multiple antennae and/or oversampling are employed at the receiver.

The approach pursued in this section examines instead how the impulse response and noise variance of the channel can be estimated as part of the iterative decoding procedure [18], using the expectation maximization algorithm [55, 56]. This approach has the advantage of operating concurrently with the turbo equalizer using essentially the same operations. The training methods or blind methods referenced above, by contrast, require the receiver to switch to a different mode of operation. They can, however, be considered as candidate methods for deducing initial channel estimates for the more refined identification method pursued here.

To begin, we first examine how the noise-free channel output relates to the channel impulse response $\mathbf{h}$. Since the channel is modeled as a tapped delay line, the noise-free channel outputs may be placed in one-to-one correspondence with the configurations of that delay line. With L denoting the channel order (or number of delay elements in the tapped delay line), introduce an augmented state vector

$$\xi_i \triangleq \begin{bmatrix} d_i \\ \mathbf{x}_{i-1} \end{bmatrix} = \begin{bmatrix} \mathbf{x}_i \\ d_{i-L} \end{bmatrix} = \begin{bmatrix} d_i \\ d_{i-1} \\ \vdots \\ d_{i-L} \end{bmatrix}.$$

For the example three-tap channel of Figure 3.7 (which has $L=2$ delay elements), this augmented vector appears as

$$\xi_i = \begin{array}{c} \\ \mathbf{x}_{i-1}\left\{ \right. \end{array} \hspace{-0.5em} \begin{bmatrix} d_i \\ d_{i-1} \\ d_{i-2} \end{bmatrix} \hspace{-0.3em} \begin{array}{l} \left. \right\} \mathbf{x}_i \\ \\ \end{array}.$$

We observe that, as ξ_i encompasses both $\mathbf{x}_{i-1}$ and $\mathbf{x}_i$, it encompasses the state transition from $\mathbf{x}_{i-1}$ to $\mathbf{x}_i$. As it is comprised of $L+1$ binary elements $d_i, \ldots, d_{i-L}$, it has 2^{L+1}

configurations, which we denote as $\Sigma_0, \ldots, \Sigma_{2^L-1}$. These relate directly to the 2^{L+1} candidate transitions at each stage of the trellis diagram of the channel. We denote by $\mathcal{H}(\Sigma_j)$ the noise-free channel output induced by the transition captured by Σ_j.

■ EXAMPLE 3.9

Noise-Free Channel Outputs. Returning to the trellis diagram of Figure 3.8, the augmented state vector assumes seven candidate configurations

$$\Sigma_0 = [+1 \quad +1 \quad +1], \quad \Sigma_1 = [-1 \quad +1 \quad +1], \quad \cdots \quad \Sigma_7 = [-1 \quad -1 \quad -1].$$

The noise-free channel outputs that the channel can produce may be enumerated as

$$\begin{aligned}
\mathcal{H}(\Sigma_0) &= \mathbf{h}^t \xi_i \big|_{\xi_i = \Sigma_0} = h_0 + h_1 + h_2 \\
\mathcal{H}(\Sigma_1) &= \mathbf{h}^t \xi_i \big|_{\xi_i = \Sigma_1} = -h_0 + h_1 + h_2 \\
&\vdots \\
\mathcal{H}(\Sigma_7) &= \mathbf{h}^t \xi_i \big|_{\xi_i = \Sigma_7} = -h_0 - h_1 - h_2
\end{aligned}$$

in one-to-one correspondence with the state transitions, and comprise all combinations of sums and differences of the channel coefficients.

Note that knowledge of the channel input sequence (d_i) implies knowledge of the state transition sequence (ξ_i), and vice-versa. A method of channel identification then consists of finding the coefficients $\mathbf{h}$ which are best compatible with the observed channel output sequence [18]. To this end, let

$$\theta = [\mathcal{H}(\Sigma_0), \ldots, \mathcal{H}(\Sigma_{2^M-1}), \sigma]$$

denote the unknown parameters for the channel, consisting of the noise-free channel output constellation values $\{\mathcal{H}(\Sigma_j)\}$ and the channel noise standard deviation σ. Consider the likelihood function $\Pr(\mathbf{y}|\theta)$ for the given received sequence $\mathbf{y} = (y_1, y_2, \ldots)$. If we maximize this function versus the parameters θ

$$\hat{\theta} = \arg\max_{\theta} \Pr(\mathbf{y}|\theta)$$

the estimate $\hat{\theta}$ is then optimal in the maximum likelihood sense. Direct optimization of the likelihood function $\Pr(\mathbf{y}|\theta)$ versus θ is usually computationally difficult, and so iterative techniques are employed.

Let $\Xi = (\xi_1, \xi_2, \ldots)$ be a valid state transition sequence through the trellis, and introduce a joint likelihood function $\Pr(\mathbf{y}, \Xi|\theta)$. We can then consider $\Pr(\mathbf{y}|\theta)$ as a marginal function obtained by summing over all valid state transition sequences

$$\Pr(y|\theta) = \sum_{\Xi} \Pr(\mathrm{y}, \Xi|\theta).$$

Let θ and θ' denote two choices for the parameter vector, and introduce the function

$$Q(\theta, \theta') = \sum_{\Xi} \Pr(y, \Xi|\theta) \log \Pr(\mathbf{y}, \Xi|\theta').$$

If we fix θ and maximize this with respect to θ', then the following standard lemma applies [18, 55].

Lemma 1 *The inequality* $Q(\theta, \theta') \geq Q(\theta, \theta)$ *implies* $\Pr(y|\theta') \geq \Pr(y|\theta)$, *with equality if and only if* $Pr(\mathbf{y}, \Xi|\theta') = Pr(\mathbf{y}, \Xi|\theta)$ *for each state transition sequence* Ξ.

For the verification, observe first that

$$Q(\theta, \theta') - Q(\theta, \theta) = \sum_{\Xi} \Pr(y, \Xi|\theta) \log \frac{\Pr(y, \Xi|\theta')}{\Pr(y, \Xi|\theta)}.$$

As $\log x$ is a concave function, we have the inequality $\log x \leq x - 1$ for all positive x, with equality if, and only if, $x = 1$. Applying this inequality to each term in the sum for $Q(\theta, \theta') - Q(\theta, \theta)$ gives

$$\begin{aligned} Q(\theta, \theta') - Q(\theta, \theta) &\leq \sum_{\Xi} \Pr(y, \Xi|\theta) \left(\frac{\Pr(y, \Xi|\theta')}{\Pr(y, \Xi|\theta)} - 1 \right) \\ &= \sum_{\Xi} \Pr(y, \Xi|\theta') - \sum_{\Xi} \Pr(y, \Xi|\theta) \\ &= \Pr(y|\theta') - \Pr(y|\theta). \end{aligned}$$

Equality holds if, and only if, the argument to each logarithm equals one, or $\Pr(\mathbf{y}, \Xi|\theta') = \Pr(\mathbf{y}, \Xi|\theta)$ for each valid state transition sequence Ξ.

From this result, the following iterative algorithm, called the expectation maximization algorithm, ensures a monotonic increase of the likelihood function $\Pr(\mathbf{y}|\theta)$, from any initial estimate $\theta^{(0)}$.

1. (*Expectation step*) Given the parameter estimate $\theta^{(m)}$ at iteration m, calculate the expectation

$$Q(\theta^{(m)}; \theta') = \sum_{\Xi} \Pr(\mathbf{y}\Xi|\theta^{(m)}) \log \Pr(y, \Xi|\theta')$$

as a function of θ'.

2. (*Maximization step*) Choose $\theta^{(m+1)}$ by maximizing the expectation:

$$\theta^{(m+1)} = \arg\max_{\theta'} Q(\theta^{(m)}, \theta')$$

The maximization step can be solved using the forward-backward algorithm, as we verify presently. First, let us develop

$$\log \Pr(y, \Xi | \theta') = \log \Pr(y | \Xi, \theta') + \log \Pr(\Xi | \theta').$$

Here we note that, given the state transition sequence Ξ and parameter vector θ', successive channel outputs (y_i) are conditionally independent because the channel noise is white and Gaussian. The mean of y_i, denoted by $\mathcal{H}'(\xi_i)$, is the noise-free channel output at time i using channel coefficients $\mathbf{h}'$ for the given extended state configuration ξ_i at time i. Thus we have

$$\Pr(y | \Xi, \theta') = \frac{1}{(\sqrt{2\pi}\sigma')^N} \prod_{i=1}^{N} \exp\left(-\frac{[y_i - \mathcal{H}'(\xi_i)]^2}{2\sigma'^2}\right).$$

We note also that $\Pr(\Xi | \theta') = \Pr(\Xi)$ since the state transition sequence Ξ depends on the channel input sequence (d_i), but not on the channel coefficients. Our development for $\log \Pr(y, \Xi | \theta')$ thus reads as

$$\begin{aligned} \log \Pr(y, \Xi | \theta') &= \log \Pr(y | \Xi, \theta') + \log \Pr(\Xi | \theta') \\ &= -\sum_i \left(\frac{[y_i - \mathcal{H}'(\xi_i)]^2}{2\sigma'^2} + \log \sigma' + \frac{\log(2\pi)}{2}\right) + \log \Pr(\Xi). \end{aligned}$$

Inserting this development into the sum for $Q(\theta^{(m)}, \theta')$ then gives

$$\begin{aligned} Q(\theta^{(m)}, \theta') = &-\sum_{\Xi} \Pr(y, \Xi | \theta^{(m)}) \sum_i \left(\frac{[y_i - H'(\xi_i)]^2}{2\sigma'^2} + \log \sigma' + \frac{\log(2\pi)}{2}\right) \\ &+ \sum_{\Xi} \Pr(y, \Xi | \theta^{(m)}) \log \Pr(\Xi) \\ = &-\sum_{i,j} \left(\frac{[y_i - \mathcal{H}'(\xi_i)]^2}{2\sigma'^2}\right) \Pr(y, \xi_i = \Sigma_j | \theta^{(m)}) \\ &+ N\left(\log \sigma' + \frac{\log(2\pi)}{2}\right) \sum_{i,j} \Pr(y, \xi_i = \Sigma_j | \theta^{(m)}) \\ &+ \sum_{\xi} \Pr(y, \Xi | \theta^{(m)}) \log \Pr(\Xi) \end{aligned}$$

which is seen to expose the marginal evaluations $\Pr(\mathbf{y}, \xi_i = \Sigma_j | \theta^{(m)})$ with respect to Ξ.

By setting derivatives with respect to the unknowns $\theta' = (\{\mathcal{H}'(\Sigma_j)\}, \sigma')$ to zero and solving, the updated parameters are found as

$$\begin{aligned}\mathcal{H}^{(m+1)}(\Sigma_j) &= \frac{\sum_i y_i \Pr(y, \xi_i = \Sigma_j | \theta^{(m)})}{\sum_i \Pr(y, \xi_i = \Sigma_j | \theta^{(m)})} \\ (\sigma'^2)^{(m+1)} &= \frac{\sum_{i,j} [y_i - \mathcal{H}^{(m+1)}(\Sigma_j)]^2 \Pr(y, \xi_i = \Sigma_j | \theta^{(m)})}{N \sum_{i,j} \Pr(y, \xi_i = \Sigma_j | \theta^{(m)})}.\end{aligned} \tag{3.7}$$

Finally, note that if each marginal likelihood evaluation $\Pr(\mathbf{y}, \xi_i = \Sigma_j | \theta^{(m)})$ is scaled by the same factor, the expressions for the updated parameters do not change. Choosing the scale factor as $1/\Pr(\mathbf{y})$ then gives

$$\frac{\Pr(y, \xi_i = \Sigma_j | \theta^{(m)})}{\Pr(y)} = \Pr(\xi_i = \Sigma_j | \mathbf{y}, \theta^{(m)}).$$

This is the state transition probability encompassed by ξ_i, and is available from the forward-backward algorithm applied to the trellis diagram for the channel. Using the three-tap channel from Figure 3.8, for example, these state transition probabilities become

$$\begin{aligned}
\Pr(\xi_i = \Sigma_0 | \mathbf{y}, \theta) &\propto \alpha_{i-1}(0)\, \gamma_i(0, 0)\, \beta_i(0) \\
\Pr(\xi_i = \Sigma_1 | \mathbf{y}, \theta) &\propto \alpha_{i-1}(0)\, \gamma_i(0, 1)\, \beta_i(1) \\
\Pr(\xi_i = \Sigma_2 | \mathbf{y}, \theta) &\propto \alpha_{i-1}(1)\, \gamma_i(1, 2)\, \beta_i(2) \\
\Pr(\xi_i = \Sigma_3 | \mathbf{y}, \theta) &\propto \alpha_{i-1}(1)\, \gamma_i(1, 3)\, \beta_i(3) \\
\Pr(\xi_i = \Sigma_4 | \mathbf{y}, \theta) &\propto \alpha_{i-1}(2)\, \gamma_i(2, 0)\, \beta_i(0) \\
\Pr(\xi_i = \Sigma_5 | \mathbf{y}, \theta) &\propto \alpha_{i-1}(2)\, \gamma_i(2, 1)\, \beta_i(1) \\
\Pr(\xi_i = \Sigma_6 | \mathbf{y}, \theta) &\propto \alpha_{i-1}(3)\, \gamma_i(3, 2)\, \beta_i(2) \\
\Pr(\xi_i = \Sigma_7 | \mathbf{y}, \theta) &\propto \alpha_{i-1}(3)\, \gamma_i(3, 3)\, \beta_i(3)
\end{aligned}$$

using the terms α, β and γ as per Section 3.5.1.

The modifications to the turbo equalizer are now straightforward. At each iteration, the means $\{\mathcal{H}^{(m+1)}(\Sigma_j)\}$ and variance σ'^2 are calculated from the inner decoder using (3.7) (along with the extrinsic values to be passed to outer decoder, as usual), and these values are used for the likelihood calculations γ_i of the inner decoder for the next iteration [as per (3.4)]. In this way, a blind turbo equalizer is obtained.

■ EXAMPLE 3.10

Obtaining the Channel Coefficients. In the parameter estimation step we contented ourselves with finding the noise-free output symbols $\mathcal{H}(\Sigma_j)$ rather than the channel coefficients themselves, as this is all that is necessary for the likelihood evaluations for the next iteration [following (3.4)]. In case the channel impulse response coefficients are desired, they may be obtained easily. We illustrate here for the case of a three-tap channel.

The noise-free channel output symbols ideally satisfy the linear relation

$$\underbrace{\begin{bmatrix} \mathcal{H}(\Sigma_0) \\ \mathcal{H}(\Sigma_1) \\ \mathcal{H}(\Sigma_2) \\ \mathcal{H}(\Sigma_3) \\ \mathcal{H}(\Sigma_4) \\ \mathcal{H}(\Sigma_5) \\ \mathcal{H}(\Sigma_6) \\ \mathcal{H}(\Sigma_7) \end{bmatrix}}_{\mathcal{H}} = \underbrace{\begin{bmatrix} 1 & 1 & 1 \\ -1 & 1 & 1 \\ 1 & -1 & 1 \\ -1 & -1 & 1 \\ 1 & 1 & -1 \\ -1 & 1 & -1 \\ 1 & -1 & -1 \\ -1 & -1 & -1 \end{bmatrix}}_{\mathbf{F}} \underbrace{\begin{bmatrix} h_0 \\ h_1 \\ h_2 \end{bmatrix}}_{\mathbf{h}}.$$

The estimated values of the means from (3.7), collected into a vector $\widehat{\mathcal{H}}$, may not satisfy exactly this relation for any $\mathbf{h}$. To compensate for possible errors in $\widehat{\mathcal{H}}$, a least-squares estimate of $\mathbf{h}$ may be computed, using the formula

$$\mathbf{h}_{LS} = (\mathbf{F}^T\mathbf{F})^{-1}\mathbf{F}^T\widehat{\mathcal{H}}.$$

We observe here that $\mathbf{F}^T\,\mathbf{F} = 8\mathbf{I}$ (with $\mathbf{I}$ the identity matrix). More generally, using a channel with degree L (having thus $L+1$ coefficients), we find that $\mathbf{F}^T\,\mathbf{F} = 2^{L+1}\mathbf{I}$, so that the least-squares solution for $\mathbf{h}$ simplifies to

$$\mathbf{h}_{LS} = 2^{-(L+1)}\mathbf{F}^T\widehat{\mathcal{H}}$$

whose calculation requires only sums, differences, and a scale factor.

■ EXAMPLE 3.11

Simulation Example for Estimating Channel Impulse Response. Figure 3.13 shows the channel impulse response estimates from a single run of the blind turbo equalizer applied to the same Proakis A channel setting as used in Example 3.6, with a rate-adjusted SNR of $E_b/\sigma^2 = 6$ dB, and an exponential distribution for the initial channel coefficient values, so that the various sums and differences of the channel coefficients (which give the means used in the first channel likelihood evaluations) are uniformly distributed. The estimated channel coefficient values are observed to converge acceptably close to their true values, indicated by the horizontal dashed lines. For this particular run, the blind turbo equalizer successfully restores the transmitted information sequence.

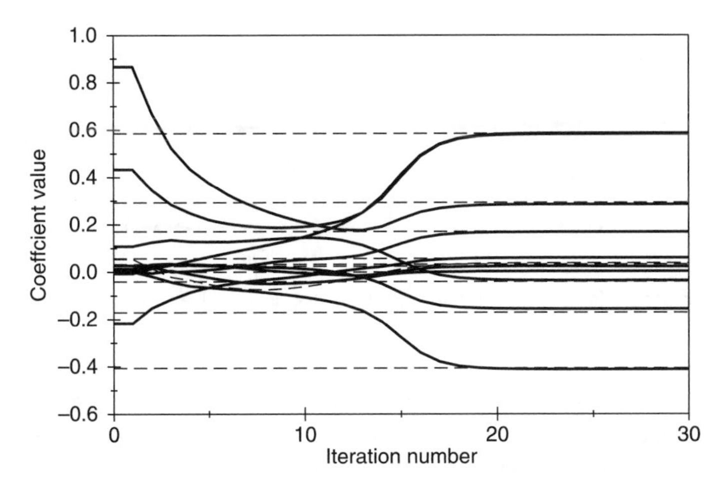

Figure 3.13 Showing the channel impulse response coefficient estimates from a single run of the blind turbo equalizer.

3.8.1 Differential Encoding

A fundamental ambiguity in blind channel estimation is that delay and/or phase shifts cannot be detected, owing to the absence of an absolute time reference. We review here how differential encoding can be exploited to remove the phase uncertainty for the case of antipodal signaling. Differential encoding may be viewed as a rate-one convolutional code operation, whose corresponding decoder may be absorbed into the turbo loop [57]. The resulting scheme gives a special case of nested iterative loops, introduced earlier in a serial coding context in [3].

The basic sign ambiguity problem of antipodal signaling is that $\mathbf{h}$ and $-\mathbf{h}$ produce the same output constellation, consisting of sums and differences of the channel coefficients. Thus if a given solution $\mathbf{h}$ maximizes the likelihood function of the previous section, so does $-\mathbf{h}$. The channel input values $(\hat{d}_i)$ inferred from the inner decoder for these two choices of the channel coefficients do not agree; rather, one sequence is the negative of the other. The principle of differential encoding is to render this sign ambiguity innocuous, by coding the relevant information into the difference of two successive channel symbols. In this way, if all channel symbols are negated, the difference from one to the next is preserved.

A differential encoder is readily absorbed into the communication chain, and effects the transformation

$$e_i = d_i\, e_{i-1}, \qquad e_0 = +1$$

where the seed value e_0 could have equally well been chosen as $e_0 = -1$. In the absence of noise, the sequence (d_i) can be recovered from the sequence (e_i) according to

$$d_i = \frac{e_i}{e_{i-1}} = e_i\, e_{i-1}$$

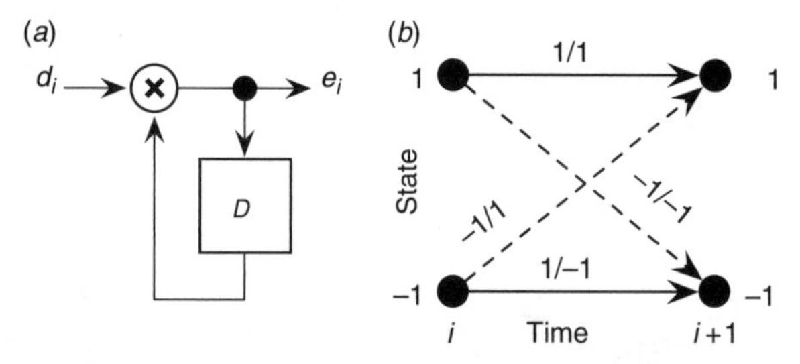

Figure 3.14 (*a*) Differential encoder; (*b*) One section of the trellis diagram, denoting transition branches as d_i/e_i.

in which the latter equality holds because each e_i is either $+1$ or -1. We observe that a sign change applied to the sequence (e_i) cancels in each term of the product $e_i\, e_{i-1}$, and so does not affect (d_i). This is the desired behavior.

In the presence of channel noise, the decoding formula $d_i = e_i\, e_{i-1}$ is less than reliable. Instead, the operation of the differential encoder may be viewed as a rate-one convolutional encoder [57], as sketched in Figure 3.14(*a*). A single section of the trellis diagram for the encoder is shown in Figure 3.14(*b*), to which the forward-backward algorithm may be applied for the purposes of decoding. The forward recursion takes the form

$$\begin{aligned}\alpha_i(+1) &\propto \alpha_{i-1}(+1)\,\Pr(e_i = 1\,|\,d_i = 1)\,\Pr(d_i = 1)\\ &\quad + \alpha_{i-1}(-1)\,\Pr(e_i = 1\,|\,d_i = -1)\,\Pr(d_i = -1)\\ \alpha_i(-1) &\propto \alpha_{i-1}(+1)\,\Pr(e_i = -1\,|\,d_i = -1)\,\Pr(d_i = -1)\\ &\quad + \alpha_{i-1}(-1)\,\Pr(e_i = -1\,|\,d_i = 1)\,\Pr(d_i = 1)\end{aligned} \tag{3.8}$$

while the backward recursion becomes

$$\begin{aligned}\beta_{i-1}(+1) &\propto \beta_i(+1)\,\Pr(e_i = 1\,|\,d_i = 1)\,\Pr(d_i = 1)\\ &\quad + \beta_i(-1)\,\Pr(e_i = -1\,|\,d_i = -1)\,\Pr(d_i = -1)\\ \beta_{i-1}(-1) &\propto \beta_i(+1)\,\Pr(e_i = 1\,|\,d_i = -1)\,\Pr(d_i = -1)\\ &\quad + \beta_i(-1)\,\Pr(e_i = -1\,|\,d_i = 1)\,\Pr(d_i = 1)\end{aligned} \tag{3.9}$$

using the boundary values

$$\begin{aligned}\alpha_0(+1) &= \Pr(e_0 = +1) \quad & \beta_n(+1) &= \Pr(e_n = +1)\\ \alpha_0(-1) &= \Pr(e_0 = -1) \quad & \beta_n(-1) &= \Pr(e_n = -1).\end{aligned}$$

These probabilities may be estimated from the first and last received symbols. Note in particular that this requires transmitting the seed value e_0 of the differential encoder.

Integrating this element into the overall communication chain then appears as in Figure 3.15 at the transmitter, and Figure 3.16 at the receiver. Observe that we now

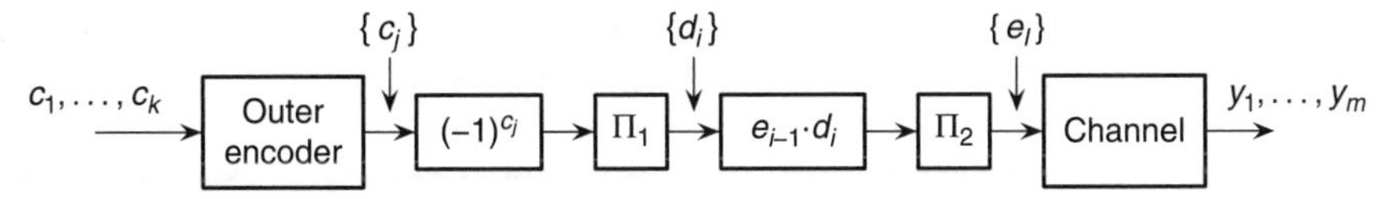

Figure 3.15 Integrating a differential encoder at the transmitter.

have a doubly-concatenated decoder [3]. The differential decoder takes as inputs the extrinsic probabilities from the channel and outer decoders, and replaces the prior and conditional probabilities according to

$$\begin{aligned} \Pr(e_i = +1 \mid d_i) &\leftarrow T_i(+1) \\ \Pr(e_i = -1 \mid d_i) &\leftarrow T_i(-1), \quad i = \Pi_2^{-1}(l) \\ \Pr(d_i = +1) &\leftarrow U_i(+1) \\ \Pr(d_i = -1) &\leftarrow U_i(-1), \quad i = \Pi_1(j) \end{aligned}$$

in running the forward-backward algorithm in (3.8) and (3.9). In addition, the boundary values for $\alpha_0(e_0)$ and $\beta_n(e_n)$ are set to $T_0(e_0)$ and $T_n(e_n)$, respectively. The differential decoder also returns two sets of extrinsic probabilities, with respect to T_j and U_j, respectively.

$$\begin{aligned} V_i(+1) &\propto \alpha_{i-1}(+1)\, U_i(+1)\, \beta_i(+1) + \alpha_{i-1}(-1)\, U_i(-1)\, \beta_i(+1) \\ V_i(-1) &\propto \alpha_{i-1}(+1)\, U_i(-1)\, \beta_i(-1) + \alpha_{i-1}(-1)\, U_i(+1)\, \beta_i(-1) \\ W_i(+1) &\propto \alpha_{i-1}(+1)\, T_i(+1)\, \beta_i(+1) + \alpha_{i-1}(-1)\, T_i(-1)\, \beta_i(-1) \\ W_i(-1) &\propto \alpha_{i-1}(+1)\, T_i(-1)\, \beta_i(-1) + \alpha_{i-1}(-1)\, T_i(+1)\, \beta_i(+1). \end{aligned}$$

The values $V_i(e_i)$, once interleaved, replace the pseudo priors in the channel decoder, while the values $W_i(d_i)$, once de-interleaved, replace the likelihood terms in the outer decoder.

Many variants in the scheduling of the two loops may be envisaged [3]. For example, one may let the inner loop run a few iterations before passing values to the outer loop, or vice-versa. The convergence behavior, however, requires good

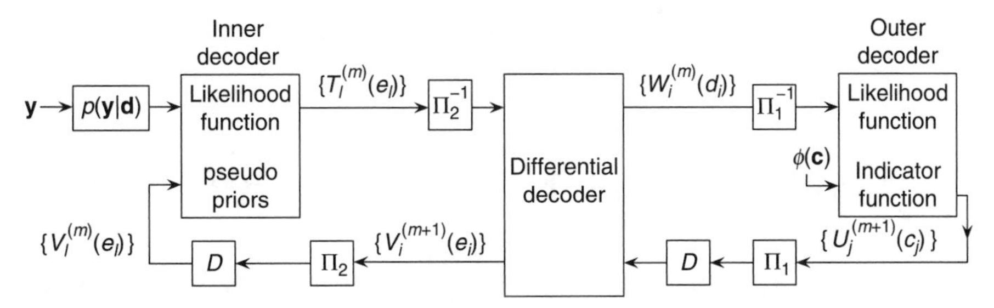

Figure 3.16 Double concatenated decoder, using a differential decoder as the middle element.

initial estimates of the channel coefficients to converge to a correct decoding solution. This would appear an inherent shortcoming of such blind schemes, that is, local convergence is often the best that can be achieved [58].

3.9 CONVERGENCE

The feedback behavior of turbo loops induces nonlinear interactions [31], leading to potentially complicated phenomena such as limit cycles or even chaos in some cases [59]. Experience shows, however, that iterative decoding usually converges under reasonable conditions under which the SNR and block length are large enough. Existing analysis methods to study convergence include those derived from information geometry [31, 60–67], numerical analysis [68–70], and density evolution [71–73]. The information geometry approach views successive iterations of the turbo loops as projectors in appropriate spaces, but is confounded by the absence of key invariants in extrinsic information extraction. The numerical analysis approaches avoid this shortcoming, but give sufficient conditions for convergence that are rather algebraic and do not readily translate into usable design criteria for engineers. The density evolution approach treats the extrinsic information probabilities as independent, identically distributed random variables whose probability density function evolves with successive iterations. A popular version of this approach uses extrinsic information transfer (EXIT) charts, which affords a graphical analysis of the convergence behavior of iterative decoding, and whose results show good agreement with the experimentally observed convergence behavior of iterative decoding. Its shortcoming is that a formal justification (e.g. [71, 72]) appeals to asymptotic approximations which are valid for long block lengths, but which break down for shorter block lengths. Accordingly, we assume in this section that the block length N is sufficiently long. Our presentation begins with an adaptation of EXIT analysis to turbo equalization [39, 74], and then provides a brief overview of some variants of this analysis for the interference canceler configuration [75], [69].

To begin, we first transform the extrinsic probabilities into the following log extrinsic probability ratios

$$\tau_i \triangleq \log \frac{T_i(d_i = +1)}{T_i(d_i = -1)}$$

$$\Upsilon_i \triangleq \log \frac{U_i(d_i = +1)}{U_i(d_i = -1)}.$$

The basic modeling assumption which underlies this analysis is summarized as follows.

Assumption 1 *The variable τ_i (resp. Υ_i) is modeled as the output of a virtual channel with input $d_{i,}$ a gain of $0.5\varsigma_\tau^2$ (resp. $0.5\varsigma_\Upsilon^2$), and additive white Gaussian noise with variance ς_τ^2 (resp. ς_Υ^2).*

In essence, the conditional mean (with respect to d_i) is half the conditional variance. The Gaussian model is motivated by the central limit theorem which applies when the block length N is sufficiently large, and is supported by empirical verification [73] (see also Problem 3.2). From this model, the log extrinsic probability ratios follow the conditional distributions [73]

$$\Pr(\tau_i|d_i = \pm 1) = \frac{1}{\sqrt{2\pi}\,\varsigma_\tau} \exp\left(-\frac{(\tau_i \mp 0.5\varsigma_\tau^2)^2}{2\varsigma_\tau^2}\right) \tag{3.10}$$

$$\Pr(\Upsilon_i|d_i = \pm 1) = \frac{1}{\sqrt{2\pi}\,\varsigma_\Upsilon} \exp\left(-\frac{(\Upsilon_i \mp 0.5\varsigma_\Upsilon^2)^2}{2\varsigma_\Upsilon^2}\right) \tag{3.11}$$

either of which is determined by a single parameter ς_τ or ς_Υ.

Let us return to the turbo decoder of Figure 3.2. This is initialized with $U_i^{(0)}(d_i) = 0.5$ for all i, corresponding to $\Upsilon_i^{(0)} = 0$ for all i. This fits the conditional Gaussian distribution (3.11) with $\varsigma_\Upsilon = 0$. The inner decoder then produces a set of extrinsic probabilities $\{T_i^{(0)}(d_i)\}$, whose log ratios follow the conditional distribution (3.10) characterized by a specific value of ς_Υ (according to the modeling assumption). These extrinsic probabilities are then mapped to a new set of priors $\{U_i^{(1)}(d_i)\}$ via the outer decoder; their log ratios $\{\Upsilon_i^{(1)}\}$ follow the conditional distribution (3.11) described by a new value of the parameter ς_Υ. The process then iterates, and we may thus track the values of the parameters ς_τ and ς_Υ through successive iterations as

$$\varsigma_\Upsilon^{(0)} = 0 \xrightarrow[\text{decoder}]{\text{inner}} \varsigma_\tau^{(0)} \xrightarrow[\text{decoder}]{\text{outer}} \varsigma_\Upsilon^{(1)} \xrightarrow[\text{decoder}]{\text{inner}} \varsigma_\tau^{(1)} \xrightarrow[\text{decoder}]{\text{outer}} \varsigma_\Upsilon^{(2)} \cdots$$

The technique of EXIT [73] analysis is to infer two transfer functions—the first maps ς_Υ to ς_τ in the inner decoder, while the second maps ς_τ to ς_Υ in the outer decoder

$$\varsigma_\tau = f_{\text{inner}}(\varsigma_\Upsilon)$$
$$\varsigma_\Upsilon = f_{\text{outer}}(\varsigma_\tau).$$

The successive values of the parameters ς_τ and ς_Υ then follow the iterations

$$\varsigma_\tau^{(m)} = f_{\text{inner}}(\varsigma_\Upsilon^{(m)}), \qquad \varsigma_\Upsilon^{(m+1)} = f_{\text{outer}}(\varsigma_\tau^{(m)}).$$

A recognized complication of this analysis technique is that closed-form expressions for the transfer functions $f_{\text{inner}}(\varsigma_\Upsilon)$ and $f_{\text{outer}}(\varsigma_\tau)$ are often unavailable; rather, they must be estimated through extensive simulations. In essence, one generates many realizations of the extrinsic probabilities $\{T_i(d_i)\}$ for a particular value of ς_τ, and numerically determines the parameter ς_Υ which best fits the variables $\{U_i(d_i)\}$ which result from the application of the outer decoder. By repeating this experiment for a range of values for ς_τ, the function $f_{\text{outer}}(\varsigma_\tau)$ is estimated. A similar procedure is carried out using the inner decoder to estimate $f_{\text{inner}}(\varsigma_\Upsilon)$.

For analysis purposes, it proves convenient to map the parameters ς_τ and ς_Y to mutual information values $I(d_i, \tau_i)$ and $I(d_i, Y_i)$, respectively; these relate to the conditional probability distribution functions $\Pr(\tau_i|d_i)$ and $\Pr(Y_i|d_i)$ from (3.10) and (3.11) according to [39, 73]

$$I(d_i, \tau_i) = \frac{1}{2}\left(\int_{\tau_i} \Pr(\tau_i|+1)\log\frac{2\Pr(\tau_i|+1)}{\Pr(\tau_i|+1)+\Pr(\tau_i|-1)}\,d\tau_i\right.$$
$$\left. + \int_{\tau_i} \Pr(\tau_i|-1)\log\frac{2\Pr(\tau_i|-1)}{\Pr(\tau_i|+1)+\Pr(\tau_i|-1)}\,d\tau_i\right)$$
$$I(d_i, Y_i) = \frac{1}{2}\left(\int_{Y_i} \Pr(Y_i|+1)\log\frac{2\Pr(Y_i|+1)}{\Pr(Y_i|+1)+\Pr(Y_i|-1)}\,dY_i\right.$$
$$\left. + \int_{\tau_i} \Pr(Y_i|-1)\log\frac{2\Pr(Y_i|-1)}{\Pr(Y_i|+1)+\Pr(Y_i|-1)}\,d\tau_i\right)$$

assuming $\Pr(d_i = 1) = \Pr(d_i = -1) = \frac{1}{2}$.

Although seemingly complicated at first sight, the key features of these mutual information expressions are readily summarized. First, the mutual information $I(d_i, \tau_i)$ is a monotonically increasing function of ς_τ, and bounded as $0 \le I(d_i, \tau_i) \le 1$, as graphed in Figure 3.17. Second, the lower bound $0 = I(d_i, \tau_i)$ is attained at $\varsigma_\tau = 0$, and implies statistical independence of d_i and τ_i. Finally, the upper bound $I(d_i, \tau_i) = 1$ is attained in the limit as $\varsigma_\tau \to \infty$, and implies $T_i(1) = d_i$. Thus, values of $I(d_i, \tau_i)$ near 1 imply a low probability of error in decoding.

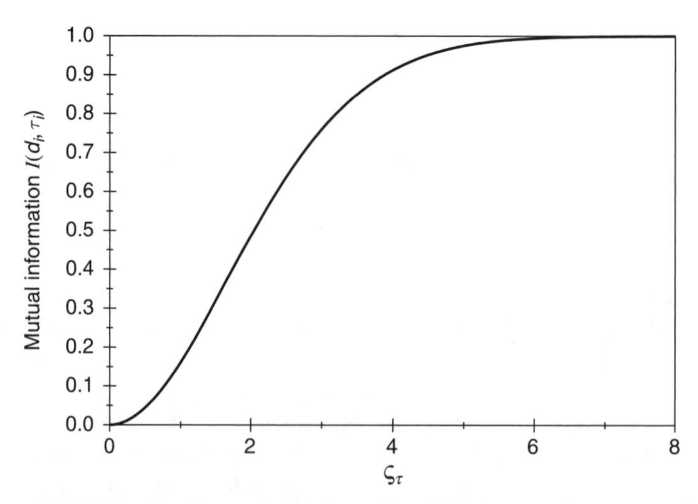

Figure 3.17 Showing the mutual information $I(d_i, \tau_i)$ versus the parameter ς_τ.

Now, since ς_τ determines $\Pr(\tau_i|d_i)$ and thus $I(d_i, \tau_i)$, and similarly ς_Υ determines $\Pr(\Upsilon_i|d_i)$ and thus $I(d_i, \Upsilon_i)$, the transfer function $\varsigma_\tau = f_{\text{inner}}(\varsigma_\Upsilon)$ can be rephrased as $I(d_i, \tau_i) = g_{\text{inner}}[I(d_i, \Upsilon_i)]$. Similarly, the transfer function $\varsigma_\Upsilon = f_{\text{outer}}(\varsigma_\tau)$ can be rephrased as $I(d_i, \Upsilon_i) = g_{\text{outer}}[I(d_i, \tau_i)]$, and successive iterations are then described as

$$I(d_i, \tau_i^{(m)}) = g_{\text{inner}}[I(d_i, \Upsilon_i^{(m)})]$$
$$I(d_i, \Upsilon_i^{(m+1)}) = g_{\text{outer}}[I(d_i, \tau_i^{(m)})]$$

■ EXAMPLE 3.12

EXIT Chart Construction. Figure 3.18 shows the extrinsic information transfer function

$$I(d_i, \Upsilon_i) = g_{\text{outer}}[I(d_i, \tau_i)]$$

for the decoder corresponding to the rate 1/2 encoder from Figure 3.4. The curve is obtained by randomly generating pseudo priors $U_i(d_i)$ such that their log ratios $\Upsilon_i = \log[U_i(1)/U_i(0)]$ follow the conditional Gaussian distribution from (3.11), for a given value of ς_Υ. These values, when fed to the decoder, give extrinsic information values T_i. The histogram of their log ratios $\tau_i = \log[T_i(1)/T_i(0)]$ was empirically verified to fit the conditional distribution (3.10), and the value ς_τ can be estimated from the mean of the values τ_i. By repeating this experiment for a range of values for ς_Υ, and transforming to mutual information values $I(d_i, \Upsilon_i)$ and $I(d_i, \tau_i)$, the plot of Figure 3.18 results.

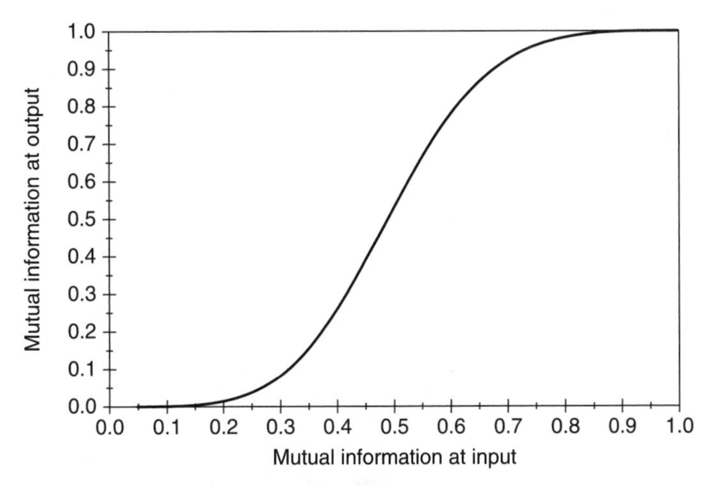

Figure 3.18 Extrinsic information transfer function for the outer decoder corresponding to the rate 1/2 encoder of Figure 3.4.

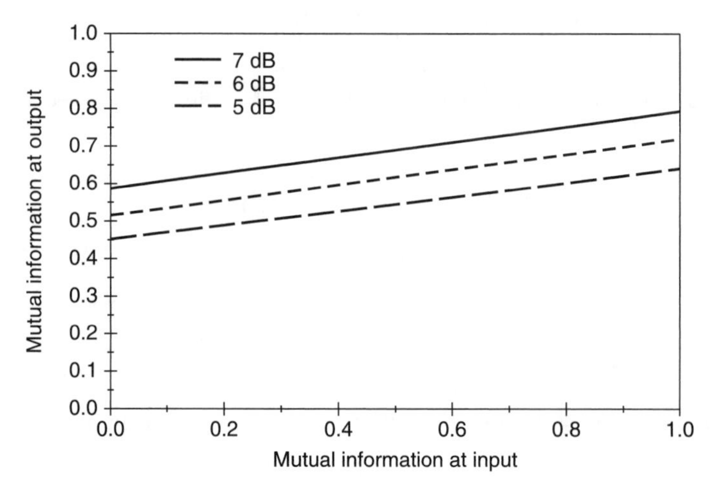

Figure 3.19 Extrinsic information transfer function for the inner (channel) decoder corresponding to the channel of Example 3.7, for three (rate-adjusted) SNRs.

Figure 3.19 shows the extrinsic information transfer function

$$I(d_i, \tau_i) = g_{\text{inner}}[I(d_i, \Upsilon_i)]$$

using the forward-backward channel decoder corresponding to the Proakis B channel used in Example 3.7, for three (rate adjusted) SNRs in the received signal that provides the initial channel likelihood function to the decoder.

The two graphs can be superimposed after exchanging the ordinate and abscissa of the outer decoder's transfer function, giving the result of Figure 3.20. The

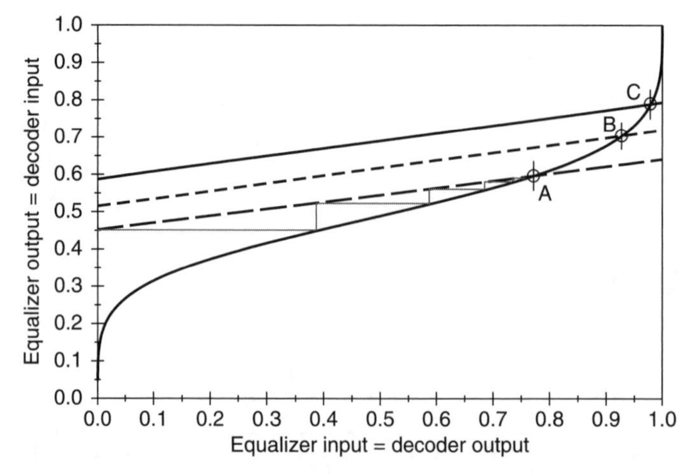

Figure 3.20 Exit chart obtained by combining the extrinsic information transfer graphs for the inner and outer decoders. The intersection of the graphs gives the mutual information values obtained after convergence, for three (rate-adjusted) SNRs.

staircase line also indicates the evolution of the mutual information from successive iterations for the case of a 5 dB (rate-adjusted) SNR. The intersection of graphs indicates the mutual information values obtained after convergence.

3.9.1 Bit Error Probability

A nice by-product of the EXIT analysis is that, subject to Assumption 1 above, the bit error probability can be related analytically to the two parameters ς_T and ς_Y, as follows.

Theorem 2 *Let ς_T and ς_Y be the parameters of the conditional distributions of the log extrinsic probability ratios. The bit error probability may be expressed as*

$$\mathrm{BER}(\varsigma_T, \varsigma_Y) = \int_0^1 g(t, \varsigma_T) f(t, \varsigma_Y)\, dt$$

in which

$$g(t, \varsigma_T) = \frac{1}{\sqrt{2\pi}\varsigma_T t(1-t)} \exp\left[-\left(\log \frac{t}{1-t} + \frac{\varsigma_T^2}{2} \right)^2 \Big/ 2\varsigma_T^2 \right]$$

is the probability distribution function of the extrinsic information $t = T_i(1)$, and

$$f(t, \varsigma_Y) = \frac{1}{2}\left[1 - \mathrm{erf}\left(\frac{\beta(t)}{\sqrt{2}\varsigma_Y} - \frac{\varsigma_Y}{2\sqrt{2}} \right) \right], \quad \text{with } \beta(t) = \log \frac{1-t}{t}$$

is the probability of error induced by $U_i(1)$ for a fixed $t = T_i(1)$, and

$$\mathrm{erf}(x) = \frac{2}{\sqrt{\pi}} \int_0^x \exp(-\xi^2)\, d\xi$$

is the Gaussian error function.

For the proof, we assume without loss of generality that $c_j = 0$ so that $d_i = (-1)^{c_j} = 1$ [with $j = \Pi(i)$], in which case the log extrinsic ratios are distributed according to

$$\log \frac{T_i(1)}{T_i(0)} \sim N\left(\frac{-\varsigma_T^2}{2}, \varsigma_T^2 \right)$$

$$\log \frac{U_i(1)}{U_i(0)} \sim N\left(\frac{-\varsigma_Y^2}{2}, \varsigma_Y^2 \right).$$

Now, the *a posteriori* bit estimate $\hat{c}_j$ is given as

$$\hat{c}_j = \frac{T_i(1)\,U_i(1)}{T_i(1)\,U_i(1) + [1 - T_i(1)][1 - U_i(1)]}, \qquad \text{with } j = \Pi(i).$$

An error occurs when $\hat{c}_j > 0.5$. It is easy to check (cf. Problem 3.1) that

$$\hat{c}_j > 0.5 \iff T_i(1) + U_i(1) > 1.$$

It thus suffices to deduce the probability that $T_i(1) + U_i(1) > 1$. Now, if we fix $T_i(1)$ at a particular value t, then

$$\Pr[U_i(1) > 1 - t] = \Pr\left(\log\frac{U_i(1)}{U_i(0)} > \log\frac{1-t}{t}\right).$$

Since $\log[U_i(1)/U_i(0)]$ follows a Gaussian distribution, this probability reduces to the error function expression for $f(t, \varsigma_Y)$ above.

It suffices now to average this over the probability density function for t. Since $\log[t/(1-t)] = \log[T_i(1)/T_i(0)]$ follows a Gaussian distribution, the probability density function for t is readily found as the expression for $g(t,\ \varsigma_T)$ in the theorem statement.

■ EXAMPLE 3.13

Performance Prediction From EXIT Chart. Consider the intersection labeled "A" in Figure 3.20, corresponding to a rate adjusted signal to noise ratio of $10\log_{10} E_b/\sigma^2 = 5$ dB. The parameters at the intersection give

$$\varsigma_Y = 2.8, \qquad \varsigma_T = 2.3.$$

Inserting these values into the formula from Theorem 2 results in

$$\text{BER}(2.3,\ 2.8) = 0.035$$

in agreement with the simulation results of Example 3.7. Similarly, at the intersections labeled "B" and "C" (corresponding to 6 and 7 dB, respectively, for the rate-adjusted SNR), we have

$$\text{B:}\quad \left.\begin{aligned} \varsigma_Y &= 4.17 \\ \varsigma_T &= 2.76 \end{aligned}\right\} \rightarrow \text{BER} = 0.0062$$

$$\text{C:}\quad \left.\begin{aligned} \varsigma_Y &= 5.14 \\ \varsigma_T &= 3.14 \end{aligned}\right\} \rightarrow \text{BER} = 0.0012$$

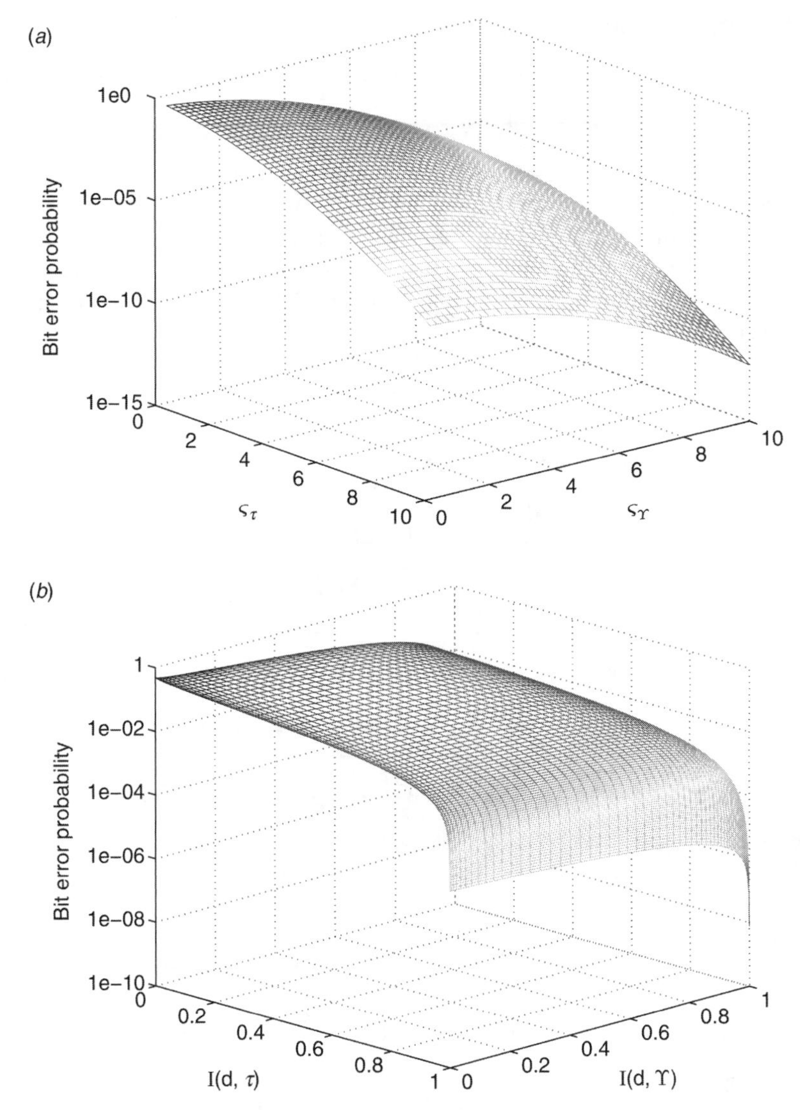

Figure 3.21 Bit error probability versus the parameters ς_τ and ς_Υ (*a*) or versus the two mutual information values used in the EXIT function analysis (*b*).

again in agreement with the values observed from the simulation results of Example 3.7.

The bit error probability as a function of the parameters σ_τ and σ_Υ appears as Figure 3.21, which may also be plotted versus the mutual information values $I(d_i, \tau_i)$ and $I(d_i, \Upsilon_i)$. The plot confirms that as these mutual information values approaches their upper limit of unity, the bit error probability becomes vanishingly small.

3.9.2 Other Encoder Variants

Although we have focused on the (5,7) encoder of Figure 3.3 for ease of exposition, many other possibilities, of course exist [5, 7, 9, 76–78]. The more commonly used variants [79] include (13, 15) and (23, 35) systematic encoders of Figure 3.22, both of rate $1/2$. A comparison of their extrinsic information transfer functions of their corresponding decoders appears as Figure 3.23, in which we observe that the coders of increasing complexity have transfer curves inching closer to a step function. Figure 3.24 shows the flow graphs of some popular rate $2/3$ encoders [79], along with the extrinsic information transfer functions in Figure 3.25. Further possibilities may be explored in [5, 7, 76, 79].

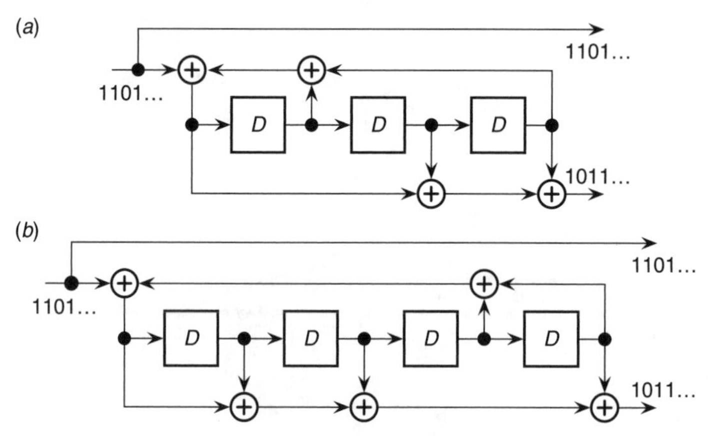

Figure 3.22 Popular rate-1/2 systematic convolutional encoders: (*a*) (13, 15); (*b*) (23, 35).

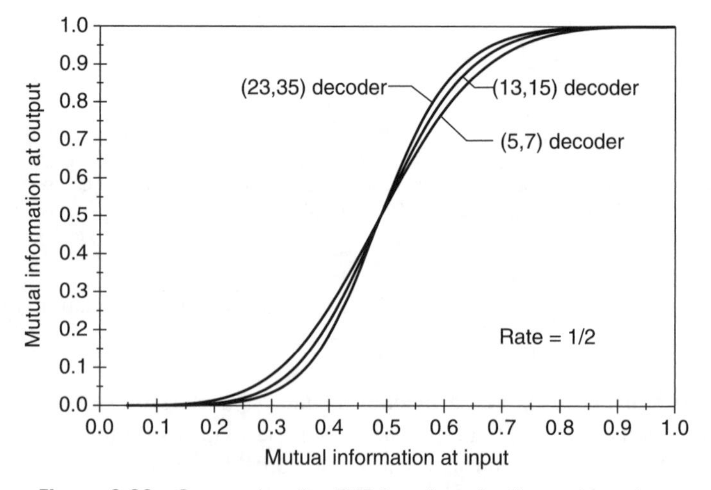

Figure 3.23 Comparing the EXIT functions for three decoders.

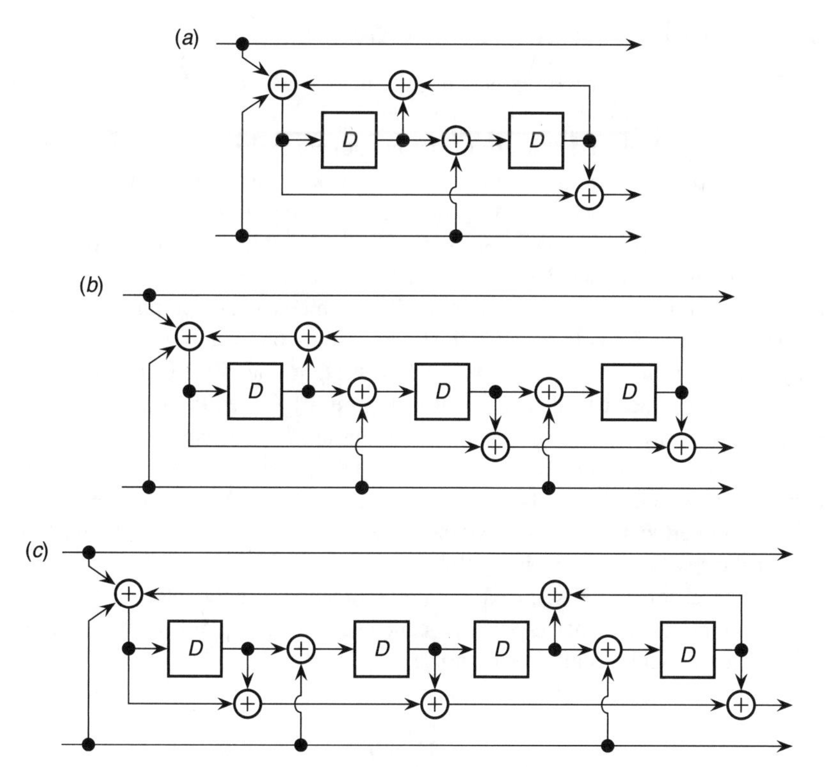

Figure 3.24 Popular rate-2/3 systematic encoders of increasing complexity. (*a*) (5,7); (*b*) (13,15); (*c*) (23,35).

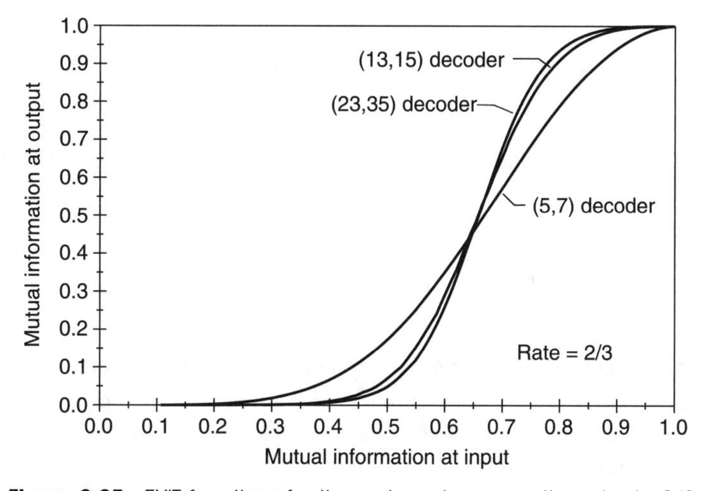

Figure 3.25 EXIT functions for three decoders operating at rate 2/3.

3.9.3 EXIT Chart for Interference Canceler

The EXIT analysis can also be extended to the interference canceler configuration [38, 39, 74], with an additional complication: The output of the interference canceler, once converted to log likelihood ratios (τ_i), does not adhere to the modeling Assumption 1, except when perfect symbol estimates are fed back to cancel the intersymbol interference. In effect, the residual intersymbol interference in the presence of imperfect feedback is decidedly non-Gaussian, such that the conditional probabilities $\Pr(\tau_i|d_i)$ assume a more complicated form. Accurate extrinsic transfer function estimation then requires delicate histogram estimation [39], which can be computationally intensive. An alternative is to observe that the transfer function relating mutual information through the interference canceler appears as a straight line in the examples from [38, 39, 74]. (This character is also apparent in Figure 3.19, even though that plot corresponds to a forward-backward equalizer, not an interference canceler.) A simplified approach thus consists in evaluating the extrinsic information transfer function of the interference canceler at its extreme points of $I(d_i, \Upsilon_i) = 0$ (no feedback) and $I(d_i, \Upsilon_i) = 1$ (perfect feedback), and then connecting the resulting output mutual information values with a straight line [39].

To illustrate, consider first the case of perfect feedback to the interference canceler. The intersymbol interference is perfectly canceled, and the equalizer output consists of a delayed input symbol plus filtered noise

$$v_i = r_0\, d_{i-L} + \sum_{l=0}^{L} p_l b_{i-l}, \qquad i > L$$

where $r_0 = \sum h_l^2 = \sum p_l^2$. For simplicity, we assume that $r_0 = 1$; otherwise we need only replace v_i by v_i/r_0 in what follows, to reach the same conclusions. Thus with $r_0 = 1$, the filtered noise term $\sum_l p_l\, b_{i-l}$ remains Gaussian, with variance σ^2. Thus, the conditional distribution of v_i remains Gaussian

$$\Pr(v_i|d_{i-L} = \pm 1) = \frac{1}{\sqrt{2\pi}\,\sigma} \exp\left(-\frac{(v_i \mp 1)^2}{2\sigma^2} \right).$$

Accordingly, the log likelihood ratios τ_j (obtained after an L-sample offset and interleaving) become

$$\tau_j = \log \frac{\Pr(v_{i+L}|d_i = +1)}{\Pr(v_{i+L}|d_i = -1)} = \frac{2v_{i+L}}{\sigma^2} \qquad j = \Pi(i).$$

The variable τ_j thus remains conditionally Gaussian given d_i [with $j = \Pi(i)$], with conditional mean $-2d_i/\sigma^2$ and conditional variance $\varsigma_\tau^2 = 4/\sigma^2$. As such, the output mutual information $I(d_j, \tau_j)$ at perfect feedback may be found from the graphical relation of Figure 3.17, using $\varsigma_\tau = 2/\sigma$.

At the other extreme of zero feedback, we have $I(d_j, Y_j) = 0$. The interference canceler output then appears as

$$v_{i+L} = d_i + \underbrace{\sum_{l=1}^{L} r_l(d_{i-l} + d_{i+l})}_{\substack{\text{intersymbol}\\ \text{interference } \mu}} + \underbrace{\sum_{l=0}^{L} p_l\, b_{i+L-l}}_{\substack{\text{filtered}\\ \text{noise}}}.$$

Although the filtered noise term remains Gaussian, the intersymbol interference term is a discrete random variable assuming values determined from various sums and differences of the channel autocorrelation coefficients r_l, $l \geq 1$. For example, with $L = 2$, the intersymbol interference term would take the following set of values

Value μ_k	Probability ν_k
0	$1/4$
r_1	$1/8$
$-r_1$	$1/8$
r_2	$1/8$
$-r_2$	$1/8$
$r_1 + r_2$	$1/16$
$r_1 - r_2$	$1/16$
$-r_1 + r_2$	$1/16$
$-r_1 - r_2$	$1/16$

The conditional probability $\Pr(v_{i+L}|d_i)$ thus assumes the form of a sum of Gaussians

$$\Pr(v_{i+L}|d_i = \pm 1) = \sum_k \nu_k N(\mu_k \pm 1, \sigma^2)$$

in which ν_k is the probability of the intersymbol interference term assuming the value μ_k, and

$$\mathcal{N}(\mu, \sigma^2) = \frac{1}{\sqrt{2\pi}\,\sigma} \exp\left(-\frac{(v-\mu)^2}{2\sigma^2}\right)$$

denotes a Gaussian distribution with mean μ and variance σ^2. The log likelihood ratio τ_j (obtained after interleaving) then becomes

$$\tau_j = \log \frac{\Pr(v_{i+L}|d_i = +1)}{\Pr(v_{i+L}\,|\,d_i = -1)} = \log \frac{\sum_k \nu_k \mathcal{N}(\mu_k + 1, \sigma^2)}{\sum_k \nu_k \mathcal{N}(\mu_k - 1, \sigma^2)} \qquad j = \Pi(i) \qquad (3.12)$$

■ **EXAMPLE 3.14**

EXIT Chart for Interference Canceler. Here we revisit the simulation setting of Example 3.7, for the which the channel autocorrelation function takes the values

$$r_1 = \frac{2}{3}, \qquad r_2 = \frac{1}{6}.$$

In the absence of feedback, the map (3.12) relating v_{i+L} to τ_j [with $j = \Pi(i)$], is monotonic for this autocorrelation function (although it need not be for other channel choices). As such, one may obtain the conditional distributions of τ as

$$\Pr(\tau_j|d_i = +1) = \frac{\Pr(v_{i+L}|d_i = +1)}{|d\tau_j/dv_{i+L}|} \qquad \Pr(\tau_j|d_i = -1) = \frac{\Pr(v_{i+L}|d_i = -1)}{|d\tau_j/dv_{i+L}|}.$$

From these, one may calculate the mutual information $I(d_i, \tau_j)$ [with $j = \Pi(i)$], when $I(d_i, \Upsilon_j) = 0$. The value when $I(d_i, \Upsilon_j) = 1$ is obtained from $\varsigma_\tau = 2/\sigma$, and the two endpoints may be connected by a straight line. Figure 3.26 shows the resulting EXIT chart for successive values of the rate-adjusted SNR (from 5 to 8 dB). The bit error probabilities inferred from Theorem 1 at the curve intersections show acceptable agreement with those of the simulations from Example 3.7, given that the extrinsic information transfer function inferred for the interference canceler is only approximate.

3.9.4 Related Analyses

The seemingly forced nature of examining mutual information transfer across the interference canceler motivates consideration of other performance-related parameters.

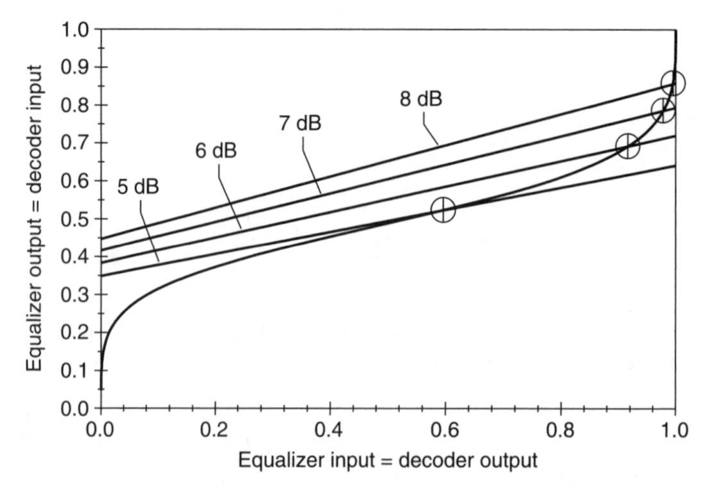

Figure 3.26 EXIT chart for the simulation of Example 3.7 using the interference canceler, for different values of the rate-adjusted SNR.

As the interference canceler appeals to linear filtering, a natural performance measure is the mean-square error of the transmitted symbol estimates [39, 75].

The basic idea is to introduce the symbol estimation errors at the outputs of the outer decoder and interference canceler as

$$e_i^{\mathrm{Y}} = d_i - \hat{d}_i \qquad \text{(symbol error from outer decoder)}$$
$$e_i^{\tau} = d_i - v_{i+L} \qquad \text{(symbol error from interference canceler).}$$

One may then examine how the error variances

$$\sigma_{\mathrm{Y}}^2 \triangleq E[(e_i^{\mathrm{Y}})^2]$$
$$\sigma_{\tau}^2 \triangleq E[(e_i^{\tau})^2]$$

evolve by considering the interconnection of the following two error variance transfer functions

$$\sigma_{\tau}^2 = f_{\mathrm{IC}}(\sigma_{\mathrm{Y}}^2) \qquad \text{(interference canceler)}$$
$$\sigma_{\mathrm{Y}}^2 = f_{\mathrm{outer}}(\sigma_{\tau}^2) \qquad \text{(outer decoder).}$$

The rationale is essentially the same as with EXIT charts, by tracking the evolution of these parameters through successive iterations [39, 75].

If the error term e_i^{Y} is modeled as independent of d_i, then the transfer function of the interference canceler is readily found as

$$\begin{aligned} \sigma_{\tau}^2 &= f_{\mathrm{IC}}(\sigma_{\mathrm{Y}}^2) \\ &= \sigma_{\mathrm{Y}}^2 \sum_{k=0}^{2L} q_k^2 = 2\sigma_{\mathrm{Y}}^2 \sum_{k=1}^{L} r_k^2 \end{aligned}$$

in which we recall that $q_L = 0$ and $r_k = r_{-k} = q_{k+L}$ for $k \neq 0$ in the final line. The effective variance gain $2\sum_{k\geq 1} r_k^2$ indicates that channels having larger autocorrelation terms r_k (with $k \geq 1$) induce a greater sensitivity to errors in the symbol estimates that are fed back to the interference canceler.

The error variance transfer function through the outer decoder, by contrast, does not admit a simple analytic form and, similar to the EXIT analysis, must be obtained through extensive simulations. More detail may be found in [75].

3.10 MULTICHANNEL AND MULTIUSER SETTINGS

Our attention has focused on single user channels for specificity, and we review now how the basic schemes may be modified to handle multichannel and multiuser settings [80–85]. The principles are essentially the same—the changes amount to adopting vector and matrix notations in the channel. We consider first the multichannel case adapted to a single user, and then summarize the basic multiuser case.

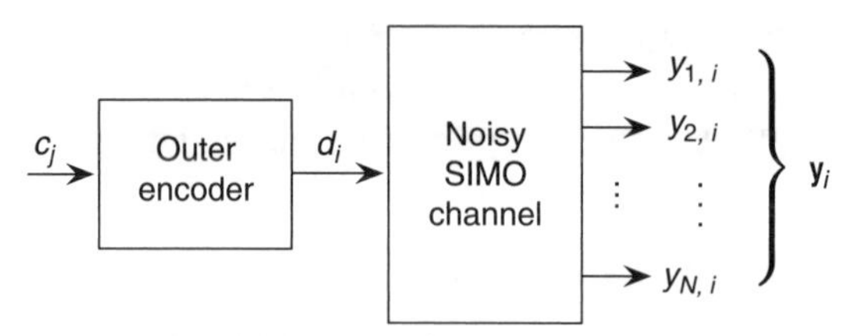

Figure 3.27 Illustrating a noisy SIMO channel.

The basic single-input/multi-output (SIMO) configuration is sketched in Figure 3.27. The multiple outputs may be obtained using multiple receiver antennas, and/or oversampling the received signal with respect to the symbol rate. Each channel output $\mathbf{y}_i$ is now vector valued (containing $\mathcal{N}$ samples), and assumes the form

$$\mathbf{y}_i = \sum_{l=0}^{L} \mathbf{h}_l \, d_{i-l} + \mathbf{b}_i$$

in which each impulse response term $\mathbf{h}_l$ and each noise term $\mathbf{b}_i$ is now a column vector of dimensions $\mathcal{N} \times 1$.

3.10.1 Forward-Backward Equalizer

The forward-backward equalizer is developed with the same methodology as in the single output case, as the channel still effects a convolution which induces a trellis diagram leading to an efficient decoding rule. The only change is that each channel output is now vector valued rather than scalar valued. Accordingly, the channel likelihood evaluation for a candidate state transition in the trellis assumes the form

$$\begin{aligned} \gamma_i(m', m) &= \Pr(\mathbf{x}_i = S_m; \mathbf{y}_i | \mathbf{x}_{i-1} = S_{m'}) \\ &\propto \Pr(d_i) \exp\left(-\frac{[y_i - \mathcal{H}(\Sigma_{m',m})]^T \mathbf{B}^{-1} [y_i - \mathcal{H}(\Sigma_{m',m})]}{2} \right). \end{aligned} \tag{3.13}$$

Here $\Pr(d_i)$ is an *a priori* probability for the input bit which provokes the transition (to be replaced by an extrinsic probability from the outer decoder), $\Sigma_{m',m}$ is an augmented state which accounts for the state transition from $\mathbf{x}_{i-1} = S_{m'}$ to $\mathbf{x}_i = S_m$, with $\mathcal{H}(\Sigma_{m',m})$ denoting the vector-valued noise-free output produced from this transition, and $\mathbf{B} = E(\mathbf{b}_i \mathbf{b}_i^T)$ is the noise autocorrelation matrix. (Often one takes $\mathbf{B} = \sigma^2 \mathbf{I}$ by assuming the noise is spatially uncorrelated.) As the channel now has $\mathcal{N}$ outputs, it may be viewed as a rate $1/\mathcal{N}$ parasitic encoder, offering potentially improved coding gain compared to the single-output case which appears as a rate 1 encoder. With the channel likelihood evaluations so modified, the remainder of the forward-backward algorithm is identical, as is its coupling to the outer decoder.

3.10.2 Interference Canceler

The interference canceler now replaces $P(z)$ by a bank of filters, as sketched in Figure 3.28. This can equivalently be viewed as an $\mathcal{N}$-input/single-output FIR filter, effecting the operations

$$v_i = \sum_{k=0}^{L} \mathbf{p}_k \, \mathbf{y}_{i-k} - \sum_{k=0}^{2L} q_k \, \hat{d}_{i-k}.$$

Here $\mathbf{p}_k$ has dimensions $1 \times \mathcal{N}$, and collects the k-th impulse response term from each filter in the receiver bank, and relates to the individual filters $P_k(z)$ according to

$$[P_1(z) \quad P_2(z) \quad \cdots \quad P_{\mathcal{N}}(z)] = \sum_{k=0}^{L} \mathbf{p}_k \, z^{-k}.$$

The optimal choice for the filter coefficients is a straightforward generalization of that for the single-output channel case (see Problem 3.6):

$$\begin{aligned} \mathbf{p}_k &= \mathbf{h}_{L-k}^T, & k &= 0, 1, \ldots, L \\ q_{k+L} &= \sum_l \mathbf{h}_l^T \mathbf{h}_{k+l}, & k &= \pm 1, \pm 2, \ldots, \pm L. \\ q_L &= 0 \end{aligned}$$

This choice maximizes the SNR at the output of the interference canceler, subject to an interference cancelation constraint if the symbols fed back from the outer decoder are correct.

Estimating the filter coefficients ($\mathbf{h}_k$) of the channel is actually simpler in the multi-output case, provided the following irreducibility [86] (or minimum phase) condition on the channel transfer function holds

$$\mathbf{h}(z) = \sum_{l=0}^{L} \mathbf{h}_l \, z^{-l} \neq 0, \qquad \text{for all } z.$$

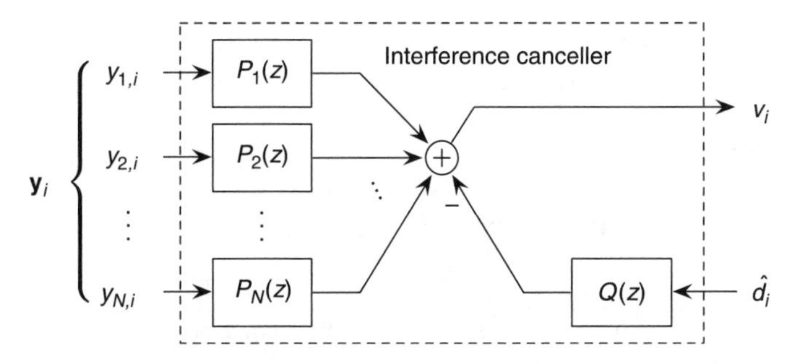

Figure 3.28 Interference canceler adapted to a multiple output channel.

In the simplest terms, this says that the $\mathcal{N}$ transfer functions contained in the vector $\mathbf{h}(z)$ have no common zero, which may be expected to hold with high probability. (If such a common zero existed, then moving one of the antennas slightly, say, would shift the zero location in the corresponding entry of $\mathbf{h}(z)$, in which case it would no longer be a common zero.) Provided the length L of the channel impulse response can be determined, then various blind methods (e.g. [49, 50, 52, 54]), using only second-order statistics of the received sequence ($\mathbf{y}_i$), can claim to identify the channel impulse response. In practice, estimating the true channel length is not always a well conditioned problem [87], and various conditioning aspects of these blind channel identification schemes can be unduly aggravated from incorrect channel length assumptions [51, 53, 88].

3.10.3 Multiuser Case

The next extension is to accommodate multiple users in a multiantenna system, for which the basic relations are summarized here.

The channel output in a typical multiuser setting becomes

$$\mathbf{y}_i = \sum_{l=0}^{L} \mathbf{H}_l \, \mathbf{d}_{i-l} + \mathbf{b}_i$$

where now $\mathbf{d}_i$ is a vector of dimension $\mathcal{M}$, containing the i-th coded symbol from each of the $\mathcal{M}$ users, and $\{\mathbf{H}_l\}_{l=0}^{L}$ is an $\mathcal{N} \times \mathcal{M}$ matrix impulse response sequence, linking the $\mathcal{M}$ users to the $\mathcal{N}$ receiver antennas.

The basic receiver configuration is sketched in Figure 3.29, comprising a user separation block followed by individual user decoders. The outputs of the decoding blocks are then fed back to the separation block for further refinement, giving the basic iterative structure.

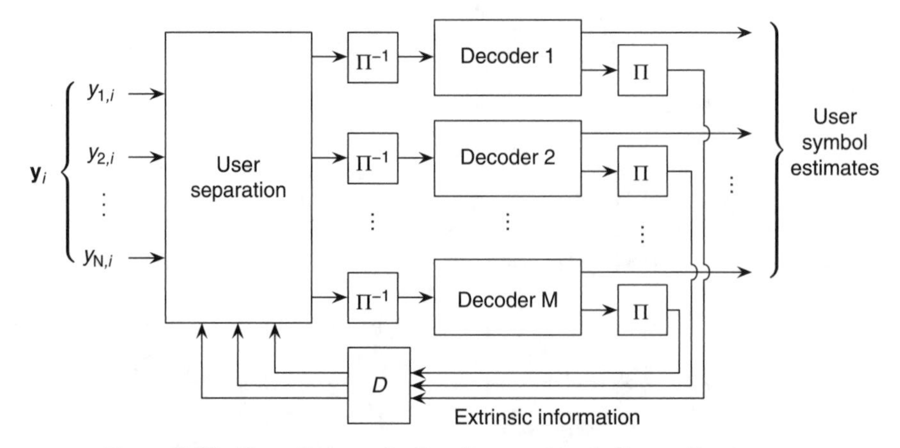

Figure 3.29 Flow diagram for iterative receiver in the multiuser case.

The user separation block when using the forward-backward algorithm again follows the same logic as in the single-user case. The state dimension of the trellis diagram, however, now grows to $2^{\mathcal{M}L}$ for $\mathcal{M}$ users and a channel of degree L. The state transition probabilities still are given by (3.13), except that the pseudo prior $\Pr(d_i)$ from the single-user case is replaced by its vector counterpart $\prod_{m=1}^{\mathcal{M}} \Pr(d_{m,i})$ to account for multiple users. The extrinsic information values from the user separation block are calculated in a per-user basis, and fed to the individual user decoding blocks. These in turn extract their extrinsic information in the usual manner, to be fed back to the user separation block as pseudo priors for the next iteration.

As the state dimension of a trellis diagram grows exponentially with the number of users $\mathcal{M}$ and the channel impulse response length L, a linear feedback equalizer presents a reduced complexity implementation, as its computation and storage requirements grow only linearly with the product $\mathcal{M}L$. The filters $P(z)$ and $Q(z)$ now have matrix-valued impulse response terms, given as

$$
\begin{aligned}
\mathbf{P}_k &= \mathbf{H}_{L-k}^T, && k = 0, 1, \ldots, L \\
\mathbf{Q}_{k+L} &= \sum_l \mathbf{H}_l^T \mathbf{H}_{k+l}, && k = \pm 1, \pm 2, \ldots, \pm L. \\
\mathbf{Q}_L &= 0
\end{aligned}
$$

If perfect symbol estimates are fed back, the interference canceler output becomes

$$
\mathbf{v}_i = \mathbf{R}_0 \mathbf{d}_{i-L} + \sum_{l=0}^{L} \mathbf{H}_{L-l}^T \mathbf{b}_{i-l}, \qquad \text{with } \mathbf{R}_0 = \sum_{l=0}^{L} \mathbf{H}_l^T \mathbf{H}_l
$$

which reveals that, although intersymbol interference is indeed eliminated, the various users are still mixed via the matrix $\mathbf{R}_0$. Since $\mathbf{R}_0$ will be positive definite (save for degenerate channels), its inverse $\mathbf{R}_0^{-1}$ exists and may be applied to $\mathbf{v}_i$ to achieve user separation, at the risk of some noise enhancement. In practice, inverting $\mathbf{R}_0$ is an expensive proposition, and blind source separation techniques [91, 92] may be envisaged here.

In a practical multiuser set-up, some means of coding is used to distinguish the users, such as code division multiple access (CDMA) [89, 90]. The spreading codes may be viewed as part of the channel impulse response, and/or treated as another serial concatenation element. The special structure induced by the spreading codes leads to further performance enhancements within the iterative loop. Further details are available in works treating turbo CDMA [80–85].

3.11 CONCLUDING REMARKS

The cooperative interconnection of equalization and decoding brings performance enhancements beyond that achievable by treating receiver components separately. We have focused on the basic variants in turbo equalization for clarity, although countless other variants may be envisaged based on the form of information exchange between components, the form of channel state information estimators employed, and/or the incorporation of other receiver functions, such as synchronization or encryption at the physical layer, or even routing and task assignment at the resource allocation layers.

The various analysis techniques for convergence overviewed in Section 3.9 allow reliable prediction of actual performance under reasonable assumptions on the block length and signal to noise ratio. Nonetheless, with the explosion of candidate iterative receiver configurations, a grand unified theory of turbo convergence applicable to all settings remains beyond immediate reach. This shortcoming is all the more relevant in latency constrained applications using multiple transmitters and receivers, in which short block lengths with asynchronous transmission conditions prevail. Further application development will thus benefit greatly from progress in understanding these more advanced settings.

3.12 PROBLEMS

3.1 Let T and U be bounded pseudo probabilities

$$0 < T < 1 \qquad \text{and} \qquad 0 < U < 1.$$

The pseudo posterior bit estimates of the turbo decoder each assume the form

$$\hat{c} = \frac{TU}{TU + (1-T)(1-U)}$$

for appropriate pseudo prior and extrinsic probabilities T and U.

(a) Verify that

$$\hat{c} - 0.5 = 0.5\,\frac{T + U - 1}{TU + (1-T)(1-U)}.$$

(b) Confirm, therefore, that $\hat{c} > 0.5$ if and only if $T + U > 1$

3.2 Consider a simple additive white Gaussian noise channel

$$y_i = d_i + b_i.$$

Here d_i is either $+1$ or -1, and b_i is zero mean Gaussian noise term with variance σ_2. The conditional probability function for y_i thus becomes

$$\Pr(y_i|d_i = \pm 1) = \frac{1}{\sqrt{2\pi}\sigma}\exp\left(-\frac{(y \mp 1)^2}{2\sigma^2}\right).$$

(a) Introduce the log likelihood ratio η as

$$\eta = \log\frac{\Pr(y_i|d_i = +1)}{\Pr(y_i|d_i = -1)}$$

Show that

$$\eta = \frac{2y_i}{\sigma^2} = \frac{2d_i}{\sigma^2} + \frac{2b_i}{\sigma^2}.$$

(b) Introduce the conditional means μ_+ and μ_- as

$$\mu_+ = E(\eta|d_i = +1), \quad \mu_- = E(\eta|d_i = -1)$$

where the expectation is with respect to the probability density function of b_i. Show that

$$\mu_+ = -\mu_- = \frac{2}{\sigma^2}.$$

(c) Introduce the conditional variances ς_+^2 and ς_-^2 as

$$\varsigma_+^2 = E[(\eta - \mu_+)^2 \,|\, d_i = +1], \; \varsigma_1^2 = E[(\eta - \mu_-)^2 \,|\, d_i = -1].$$

Show that

$$\varsigma_+^2 = \varsigma_-^2 = \frac{4}{\sigma^2}$$

so that the conditional mean is half the conditional variance (give or take a sign factor). What happens to the conditional means and variances as $\sigma^2 \to 0$?

3.3 Consider the log likelihood ratio as in Problem 3.2 (dropping the subscript i for convenience)

$$\eta = \log \frac{\Pr(y|d = +1)}{\Pr(y|d = -1)} = \frac{2y}{\sigma^2}.$$

The variable η is then characterized by a conditional variance parameter $\varsigma_\eta^2 = 4/\sigma^2$. Assuming d is binary and equiprobable $[\Pr(d = +1) = \Pr(d = -1) = \frac{1}{2}]$, the mutual information between d and y is expressed in terms of the conditional distribution $\Pr(y|d)$ as

$$I(d, y) = \frac{1}{2}\left(\int_y \Pr(y|+1) \log \frac{2\Pr(y|+1)}{\Pr(y|+1) + \Pr(y|-1)}\, dy \right.$$
$$\left. + \int_y \Pr(y|-1) \log \frac{2\Pr(y|-1)}{\Pr(y|+1) + \Pr(y|-1)}\, dy \right).$$

An analogous expression gives $I(d, \eta)$ in terms of $\Pr(\eta|d)$. Show that

$$I(d, \eta) = I(d, y).$$

3.4 Let τ be a log extrinsic ratio whose probability distribution, conditioned on $d = -1$, is:

$$\Pr(\tau|d = -1) = \frac{1}{\sqrt{2\pi}\,\varsigma}\exp\left(-\frac{(\tau + 0.5\varsigma^2)^2}{2\varsigma^2}\right).$$

Let t be the corresponding extrinsic probability

$$t = \frac{e^\tau}{1 + e^\tau}.$$

(a) Show that t is a bounded variable

$$0 \leq t \leq 1.$$

(b) Show that t increases monotonically with τ

$$\frac{dt}{d\tau} > 0.$$

(c) Show that the probability distribution function for t (conditioned on $d = -1$) is given by

$$\Pr(t|d = -1) = \frac{1}{\sqrt{2\pi}\,\varsigma\, t(1 - t)}\exp\left[-\left(\log\frac{t}{1 - t} + \frac{\varsigma^2}{2}\right)^2 \Big/ 2\varsigma^2\right].$$

Hint: Use the fact that if t is a unique function of τ, that is, $t = f(\tau)$ where $f(\cdot)$ is monotonic, then $\Pr(t|d) = \Pr(\tau|d)/|df/d\tau|$.

3.5 Consider the simplified schematic of Figure 3.30 in which the sequence (y_i) is obtained from an additive white Gaussian noise channel

$$y_i = d_i + b_i, \qquad i = 1, 2, \ldots, N.$$

Here b_i has zero mean and variance σ^2, the bits $d_1, \ldots, d_K$ are the antipodal forms of the information bits, and the remaining terms $d_{K+1}, \ldots, d_N$ are the antipodal forms of the parity-check bits, and are thus uniquely determined from $d_1, \ldots, d_K$. The log extrinsic ratios (τ_i) are produced by the turbo decoder, whose input is the sequence (y_i) alone. The variables (y_i) and (τ_i) are assumed

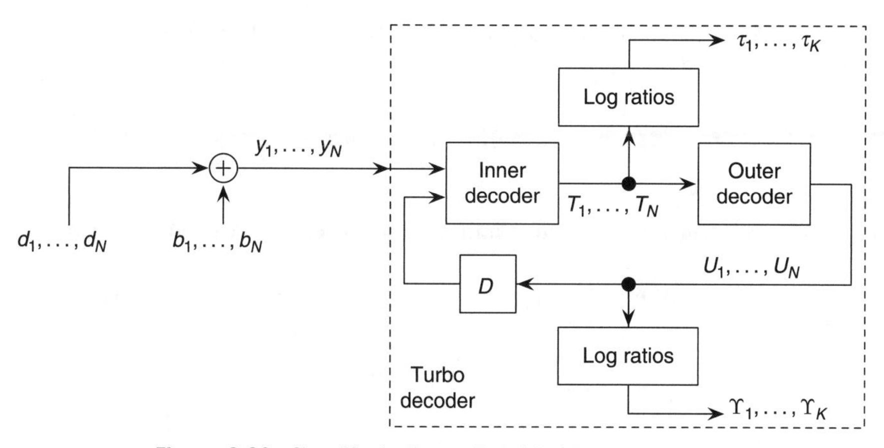

Figure 3.30 Simplified schematic of the turbo decoder.

conditionally independent and identically distributed given (d_i). (This is true for $\{y_i\}$ if the noise samples $\{b_i\}$ are independent, but not strictly true for $\{\tau_i\}$ without additional mathematical machinery invoking interleavers.) For notational convenience, introduce the vectors

$$\mathbf{d} = [d_1, \ldots, d_K]$$
$$\mathbf{y} = [y_1, \ldots, y_N]$$
$$\tau = [\tau_1, \ldots, \tau_K].$$

(a) Since the variables $\{\tau_i\}$ can be reduced to a (somewhat complicated) function of the $\{y_i\}$ alone, argue that

$$\Pr(\tau|\mathbf{y}, \mathbf{d}) = \Pr(\tau|\mathbf{y}).$$

From this, verify that

$$\Pr(\mathbf{d}, \mathbf{y}, \tau) = \Pr(\mathbf{d}) \times \Pr(y|\mathbf{d}) \times \Pr(\tau|\mathbf{y}).$$

The variables $\mathbf{d}$, $\mathbf{y}$, and τ are then said to form a Markov chain, denoted $\mathbf{d} \to \mathbf{y} \to \tau$.

(b) The data processing inequality from information theory [44] asserts that mutual information can only decrease through a Markov chain

$$\text{If } \mathbf{d} \to \mathbf{y} \to \tau, \quad \text{then } I(\mathbf{d}, \tau) \leq I(\mathbf{d}, \mathbf{y}).$$

Let η be a vector collecting the log likelihood ratios (η_i) obtained from the samples (y_i) as in Problem 3.3. Show that

$$I(\mathbf{d}, \tau) \leq I(\mathbf{d}, \eta).$$

Note that if the variables (τ_i) are indeed conditionally independent given the values (d_i), then

$$I(\mathbf{d};\, \tau) = \sum_{i=1}^{K} I(d_i;\, \tau_i).$$

(c) The chain rule for mutual information [44] can be invoked to show that

$$\begin{aligned} I(\mathbf{d};\, \mathbf{y}) &= I(\mathbf{d};\, y_1, \ldots, y_K) + I(\mathbf{d};\, y_{K+1}, \ldots, y_N \,|\, y_1, \ldots, y_K) \\ &= \sum_{i=1}^{K} I(d_i;\, y_i) + I(\mathbf{d};\, y_{K+1}, \ldots, y_N \,|\, y_1, \ldots, y_K). \end{aligned}$$

In the absence of coding, the variables $y_{K+1}, \ldots, y_N$ would be absent. Show that, in this case, one would have $I(d_i;\, \tau_i) \leq I(d_i;\, y_i)$ and thus

$$\varsigma_\tau^2 \leq \frac{4}{\sigma^2} \qquad \text{(without coding).}$$

3.6 Consider a single-input/multiple-output channel, whose received (vector-valued) sequence is

$$\mathbf{y}_i = \sum_{k=0}^{L} \mathbf{h}_k\, d_{i-k} + \mathbf{b}_i$$

where $\mathbf{y}_i$, $\mathbf{h}_k$ and $\mathbf{b}_i$ are all column vectors having $\mathcal{N}$ elements, and the sequence $\{\mathbf{b}_k\}$ is spatially and temporally white

$$E(\mathbf{b}_j\, \mathbf{b}_k^T) = \begin{cases} \sigma^2 \mathbf{I}, & j = k; \\ 0, & \text{otherwise.} \end{cases}$$

The output of the interference canceler is

$$\begin{aligned} v_i &= \sum_{k=0}^{L} \mathbf{p}_k\, \mathbf{y}_{i-k} - \sum_{k=0}^{2L} q_k\, \hat{d}_{i-k} \\ &= \sum_{k=0}^{2L} \rho_k\, d_{i-k} - \sum_{k=0}^{2L} q_k\, \hat{d}_{i-k} + \sum_{k=0}^{L} \mathbf{p}_k\, \mathbf{b}_{i-k} \end{aligned}$$

where each impulse response term $\mathbf{p}_k$ has dimensions $1 \times \mathcal{N}$, and where (ρ_k) is the combined impulse response, obtained by convolving $(\mathbf{p}_k)$ and $(\mathbf{h}_k)$

$$\rho_k = \sum_{l} \mathbf{p}_l\, \mathbf{h}_{k-l}, \qquad k = 0, 1, \ldots, 2L.$$

(a) Verify that the combined response sequence (ρ_k) can be expressed in matrix form as

$$\begin{bmatrix} \rho_0 \\ \rho_1 \\ \vdots \\ \rho_L \\ \rho_{L+1} \\ \vdots \\ \rho_{2L} \end{bmatrix} = \begin{bmatrix} \mathbf{p}_0 & 0 & \cdots & 0 \\ \mathbf{p}_1 & \mathbf{p}_0 & \ddots & \vdots \\ \vdots & \ddots & \ddots & 0 \\ \mathbf{p}_L & \cdots & \mathbf{p}_1 & \mathbf{p}_0 \\ 0 & \mathbf{p}_L & \cdots & \mathbf{p}_1 \\ \vdots & \ddots & \ddots & \vdots \\ 0 & \cdots & 0 & \mathbf{p}_L \end{bmatrix} \begin{bmatrix} \mathbf{h}_0 \\ \mathbf{h}_1 \\ \vdots \\ \mathbf{h}_M \end{bmatrix}$$

(b) Suppose we constrain $\rho_L = 1$, and $q_L = 0$, and feed back perfect symbol estimates so that $\hat{d}_i = d_i$. Show that the interference canceler output becomes

$$v_i = d_{i-L} + \underbrace{\sum_{\substack{k=0 \\ k\neq L}}^{2L} (\rho_k - q_k)\, d_{i-k}}_{\substack{\text{intersymbol} \\ \text{interference}}} + \underbrace{\sum_{k=0}^{L} \mathbf{p}_k\, \mathbf{b}_{i-k}}_{\substack{\text{filtered} \\ \text{noise: } \hat{b}_i}}.$$

Deduce the formula for the coefficients (q_k) which ensure cancellation of the intersymbol interference.

(c) Show that the filtered noise term has variance

$$E(\hat{b}_i^2) = \sigma^2 \sum_{k=1}^{L} \|\mathbf{p}_k\|^2$$

where $\|\mathbf{p}_k\|^2 = \mathbf{p}_k \mathbf{p}_k^T$.

(d) Rewrite the constraint $\rho_L = 1$ as

$$\underbrace{\begin{bmatrix} \mathbf{p}_L & \mathbf{p}_{L-1} & \cdots & \mathbf{p}_1 & \mathbf{p}_0 \end{bmatrix}}_{\mathbf{P}^T} \underbrace{\begin{bmatrix} \mathbf{h}_0 \\ \mathbf{h}_1 \\ \vdots \\ \mathbf{h}_{L-1} \\ \mathbf{h}_L \end{bmatrix}}_{\mathbf{H}} = 1$$

with the vectors $\mathbf{P}$ and $\mathbf{H}$ so defined. The Cauchy–Schwarz inequality asserts that $|\mathbf{P}^T\mathbf{H}| \leq \|\mathbf{P}\| \cdot \|\mathbf{H}\|$, with equality if, and only if, the vectors $\mathbf{P}$ and $\mathbf{H}$ are colinear, that is, $\mathbf{P} = a\mathbf{H}$ for some scalar a. From this, show that the filtered noise power is lower bounded as

$$E(\hat{b}_i^2) \geq \frac{\sigma^2}{\|\mathbf{H}\|^2} = \frac{\sigma^2}{\sum_{k=0}^{L} \|\mathbf{h}_k\|^2}.$$

Deduce the optimal coefficients ($\mathbf{p}_k$) to minimize the filtered noise power.

REFERENCES

1. C. Berrou and A. Glavieux, Near optimum error correction coding and decoding: Turbo codes, *IEEE Trans. Communications*, 44, pp. 1262–1271, (1996).
2. S. Benedetto and G. Montorsi, Iterative decoding of serially concatenated convolutional codes, *Electronics Letters*, 32, pp. 1186–1188, (1996).
3. S. Benedetto, D. Divsalar, G. Montorsi, and F. Pollara, Analysis, design, and iterative decoding of double serial concatenated codes with interleaves, *IEEE Selected Areas of Communications*, 16, pp. 231–244, (1998).
4. J. Haguenauer, E. Offer, and L. Papke, Iterative decoding of binary block and convolutional codes, *IEEE Trans. Information Theory*, 42, pp. 429–445, (1996).
5. L. Hanzo, T. H. Liew, and B. L. Yeap, *Turbo Coding, Turbo Equalisation and Space-Time Coding*. Chichester, UK: Wiley, 2002.
6. C. Heegard and S. B. Wicker, *Turbo Coding*. Boston, MA: Kluwer, 1999.
7. B. Vucetic and J. Yuan, *Turbo Codes: Principles and Applications*. Boston, MA: Kluwer, 2000.
8. C. E. Shannon, A mathematical theory of communication, *Bell System Technical J.*, 27, pp. 379–423, 623–656, (1948).
9. J. G. Proakis, *Digital Communications*. Boston, MA: McGraw-Hill, 2001.
10. N. Blaustein and C. G. Christodoulou, *Radio Propagation and Adaptive Antennas for Wireless Communication Links*. Hoboken, NJ: Wiley, 2007.
11. S. Haykin, *Adaptive Filter Theory*. Upper Saddle River, NJ: Prentice-Hall, 4th ed., 2001.
12. A. H. Sayed, *Fundamentals of Adaptive Filtering*. Hoboken, NJ: Wiley, 2003.
13. G. D. Forney, Jr., The Viterbi algorithm, *Proc. IEEE*, 61, pp. 268–278, (1973).
14. R. W. Chang and J. C. Hancock, On receiver structures for channels having memory, *IEEE Trans. Information Theory*, 12, pp. 463–468, (1966).
15. L. E. Baum and T. Petrie, Statistical inference for probabilistic functions of finite state Markov chains, *Ann. Math. Statist.*, 37, pp. 1554–1563, (1966).
16. G. Ungerboeck, Nonlinear equalization of binary signals in Gaussian noise, *IEEE Trans. Communication Technology*, 19, pp. 1128–1137, (1971).
17. L. R. Bahl, J. Cocke, F. Jelinek, and J. Raviv, Optimal decoding of linear codes for minimizing symbol error rate, *IEEE Trans. Information Theoiy*, 20, No. 3, pp. 284–287, (1974).
18. G. K. Kaleh and R. Vallet, Joint parameter estimation and symbol detection for linear or nonlinear unknown channels, *IEEE Trans. Communications*, 42, pp. 2406–2413, (1994).
19. S. Chen, G. J. Gibson, C. F. N. Cowan, and P. M. Grant, Adaptive equalization of finite nonlinear channels using multilayer perceptions, *Signal Processing*, 10, pp. 107–119, (1990).

20. G. J. Gibson, S. Siu, and C. F. N. Cowan, The application of nonlinear structures to the reconstruction of binary signals, *IEEE Trans. Signal Processing*, 39, pp. 1877–1884, (1991).
21. M. Meyer and G. Pfeiffer, Multilayer perceptron based decision feedback equaliser for channels with intersymbol interference, *IEE Proceedings I: Communications, Speech and Vision*, 140, pp. 420–424, (1993).
22. T. X. Brown, "Neural networks for adaptive equalization," in J. Alspector, R. Goodman, and T. X. Brown, Eds., *Applications of Neural Networks in Telecommunications*, Hillsdale, NJ: Erlbaum, 1993.
23. G. Kechriotis, E. Zervas, and E. S. Manolakos, Using recurrent neural networks for adaptive communication channel equalization, *IEEE Trans. Neural Networks*, 5, pp. 267–278, (1994).
24. J. Cid-Sueiro, A. Artès-Rodriguez, and A. R. Figueiras-Vidal, Recurrent radial basis function networks for optimal symbol-by-symbol equalization, *Signal Processing*, 40, pp. 53–63, (1994).
25. S. Chen, S. McLaughlin, and B. Mulgrew, Complex-valued radial basis function network, Part II: Application to digital communications channel equalization, *Signal Processing*, 35, pp. 19–31, (1994).
26. I. Cha and S. A. Kassam, Channel equalization using adaptive complex radial basis functions, *IEEE J. Selected Areas in Communications*, 13, pp. 122–131, (1995).
27. S. Theodoridis, C. F. N. Cowan, C. P. Callender, and C- M. S. See, Schemes for equalisation of communication channels with nonlinear impairments, *IEE Proc. Communications*, 142, No. 3, pp. 165–171, (1995).
28. T. Adalı, X. Liu, and K. Sönmez, Conditional distribution learning with neural networks and its application to channel equalization, *IEEE Trans. Signal Processing*, 45, pp. 1051–1064, (1997).
29. Y. Kopsinis, S. Theodoridis, and E. Kofidis, An efficient low complexity cluster-based MLSE equalizer for frequency-selective fading channels, *IEEE Trans. Wireless Communications*, 5, pp. 705–711, (2006).
30. C. Douillard, M. Jezequel, C. Berrou, P. Picart, P. Didier, and A. Glavieux, Iterative correction of intersymbol interference: Turbo equalization, *European Telecommunications Transactions*, 6, No. 5, pp. 507–512, (1995).
31. T. Richardson, The geometry of turbo-decoding dynamics, *IEEE Trans. Information Theory*, 46, pp. 9–23, (2000).
32. P. A. Regalia, Iterative decoding of concatenated codes: A tutorial, *EURASIP J. Applied Signal Processing*, 2005, pp. 762–774, (2005).
33. T. L. Saaty and J. Bram, *Nonlinear Mathematics*. New York: McGraw-Hill, 1964.
34. S. Verdù and H. V. Poor, Abstract dynamic programming models under commutativity conditions, *SUM J. Control and Optimization*, 25, pp. 990–1006, (1987).
35. L. Rabiner, A tutorial on hidden Markov models and selected applications in speech recognition, *Proc. IEEE*, 77, pp. 257–285, (1989).
36. C. Laot, A. Glavieux, and J. Labat, Turbo equalization: Adaptive equalization and channel decoding jointly optimized, *IEEE J. Selected Areas in Communications*, 19, pp. 673–680, (2001).

37. M. Tüchler, R. Kotter, and A. Singer, Turbo equalization: Principles and new results, *IEEE Trans. Communications*, 50, pp. 754–767, (2002).
38. R. Kötter, A. C. Singer, and M. Tuchler, Turbo equalization, *IEEE Signal Processing Magazine*, pp. 67–80, (2004).
39. S.-K. Lee, A. C. Singer, and N. R. Shanbhag, Linear turbo equalization analysis via BER transfer and EXIT charts, *IEEE Trans. Signal Processing*, 53, pp. 2883–2897, (2005).
40. P. Supnithi, R. Lopes, and S. W. McLaughlin, Reduced-complexity turbo equalization for high-density magnetic recording systems, *IEEE Trans. Magnetics*, 39, pp. 2585–2587, (2003).
41. S. Jiang, L. Ping, H. Sun, and C. S. Leung, Modified LMMSE turbo equalization, *IEEE Communications Letters*, 8, pp. 174–176, (2004).
42. F. R. Rad and J. Moon, Turbo equalization utilizing soft decision feedback, *IEEE Trans. Magnetics*, 41, pp. 2998–3000, (2005).
43. R. R. Lopes and J. R. Barry, The soft-feedback equalizer for turbo equalization of highly dispersive channels, *IEEE Trans, Communications*, 54, pp. 783–788, (2006).
44. T. M. Cover and J. A. Thomas, *Elements of Information Theory*. New York: Wiley, 2nd ed., 2007.
45. W. L. Root and P. P. Varaiya, Capacity of classes of Gaussian channels, *SIAM J. Applied Mathematics*, 16, pp. 1350–1393, (1968).
46. G. J. Foschsini and R. K. Müller, The capacity of linear channels with additive Gaussian noise, *Bell System Technical J.*, pp. 81–94, (1970).
47. R. A. Horn and C. R. Johnson, *Matrix Analysis*. Cambridge University Press, 1985.
48. I. Fijalkow, A. Roumy, S. Ronger, D. Perez, and P. Vila, "Improved interference cancellation for turbo equalization," in *Proc. Int. Conf. Acoustics, Speech and Signal Processing*, (Istanbul, Turkey), pp. 419–422, (2000).
49. G. Xu, H. Liu, L. Tong, and T. Kailath, A least-squares approach to blind channel identification, *IEEE Trans. Signal Processing*, 43, pp. 2981–2993, (1995).
50. Z. Ding, Matrix outer-product decomposition method for blind multiple channel identification, *IEEE Trans. Signal Processing*, 45, pp. 3053–3061, (1997).
51. A. P. Liavas, P. A. Regalia, and J.-P. Delmas, On the robustness of the linear prediction method for blind channel identification with respect to effective channel undermodeling/overmodeling, *IEEE Trans. Signal Processing*, 48, pp. 1477–1481, (2000).
52. K. Abed-Meraim, E. Moulines, and P. Loubaton, Prediction error method for second-order blind identification, *IEEE Trans. Signal Processing*, 45, pp. 694–705, (1997).
53. J.-P. Delmas, H. Gazzah, A. P. Liavas, and P. A. Regalia, Statistical analysis of some second-order methods for blind channel identification/equalization with respect to channel undermodeling, *IEEE Trans. Signal Processing*, 48, pp. 1984–1998, (2000).
54. E. Moulines, P. Duhamel, J.-F. Cardoso, and S. Mayrargue, Subspace methods for the blind identification of multichannel FIR filters, *IEEE Trans. Signal Processing*, 43, pp. 516–525, (1995).
55. L. A. Liporace, Maximum likelihood estimation for multivariate observations of Markov sources, *IEEE Trans. Information Theory*, 28, pp. 729–734, (1982).
56. S. Amari, The EM algorithm and information geometry in neural network learning, *Neural Computation*, 7, pp. 13–18, (1994).

57. K. R. Narayanan and G. L. Stüber, A serial concatenation approach to iterative demodulation and decoding, *IEEE Trans. Communications*, 47, pp. 956–961, (1999).
58. X. Wang and R. Chen, Blind turbo equalization in Gaussian and impulse noise, *IEEE Trans. Vehicular Technology*, 50, pp. 1092–1105, (2001).
59. L. Kocarev, F. Lehmann, G. M. Maggio, B. Scanvino, Z. Tasev, and A. Vardy, Nonlinear dynamics of iterative decoding systems: Analysis and applications, *IEEE Trans. Information Theory*, 52, pp. 1366–1384, (2006).
60. M. Moher and T. A. Gulliver, Cross entropy and iterative decoding, *IEEE Trans. Information Theory*, 44, pp. 3097–3104, (1998).
61. A. Montanari and N. Sourlas, The statistical mechanics of turbo codes, *European Physics J. B*, 18, pp. 107–119, (2000).
62. B. Muquet, P. Duhamel, and M. de Courville, "Geometric interpretations of iterative 'turbo' decoding," in *Proc Int. Symp. Information Theory*, June 2002.
63. S. Ikeda, T. Tanaka, and S. Amari, Stochastic reasoning, free energy and information geometry, *Neural Computataion*, 16, pp. 1779–1810, (2004).
64. S. Ikeda, T. Tanaka, and S. Amari, Information geometry of turbo and low-density parity-check codes, *IEEE Trans. Information Theory*, 50, pp. 1097–1114, (2004).
65. J. Yedidia, W. Freeman, and Y. Weiss, Constructing free-energy approximations and generalized belief propagation algorithms, *IEEE Trans. Information Theory*, 51, pp. 2282–2312, (2005).
66. J. M. Walsh, P. A. Regalia, and C. R. Johnson, Jr., "A refined information geometric interpretation of turbo decoding," in *Proc. Int. Conf. Acoustics, Speech and Signal Processing*, (Philadelphia, PA), May 2005.
67. P. A. Regalia and J. M. Walsh, Optimality and duality of the turbo decoder, *Proc. IEEE*, 95, pp. 1362–1377, (2007).
68. P. Moquist and T. M. Aulin, "Turbo decoding as a numerical analysis problem," in *Int. Symp. Information Theory*, (Sorrento, Italy), p. 485, June 2000.
69. P. A. Regalia, "Contractivity in turbo iterations," in *Proc. Int. Conf. Acoustics, Speech and Signal Processing*, vol. 4, (Montreal, Canada), pp. 637–640, May 2004.
70. J. M. Walsh, P. A. Regalia, and C. R. Johnson, Jr., "A convergence proof for the turbo decoder as an instance of the Gauss-Seidel iteration," in *Int. Symp. Information Theory*, (Adelaide, Australia), pp. 734–738, Sept. 2005.
71. D. Divsalar, S. Dolinar, and F. Pollara, Iterative turbo decoder analysis based on density evolution, *IEEE J. Selected Areas in Communications*, 19, pp. 891–907, (2001).
72. H. El Gamal and A. R. Hommons, Analyzing the turbo decoder using the Gaussian approximation, *IEEE Trans. Information Theory*, 47, pp. 671–686, (2001).
73. S. ten Brink, Convergence behavior of iteratively decoded parallel concatenated codes, *IEEE Trans. Communications*, 49, pp. 1727–1737, (2001).
74. R. L. Bidan, C. Laot, D. LeRoux, and A. Glavieux, "Analyse de convergence en turbo détection," in *Proc. GRETSI-2001*, (Toulouse, France), September 2001.
75. A. Roumy, A. J. Grant, I. Fijalkow, P. D. Alexander, and D. Perez, "Turbo equalization: Convergence analysis," in *Proc. Int. Conf. Acoustics, Speech and Signal Processing*, (Salt Lake City, UT), pp. 2645–2648, 2001.
76. G. D. Forney, Jr., *Concatenated Codes*. Cambridge, MA: MIT Press, 1966.

77. J. Costello, Free distance bounds for convolutional codes, *IEEE Trans. Information Theory*, 20, pp. 356–365, (1974).
78. S. G. Wilson, *Digital Modulation and Coding*. Upper Saddle River, NJ: Prentice-Hall, 1996.
79. C. Berrou, The ten-year-old turbo codes are entering into service, *IEEE Communications Magazine*, 41, pp. 110–116, (2003).
80. X. Wang and H. V. Poor, Iterative (turbo) soft interference cancellation and decoding for coded CDMA, *IEEE Trans. Communications*, 47, pp. 1046–1061, (1999).
81. X. Wang and H. V. Poor, Blind joint equalization and multiuser detection for DS-CDMA in unknown correlated noise, *IEEE Trans. Circuits and Systems II*, 46, pp. 886–895, (1999).
82. P. D. Alexander, M. C. Reed, J. A. Asenstorfer, and C. B. Schlegel, Iterative multiuser interference reduction: Turbo CDMA, *IEEE Trans. Communications*, 47, pp. 1008–1014, (1999).
83. A. Roumy, I. Fijalkow, D. Perez, and P. Duvaut, "Interative multi-user algorithm for convolutionally coded asynchronous DS-CDMA systems: Turbo CDMA," in *Proc. Int. Conf. Acoustics, Speech and Signal Processing*, vol. 5, (Istanbul, Turkey), pp. 2893–2896, June 2000.
84. J. Boutros and G. Caire, Iterative multiuser joint decoding: Unified framework and asymptotic analysis, *IEEE Trans. Information Theory*, 48, pp. 1772–1793, (2002).
85. Z. Shi and C. Schlegel, Iterative multiuser detection and error control code decoding in random CDMA, *IEEE Trans. Signal Processing*, 54, pp. 1886–1895, (2006).
86. T. Kailath, *Linear Systems*. Upper Saddle River, NJ: Prentice-Hall, 1980.
87. A. P. Liavas, P. A. Regalia, and J.-P. Delmas, Blind channel approximation: Effective channel order determination, *IEEE Trans. Signal Processing*, 47, pp. 3336—3344, (1999).
88. A. P. Liavas, P. A. Regalia, and J.-P. Delrnas, Robustness of least-squares and subspace methods for blind channel identification/equalization with respect to channel undermodeling, *IEEE Trans. Signal Processing*, 47, pp. 1636–1645, (1999).
89. V. P. Ipatov, *Spread Spectrum and CDMA: Principles and Applications*. New York: Wiley, 2005.
90. H. Schulze and C. Lueders, *Theory and Applications of OFDM and CDMA: Wideband Wireless Communications*. New York: Wiley, 2005.
91. Y.-O. Li, T. Adalı W. Wang, and V. D. Calhoun, Joint blind source separation by multiset canonical correlation analysis, *IEEE Trans. Signal Processing*, 57, pp. 3918–3929, (2009).
92. T. Mei, F. Yin, and J. Wang, Blind source separation based on cumulants with time and frequency non-properties, *IEEE Trans. Audio, Speech and Language Processing*, 17, pp. 1099–1108, (2009).

4

SUBSPACE TRACKING FOR SIGNAL PROCESSING

Jean Pierre Delmas
TELECOM SudParis, Evry, France

4.1 INTRODUCTION

Research in subspace and component-based techniques originated in statistics in the middle of the last century through the problem of linear feature extraction solved by the Karhunen–Loeve Transform (KLT). It began to be applied to signal processing 30 years ago and considerable progress has since been made. Thorough studies have shown that the estimation and detection tasks in many signal processing and communications applications such as data compression, data filtering, parameter estimation, pattern recognition, and neural analysis can be significantly improved by using the subspace and component-based methodology. Over the past few years new potential applications have emerged, and subspace and component methods have been adopted in several diverse new fields such as smart antennas, sensor arrays, multiuser detection, time delay estimation, image segmentation, speech enhancement, learning systems, magnetic resonance spectroscopy, and radar systems, to mention only a few examples. The interest in subspace and component-based methods stems from the fact that they consist in splitting the observations into a set of desired and a set of disturbing components. They not only provide new insights into many such problems, but they also offer a good tradeoff between achieved performance and computational complexity. In most cases they can be considered to be low-cost alternatives to computationally intensive maximum-likelihood approaches.

Adaptive Signal Processing: Next Generation Solutions. Edited by Tülay Adalı and Simon Haykin
Copyright © 2010 John Wiley & Sons, Inc.

In general, subspace and component-based methods are obtained by using batch methods, such as the eigenvalue decomposition (EVD) of the sample covariance matrix or the singular value decomposition (SVD) of the data matrix. However, these two approaches are not suitable for adaptive applications for tracking nonstationary signal parameters, where the required repetitive estimation of the subspace or the eigenvectors can be a real computational burden because their iterative implementation needs $O(n^3)$ operations at each update, where n is the dimension of the vector-valued data sequence. Before proceeding with a brief literature review of the main contributions of adaptive estimation of subspace or eigenvectors, let us first classify these algorithms with respect to their computational complexity. If r denotes the rank of the principal or dominant or minor subspace we would like to estimate, since usually $r \ll n$, it is traditional to refer to the following classification. Algorithms requiring $O(n^2r)$ or $O(n^2)$ operations by update are classified as high complexity; algorithms with $O(nr^2)$ operations as medium complexity and finally, algorithms with $O(nr)$ operations as low complexity. This last category constitutes the most important one from a real time implementation point of view, and schemes belonging to this class are also known in the literature as fast subspace tracking algorithms. It should be mentioned that methods belonging to the high complexity class usually present faster convergence rates compared to the other two classes. From the paper by Owsley [56], that first introduced an adaptive procedure for the estimation of the signal subspace with $O(n^2r)$ operations, the literature referring to the problem of subspace or eigenvectors tracking from a signal processing point of view is extremely rich. The survey paper [20] constitutes an excellent review of results up to 1990, treating the first two classes, since the last class was not available at the time it was written. The most popular algorithm of the medium class was proposed by Karasalo in [40]. In [20], it is stated that this dominant subspace algorithm offers the best performance to cost ratio and thus serves as a point of reference for subsequent algorithms by many authors. The merger of signal processing and neural networks in the early 1990s [39] brought much attention to a method originated by Oja [50] and applied by many others. The Oja method requires only $O(nr)$ operations at each update. It is clearly the continuous interest in the subject and significant recent developments that gave rise to this third class. It is outside of the scope of this chapter to give a comprehensive survey of all the contributions, but rather to focus on some of them. The interested reader may refer to [29, pp. 30–43] for an exhaustive literature review and to [8] for tables containing exact computational complexities and ranking with respect to the convergence of recent subspace tracking algorithms. In the present work, we mainly emphasize the low complexity class for both dominant and minor subspace, and dominant and minor eigenvector tracking, while we briefly address the most important schemes of the other two classes. For these algorithms, we will focus on their derivation from different iterative procedures coming from linear algebra and on their theoretical convergence and performance in stationary environments. Many important issues such as the finite precision effects on their behavior (e.g. possible numerical instabilities due to roundoff error accumulation), the different adaptive stepsize strategies and the tracking capabilities of these algorithms in nonstationary environments will be left aside. The interested reader may refer to the simulation sections of the different papers that deal with these issues.

The derivation and analysis of algorithms for subspace tracking requires a minimum background in linear algebra and matrix analysis. This is the reason why in Section 2, standard linear algebra materials necessary for this chapter are recalled. This is followed in Section 3 by the general studied observation model to fix the main notations and by the statement of the adaptive and tracking of principal or minor subspaces (or eigenvectors) problems. Then, Oja's neuron is introduced in Section 4 as a preliminary example to show that the subspace or component adaptive algorithms are derived empirically from different adaptations of standard iterative computational techniques issued from numerical methods. In Sections 5 and 6 different adaptive algorithms for principal (or minor) subspace and component analysis are introduced respectively. As for Oja's neuron, the majority of these algorithms can be viewed as some heuristic variations of the power method. These heuristic approaches need to be validated by convergence and performance analysis. Several tools such as the stability of the ordinary differential equation (ODE) associated with a stochastic approximation algorithm and the Gaussian approximation to address these points in a stationary environment are given in Section 7. Some illustrative applications of principal and minor subspace tracking in signal processing are given in Section 8. Section 9 contains some concluding remarks. Finally, some exercises are proposed in Section 10, essentially to prove some properties and relations introduced in the other sections.

4.2 LINEAR ALGEBRA REVIEW

In this section several useful notions coming from linear algebra as the EVD, the QR decomposition and the variational characterization of eigenvalues/eigenvectors of real symmetric matrices, and matrix analysis as a class of standard subspace iterative computational techniques are recalled. Finally a characterization of the principal subspace of a covariance matrix derived from the minimization of a mean square error will complete this section.

4.2.1 Eigenvalue Value Decomposition

Let $\mathbf{C}$ be an $n \times n$ real symmetric [resp. complex Hermitian] matrix, which is also *non-negative definite* because $\mathbf{C}$ will represent throughout this chapter a covariance matrix. Then, there exists (see e.g. [37, Sec. 2.5]) an orthonormal [resp. unitary] matrix $\mathbf{U} = [\mathbf{u}_1, \ldots, \mathbf{u}_n]$ and a real diagonal matrix $\boldsymbol{\Delta} = \text{Diag}(\lambda_1, \ldots, \lambda_n)$ such that $\mathbf{C}$ can be decomposed[1] as follows

$$\mathbf{C} = \mathbf{U}\boldsymbol{\Delta}\mathbf{U}^T = \sum_{i=1}^{n} \lambda_i \mathbf{u}_i \mathbf{u}_i^T, \quad \left[\text{resp., } \mathbf{U}\boldsymbol{\Delta}\mathbf{U}^H = \sum_{i=1}^{n} \lambda_i \mathbf{u}_i \mathbf{u}_i^H\right]. \tag{4.1}$$

[1]Note that for nonnegative real symmetric or complex Hermitian matrices, this EVD is identical to the SVD where the associated left and right singular vectors are identical.

The diagonal elements of $\mathbf{\Delta}$ are called *eigenvalues* and arranged in decreasing order, satisfy $\lambda_1 \geq \cdots \geq \lambda_n > 0$, while the orthogonal columns $(\mathbf{u}_i)_{i=1,\ldots,n}$ of $\mathbf{U}$ are the corresponding unit 2-norm *eigenvectors* of $\mathbf{C}$.

For the sake of simplicity, only real-valued data will be considered from the next subsection and throughout this chapter. The extension to complex-valued data is often straightforward by changing the transposition operator to the conjugate transposition one. But we note two difficulties. First, for simple[2] eigenvalues, the associated eigenvectors are unique up to a multiplicative sign in the real case, but only to a unit modulus constant in the complex case, and consequently a constraint ought to be added to fix them to avoid any discrepancies between the statistics observed in numerical simulations and the theoretical formulas. The reader interested by the consequences of this nonuniqueness on the derivation of the asymptotic variance of estimated eigenvectors from sample covariance matrices (SCMs) can refer to [34], (see also Exercise 4.1). Second, in the complex case, the second-order properties of multidimensional zero-mean random variables $\mathbf{x}$ are not characterized by the complex Hermitian covariance matrix $\mathrm{E}(\mathbf{x}\mathbf{x}^H)$ only, but also by the complex symmetric complementary covariance [58] matrix $\mathrm{E}(\mathbf{x}\mathbf{x}^T)$.

The computational complexity of the most efficient existing iterative algorithms that perform EVD of real symmetric matrices is cubic by iteration with respect to the matrix dimension (more details can be sought in [35, Chap. 8]).

4.2.2 QR Factorization

The QR factorization of an $n \times r$ real-valued matrix $\mathbf{W}$, with $n \geq r$ is defined as (see e.g. [37, Sec. 2.6])

$$\mathbf{W} = \mathbf{Q}\mathbf{R} = \mathbf{Q}_1\mathbf{R}_1 \tag{4.2}$$

where $\mathbf{Q}$ is an $n \times n$ orthonormal matrix, $\mathbf{R}$ is an $n \times r$ upper triangular matrix, and $\mathbf{Q}_1$ denotes the first r columns of $\mathbf{Q}$ and $\mathbf{R}_1$ the $r \times r$ matrix constituted with the first r rows of $\mathbf{R}$. If $\mathbf{W}$ is of full column rank, the columns of $\mathbf{Q}_1$ form an orthonormal basis for the range of $\mathbf{W}$. Furthermore, in this case the skinny factorization $\mathbf{Q}_1\mathbf{R}_1$ of $\mathbf{W}$ is unique if $\mathbf{R}_1$ is constrained to have positive diagonal entries. The computation of the QR decomposition can be performed in several ways. Existing methods are based on Householder, block Householder, Givens or fast Givens transformations. Alternatively, the Gram–Schmidt orthonormalization process or a more numerically stable variant called modified Gram–Schmidt can be used. The interested reader can seek details for the aforementioned QR implementations in [35, pp. 224–233]), where the complexity is of the order of $O(nr^2)$ operations.

[2]This is in contrast to multiple eigenvalues for which only the subspaces generated by the eigenvectors associated with these multiple eigenvalues are unique.

4.2.3 Variational Characterization of Eigenvalues/Eigenvectors of Real Symmetric Matrices

The eigenvalues of a general $n \times n$ matrix $\mathbf{C}$ are only characterized as the roots of the associated characteristic equation. But for real symmetric matrices, they can be characterized as the solutions to a series of optimization problems. In particular, the largest λ_1 and the smallest λ_n eigenvalues of $\mathbf{C}$ are solutions to the following constrained maximum and minimum problem (see e.g. [37, Sec. 4.2]).

$$\lambda_1 = \max_{\|\mathbf{w}\|_2=1,\ \mathbf{w}\in\mathcal{R}^n} \mathbf{w}^T\mathbf{C}\mathbf{w} \quad \text{and} \quad \lambda_n = \min_{\|\mathbf{w}\|_2=1,\ \mathbf{w}\in\mathcal{R}^n} \mathbf{w}^T\mathbf{C}\mathbf{w}. \tag{4.3}$$

Furthermore, the maximum and minimum are attained by the unit 2-norm eigenvectors $\mathbf{u}_1$ and $\mathbf{u}_n$ associated with λ_1 and λ_n respectively, which are unique up to a sign for simple eigenvalues λ_1 and λ_n. For nonzero vectors $\mathbf{w} \in \mathcal{R}^n$, the expression $\frac{\mathbf{w}^T\mathbf{C}\mathbf{w}}{\mathbf{w}^T\mathbf{w}}$ is known as the *Rayleigh's quotient* and the constrained maximization and minimization (4.3) can be replaced by the following unconstrained maximization and minimization

$$\lambda_1 = \max_{\mathbf{w}\neq\mathbf{0},\ \mathbf{w}\in\mathcal{R}^n} \frac{\mathbf{w}^T\mathbf{C}\mathbf{w}}{\mathbf{w}^T\mathbf{w}} \quad \text{and} \quad \lambda_n = \min_{\mathbf{w}\neq\mathbf{0},\ \mathbf{w}\in\mathcal{R}^n} \frac{\mathbf{w}^T\mathbf{C}\mathbf{w}}{\mathbf{w}^T\mathbf{w}}. \tag{4.4}$$

For simple eigenvalues $\lambda_1, \lambda_2, \ldots, \lambda_r$ or $\lambda_n, \lambda_{n-1}, \ldots, \lambda_{n-r+1}$, (4.3) extends by the following iterative constrained maximizations and minimizations (see e.g. [37, Sec. 4.2])

$$\lambda_k = \max_{\|\mathbf{w}\|_2=1,\ \mathbf{w}\perp\mathbf{u}_1,\mathbf{u}_2,\ldots,\mathbf{u}_{k-1},\ \mathbf{w}\in\mathcal{R}^n} \mathbf{w}^T\mathbf{C}\mathbf{w}, \quad k = 2,\ldots,r \tag{4.5}$$

$$= \min_{\|\mathbf{w}\|_2=1,\ \mathbf{w}\perp\mathbf{u}_n,\mathbf{u}_{n-1},\ldots,\mathbf{u}_{k+1},\ \mathbf{w}\in\mathcal{R}^n} \mathbf{w}^T\mathbf{C}\mathbf{w}, \quad k = n-1,\ldots,n-r+1 \tag{4.6}$$

and the constrained maximum and minimum are attained by the unit 2-norm eigenvectors $\mathbf{u}_k$ associated with λ_k which are unique up to a sign.

Note that when $\lambda_r > \lambda_{r+1}$ or $\lambda_{n-r} > \lambda_{n-r+1}$, the following global constrained maximizations or minimizations (denoted *subspace criterion*)

$$\max_{\mathbf{W}^T\mathbf{W}=\mathbf{I}_r} \mathrm{Tr}(\mathbf{W}^T\mathbf{C}\mathbf{W}) = \max_{\mathbf{W}^T\mathbf{W}=\mathbf{I}_r} \sum_{k=1}^{r} \mathbf{w}_k^T\mathbf{C}\mathbf{w}_k$$

or

$$\min_{\mathbf{W}^T\mathbf{W}=\mathbf{I}_r} \mathrm{Tr}(\mathbf{W}^T\mathbf{C}\mathbf{W}) = \min_{\mathbf{W}^T\mathbf{W}=\mathbf{I}_r} \sum_{k=1}^{r} \mathbf{w}_k^T\mathbf{C}\mathbf{w}_k \tag{4.7}$$

where $\mathbf{W} = [\mathbf{w}_1, \ldots, \mathbf{w}_r]$ is an arbitrary $n \times r$ matrix, have four solutions (see e.g. [70] and Exercise 4.6), $\mathbf{W} = [\mathbf{u}_1, \ldots, \mathbf{u}_r]\mathbf{Q}$ or $\mathbf{W} = [\mathbf{u}_{n-r+1}, \ldots, \mathbf{u}_n]\mathbf{Q}$ respectively, where $\mathbf{Q}$ is an arbitrary $r \times r$ orthogonal matrix. Thus, subspace criterion (4.7) determines the subspace spanned by $\{\mathbf{u}_1, \ldots, \mathbf{u}_r\}$ or $\{\mathbf{u}_{n-r+1}, \ldots, \mathbf{u}_n\}$, but does not specify the basis of this subspace at all.

Finally, when now, $\lambda_1 > \lambda_2 > \cdots > \lambda_r > \lambda_{r+1}$ or $\lambda_{n-r} > \lambda_{n-r+1} > \cdots > \lambda_{n-1} > \lambda_n$,[3] if $(\omega_k)_{k=1,\ldots,r}$ denotes r arbitrary positive and different real numbers such that $\omega_1 > \omega_2 > \cdots > \omega_r > 0$, the following modification of subspace criterion (4.7) denoted *weighted subspace criterion*

$$\max_{\mathbf{W}^T\mathbf{W}=\mathbf{I}_r} \operatorname{Tr}(\mathbf{\Omega}\mathbf{W}^T\mathbf{C}\mathbf{W}) = \max_{\mathbf{W}^T\mathbf{W}=\mathbf{I}_r} \sum_{k=1}^{r} \omega_k \mathbf{w}_k^T\mathbf{C}\mathbf{w}_k$$

or

$$\min_{\mathbf{W}^T\mathbf{W}=\mathbf{I}_r} \operatorname{Tr}(\mathbf{\Omega}\mathbf{W}^T\mathbf{C}\mathbf{W}) = \min_{\mathbf{W}^T\mathbf{W}=\mathbf{I}_r} \sum_{k=1}^{r} \omega_k \mathbf{w}_k^T\mathbf{C}\mathbf{w}_k \tag{4.8}$$

with $\mathbf{\Omega} = \operatorname{Diag}(\omega_1, \ldots, \omega_r)$, has [54] the unique solution $\{\pm\mathbf{u}_1, \ldots, \pm\mathbf{u}_r\}$ or $\{\pm\mathbf{u}_{n-r+1}, \ldots, \pm\mathbf{u}_n\}$, respectively.

4.2.4 Standard Subspace Iterative Computational Techniques

The first subspace problem consists in computing the eigenvector associated with the largest eigenvalue. The *power method* presented in the sequel is the simplest iterative techniques for this task. Under the condition that λ_1 is the unique dominant eigenvalue associated with $\mathbf{u}_1$ of the real symmetric matrix $\mathbf{C}$, and starting from arbitrary unit 2-norm $\mathbf{w}_0$ not orthogonal to $\mathbf{u}_1$, the following iterations produce a sequence $(\alpha_i, \mathbf{w}_i)$ that converges to the largest eigenvalue λ_1 and its corresponding eigenvector unit 2-norm $\pm\mathbf{u}_1$.

$$\begin{aligned}
&\mathbf{w}_0 \text{ arbitrary such that } \mathbf{w}_0^T\mathbf{u}_1 \neq 0\\
\text{for } i = 0, 1, \ldots \quad &\mathbf{w}'_{i+1} = \mathbf{C}\mathbf{w}_i\\
&\mathbf{w}_{i+1} = \mathbf{w}'_{i+1}/\|\mathbf{w}'_{i+1}\|_2\\
&\alpha_{i+1} = \mathbf{w}_{i+1}^T\mathbf{C}\mathbf{w}_{i+1}.
\end{aligned} \tag{4.9}$$

The proof can be found in [35, p. 406], where the definition and the speed of this convergence are specified in the following. Define $\theta_i \in [0, \pi/2]$ by $\cos(\theta_i) \stackrel{\text{def}}{=} |\mathbf{w}_i^T\mathbf{u}_1|$ satisfying $\cos(\theta_0) \neq 0$, then

$$|\sin(\theta_i)| \leq \tan(\theta_0)\left|\frac{\lambda_2}{\lambda_1}\right|^i \quad \text{and} \quad |\alpha_i - \lambda_1| \leq |\lambda_1 - \lambda_n|\tan^2(\theta_0)\left|\frac{\lambda_2}{\lambda_1}\right|^{2i}. \tag{4.10}$$

Consequently the convergence rate of the power method is exponential and proportional to the ratio $\left|\frac{\lambda_2}{\lambda_1}\right|^i$ for the eigenvector and to $\left|\frac{\lambda_2}{\lambda_1}\right|^{2i}$ for the associated eigenvalue. If $\mathbf{w}_0$ is selected randomly, the probability that this vector is orthogonal to $\mathbf{u}_1$ is equal to

[3] Or simply $\lambda_1 > \lambda_2 > \cdots > \cdots \lambda_n$ when $r = n$, if we are interested by all the eigenvectors.

zero. Furthermore, if $\mathbf{w}_0$ is deliberately chosen orthogonal to $\mathbf{u}_1$, the effect of finite precision in arithmetic computations will introduce errors that will finally provoke the loss of this orthogonality and therefore convergence to $\pm\mathbf{u}_1$.

Suppose now that $\mathbf{C}$ nonnegative. A straightforward generalization of the power method allows for the computation of the r eigenvectors associated with the r largest eigenvalues of $\mathbf{C}$ when its first $r+1$ eigenvalues are distinct, or of the subspace corresponding to the r largest eigenvalues of $\mathbf{C}$ when $\lambda_r > \lambda_{r+1}$ only. This method can be found in the literature under the name of orthogonal iteration, for example, in [35], subspace iteration, for example, in [57] or simultaneous iteration method, for example, in [64]. First, consider the case where the $r+1$ largest eigenvalues of $\mathbf{C}$ are distinct. With $\mathbf{U}_r \stackrel{\text{def}}{=} [\mathbf{u}_1, \ldots, \mathbf{u}_r]$ and $\mathbf{\Delta}_r = \text{Diag}(\lambda_1, \ldots, \lambda_r)$, the following iterations produce a sequence $(\mathbf{\Lambda}_i, \mathbf{W}_i)$ that converges to $(\mathbf{\Delta}_r, [\pm\mathbf{u}_1, \ldots, \pm\mathbf{u}_r])$.

$$\begin{aligned} &\mathbf{W}_0 \text{ arbitrary } n \times r \text{ matrix such that } \mathbf{W}_0^T\mathbf{U}_r \text{ not singular} \\ \text{for } i = 0, 1, \ldots \; \mathbf{W}'_{i+1} &= \mathbf{CW}_i \\ \mathbf{W}'_{i+1} &= \mathbf{W}_{i+1}\mathbf{R}_{i+1} \text{ skinny QR factorization} \\ \mathbf{\Lambda}_{i+1} &= \text{Diag}(\mathbf{W}_{i+1}^T\mathbf{CW}_{i+1}). \end{aligned} \tag{4.11}$$

The proof can be found in [35, p. 411]. The definition and the speed of this convergence are similar to those of the power method, it is exponential and proportional to $\left(\frac{\lambda_{r+1}}{\lambda_r}\right)^i$ for the eigenvectors and to $\left(\frac{\lambda_{r+1}}{\lambda_r}\right)^{2i}$ for the eigenvalues. Note that if $r=1$, then this is just the power method. Moreover for arbitrary r, the sequence formed by the first column of $\mathbf{W}_i$ is precisely the sequence of vectors produced by the power method with the first column of $\mathbf{W}_0$ as starting vector.

Consider now the case where $\lambda_r > \lambda_{r+1}$. Then the following iteration method

$$\begin{aligned} &\mathbf{W}_0 \text{ arbitrary } n \times r \text{ matrix such that } \mathbf{W}_0^T\mathbf{U}_r \text{ not singular} \\ \text{for } i = 0, 1, \ldots \; \mathbf{W}_{i+1} &= \text{Orthonorm}\{\mathbf{CW}_i\} \end{aligned} \tag{4.12}$$

where the orthonormalization (Orthonorm) procedure not necessarily given by the QR factorization, generates a sequence $\mathbf{W}_i$ that converges to the dominant subspace generated by $\{\mathbf{u}_1, \ldots, \mathbf{u}_r\}$ only. This means precisely that the sequence $\mathbf{W}_i\mathbf{W}_i^T$ (which here is a projection matrix because $\mathbf{W}_i^T\mathbf{W}_i = \mathbf{I}_r$) converges to the projection matrix $\mathbf{\Pi}_r \stackrel{\text{def}}{=} \mathbf{U}_r\mathbf{U}_r^T$. In the particular case where the QR factorization is used in the orthonormalization step, the speed of this convergence is exponential and proportional to $\left(\frac{\lambda_{r+1}}{\lambda_r}\right)^i$, that is, more precisely [35, p. 411]

$$\|\mathbf{W}_i\mathbf{W}_i^T - \mathbf{\Pi}_r\|_2 \leq \tan(\theta)\left(\frac{\lambda_{r+1}}{\lambda_r}\right)^i$$

where $\theta \in [0, \pi/2]$ is specified by $\cos(\theta) = \min_{\mathbf{u}\in \text{Span}(\mathbf{W}_0), \mathbf{v}\in \text{Span}(\mathbf{U}_r)} \frac{|\mathbf{u}^T\mathbf{v}|}{\|\mathbf{u}\|_2\|\mathbf{v}\|_2} > 0$. This type of convergence is very specific. The r orthonormal columns of $\mathbf{W}_i$ do not necessary converge to a particular orthonormal basis of the dominant subspace generated by $\mathbf{u}_1, \ldots, \mathbf{u}_r$, but may eventually rotate in this dominant subspace as i increases. Note that the orthonormalization step (4.12) can be realized by other means than the QR decomposition. For example, extending the $r = 1$ case

$$\mathbf{w}_{i+1} = \mathbf{C}\mathbf{w}_i/\|\mathbf{C}\mathbf{w}_i\|_2 = \mathbf{C}\mathbf{w}_i(\mathbf{w}_i^T\mathbf{C}^2\mathbf{w}_i)^{-1/2}$$

to arbitrary r, yields

$$\mathbf{W}_{i+1} = \mathbf{C}\mathbf{W}_i(\mathbf{W}_i^T\mathbf{C}^2\mathbf{W}_i)^{-1/2} \tag{4.13}$$

where the square root inverse of the matrix $\mathbf{W}_i^T\mathbf{C}^2\mathbf{W}_i$ is defined by the EVD of the matrix with its eigenvalues replaced by their square root inverses. The speed of convergence of the associated algorithm is exponential and proportional to $\left(\frac{\lambda_{r+1}}{\lambda_r}\right)^i$ as well [38].

Finally, note that the power and the orthogonal iteration methods can be extended to obtain the minor subspace or eigenvectors by replacing the matrix $\mathbf{C}$ by $\mathbf{I}_n - \mu\mathbf{C}$ where $0 < \mu < 1/\lambda_1$ such that the eigenvalues $1 - \mu\lambda_n > \cdots \geq 1 - \mu\lambda_1 > 0$ of $\mathbf{I}_n - \mu\mathbf{C}$ are strictly positive.

4.2.5 Characterization of the Principal Subspace of a Covariance Matrix from the Minimization of a Mean Square Error

In the particular case where the matrix $\mathbf{C}$ is the covariance of the zero-mean random variable $\mathbf{x}$, consider the scalar function $J(\mathbf{W})$ where $\mathbf{W}$ denotes an arbitrary $n \times r$ matrix

$$J(\mathbf{W}) \stackrel{\text{def}}{=} \mathrm{E}\left(\|\mathbf{x} - \mathbf{W}\mathbf{W}^T\mathbf{x}\|^2\right). \tag{4.14}$$

The following two properties are proved (e.g. see [71] and Exercises 4.7 and 4.8).

First, the stationary points $\mathbf{W}$ of $J(\mathbf{W})$ [i.e., the points $\mathbf{W}$ that cancel $J(\mathbf{W})$] are given by $\mathbf{W} = \mathbf{U}_r\mathbf{Q}$ where the r columns of $\mathbf{U}_r$ denotes here arbitrary r distinct unit-2 norm eigenvectors among $\mathbf{u}_1, \ldots, \mathbf{u}_n$ of $\mathbf{C}$ and where $\mathbf{Q}$ is an arbitrary $r \times r$ orthogonal matrix. Furthermore at each stationary point, $J(\mathbf{W})$ equals the sum of eigenvalues whose eigenvectors are not included in $\mathbf{U}_r$.

Second, in the particular case where $\lambda_r > \lambda_{r+1}$, all stationary points of $J(\mathbf{W})$ are saddle points except the points $\mathbf{W}$ whose associated matrix $\mathbf{U}_r$ contains the r dominant eigenvectors $\mathbf{u}_1, \ldots, \mathbf{u}_r$ of $\mathbf{C}$. In this case $J(\mathbf{W})$ attains the global minimum $\sum_{i=r+1}^n \lambda_i$. It is important to note that at this global minimum, $\mathbf{W}$ does not necessarily contain the r dominant eigenvectors $\mathbf{u}_1, \ldots, \mathbf{u}_r$ of $\mathbf{C}$, but rather an arbitrary orthogonal basis of the associated dominant subspace. This is not surprising because

$$J(\mathbf{W}) = \mathrm{Tr}(\mathbf{C}) - 2\mathrm{Tr}(\mathbf{W}^T\mathbf{C}\mathbf{W}) + \mathrm{Tr}(\mathbf{W}\mathbf{W}^T\mathbf{C}\mathbf{W}\mathbf{W}^T)$$

with $\mathrm{Tr}(\mathbf{W}^T\mathbf{C}\mathbf{W}) = \mathrm{Tr}(\mathbf{C}\mathbf{W}\mathbf{W}^T)$ and thus $J(\mathbf{W})$ is expressed as a function of $\mathbf{W}$ through $\mathbf{W}\mathbf{W}^T$ which is invariant with respect to rotation $\mathbf{W}\mathbf{Q}$ of $\mathbf{W}$. Finally, note that when $r = 1$ and $\lambda_1 > \lambda_2$, the solution of the minimization of $J(\mathbf{w})$ (4.14) is given by the unit 2-norm dominant eigenvector $\pm\mathbf{u}_1$.

4.3 OBSERVATION MODEL AND PROBLEM STATEMENT

4.3.1 Observation Model

The general iterative subspace determination problem described in the previous section, will be now specialized to a class of matrices $\mathbf{C}$ computed from observation data. In typical applications of subspace-based signal processing, a sequence[4] of data vectors $\mathbf{x}(k) \in \mathcal{R}^n$ is observed, satisfying the following very common observation signal model

$$\mathbf{x}(k) = \mathbf{s}(k) + \mathbf{n}(k), \tag{4.15}$$

where $\mathbf{s}(k)$ is a vector containing the information signal lying on an r-dimensional linear subspace of $\mathcal{R}^n$ with $r < n$, while $\mathbf{n}(k)$ is a zero-mean additive random white noise (AWN) random vector, uncorrelated from $\mathbf{s}(k)$. Note that $\mathbf{s}(k)$ is often given by $\mathbf{s}(k) = \mathbf{A}(k)\mathbf{r}(k)$ where the full rank $n \times r$ matrix $\mathbf{A}(k)$ is deterministically parameterized and $\mathbf{r}(k)$ is a r-dimensional zero-mean full random vector (i.e., with $\mathrm{E}[\mathbf{r}(k)\mathbf{r}^T(k)]$ nonsingular). The signal part $\mathbf{s}(k)$ may also randomly select among r deterministic vectors. This random selection does not necessarily result in a zero-mean signal vector $\mathbf{s}(k)$.

In these assumptions, the covariance matrix $\mathbf{C}_s(k)$ of $\mathbf{s}(k)$ is r-rank deficient and

$$\mathbf{C}_x(k) \stackrel{\text{def}}{=} \mathrm{E}[\mathbf{x}(k)\mathbf{x}^T(k)] = \mathbf{C}_s(k) + \sigma_n^2(k)\mathbf{I}_n \tag{4.16}$$

where $\sigma_n^2(k)$ denotes the AWN power. Taking into account that $\mathbf{C}_s(k)$ is of rank r and applying the EVD (4.1) on $\mathbf{C}_x(k)$ yields

$$\mathbf{C}_x(k) = [\mathbf{U}_s(k), \mathbf{U}_n(k)]\begin{bmatrix} \boldsymbol{\Delta}_s(k) + \sigma_n^2(k)\mathbf{I}_r & \mathbf{O} \\ \mathbf{O} & \sigma_n^2(k)\mathbf{I}_{n-r} \end{bmatrix}\begin{bmatrix} \mathbf{U}_s^T(k) \\ \mathbf{U}_n^T(k) \end{bmatrix} \tag{4.17}$$

where the $n \times r$ and $n \times (n-r)$ matrices $\mathbf{U}_s(k)$ and $\mathbf{U}_n(k)$ are orthonormal bases for the denoted signal or dominant and noise or minor subspace of $\mathbf{C}_x(k)$ and $\boldsymbol{\Delta}_s(k)$ is a $r \times r$ diagonal matrix constituted by the r nonzero eigenvalues of $\mathbf{C}_s(k)$. We note that the column vectors of $\mathbf{U}_s(k)$ are generally unique up to a sign, in contrast to the column

[4]Note that k generally represents successive instants, but it can also represent successive spatial coordinates (e.g. in [11]) where k denotes the position of the secondary range cells in Radar.

vectors of $\mathbf{U}_n(k)$ for which $\mathbf{U}_n(k)$ is defined up to a right multiplication by a $(n-r) \times (n-r)$ orthonormal matrix $\mathbf{Q}$. However, the associated orthogonal projection matrices $\mathbf{\Pi}_s(k) \stackrel{\text{def}}{=} \mathbf{U}_s(k)\mathbf{U}_s^T(k)$ and $\mathbf{\Pi}_n(k) \stackrel{\text{def}}{=} \mathbf{U}_n(k)\mathbf{U}_n^T(k)$ respectively denoted signal or dominant projection matrices and noise or minor projection matrices that will be introduced in the next sections are both unique.

4.3.2 Statement of the Problem

A very important problem in signal processing consists in continuously updating the estimate $\mathbf{U}_s(k)$, $\mathbf{U}_n(k)$, $\mathbf{\Pi}_s(k)$ or $\mathbf{\Pi}_n(k)$ and sometimes with $\mathbf{\Delta}_s(k)$ and $\sigma_n^2(k)$, assuming that we have available consecutive observation vectors $\mathbf{x}(i)$, $i = \ldots, k-1, k, \ldots$ when the signal or noise subspace is slowly time-varying compared to $\mathbf{x}(k)$. The dimension r of the signal subspace may be known *a priori* or estimated from the observation vectors. A straightforward way to come up with a method that solves these problems is to provide efficient adaptive estimates $\mathbf{C}(k)$ of $\mathbf{C}_x(k)$ and simply apply an EVD at each time step k. Candidates for this estimate $\mathbf{C}(k)$ are generally given by sliding windowed sample data covariance matrices when the sequence of $\mathbf{C}_x(k)$ undergoes relatively slow changes. With an exponential window, the estimated covariance matrix is defined as

$$\mathbf{C}(k) = \sum_{i=0}^{k} \beta^{k-i}\mathbf{x}(i)\mathbf{x}^T(i) \tag{4.18}$$

where $0 < \beta < 1$ is the forgetting factor. Its use is intended to ensure that the data in the distant past are downweighted in order to afford the tracking capability when we operate in a nonstationary environment. $\mathbf{C}(k)$ can be recursively updated according to the following scheme

$$\mathbf{C}(k) = \beta\mathbf{C}(k-1) + \mathbf{x}(k)\mathbf{x}^T(k). \tag{4.19}$$

Note that

$$\begin{aligned}\mathbf{C}(k) &= (1-\beta')\mathbf{C}(k-1) + \beta'\mathbf{x}(k)\mathbf{x}^T(k) \\ &= \mathbf{C}(k-1) + \beta'[\mathbf{x}(k)\mathbf{x}^T(k) - \mathbf{C}(k-1)]\end{aligned} \tag{4.20}$$

is also used. These estimates $\mathbf{C}(k)$ tend to smooth the variations of the signal parameters and so are only suitable for slowly changing signal parameters. For sudden signal parameter changes, the use of a truncated window may offer faster tracking. In this case, the estimated covariance matrix is derived from a window of length l

$$\mathbf{C}(k) = \sum_{i=k-l+1}^{k} \beta^{k-i}\mathbf{x}(i)\mathbf{x}^T(i) \tag{4.21}$$

where $0 < \beta \leq 1$. The case $\beta = 1$ corresponds to a rectangular window. This matrix can be recursively updated according to the following scheme

$$\mathbf{C}(k) = \beta\mathbf{C}(k-1) + \mathbf{x}(k)\mathbf{x}^T(k) - \beta^l\mathbf{x}(k-l)\mathbf{x}^T(k-l). \tag{4.22}$$

Both versions require $O(n^2)$ operations with the first having smaller computational complexity and memory needs. Note that for $\beta = 0$, (4.22) gives the coarse estimate $\mathbf{x}(k)\mathbf{x}^T(k)$ of $\mathbf{C}_x(k)$ as used in the least mean square (LMS) algorithms for adaptive filtering (see e.g. [36]).

Applying an EVD on $\mathbf{C}(k)$ at each time k is of course the best possible way to estimate the eigenvectors or subspaces we are looking for. This approach is known as direct EVD and has high complexity which is $O(n^3)$. This method usually serves as point of reference when dealing with the different less computationally demanding approaches described in the next sections. These computationally efficient algorithms will compute signal or noise eigenvectors (or signal or noise projection matrices) at the time instant $k+1$ from the associated estimate at time k and the new arriving sample vector $\mathbf{x}(k)$.

4.4 PRELIMINARY EXAMPLE: OJA'S NEURON

Let us introduce these adaptive procedures by a simple example: The following Oja's neuron originated by Oja [50] and then applied by many others that estimates the eigenvector associated with the unique largest eigenvalue of a covariance matrix of the stationary vector $\mathbf{x}(k)$.

$$\mathbf{w}(k+1) = \mathbf{w}(k) + \mu\{[\mathbf{I}_n - \mathbf{w}(k)\mathbf{w}^T(k)]\mathbf{x}(k)\mathbf{x}^T(k)\mathbf{w}(k)\}. \tag{4.23}$$

The first term on the right side is the previous estimate of $\pm\mathbf{u}_1$, which is kept as a memory of the iteration. The whole term in brackets is the new information. This term is scaled by the stepsize μ and then added to the previous estimate $\mathbf{w}(k)$ to obtain the current estimate $\mathbf{w}(k+1)$. We note that this new information is formed by two terms. The first one $\mathbf{x}(k)\mathbf{x}^T(k)\mathbf{w}(k)$ contains the first step of the power method (4.9) and the second one is simply the previous estimate $\mathbf{w}(k)$ adjusted by the scalar $\mathbf{w}^T(k)\mathbf{x}(k)\mathbf{x}^T(k)\mathbf{w}(k)$ so that these two terms are on the same scale. Finally, we note that if the previous estimate $\mathbf{w}(k)$ is already the desired eigenvector $\pm\mathbf{u}_1$, the expectation of this new information is zero, and hence, $\mathbf{w}(k+1)$ will be hovering around $\pm\mathbf{u}_1$. The step size μ controls the balance between the past and the new information. Introduced in the neural networks literature [50] within the framework of a new synaptic modification law, it is interesting to note that this algorithm can be derived from different heuristic variations of numerical methods introduced in Section 4.2.

First consider the variational characterization recalled in Subsection 4.2.3. Because $\nabla_{\mathbf{w}}(\mathbf{w}^T\mathbf{C}_x\mathbf{w}) = 2\mathbf{C}_x\mathbf{w}$, the constrained maximization (4.3) or (4.7) can be solved using the following constrained gradient-search procedure

$$\mathbf{w}'(k+1) = \mathbf{w}(k) + \mu \mathbf{C}_x(k)\mathbf{w}(k)$$
$$\mathbf{w}(k+1) = \mathbf{w}'(k+1)/\|\mathbf{w}'(k+1)\|_2$$

in which the stepsize μ is sufficiency small enough. Using the approximation $\mu^2 \ll \mu$ yields

$$\begin{aligned}\mathbf{w}'(k+1)/\|\mathbf{w}'(k+1)\|_2 &= [\mathbf{I}_n + \mu\mathbf{C}_x(k)]\mathbf{w}(k)/\{\mathbf{w}^T(k)[\mathbf{I}_n + \mu\mathbf{C}_x(k)]^2\mathbf{w}(k)\}^{1/2} \\ &\approx [\mathbf{I}_n + \mu\mathbf{C}_x(k)]\mathbf{w}(k)/[1 + 2\mu\mathbf{w}^T(k)\mathbf{C}_x(k)\mathbf{w}(k)]^{1/2} \\ &\approx [\mathbf{I}_n + \mu\mathbf{C}_x(k)]\mathbf{w}(k)[1 - \mu\mathbf{w}^T(k)\mathbf{C}_x(k)\mathbf{w}(k)] \\ &\approx \mathbf{w}(k) + \mu[\mathbf{I}_n - \mathbf{w}(k)\mathbf{w}^T(k)]\mathbf{C}_x(k)\mathbf{w}(k).\end{aligned}$$

Then, using the instantaneous estimate $\mathbf{x}(k)\mathbf{x}^T(k)$ of $\mathbf{C}_x(k)$, Oja's neuron (4.23) is derived.

Consider now the power method recalled in Subsection 4.2.4. Noticing that $\mathbf{C}_x$ and $\mathbf{I}_n + \mu\mathbf{C}_x$ have the same eigenvectors, the step $\mathbf{w}'_{i+1} = \mathbf{C}_x\mathbf{w}_i$ of (4.9) can be replaced by $\mathbf{w}'_{i+1} = (\mathbf{I}_n + \mu\mathbf{C}_x)\mathbf{w}_i$ and using the previous approximations yields Oja's neuron (4.23) anew.

Finally, consider the characterization of the eigenvector associated with the unique largest eigenvalue of a covariance matrix derived from the mean square error $\mathrm{E}\left(\|\mathbf{x} - \mathbf{w}\mathbf{w}^T\mathbf{x}\|^2\right)$ recalled in Subsection 4.2.5. Because

$$\nabla_{\mathbf{w}}\mathrm{E}\left(\|\mathbf{x} - \mathbf{w}\mathbf{w}^T\mathbf{x}\|^2\right) = 2\left(-2\mathbf{C}_x + \mathbf{C}_x\mathbf{w}\mathbf{w}^T + \mathbf{w}\mathbf{w}^T\mathbf{C}_x\right)\mathbf{w}$$

an unconstrained gradient-search procedure yields

$$\mathbf{w}(k+1) = \mathbf{w}(k) - \mu[-2\mathbf{C}_x(k) + \mathbf{C}_x(k)\mathbf{w}(k)\mathbf{w}^T(k) + \mathbf{w}(k)\mathbf{w}^T(k)\mathbf{C}_x(k)]\mathbf{w}(k).$$

Then, using the instantaneous estimate $\mathbf{x}(k)\mathbf{x}^T(k)$ of $\mathbf{C}_x(k)$ and the approximation $\mathbf{w}^T(k)\mathbf{w}(k) = 1$ justified by the convergence of the deterministic gradient-search procedure to $\pm\mathbf{u}_1$ when $\mu \to 0$, Oja's neuron (4.23) is derived again.

Furthermore, if we are interested in adaptively estimating the associated single eigenvalue λ_1, the minimization of the scalar function $J(\lambda) = (\lambda - \mathbf{u}_1^T\mathbf{C}_x\mathbf{u}_1)^2$ by a gradient-search procedure can be used. With the instantaneous estimate $\mathbf{x}(k)\mathbf{x}^T(k)$ of $\mathbf{C}_x(k)$ and with the estimate $\mathbf{w}(k)$ of $\mathbf{u}_1$ given by (4.23), the following stochastic gradient algorithm is obtained.

$$\lambda(k+1) = \lambda(k) + \mu[\mathbf{w}^T(k)\mathbf{x}(k)\mathbf{x}^T(k)\mathbf{w}(k) - \lambda(k)]. \tag{4.24}$$

We note that the previous two heuristic derivations could be extended to the adaptive estimation of the eigenvector associated with the unique smallest eigenvalue of $\mathbf{C}_x(k)$. Using the constrained minimization (4.3) or (4.7) solved by a constrained gradient-search procedure or the power method (4.9) where the step $\mathbf{w}'_{i+1} = \mathbf{C}_x\mathbf{w}_i$ of (4.9) is replaced by $\mathbf{w}'_{i+1} = (\mathbf{I}_n - \mu\mathbf{C}_x)\mathbf{w}_i$ (where $0 < \mu < 1/\lambda_1$) yields (4.23) after the same derivation, but where the sign of the stepsize μ is reversed.

$$\mathbf{w}(k+1) = \mathbf{w}(k) - \mu\{[\mathbf{I}_n - \mathbf{w}(k)\mathbf{w}^T(k)]\mathbf{x}(k)\mathbf{x}^T(k)\mathbf{w}(k)\}. \tag{4.25}$$

The associated eigenvalue λ_n could be also derived from the minimization of $J(\lambda) = (\lambda - \mathbf{u}_n^T\mathbf{C}_x\mathbf{u}_n)^2$ and consequently obtained by (4.24) as well, where $\mathbf{w}(k)$ is issued from (4.25).

These heuristic approaches are derived from iterative computational techniques issued from numerical methods recalled in Section 4.2, and need to be validated by convergence and performance analysis for stationary data $\mathbf{x}(k)$. These issues will be considered in Section 4.7. In particular it will be proved that the coupled stochastic approximation algorithms (4.23) and (4.24) in which the stepsize μ is decreasing, converge to the pair $(\pm\mathbf{u}_1, \lambda_1)$, in contrast to the stochastic approximation algorithm (4.25) that diverges. Then, due to the possible accumulation of rounding errors, the algorithms that converge theoretically must be tested through numerical experiments to check their numerical stability in stationary environments. Finally extensive Monte Carlo simulations must be carried out with various stepsizes, initialization conditions, SNRs, and parameters configurations in nonstationary environments.

4.5 SUBSPACE TRACKING

In this section, we consider the adaptive estimation of dominant (signal) and minor (noise) subspaces. To derive such algorithms from the linear algebra material recalled in Subsections 4.2.3, 4.2.4, and 4.2.5 similarly as for Oja's neuron, we first note that the general orthogonal iterative step (4.12): $\mathbf{W}_{i+1} = \text{Orthonorm}\{\mathbf{C}\mathbf{W}_i\}$ allows for the following variant for adaptive implementation

$$\mathbf{W}_{i+1} = \text{Orthonorm}\{(\mathbf{I}_n + \mu\mathbf{C})\mathbf{W}_i\}$$

where $\mu > 0$ is a small parameter known as stepsize, because $\mathbf{I}_n + \mu\mathbf{C}$ has the same eigenvectors as $\mathbf{C}$ with associated eigenvalues $(1 + \mu\lambda_i)_{i=1,\ldots,n}$. Noting that $\mathbf{I}_n - \mu\mathbf{C}$ has also the same eigenvectors as $\mathbf{C}$ with associated eigenvalues $(1 - \mu\lambda_i)_{i=1,\ldots,n}$, arranged exactly in the opposite order as $(\lambda_i)_{i=1,\ldots,n}$ for μ sufficiently small ($\mu < 1/\lambda_1$), the general orthogonal iterative step (4.12) allows for the following second variant of this iterative procedure to converge to the r-dimensional minor subspace of $\mathbf{C}$ if $\lambda_{n-r} > \lambda_{n-r+1}$.

$$\mathbf{W}_{i+1} = \text{Orthonorm}\{(\mathbf{I}_n - \mu\mathbf{C})\mathbf{W}_i\}.$$

When the matrix $\mathbf{C}$ is unknown and, instead we have sequentially the data sequence $\mathbf{x}(k)$, we can replace $\mathbf{C}$ by an adaptive estimate $\mathbf{C}(k)$ (see Section 4.3.2). This leads to the adaptive orthogonal iteration algorithm

$$\mathbf{W}(k+1) = \text{Orthonorm}\{[\mathbf{I}_n \pm \mu_k \mathbf{C}(k)]\mathbf{W}(k)\} \tag{4.26}$$

where the + sign generates estimates for the signal subspace (if $\lambda_r > \lambda_{r+1}$) and the − sign for the noise subspace (if $\lambda_{n-r} > \lambda_{n-r+1}$). Depending on the choice of the estimate $\mathbf{C}(k)$ and of the orthonormalization (or approximate orthonormalization), we can obtain alternative subspace tracking algorithms.

We note that maximization or minimization in (4.7) of $J(\mathbf{W}) \stackrel{\text{def}}{=} \text{Tr}(\mathbf{W}^T\mathbf{C}\mathbf{W})$ subject to the constraint $\mathbf{W}^T\mathbf{W} = \mathbf{I}_r$ can be solved by a constrained gradient-descent technique. Because $\nabla_{\mathbf{W}} J = 2\mathbf{C}(k)\mathbf{W}$, we obtain the following Rayleigh quotient-based algorithm

$$\mathbf{W}(k+1) = \text{Orthonorm}\{\mathbf{W}(k) \pm \mu_k \mathbf{C}(k)\mathbf{W}(k)\} \tag{4.27}$$

whose general expression is the same as general expression (4.26) derived from the orthogonal iteration approach. We will denote this family of algorithms as the power-based methods. It is interesting to note that a simple sign change enables one to switch from the dominant to the minor subspaces. Unfortunately, similarly to Oja's neuron, many minor subspace algorithms will be unstable or stable but non-robust (i.e., numerically unstable with a tendency to accumulate round-off errors until their estimates are meaningless), in contrast to the associated majorant subspace algorithms. Consequently, the literature of minor subspace tracking techniques is very limited as compared to the wide variety of methods that exists for the tracking of majorant subspaces.

4.5.1 Subspace Power-based Methods

Clearly the simplest selection for $\mathbf{C}(k)$ is the instantaneous estimate $\mathbf{x}(k)\mathbf{x}^T(k)$, which gives rise to the Data Projection Method (DPM) first introduced in [70] where the orthonormalization is performed using the Gram–Schmidt procedure.

$$\mathbf{W}(k+1) = \text{GS Orth.}\{\mathbf{W}(k) \pm \mu_k \mathbf{x}(k)\mathbf{x}^T(k)\mathbf{W}(k)\}. \tag{4.28}$$

In nonstationary situations, estimates (4.19) or (4.20) of the covariance $\mathbf{C}_x(k)$ of $\mathbf{x}(k)$ at time k have been tested in [70]. For this algorithm to converge, we need to select a stepsize μ such that $\mu \ll 1/\lambda_1$ (see e.g. [29]). To satisfy this requirement (in nonstationary situations included) and because most of the time we have $\text{Tr}[\mathbf{C}_x(k)] \gg \lambda_1(k)$, the following two normalized step sizes have been proposed in [70]

$$\mu_k = \frac{\mu}{\|\mathbf{x}(k)\|^2} \quad \text{and} \quad \mu_k = \frac{\mu}{\sigma_x^2(k)} \quad \text{with } \sigma_x^2(k+1) = \nu\sigma_x^2(k) + (1-\nu)\|\mathbf{x}(k)\|^2$$

where μ may be close to unity and where the choice of $\nu \in (0, 1)$ depends on the rapidity of the change of the parameters of the observation signal model (4.15).

Note that a better numerical stability can be achieved [5] if μ_k is chosen, similar to the normalized LMS algorithm [36], as $\mu_k = \dfrac{\mu}{\|\mathbf{x}(k)\|^2 + \alpha}$ where α is a very small positive constant. Obviously, this algorithm (4.28) has very high computational complexity due to the Gram–Schmidt orthonormalization step.

To reduce this computational complexity, many algorithms have been proposed. Going back to the DPM algorithm (4.28), we observe that we can write

$$\mathbf{W}(k+1) = \{\mathbf{W}(k) \pm \mu_k \mathbf{x}(k)\mathbf{x}^T(k)\mathbf{W}(k)\}\mathbf{G}(k+1) \tag{4.29}$$

where the matrix $\mathbf{G}(k+1)$ is responsible for performing exact or approximate orthonormalization while preserving the space generated by the columns of $\mathbf{W}'(k+1) \stackrel{\text{def}}{=} \mathbf{W}(k) \pm \mu_k \mathbf{x}(k)\mathbf{x}^T(k)\mathbf{W}(k)$. It is the different choices of $\mathbf{G}(k+1)$ that will pave the way for alternative less computationally demanding algorithms. Depending on whether this orthonormalization is exact or approximate, two families of algorithms have been proposed in the literature.

The Approximate Symmetric Orthonormalization Family The columns of $\mathbf{W}'(k+1)$ can be approximately orthonormalized in a symmetrical way. Since $\mathbf{W}(k)$ has orthonormal columns, for sufficiently small μ_k the columns of $\mathbf{W}'(k+1)$ will be linearly independent, although not orthonormal. Then $\mathbf{W}'^T(k+1)\mathbf{W}'(k+1)$ is positive definite, and $\mathbf{W}(k+1)$ will have orthonormal columns if $\mathbf{G}(k+1) = \{\mathbf{W}'^T(k+1)\mathbf{W}'(k+1)\}^{-1/2}$ (unique if $\mathbf{G}(k+1)$ is constrained to be symmetric). A stochastic algorithm denoted Subspace Network Learning (SNL) and later Oja's algorithm have been derived in [53] to estimate dominant subspace. Assuming μ_k is sufficiency enough, $\mathbf{G}(k+1)$ can be expanded in μ_k as follows

$$\begin{aligned}
\mathbf{G}(k+1) &= \{[\mathbf{W}(k) + \mu_k \mathbf{x}(k)\mathbf{x}^T(k)\mathbf{W}(k)]^T[\mathbf{W}(k) + \mu_k \mathbf{x}(k)\mathbf{x}^T(k)\mathbf{W}(k)]\}^{-1/2} \\
&= \{\mathbf{I}_r + 2\mu_k \mathbf{W}^T(k)\mathbf{x}(k)\mathbf{x}^T(k)\mathbf{W}(k) + O(\mu_k^2)\}^{-1/2} \\
&= \mathbf{I}_r - \mu_k \mathbf{W}^T(k)\mathbf{x}(k)\mathbf{x}^T(k)\mathbf{W}(k) + O(\mu_k^2).
\end{aligned}$$

Omitting second-order terms, the resulting algorithm reads[5]

$$\mathbf{W}(k+1) = \mathbf{W}(k) + \mu_k[\mathbf{I}_n - \mathbf{W}(k)\mathbf{W}^T(k)]\mathbf{x}(k)\mathbf{x}^T(k)\mathbf{W}(k). \tag{4.30}$$

The convergence of this algorithm has been earlier studied in [78] and then in [69], where it was shown that the solution $\mathbf{W}(t)$ of its associated ODE (see Subsection 4.7.1) need not tend to the eigenvectors $\{\mathbf{v}_1, \ldots, \mathbf{v}_r\}$, but only to a rotated basis $\mathbf{W}_*$ of the subspace spanned by them. More precisely, it has been proved in [16] that under the assumption that $\mathbf{W}(0)$ is of full column rank such that its projection to the

[5]Note that this algorithm can be directly deduced from the optimization of the cost function $J(\mathbf{W}) = \mathrm{Tr}[\mathbf{W}^T\mathbf{x}(k)\mathbf{x}^T(k)\mathbf{W}]$ defined on the set of $n \times r$ orthogonal matrices $\mathbf{W}(\mathbf{W}^T\mathbf{W} = \mathbf{I}_r)$ with the help of continuous-time matrix algorithms [21, Ch. 7.2] (see also (4.93) in Exercise 4.15).

signal subspace of $\mathbf{C}_x$ is linearly independent, there exists a rotated basis $\mathbf{W}_*$ of this signal subspace such that $\|\mathbf{W}(t) - \mathbf{W}_*\|_{\text{Fro}} = O(e^{-(\lambda_r - \lambda_{r+1})t})$. A performance analysis has been given in [24, 25]. This issue will be used as an example analysis of convergence and performance in Subsection 4.7.3. Note that replacing $\mathbf{x}(k)\mathbf{x}^T(k)$ by $\beta\mathbf{I}_n \pm \mathbf{x}(k)\mathbf{x}^T(k)$ (with $\beta > 0$) in (4.30), leads to a modified Oja's algorithm [15], which, not affecting its capability of tracking a signal subspace with the sign +, can track a noise subspace by changing the sign (if $\beta > \lambda_1$). Of course, these modified Oja's algorithms enjoy the same convergence properties as Oja's algorithm (4.30).

Many other modifications of Oja's algorithm have appeared in the literature, particularly to adapt it to noise subspace tracking. To obtain such algorithms, it is interesting to point out that, in general, it is not possible to obtain noise subspace tracking algorithms by simply changing the sign of the stepsize of a signal subspace tracking algorithm. For example, changing the sign in (4.30) or (4.85) leads to an unstable algorithm (divergence) as will be explained in Subsection 4.7.3 for $r = 1$. Among these modified Oja's algorithms, Chen et al. [16] have proposed the following unified algorithm

$$\begin{aligned}\mathbf{W}(k+1) = \mathbf{W}(k) \pm \mu_k[&\mathbf{x}(k)\mathbf{x}^T(k)\mathbf{W}(k)\mathbf{W}^T(k)\mathbf{W}(k) \\ &- \mathbf{W}(k)\mathbf{W}^T(k)\mathbf{x}(k)\mathbf{x}^T(k)\mathbf{W}(k)]\end{aligned} \tag{4.31}$$

where the signs + and − are respectively associated with signal and noise tracking algorithms. While the associated ODE maintains $\mathbf{W}^T(t)\mathbf{W}(t) = \mathbf{I}_r$ if $\mathbf{W}^T(0)\mathbf{W}(0) = \mathbf{I}_r$ and enjoys [16] the same stability properties as Oja's algorithm, the stochastic approximation to algorithm (4.31) suffers from numerical instabilities (see e.g. the numerical simulations in [28]). Thus, its practical use requires periodic column reorthonormalization. To avoid these numerical instabilities, this algorithm has been modified [17] by adding the penalty term $\mathbf{W}(k)[\mathbf{I}_n - \mathbf{W}(k)\mathbf{W}^T(k)]$ to the field of (4.31). As far as noise subspace tracking is concerned, Douglas et al. [28] have proposed modifying the algorithm (4.31) by multiplying the first term of its field by $\mathbf{W}^T(k)\mathbf{W}(k)$ whose associated term in the ODE tends to $\mathbf{I}_r$, viz

$$\begin{aligned}\mathbf{W}(k+1) = \mathbf{W}(k) - \mu_k[&\mathbf{x}(k)\mathbf{x}^T(k)\mathbf{W}(k)\mathbf{W}^T(k)\mathbf{W}(k)\mathbf{W}^T(k)\mathbf{W}(k) \\ &- \mathbf{W}(k)\mathbf{W}^T(k)\mathbf{x}(k)\mathbf{x}^T(k)\mathbf{W}(k)].\end{aligned} \tag{4.32}$$

It is proved in [28] that the locally asymptotically stable points $\mathbf{W}$ of the ODE associated with this algorithm satisfy $\mathbf{W}^T\mathbf{W} = \mathbf{I}_r$ and $\text{Span}(\mathbf{W}) = \text{Span}(\mathbf{U}_n)$. But the solution $\mathbf{W}(t)$ of the associated ODE does not converge to a particular basis $\mathbf{W}_*$ of the noise subspace but rather, it is proved that $\text{Span}[\mathbf{W}(t)]$ tends to $\text{Span}(\mathbf{U}_n)$ (in the sense that the projection matrix associated with the subspace $\text{Span}[\mathbf{W}(t)]$ tends to $\mathbf{\Pi}_n$). Numerical simulations presented in [28] show that this algorithm is numerically more stable than the minor subspace version of algorithm (4.31).

To eliminate the instability of the noise tracking algorithm derived from Oja's algorithm (4.30) where the sign of the stepsize is changed, Abed Meraim et al. [2]

have proposed forcing the estimate $\mathbf{W}(k)$ to be orthonormal at each time step k (see Exercise 4.10) that can be used for signal subspace tracking (by reversing the sign of the stepsize) as well. But this algorithm converges with the same speed of convergence as Oja's algorithm (4.30). To accelerate its convergence, two normalized versions (denoted Normalized Oja's algorithm (NOja) and Normalized Orthogonal Oja's algorithm (NOOJa)) of this algorithm have been proposed in [4]. They can perform both signal and noise tracking by switching the sign of the stepsize for which an approximate closed-form expression has been derived. A convergence analysis of the NOja algorithm has been presented in [7] using the ODE approach. Because the ODE associated with the field of this stochastic approximation algorithm is the same as those associated with the projection approximation-based algorithm (4.43), it enjoys the same convergence properties.

The Exact Orthonormalization Family The orthonormalization (4.29) of the columns of $\mathbf{W}'(k+1)$ can be performed exactly at each iteration by the symmetric square root inverse of $\mathbf{W}'^T(k+1)\mathbf{W}'(k+1)$ due to the fact that the latter is a rank one modification of the identity matrix

$$\mathbf{W}'^T(k+1)\mathbf{W}'(k+1) = \mathbf{I}_r \pm (2\mu_k \pm \mu_k^2\|\mathbf{x}(k)\|^2)\mathbf{y}(k)\mathbf{y}^T(k) \stackrel{\text{def}}{=} \mathbf{I}_r \pm \mathbf{z}\mathbf{z}^T \tag{4.33}$$

with $\mathbf{y}(k) \stackrel{\text{def}}{=} \mathbf{W}^T(k)\mathbf{x}(k)$ and $\mathbf{z} \stackrel{\text{def}}{=} \sqrt{2\mu_k \pm \mu_k^2\|\mathbf{x}(k)\|^2}\ \mathbf{y}(k)$. Using the identity

$$\left(\mathbf{I}_r \pm \mathbf{z}\mathbf{z}^T\right)^{-1/2} = \mathbf{I}_r + \left(\frac{1}{(1 \pm \|\mathbf{z}\|^2)^{1/2}} - 1\right)\frac{\mathbf{z}\mathbf{z}^T}{\|\mathbf{z}\|^2} \tag{4.34}$$

we obtain

$$\mathbf{G}(k+1) = \{\mathbf{W}'^T(k+1)\mathbf{W}'(k+1)\}^{-1/2} = \mathbf{I}_r + \tau_k\mathbf{y}(k)\mathbf{y}^T(k) \tag{4.35}$$

with $\tau_k \stackrel{\text{def}}{=} \left(\frac{1}{(1 \pm (2\mu_k \pm \mu_k^2\|\mathbf{x}(k)\|^2)\|\mathbf{y}(k)\|^2)^{1/2}} - 1\right)\frac{1}{\|\mathbf{y}(k)\|^2}$. Substituting (4.35) into (4.29) leads to

$$\mathbf{W}(k+1) = \mathbf{W}(k) \pm \mu_k\mathbf{p}(k)\mathbf{x}^T(k)\mathbf{W}(k) \tag{4.36}$$

where $\mathbf{p}(k) \stackrel{\text{def}}{=} \pm\frac{\tau_k}{\mu_k}\mathbf{W}(k)\mathbf{y}(k) + (1 + \tau_k\|\mathbf{y}(k)\|^2)\mathbf{x}(k)$. All these steps lead to the Fast Rayleigh quotient-based Adaptive Noise Subspace algorithm (FRANS) introduced by Attallah et al. in [5]. As stated in [5], this algorithm is stable and robust in the case of signal subspace tracking (associated with the sign $+$) including initialization with a nonorthonormal matrix $\mathbf{W}(0)$. By contrast, in the case of noise subspace tracking (associated with the sign $-$), this algorithm is numerically unstable because of round-off error accumulation. Even when initialized with an orthonormal matrix, it requires periodic reorthonormalization of $\mathbf{W}(k)$ in order to maintain the orthonormality of the columns of $\mathbf{W}(k)$. To remedy this instability, another implementation of this algorithm based on the numerically well behaved Householder transform has been proposed [6]. This Householder FRANS algorithm (HFRANS) comes from (4.36) which can be rewritten after cumbersome manipulations as

$$\mathbf{W}(k+1) = \mathbf{H}(k)\mathbf{W}(k) \quad \text{with } \mathbf{H}(k) = \mathbf{I}_n - 2\mathbf{u}(k)\mathbf{u}^T(k)$$

with $\mathbf{u}(k) \stackrel{\text{def}}{=} \frac{\mathbf{p}(k)}{\|\mathbf{p}(k)\|_2}$. With no additional numerical complexity, this Householder transform allows one to stabilize the noise subspace version of the FRANS algorithm.[6] The interested reader may refer to [75] that analyzes the orthonormal error propagation [i.e., a recursion of the distance to orthonormality $\|\mathbf{W}^T(k)\mathbf{W}(k) - \mathbf{I}_r\|^2_{\text{Fro}}$ from a nonorthogonal matrix $\mathbf{W}(0)$] in the FRANS and HFRANS algorithms.

Another solution to orthonormalize the columns of $\mathbf{W}'(k+1)$ has been proposed in [29, 30]. It consists of two steps. The first one orthogonalizes these columns using a matrix $\mathbf{G}(k+1)$ to give $\mathbf{W}''(k+1) = \mathbf{W}'(k+1)\mathbf{G}(k+1)$, and the second one normalizes the columns of $\mathbf{W}''(k+1)$. To find such a matrix $\mathbf{G}(k+1)$ which is of course not unique, notice that if $\mathbf{G}(k+1)$ is an orthogonal matrix having as first column, the vector $\frac{\mathbf{y}(k)}{\|\mathbf{y}(k)\|_2}$ with the remaining $r-1$ columns completing an orthonormal basis, then using (4.33), the product $\mathbf{W}''^T(k+1)\mathbf{W}''(k+1)$ becomes the following diagonal matrix

$$\begin{aligned} \mathbf{W}''^T(k+1)\mathbf{W}''(k+1) &= \mathbf{G}^T(k+1)[\mathbf{I}_r + \delta_k \mathbf{y}(k)\mathbf{y}^T(k)]\mathbf{G}(k+1) \\ &= \mathbf{I}_r + \delta_k \|\mathbf{y}(k)\|^2 \mathbf{e}_1\mathbf{e}_1^T \end{aligned}$$

where $\delta_k \stackrel{\text{def}}{=} \pm 2\mu_k + \mu_k^2\|\mathbf{x}(k)\|^2$ and $\mathbf{e}_1 \stackrel{\text{def}}{=} [0, \ldots, 0]^T$. It is fortunate that there exists such an orthonogonal matrix $\mathbf{G}(k+1)$ with the desired properties known as a Householder reflector [35, Chap. 5], and can be very easily generated since it is of the form

$$\mathbf{G}(k+1) = \mathbf{I}_r - \frac{2}{\|\mathbf{a}(k)\|^2}\mathbf{a}(k)\mathbf{a}^T(k) \quad \text{with } \mathbf{a}(k) = \mathbf{y}(k) - \|\mathbf{y}(k)\|\mathbf{e}_1. \tag{4.37}$$

This gives the Fast Data Projection Method (FDPM)

$$\mathbf{W}(k+1) = \text{Normalize}\{[\mathbf{W}(k) \pm \mu_k \mathbf{x}(k)\mathbf{x}^T(k)\mathbf{W}(k)]\mathbf{G}(k+1)\}, \tag{4.38}$$

where Normalize$\{\mathbf{W}''(\mathrm{k}+1)\}$ stands for normalization of the columns of $\mathbf{W}''(k+1)$, and $\mathbf{G}(k+1)$ is the Householder transform given by (4.37). Using the independence assumption [36, Chap. 9.4] and the approximation $\mu_k \ll 1$, a simplistic theoretical

[6]However, if one looks very carefully at the simulation graphs representing the orthonormality error [75, Fig. 7], it is easy to realize that the HFRANS algorithm exhibits a slight linear instability.

analysis has been presented in [31] for both signal and noise subspace tracking. It shows that the FDPM algorithm is locally stable and the distance to orthonormality $E(\|\mathbf{W}^T(k)\mathbf{W}(k) - \mathbf{I}_r\|^2)$ tends to zero as $O(e^{-ck})$ where $c > 0$ does not depend on μ. Furthermore, numerical simulations presented in [29–31] with $\mu_k = \frac{\mu}{\|\mathbf{x}(k)\|^2}$ demonstrate that this algorithm is numerically stable for both signal and noise subspace tracking, and if for some reason, orthonormality is lost, or the algorithm is initialized with a matrix that is not orthonormal, the algorithm exhibits an extremely high convergence speed to an orthonormal matrix. This FDPM algorithm is to the best to our knowledge, the only power-based minor subspace tracking methods of complexity $O(nr)$ that is truly numerically stable since it does not accumulate rounding errors.

Power-based Methods Issued from Exponential or Sliding Windows Of course, all of the above algorithms that do not use the rank one property of the instantaneous estimate $\mathbf{x}(k)\mathbf{x}^T(k)$ of $\mathbf{C}_x(k)$ can be extended to the exponential (4.19) or sliding windowed (4.22) estimates $\mathbf{C}(k)$, but with an important increase in complexity. To keep the $O(nr)$ complexity, the orthogonal iteration method (4.12) must be adapted to the following iterations

$$\begin{aligned}\mathbf{W}'(k+1) &= \mathbf{C}(k)\mathbf{W}(k)\\ \mathbf{W}(k+1) &= \text{Orthonorm}\{\mathbf{W}'(k+1)\}\\ &= \mathbf{W}'(k+1)\mathbf{G}(k+1)\end{aligned}$$

where the matrix $\mathbf{G}(k+1)$ is a square root inverse of $\mathbf{W}'^T(k+1)\mathbf{W}'(k+1)$ responsible for performing the orthonormalization of $\mathbf{W}'(k+1)$. It is the choice of $\mathbf{G}(k+1)$ that will pave the way for different adaptive algorithms.

Based on the approximation

$$\mathbf{C}(k-1)\mathbf{W}(k) = \mathbf{C}(k-1)\mathbf{W}(k-1), \tag{4.39}$$

which is clearly valid if $\mathbf{W}(k)$ is slowly varying with k, an adaptation of the power method denoted natural power method 3 (NP3) has been proposed in [38] for the exponential windowed estimate (4.19) $\mathbf{C}(k) = \beta\mathbf{C}(k-1) + \mathbf{x}(k)\mathbf{x}^T(k)$. Using (4.19) and (4.39), we obtain

$$\mathbf{W}'(k+1) = \beta\mathbf{W}'(k) + \mathbf{x}(k)\mathbf{y}^T(k)$$

with $\mathbf{y}(k) \stackrel{\text{def}}{=} \mathbf{W}^T(k)\mathbf{x}(k)$. It then follows that

$$\begin{aligned}\mathbf{W}'^T(k+1)\mathbf{W}'(k+1) &= \beta^2\mathbf{W}'^T(k)\mathbf{W}'(k) + \mathbf{z}(k)\mathbf{y}^T(k) + \mathbf{y}(k)\mathbf{z}^T(k)\\ &\quad + \|\mathbf{x}(k)\|^2\mathbf{y}(k)\mathbf{y}^T(k)\end{aligned} \tag{4.40}$$

with $\mathbf{z}(k) \stackrel{\text{def}}{=} \beta \mathbf{W}'^T(k)\mathbf{x}(k)$, which implies (see Exercise 4.9) the following recursions

$$\mathbf{G}(k+1) = \frac{1}{\beta}[\mathbf{I}_n - \tau_1 \mathbf{e}_1 \mathbf{e}_1^T - \tau_2 \mathbf{e}_2 \mathbf{e}_2^T]\mathbf{G}(k) \tag{4.41}$$

$$\begin{aligned} \mathbf{W}(k+1) &= \mathbf{W}(k)[\mathbf{I}_n - \tau_1 \mathbf{e}_1 \mathbf{e}_1^T - \tau_2 \mathbf{e}_2 \mathbf{e}_2^T] \\ &\quad + \frac{1}{\beta}\mathbf{x}(k)\mathbf{y}^T(k)\mathbf{G}^T(k)[\mathbf{I}_n - \tau_1 \mathbf{e}_1 \mathbf{e}_1^T - \tau_2 \mathbf{e}_2 \mathbf{e}_2^T] \end{aligned} \tag{4.42}$$

where τ_1, τ_2 and $\mathbf{e}_1$, $\mathbf{e}_2$ are defined in Exercise 4.9.

Note that the square root inverse matrix $\mathbf{G}(k+1)$ of $\mathbf{W}'^T(k+1)\mathbf{W}'(k+1)$ is asymmetric even if $\mathbf{G}(0)$ is symmetric. Expressions (4.41) and (4.42) provide an algorithm which does not involve any matrix–matrix multiplications and in fact requires only $O(nr)$ operations.

Based on the approximation that $\mathbf{W}(k)$ and $\mathbf{W}(k+1)$ span the same r-dimensional subspace, another power-based algorithm referred to as the approximated power iteration (API) algorithm and its fast implementation (FAPI) have been proposed in [8]. Compared to the NP3 algorithm, this scheme has the advantage that it can handle the exponential (4.19) or the sliding windowed (4.22) estimates of $\mathbf{C}_x(k)$ in the same framework (and with the same complexity of $O(nr)$ operations) by writing (4.19) and (4.22) in the form

$$\mathbf{C}(k) = \beta \mathbf{C}(k-1) + \mathbf{x}'(k)\mathbf{J}\mathbf{x}'^T(k)$$

with $\mathbf{J} = 1$ and $\mathbf{x}'(k) = \mathbf{x}(k)$ for the exponential window and $\mathbf{J} = \begin{bmatrix} 1 & 0 \\ 0 & -\beta^l \end{bmatrix}$ and $\mathbf{x}'(k) = [\mathbf{x}(k), \mathbf{x}(k-l)]$ for the sliding window [see (4.22)]. Among the power-based minor subspace tracking methods issued from the exponential of sliding window, this FAPI algorithm has been considered by many practitioners (e.g. [11]) as outperforming the other algorithms having the same computational complexity.

4.5.2 Projection Approximation-based Methods

Since (4.14) describes an unconstrained cost function to be minimized, it is straightforward to apply the gradient-descent technique for dominant subspace tracking. Using expression (4.90) of the gradient given in Exercise 4.7 with the estimate $\mathbf{x}(k)\mathbf{x}^T(k)$ of $\mathbf{C}_x(k)$ gives

$$\begin{aligned} \mathbf{W}(k+1) &= \mathbf{W}(k) - \mu_k[-2\mathbf{x}(k)\mathbf{x}^T(k) + \mathbf{x}(k)\mathbf{x}^T(k)\mathbf{W}(k)\mathbf{W}^T(k) \\ &\quad + \mathbf{W}(k)\mathbf{W}^T(k)\mathbf{x}(k)\mathbf{x}^T(k)]\mathbf{W}(k). \end{aligned} \tag{4.43}$$

We note that this algorithm can be linked to Oja's algorithm (4.30). First, the term between brackets is the symmetrization of the term $-\mathbf{x}(k)\mathbf{x}^T(k) + \mathbf{W}(k)\mathbf{W}^T$

$(k)\mathbf{x}(k)\mathbf{x}^T(k)$ of Oja's algorithm (4.30). Second, we see that when $\mathbf{W}^T(k)\mathbf{W}(k)$ is approximated by $\mathbf{I}_r$ (which is justified from the stability property below), algorithm (4.43) gives Oja's algorithm (4.30). We note that because the field of the stochastic approximation algorithm (4.43) is the opposite of the derivative of the positive function (4.14), the orthonormal bases of the dominant subspace are globally asymptotically stable for its associated ODE (see Subsection 4.7.1) in contrast to Oja's algorithm (4.30), for which they are only locally asymptotically stable. A complete performance analysis of the stochastic approximation algorithm (4.43) has been presented in [24] where closed-form expressions of the asymptotic covariance of the estimated projection matrix $\mathbf{W}(k)\mathbf{W}^T(k)$ are given and commented on for independent Gaussian data $\mathbf{x}(k)$ and constant stepsize μ.

If now $\mathbf{C}_x(k)$ is estimated by the exponentially weighted sample covariance matrix $\mathbf{C}(k) = \sum_{i=0}^{k} \beta^{k-i}\mathbf{x}(i)\mathbf{x}^T(i)$ (4.18) instead of $\mathbf{x}(k)\mathbf{x}^T(k)$, the scalar function $J(\mathbf{W})$ becomes

$$J(\mathbf{W}) = \sum_{i=0}^{k} \beta^{k-i}\|\mathbf{x}(i) - \mathbf{W}\mathbf{W}^T\mathbf{x}(i)\|^2 \tag{4.44}$$

and all data $\mathbf{x}(i)$ available in the time interval $\{0, \ldots, k\}$ are involved in estimating the dominant subspace at time instant $k+1$ supposing this estimate is known at time instant k. The key issue of the projection approximation subspace tracking algorithm (PAST) proposed by Yang in [71] is to approximate $\mathbf{W}^T(k)\mathbf{x}(i)$ in (4.44), the unknown projection of $\mathbf{x}(i)$ onto the columns of $\mathbf{W}(k)$ by the expression $y(i) = \mathbf{W}^T(i)\mathbf{x}(i)$ which can be calculated for all $0 \le i \le k$ at the time instant k. This results in the following modified cost function

$$J'(\mathbf{W}) = \sum_{i=0}^{k} \beta^{k-i}\|\mathbf{x}(i) - \mathbf{W}\mathbf{y}(i)\|^2 \tag{4.45}$$

which is now quadratic in the elements of $\mathbf{W}$. This projection approximation, hence the name PAST, changes the error performance surface of $J(\mathbf{W})$. For stationary or slowly varying $\mathbf{C}_x(k)$, the difference between $\mathbf{W}^T(k)\mathbf{x}(i)$ and $\mathbf{W}^T(i)\mathbf{x}(i)$ is small, in particular when i is close to k. However, this difference may be larger in the distant past with $i \ll k$, but the contribution of the past data to the cost function (4.45) is decreasing for growing k, due to the exponential windowing. It is therefore expected that $J'(\mathbf{W})$ will be a good approximation to $J(\mathbf{W})$ and the matrix $\mathbf{W}(k)$ minimizing $J'(\mathbf{W})$ be a good estimate for the dominant subspace of $\mathbf{C}_x(k)$. In case of sudden parameter changes of the model (4.15), the numerical experiments presented in [71] show that the algorithms derived from the PAST approach still converge. The main advantage of this scheme is that the least square minimization of (4.45) whose solution is given by $\mathbf{W}(k+1) = \mathbf{C}_{x,y}(k)\mathbf{C}_y^{-1}(k)$ where $\mathbf{C}_{x,y}(k) \stackrel{\text{def}}{=} \sum_{i=0}^{k} \beta^{k-i}\mathbf{x}(i)\mathbf{y}^T(i)$ and $\mathbf{C}_y(k) \stackrel{\text{def}}{=} \sum_{i=0}^{k} \beta^{k-i}\mathbf{y}(i)\mathbf{y}^T(i)$ has been extensively studied in adaptive filtering (see e.g. [36, Chap. 13] and [68, Chap. 12]) where various recursive least square algorithms

(RLS) based on the matrix inversion lemma have been proposed.[7] We note that because of the approximation of $J(\mathbf{W})$ by $J'(\mathbf{W})$, the columns of $\mathbf{W}(k)$ are not exactly orthonormal. But this lack of orthonormality does not mean that we need to perform a reorthonormalization of $\mathbf{W}(k)$ after each update. For this algorithm, the necessity of orthonormalization depends solely on the post processing method which uses this signal subspace estimate to extract the desired signal information (see e.g. Section 4.8). It is shown in the numerical experiments presented in [71] that the deviation of $\mathbf{W}(k)$ from orthonormality is very small and for a growing sliding window ($\beta = 1$), $\mathbf{W}(k)$ converges to a matrix with exactly orthonormal columns under stationary signal. Finally, note that a theoretical study of convergence and a derivation of the asymptotic distribution of the recursive subspace estimators have been presented in [73, 74] respectively. Using the ODE associated with this algorithm (see Section 4.7.1) which is here a pair of coupled matrix differential equations, it is proved that under signal stationarity and other weak conditions, the PAST algorithm converges to the desired signal subspace with probability one.

To speed up the convergence of the PAST algorithm and to guarantee the orthonormality of $\mathbf{W}(k)$ at each iteration, an orthonormal version of the PAST algorithm dubbed OPAST has been proposed in [1]. This algorithm consists of the PAST algorithm where $\mathbf{W}(k+1)$ is related to $\mathbf{W}(k)$ by $\mathbf{W}(k+1) = \mathbf{W}(k) + \mathbf{p}(k)\mathbf{q}(k)$, plus an orthonormalization step of $\mathbf{W}(k)$ based on the same approach as those used in the FRANS algorithm (see Subsection 4.5.1) which leads to the update $\mathbf{W}(k+1) = \mathbf{W}(k) + \mathbf{p}'(k)\mathbf{q}(k)$.

Note that the PAST algorithm cannot be used to estimate the noise subspace by simply changing the sign of the stepsize because the associated ODE is unstable. Efforts to eliminate this instability were attempted in [4] by forcing the orthonormality of $\mathbf{W}(k)$ at each time step. Although there was a definite improvement in the stability characteristics, the resulting algorithm remains numerically unstable.

4.5.3 Additional Methodologies

Various generalizations of criteria (4.7) and (4.14) have been proposed (e.g. in [41]), which generally yield robust estimates of principal subspaces or eigenvectors that are totally different from the standard ones. Among them, the following novel information criterion (NIC) [48] results in a fast algorithm to estimate the principal subspace with a number of attractive properties

$$\max_{\mathbf{W}}\{J(\mathbf{W})\} \quad \text{with } J(\mathbf{W}) \stackrel{\text{def}}{=} \text{Tr}[\ln(\mathbf{W}^T\mathbf{C}\mathbf{W})] - \text{Tr}(\mathbf{W}^T\mathbf{W}) \tag{4.46}$$

given that $\mathbf{W}$ lies in the domain $\{\mathbf{W}$ such that $\mathbf{W}^T\mathbf{C}\mathbf{W} > 0\}$, where the matrix logarithm is defined, for example, in [35, Chap. 11]. It is proved in [48] (see also

[7]For possible sudden signal parameter changes (see Subsection 4.3.1), the use of a sliding exponential window (4.21) version of the cost function may offer faster convergence. In this case, $\mathbf{W}(k)$ can be calculated recursively as well [71] by applying the general form of the matrix inversion lemma $(\mathbf{A} + \mathbf{B}\mathbf{D}\mathbf{C}^T)^{-1} = \mathbf{A}^{-1} - \mathbf{A}^{-1}\mathbf{B}(\mathbf{D}^{-1} + \mathbf{C}^T\mathbf{A}^{-1}\mathbf{B})^{-1}\mathbf{C}^T\mathbf{A}^{-1}$ which requires inversion of a 2×2 matrix.

Exercises 4.11 and 4.12) that the above criterion has a global maximum that is attained when and only when $\mathbf{W} = \mathbf{U}_r\mathbf{Q}$ where $\mathbf{U}_r = [\mathbf{u}_1, \ldots, \mathbf{u}_r]$ and $\mathbf{Q}$ is an arbitrary $r \times r$ orthogonal matrix and all the other stationary points are saddle points. Taking the gradient of (4.46) [which is given explicitly by (4.92)], the following gradient ascent algorithm has been proposed in [48] for updating the estimate $\mathbf{W}(k)$

$$\mathbf{W}(k+1) = \mathbf{W}(k) + \mu_k\{\mathbf{C}(k)\mathbf{W}(k)[\mathbf{W}^T(k)\mathbf{C}(k)\mathbf{W}(k)]^{-1} - \mathbf{W}(k)\}. \tag{4.47}$$

Using the recursive estimate $\mathbf{C}(k) = \sum_{i=0}^{k} \beta^{k-i}\mathbf{x}(i)\mathbf{x}^T(i)$ (4.18), and the projection approximation introduced in [71] $\mathbf{W}^T(k)\mathbf{x}(i) = \mathbf{W}^T(i)\mathbf{x}(i)$ for all $0 \le i \le k$, the update (4.47) becomes

$$\mathbf{W}(k+1) = \mathbf{W}(k) + \mu_k\left[\left(\sum_{i=0}^{k} \beta^{k-i}\mathbf{x}(i)\mathbf{y}^T(i)\right)\left(\sum_{i=0}^{k} \beta^{k-i}\mathbf{y}(i)\mathbf{y}^T(i)\right)^{-1} - \mathbf{W}(k)\right], \tag{4.48}$$

with $\mathbf{y}(i) \stackrel{\text{def}}{=} \mathbf{W}^T(i)\mathbf{x}(i)$. Consequently, similarly to the PAST algorithms, standard RLS techniques used in adaptive filtering can be applied. According to the numerical experiments presented in [38], this algorithm performs very similarly to the PAST algorithm also having the same complexity. Finally, we note that it has been proved in [48] that the points $\mathbf{W} = \mathbf{U}_r\mathbf{Q}$ are the only asymptotically stable points of the ODE (see Subsection 4.7.1) associated with the gradient ascent algorithm (4.47) and that the attraction set of these points is the domain $\{\mathbf{W}$ such that $\mathbf{W}^T\mathbf{C}\mathbf{W} > 0\}$. But to the best of our knowledge, no complete theoretical performance analysis of algorithm (4.48) has been carried out so far.

4.6 EIGENVECTORS TRACKING

Although, the adaptive estimation of the dominant or minor subspace through the estimate $\mathbf{W}(k)\mathbf{W}^T(k)$ of the associated projector is of most importance for subspace-based algorithms, there are situations where the associated eigenvalues are simple ($\lambda_1 > \cdots > \lambda_r > \lambda_{r+1}$ or $\lambda_n < \cdots < \lambda_{n-r+1} < \lambda_{n-r}$) and the desired estimated orthonormal basis of this space must form an eigenbasis. This is the case for the statistical technique of principal component analysis in data compression and coding, optimal feature extraction in pattern recognition, and for optimal fitting in the total least square sense, or for Karhunen-Loève transformation of signals, to mention only a few examples. In these applications, $\{y_1(k), \ldots, y_r(k)\}$ or $\{y_n(k), \ldots, y_{n-r+1}(k)\}$ with $y_i(k) \stackrel{\text{def}}{=} \mathbf{w}_i^T(k)\mathbf{x}(k)$ where $\mathbf{W} = [\mathbf{w}_1(k), \ldots, \mathbf{w}_r(k)]$ or $\mathbf{W} = [\mathbf{w}_n(k), \ldots, \mathbf{w}_{n-r+1}(k)]$ are the estimated r first principal or r last minor components of the data $\mathbf{x}(k)$. To derive such adaptive estimates, the stochastic approximation algorithms that have been proposed, are issued from adaptations of the iterative constrained maximizations (4.5) and minimizations (4.6) of Rayleigh quotients; the weighted subspace criterion (4.8); the orthogonal iterations (4.11) and, finally the gradient-descent technique applied to the minimization of (4.14).

4.6.1 Rayleigh Quotient-based Methods

To adapt maximization (4.5) and minimization (4.6) of Rayleigh quotients to adaptive implementations, a method has been proposed in [61]. It is derived from a Givens parametrization of the constraint $\mathbf{W}^T\mathbf{W} = \mathbf{I}_r$, and from a gradient-like procedure. The Givens rotations approach introduced by Regalia [61] is based on the properties that any $n \times 1$ unit 2-norm vector and any orthogonal vector to this vector can be respectively written as the last column of an $n \times n$ orthogonal matrix and as a linear combination of the first $n-1$ columns of this orthogonal matrix, that is,

$$\mathbf{w}_1 = \mathbf{Q}_1 \begin{bmatrix} \mathbf{0} \\ 1 \end{bmatrix}, \mathbf{w}_2 = \mathbf{Q}_1 \begin{bmatrix} \mathbf{Q}_2 \begin{bmatrix} \mathbf{0} \\ 1 \end{bmatrix} \\ 0 \end{bmatrix}, \ldots, \mathbf{w}_r = \mathbf{Q}_1 \begin{bmatrix} \mathbf{Q}_2 \begin{bmatrix} \mathbf{Q}_r \begin{bmatrix} \mathbf{0} \\ 1 \end{bmatrix} \\ 0 \end{bmatrix} \\ 0 \end{bmatrix}$$

where $\mathbf{Q}_i$ is the following orthogonal matrix of order $n-i+1$

$$\mathbf{Q}_i = \mathbf{U}_{i,1} \cdots \mathbf{U}_{i,j} \cdots \mathbf{U}_{i,n-i} \quad \text{with } \mathbf{U}_{i,j} \stackrel{\text{def}}{=} \begin{bmatrix} \mathbf{I}_{j-1} & \mathbf{0} & \mathbf{0} & \mathbf{0} \\ \mathbf{0} & -\sin\theta_{i,j} & \cos\theta_{i,j} & \mathbf{0} \\ \mathbf{0} & \cos\theta_{i,j} & \sin\theta_{i,j} & \mathbf{0} \\ \mathbf{0} & \mathbf{0} & \mathbf{0} & \mathbf{I}_{n-i-j} \end{bmatrix}$$

and $\theta_{i,j}$ belongs to $[-\frac{\pi}{2}, +\frac{\pi}{2}]$. The existence of such a parametrization[8] for all orthonormal sets $\{\mathbf{w}_1, \ldots, \mathbf{w}_r\}$ is proved in [61]. It consists of $r(2n-r-1)/2$ real parameters. Furthermore, this parametrization is unique if we add some constraints on $\theta_{i,j}$. A deflation procedure, inspired by the maximization (4.5) and minimization (4.6) has been proposed [61]. First the maximization or minimization (4.3) is performed with the help of the classical stochastic gradient algorithm, in which the parameters are $\theta_{1,1}, \ldots, \theta_{1,n-1}$, whereas the maximization (4.5) or minimization (4.6) are realized thanks to stochastic gradient algorithms with respect to the parameters $\theta_{i,1}, \ldots, \theta_{i,n-i}$, in which the preceding parameters $\theta_{l,1}(k), \ldots, \theta_{l,n-l}(k)$ for $l = 1, \ldots, i-1$ are injected from the $i-1$ previous algorithms. The deflation procedure is achieved by coupled stochastic gradient algorithms

$$\begin{bmatrix} \boldsymbol{\theta}_1(k+1) \\ \vdots \\ \boldsymbol{\theta}_r(k+1) \end{bmatrix} = \begin{bmatrix} \boldsymbol{\theta}_1(k) \\ \vdots \\ \boldsymbol{\theta}_r(k) \end{bmatrix} \pm \mu_k \begin{bmatrix} f_1[\boldsymbol{\theta}_1(k), \mathbf{x}(k)] \\ \vdots \\ f_r[\boldsymbol{\theta}_1(k), \ldots, \boldsymbol{\theta}_r(k), \mathbf{x}(k)] \end{bmatrix} \tag{4.49}$$

[8]Note that this parametrization extends immediately to the complex case using the kernel $\begin{bmatrix} -\sin\theta_{i,j} & \cos\theta_{i,j} \\ e^{i\phi_{i,j}}\cos\theta_{i,j} & e^{i\phi_{i,j}}\sin\theta_{i,j} \end{bmatrix}$.

with $\boldsymbol{\theta}_i \stackrel{\text{def}}{=} [\theta_{i,1}, \ldots, \theta_{i,n-i}]^T$ and $f_i(\boldsymbol{\theta}_1, \ldots, \boldsymbol{\theta}_i, \mathbf{x}) \stackrel{\text{def}}{=} \nabla_{\boldsymbol{\theta}_i}(\mathbf{w}_i^T\mathbf{x}\mathbf{x}^T\mathbf{w}_i) = 2\nabla_{\boldsymbol{\theta}_i}(\mathbf{w}_i^T)\mathbf{x}$ $\mathbf{x}^T\mathbf{w}_i$, $i = 1, \ldots, r$. This rather intuitive computational process was confirmed by simulation results [61]. Later a formal analysis of the convergence and performance was performed in [23] where it has been proved that the stationary points of the associated ODE are globally asymptotically stable (see Subsection 4.7.1) and that the stochastic algorithm (4.49) converges almost surely to these points for stationary data $\mathbf{x}(k)$ when μ_k is decreasing with $\lim_{k\to\infty} \mu_k = 0 =$ and $\sum_k \mu_k = \infty$. We note that this algorithm yields exactly orthonormal r dominant or minor estimated eigenvectors by a simple change of sign in its stepsize, and requires $O(nr)$ operations at each iteration but without accounting for the trigonometric functions.

Alternatively, a stochastic gradient-like algorithm denoted direct adaptive subspace estimation (DASE) has been proposed in [62] with a direct parametrization of the eigenvectors by means of their coefficients. Maximization or minimization (4.3) is performed with the help of a modification of the classic stochastic gradient algorithm to assure an approximate unit norm of the first estimated eigenvector $\mathbf{w}_1(k)$ [in fact a rewriting of Oja's neuron (4.23)]. Then, a modification of the classical stochastic gradient algorithm using a deflation procedure, inspired by the constraint $\mathbf{W}^T\mathbf{W} = \mathbf{I}_r$ gives the estimates $[\mathbf{w}_i(k)]_{i=2,\ldots,r}$

$$\mathbf{w}_1(k+1) = \mathbf{w}_1(k) \pm \mu_k\{\mathbf{x}(k)\mathbf{x}^T(k) - [\mathbf{w}_1^T(k)\mathbf{x}(k)\mathbf{x}^T(k)\mathbf{w}_1(k)]\mathbf{I}_n\}\mathbf{w}_1(k)$$

$$\mathbf{w}_i(k+1) = \mathbf{w}_i(k) \pm \mu_k\left\{\mathbf{x}(k)\mathbf{x}^T(k) - [\mathbf{w}_i^T(k)\mathbf{x}(k)\mathbf{x}^T(k)\mathbf{w}_i(k)].\right.$$
$$\left.\times\left(\mathbf{I}_n - \sum_{j=1}^{i-1}\mathbf{w}_j(k)\mathbf{w}_j^T(k)\right)\right\}\mathbf{w}_i(k) \quad \text{for } i = 2, \ldots, r. \tag{4.50}$$

This totally empirical procedure has been studied in [63]. It has been proved that the stationary points of the associated ODE are all eigenvector bases $\{\pm\mathbf{u}_{i_1}, \ldots, \pm\mathbf{u}_{i_r}\}$. Using the eigenvalues of the derivative of the mean field (see Subsection 4.7.1), it is shown that all these eigenvector bases are unstable except $\{\pm\mathbf{u}_1\}$ for $r = 1$ associated with the sign $+$ [where algorithm (4.50) is Oja's neuron (4.23)]. But a closer, examination of these eigenvalues that are all real-valued, shows that for only the eigenbasis $\{\pm\mathbf{u}_1, \ldots, \pm\mathbf{u}_r\}$ and $\{\pm\mathbf{u}_n, \ldots, \pm\mathbf{u}_{n-r+1}\}$ associated with the sign $+$ and $-$ respectively, all the eigenvalues of the derivative of the mean field are strictly negative except for the eigenvalues associated with variations of the eigenvectors $\{\pm\mathbf{u}_1, \ldots, \pm\mathbf{u}_r\}$ and $\{\pm\mathbf{u}_n, \ldots, \pm\mathbf{u}_{n-r+1}\}$ in their directions. Consequently, it is claimed in [63] that if the norm of each estimated eigenvector is set to one at each iteration, the stability of the algorithm is ensured. The simulations presented in [62] confirm this intuition.

4.6.2 Eigenvector Power-based Methods

Note that similarly to the subspace criterion (4.7), the maximization or minimization of the weighted subspace criterion (4.8) $J(\mathbf{W}) \stackrel{\text{def}}{=} \text{Tr}[\boldsymbol{\Omega}\mathbf{W}^T\mathbf{C}(k)\mathbf{W}]$ subject to the constraint $\mathbf{W}^T\mathbf{W} = \mathbf{I}_r$ can be solved by a constrained gradient-descent technique. Clearly, the simplest selection for $\mathbf{C}(k)$ is the instantaneous estimate $\mathbf{x}(k)\mathbf{x}^T(k)$. Because in this case, $\nabla_{\mathbf{W}} J = 2\mathbf{x}(k)\mathbf{x}^T(k)\mathbf{W}\boldsymbol{\Omega}$, we obtain the following stochastic approximation algorithm that will be a starting point for a family of algorithms that have been derived to adaptively estimate major or minor eigenvectors

$$\mathbf{W}(k+1) = \{\mathbf{W}(k) \pm \mu_k \mathbf{x}(k)\mathbf{x}^T(k)\mathbf{W}(k)\boldsymbol{\Omega}\}\mathbf{G}(k+1) \tag{4.51}$$

in which $\mathbf{W}(k) = [\mathbf{w}_1(k), \ldots, \mathbf{w}_r(k)]$ and the matrix $\boldsymbol{\Omega}$ is a diagonal matrix $\text{Diag}(\omega_1, \ldots, \omega_r)$ with $\omega_1 > \cdots > \omega_r > 0$. $\mathbf{G}(k+1)$ is a matrix depending on

$$\mathbf{W}'(k+1) \stackrel{\text{def}}{=} \mathbf{W}(k) \pm \mu_k \mathbf{x}(k)\mathbf{x}^T(k)\mathbf{W}(k)\boldsymbol{\Omega}$$

which orthonormalizes or approximately orthonormalizes the columns of $\mathbf{W}'(k+1)$. Thus, $\mathbf{W}(k)$ has orthonormal or approximately orthonormal columns for all k. Depending on the form of matrix $\mathbf{G}(k+1)$, variants of the basic stochastic algorithm are obtained. Going back to the general expression (4.29) of the subspace power-based algorithm, we note that (4.51) can also be derived from (4.29), where different stepsizes $\mu_k\omega_1, \ldots, \mu_k\omega_r$ are introduced for each column of $\mathbf{W}(k)$.

Using the same approach as for deriving (4.30), that is, where $\mathbf{G}(k+1)$ is the symmetric square root inverse of $\mathbf{W}'^T(k+1)\mathbf{W}'(k+1)$, we obtain the following stochastic approximation algorithm

$$\begin{aligned}\mathbf{W}(k+1) = \mathbf{W}(k) \pm \mu_k[&\mathbf{x}(k)\mathbf{x}^T(k)\mathbf{W}(k)\boldsymbol{\Omega} - \tfrac{1}{2}\mathbf{W}(k)\boldsymbol{\Omega}\mathbf{W}^T(k)\mathbf{x}(k)\mathbf{x}^T(k)\mathbf{W}(k)\\ &- \tfrac{1}{2}\mathbf{W}(k)\mathbf{W}^T(k)\mathbf{x}(k)\mathbf{x}^T(k)\mathbf{W}(k)\boldsymbol{\Omega}].\end{aligned} \tag{4.52}$$

Note that in contrast to the Oja's algorithm (4.30), this algorithm is different from the algorithm issued from the optimization of the cost function $J(\mathbf{W}) \stackrel{\text{def}}{=} \text{Tr}[\boldsymbol{\Omega}\mathbf{W}^T\mathbf{x}(k)\mathbf{x}^T(k)\mathbf{W}]$ defined on the set of $n \times r$ orthogonal matrices $\mathbf{W}$ with the help of continuous-time matrix algorithms (see e.g. [21, Ch. 7.2], [19, Ch. 4] or (4.91) in Exercise 4.15).

$$\mathbf{W}(k+1) = \mathbf{W}(k) \pm \mu_k[\mathbf{x}(k)\mathbf{x}^T(k)\mathbf{W}(k)\boldsymbol{\Omega} - \mathbf{W}(k)\boldsymbol{\Omega}\mathbf{W}^T(k)\mathbf{x}(k)\mathbf{x}^T(k)\mathbf{W}(k)]. \tag{4.53}$$

We note that these two algorithms reduce to the Oja's algorithm (4.30) for $\boldsymbol{\Omega} = \mathbf{I}_r$ and to Oja's neuron (4.23) for $r = 1$, which of course is unstable for tracking the minorant eigenvectors with the sign $-$. Techniques used for stabilizing Oja's algorithm (4.30) for minor subspace tracking, have been transposed to stabilize the weighted Oja's algorithm for tracking the minorant eigenvectors. For example, in [9], $\mathbf{W}(k)$ is forced to be orthonormal at each time step k as in [2] (see Exercise 4.10) with the MCA-OOja algorithm and the MCA-OOjaH algorithm using Householder transforms. Note, that by proving a recursion of the distance to orthonormality $\|\mathbf{W}^T(k)\mathbf{W}(k) - \mathbf{I}_r\|^2_{\text{Fro}}$ from a nonorthogonal matrix $\mathbf{W}(0)$, it has been shown in [10], that the latter algorithm is numerically stable in contrast to the former.

Instead of deriving a stochastic approximation algorithm from a specific orthonormalization matrix $\mathbf{G}(k+1)$, an analogy with Oja's algorithm (4.30) has been used in [54] to derive the following algorithm

$$\mathbf{W}(k+1) = \mathbf{W}(k) \pm \mu_k[\mathbf{x}(k)\mathbf{x}^T(k)\mathbf{W}(k) \\ -\mathbf{W}(k)\boldsymbol{\Omega}\mathbf{W}^T(k)\mathbf{x}(k)\mathbf{x}^T(k)\mathbf{W}(k)\boldsymbol{\Omega}^{-1}]. \quad (4.54)$$

It has been proved in [55], that for tracking the dominant eigenvectors (i.e. with the sign +), the eigenvectors $\{\pm\mathbf{u}_1, \ldots, \pm\mathbf{u}_r\}$ are the only locally asymptotically stable points of the ODE associated with (4.54). But to the best of our knowledge, no complete theoretical performance analysis of these three algorithms (4.52), (4.53), and (4.54), has been carried out until now, except in [27] which gives the asymptotic distribution of the estimated principal eigenvectors.

If now the matrix $\mathbf{G}(k+1)$ performs the Gram–Schmidt orthonormalization on the columns of $\mathbf{W}'(k+1)$, an algorithm, denoted stochastic gradient ascent (SGA) algorithm, is obtained if the successive columns of matrix $\mathbf{W}(k+1)$ are expanded, assuming μ_k is sufficiently small. By omitting the $O(\mu_k^2)$ term in this expansion [51], we obtain the following algorithm

$$\mathbf{w}_i(k+1) = \mathbf{w}_i(k) + \alpha_i\mu_k\left[\mathbf{I}_n - \mathbf{w}_i(k)\mathbf{w}_i^T(k) - \sum_{j=1}^{i-1}\left(1 + \frac{\alpha_j}{\alpha_i}\right)\mathbf{w}_j(k)\mathbf{w}_j^T(k)\right] \\ \times \mathbf{x}(k)\mathbf{x}^T(k)\mathbf{w}_i(k) \qquad \text{for } i = 1, \ldots, r \quad (4.55)$$

where here $\boldsymbol{\Omega} = \text{Diag}(\alpha_1, \alpha_2, \ldots, \alpha_r)$ with α_i arbitrary strictly positive numbers.

The so called generalized Hebbian algorithm (GHA) is derived from Oja's algorithm (4.30) by replacing the matrix $\mathbf{W}^T(k)\mathbf{x}(k)\mathbf{x}^T(k)\mathbf{W}(k)$ of Oja's algorithm by its diagonal and superdiagonal only

$$\mathbf{W}(k+1) = \mathbf{W}(k) + \mu_k\{\mathbf{x}(k)\mathbf{x}^T(k)\mathbf{W}(k) - \mathbf{W}(k)\text{upper}[\mathbf{W}^T(k)\mathbf{x}(k)\mathbf{x}^T(k)\mathbf{W}(k)]\}$$

in which the operator upper sets all subdiagonal elements of a matrix to zero. When written columnwise, this algorithm is similar to the SGA algorithm (4.57) where $\alpha_i = 1$, $i = 1, \ldots, r$, with the difference that there is no coefficient 2 in the sum

$$\mathbf{w}_i(k+1) = \mathbf{w}_i(i) + \mu_k\left[\mathbf{I}_n - \sum_{j=1}^{i}\mathbf{w}_j(k)\mathbf{w}_j^T(k)\right]\mathbf{x}(k)\mathbf{x}^T(k)\mathbf{w}_i(k) \\ \text{for } i = 1, \ldots, r. \quad (4.56)$$

Oja et al. [54] proposed an algorithm denoted weighted subspace algorithm (WSA), which is similar to the Oja's algorithm, except for the scalar parameters $\beta_1, \ldots, \beta_r$

$$\mathbf{w}_i(k+1) = \mathbf{w}_i(k) + \mu_k \left[\mathbf{I}_n - \sum_{j=1}^{r} \frac{\beta_j}{\beta_i} \mathbf{w}_j(k)\mathbf{w}_j^T(k) \right] \times \mathbf{x}(k)\mathbf{x}^T(k)\mathbf{w}_i(k) \quad \text{for } i = 1, \ldots, r \tag{4.57}$$

with $\beta_1 > \cdots > \beta_r > 0$. If $\beta_i = 1$ for all i, this algorithm reduces to Oja's algorithm.

Following the deflation technique introduced in the adaptive principal component extraction (APEX) algorithm [42], note finally that Oja's neuron can be directly adapted to estimate the r principal eigenvectors by replacing the instantaneous estimate $\mathbf{x}(k)\mathbf{x}^T(k)$ of $\mathbf{C}_x(k)$ by $\mathbf{x}(k)\mathbf{x}^T(k)\left[\mathbf{I}_n - \sum_{j=1}^{i-1} \mathbf{w}_j(k)\mathbf{w}_j^T(k)\right]$ to successively estimate $\mathbf{w}_i(k)$, $i = 2, \ldots, r$

$$\mathbf{w}_i(k+1) = \mathbf{w}_i(i) + \mu_k[\mathbf{I}_n - \mathbf{w}_i(k)\mathbf{w}_i^T(k)]\mathbf{x}(k)\mathbf{x}^T(k) \times \left[\mathbf{I}_n - \sum_{j=1}^{i-1} \mathbf{w}_j(k)\mathbf{w}_j^T(k) \right] \mathbf{w}_i(k) \quad \text{for } i = 1, \ldots, r.$$

Minor component analysis was also considered in neural networks to solve the problem of optimal fitting in the total least square sense. Xu et al. [79] introduced the optimal fitting analyzer (OFA) algorithm by modifying the SGA algorithm. For the estimate $\mathbf{w}_n(k)$ of the eigenvector associated with the smallest eigenvalue, this algorithm is derived from the Oja's neuron (4.23) by replacing $\mathbf{x}(k)\mathbf{x}^T(k)$ by $\mathbf{I}_n - \mathbf{x}(k)\mathbf{x}^T(k)$, viz

$$\mathbf{w}_n(k+1) = \mathbf{w}_n(k) + \mu[\mathbf{I}_n - \mathbf{w}_n(k)\mathbf{w}_n^T(k)][\mathbf{I}_n - \mathbf{x}(k)\mathbf{x}^T(k)]\mathbf{w}_n(k)$$

and for $i = n, \ldots, n - r + 1$, his algorithm reads

$$\mathbf{w}_i(k+1) = \mathbf{w}_i(k) + \mu_k \left([\mathbf{I}_n - \mathbf{w}(k)\mathbf{w}^T(k)][\mathbf{I}_n - \mathbf{x}(k)\mathbf{x}^T(k)]. - \beta \sum_{i=k+1}^{n} \mathbf{w}_{t,i}\mathbf{w}_{t,i}^T \mathbf{x}(k)\mathbf{x}^T(k) \right) \mathbf{w}_i(k). \tag{4.58}$$

Oja [53] showed that, under the conditions that the eigenvalues are distinct, and that $\lambda_{n-r+1} < 1$ and $\beta > \frac{\lambda_{n-r+1}}{\lambda_n} - 1$, the only asymptotically stable points of the associated ODE are the eigenvectors $\{\pm \mathbf{v}_n, \ldots, \pm \mathbf{v}_{n-r+1}\}$. Note that the magnitude of the eigenvalues must be controlled in practice by normalizing $\mathbf{x}(k)$ so that the expression between brackets in (4.58) becomes homogeneous.

The derivation of these algorithms seems empirical. In fact, they have been derived from slight modifications of the ODE (4.75) associated with the Oja's neuron in order to keep adequate conditions of stability (see e.g. [53]). It was established by Oja [52], Sanger [67], and Oja et al. [55] for the SGA, GHA, and WSA algorithms respectively, that the only asymptotically stable points of their associated ODE are the eigenvectors $\{\pm \mathbf{v}_1, \ldots, \pm \mathbf{v}_r\}$. We note that the first vector ($k=1$) estimated by the SGA and GHA algorithms, and the vector ($r = k = 1$) estimated by the SNL and WSA algorithms gives the constrained Hebbian learning rule of the basic PCA neuron (4.23) introduced by Oja [50].

A performance analysis of different eigenvector power-based algorithms has been presented in [22]. In particular, the asymptotic distribution of the eigenvector estimates and of the associated projection matrices given by these stochastic algorithms with constant stepsize μ for stationary data has been derived, where closed-form expressions of the covariance of these distributions has been given and analyzed for independent Gaussian distributed data $\mathbf{x}(k)$. Closed-form expressions of the mean square error of these estimators has been deduced and analyzed. In particular, they allow us to specify the influence of the different parameters $(\alpha_2, \ldots, \alpha_r)$, $(\beta_1, \ldots, \beta_r)$ and β of these algorithms on their performance and to take into account tradeoffs between the misadjustment and the speed of convergence. An example of such derivation and analysis is given for the Oja's neuron in Subsection 4.7.3.

Eigenvector Power-based Methods Issued from Exponential Windows Using the exponential windowed estimates (4.19) of $\mathbf{C}_x(k)$, and following the concept of power method (4.9) and the subspace deflation technique introduced in [42], the following algorithm has been proposed in [38]

$$\mathbf{w}'_i(k+1) = \mathbf{C}_i(k)\mathbf{w}_i(k) \tag{4.59}$$

$$\mathbf{w}_i(k+1) = \mathbf{w}'_i(k+1)/\|\mathbf{w}'_i(k+1)\|_2 \tag{4.60}$$

where $\mathbf{C}_i(k) = \beta\mathbf{C}_i(k-1) + \mathbf{x}(k)\mathbf{x}^T(k)[\mathbf{I}_n - \sum_{j=1}^{i-1} \mathbf{w}_j(k)\mathbf{w}_j^T(k)]$ for $i = 1, \ldots, r$. Applying the approximation $\mathbf{w}'_i(k) \approx \mathbf{C}_i(k-1)\mathbf{w}_i(k)$ in (4.59) to reduce the complexity, (4.59) becomes

$$\mathbf{w}'_i(k+1) = \beta\mathbf{w}'_i(k) + \mathbf{x}(k)[g_i(k) - \mathbf{y}_i^T(k)\mathbf{c}_i(k)] \tag{4.61}$$

with $g_i(k) \stackrel{\text{def}}{=} \mathbf{x}^T(k)\mathbf{w}_i(k)$, $\mathbf{y}_i(k) \stackrel{\text{def}}{=} [\mathbf{w}_1(k), \ldots, \mathbf{w}_{i-1}(k)]^T\mathbf{x}(k)$ and $\mathbf{c}_i(k) \stackrel{\text{def}}{=} [\mathbf{w}_1(k), \ldots, \mathbf{w}_{i-1}(k)]^T\mathbf{w}_i(k)$. Equations (4.61) and (4.59) should be run successively for $i = 1, \ldots, r$ at each iteration k.

Note that up to a common factor estimate of the eigenvalues $\lambda_i(k+1)$ of $\mathbf{C}_x(k)$ can be updated as follows. From (4.59), one can write

$$\lambda_i(k+1) \stackrel{\text{def}}{=} \mathbf{w}_i^T(k)\mathbf{C}_i(k)\mathbf{w}_i(k) = \mathbf{w}_i^T(k)\mathbf{w}'_i W(k+1). \tag{4.62}$$

Using (4.61) and applying the approximations $\lambda_i(k) \approx \mathbf{w}_i^T(k)\mathbf{w}'_i(k)$ and $\mathbf{c}_i(k) \approx \mathbf{0}$, one can replace (4.62) by

$$\lambda_i(k+1) = \beta\lambda_i(k) + |g_i(k)|^2$$

that can be used to track the rank r and the signal eigenvectors, as in [72].

4.6.3 Projection Approximation-based Methods

A variant of the PAST algorithm, named PASTd and presented in [71], allows one to estimate the r dominant eigenvectors. This algorithm is based on a deflation technique that consists in estimating sequentially the eigenvectors. First the most dominant estimated eigenvector $\mathbf{w}_1(k)$ is updated by applying the PAST algorithm with $r = 1$. Then the projection of the current data $\mathbf{x}(k)$ onto this estimated eigenvector is removed from $\mathbf{x}(k)$ itself. Because now the second dominant eigenvector becomes the most dominant one in the updated data vector $(\mathrm{E}\{[\mathbf{x}(k) - \mathbf{v}_1\mathbf{v}_1^T\mathbf{x}(k)][\mathbf{x}(k) - \mathbf{v}_1\mathbf{v}_1^T\mathbf{x}(k)]^T\} = \mathbf{C}_x(k) - \lambda_1\mathbf{v}_1\mathbf{v}_1^T)$, it can be extracted in the same way as before. Applying this procedure repeatedly, all of the r dominant eigenvectors and the associated eigenvalues are estimated sequentially. These estimated eigenvalues may be used to estimate the rank r if it is not known *a priori* [72]. It is interesting to note that for $r = 1$, the PAST and the PASTd algorithms, that are identical, simplify as

$$\mathbf{w}(k+1) = \mathbf{w}(k) + \mu_k[\mathbf{I}_n - \mathbf{w}(k)\mathbf{w}^T(k)]\mathbf{x}(k)\mathbf{x}^T(k)\mathbf{w}(k) \tag{4.63}$$

where $\mu_k = \dfrac{1}{\sigma_y^2(k)}$ with $\sigma_y^2(k+1) = \beta\sigma_y^2(k) + y^2(k)$ and $y(k) \stackrel{\text{def}}{=} \mathbf{w}^T(k)\mathbf{x}(k)$. A comparison with Oja's neuron (4.23) shows that both algorithms are identical except for the stepsize. While Oja's neuron uses a fixed stepsize μ which needs careful tuning, (4.63) implies a time varying, self-tuning stepsize μ_k. The numerical experiments presented in [71] show that this deflation procedure causes a stronger loss of orthonormality between $\mathbf{w}_i(k)$ and a slight increase in the error in the successive estimates $\mathbf{w}_i(k)$. By invoking the ODE approach (see Section 4.7.1), it has been proved in [73] for stationary signals and other weak conditions, that the PASTd algorithm converges to the desired r dominant eigenvectors with probability one.

In contrast to the PAST algorithm, the PASTd algorithm can be used to estimate the minor eigenvectors by changing the sign of the stepsize with an orthonormalization of the estimated eigenvectors at each step. It has been proved [65] that for $\beta = 1$, the only locally asymptotically stable points of the associated ODE are the desired eigenvectors $\{\pm\mathbf{v}_n, \ldots, \pm\mathbf{v}_{n-r+1}\}$. To reduce the complexity of the Gram–Schmidt orthonormalization step used in [65], [9] proposed a modification of this part.

4.6.4 Additional Methodologies

Among the other approaches to adaptively estimate the eigenvectors of a covariance matrix, the maximum likelihood adaptive subspace estimation (MALASE) [18] provides a number of desirable features. It is based on the adaptive maximization of the log-likelihood of the EVD parameters associated with the covariance matrix $\mathbf{C}_x$ for Gaussian distributed zero-mean data $\mathbf{x}(k)$. Up to an additive constant, this log-likelihood is given by

$$\begin{aligned} L(\mathbf{W}, \boldsymbol{\Lambda}) &= -\ln(\det \mathbf{C}_x) - \mathbf{x}^T(k)\mathbf{C}_x^{-1}\mathbf{x}(k) \\ &= -\sum_{i=1}^{n} \ln(\lambda_i) - \mathbf{x}^T(k)\mathbf{W}\boldsymbol{\Lambda}^{-1}\mathbf{W}^T\mathbf{x}(k) \end{aligned} \tag{4.64}$$

where $\mathbf{C}_x = \mathbf{W}\boldsymbol{\Lambda}\mathbf{W}^T$ represents the EVD of $\mathbf{C}_x$ with $\mathbf{W}$ an orthogonal $n \times n$ matrix and $\boldsymbol{\Lambda} = \text{Diag}(\lambda_1, \ldots, \lambda_n)$. This is a quite natural criterion for statistical estimation purposes, even if the minimum variance property of the likelihood functional is actually an asymptotic property. To deduce an adaptive algorithm, a gradient ascent procedure has been proposed in [18] in which a new data $\mathbf{x}(k)$ is used at each time iteration k of the maximization of (4.64). Using the differential of $L(\mathbf{W}, \boldsymbol{\Lambda})$ defined on the manifold of $n \times n$ orthogonal matrices [see [21, pp. 62–63] or Exercise 4.15 (4.93)], we obtain the following gradient of $L(\mathbf{W}, \boldsymbol{\Lambda})$

$$\nabla_{\mathbf{W}} L = \mathbf{W}[\boldsymbol{\Lambda}^{-1}\mathbf{y}(k)\mathbf{y}^T(k) - \mathbf{y}(k)\mathbf{y}^T(k)\boldsymbol{\Lambda}^{-1}]$$

$$\nabla_{\boldsymbol{\Lambda}} L = -\boldsymbol{\Lambda}^{-1} + \boldsymbol{\Lambda}^{-2}\text{Diag}[\mathbf{y}(k)\mathbf{y}^T(k)]$$

where $\mathbf{y}(k) \stackrel{\text{def}}{=} \mathbf{W}^T\mathbf{x}(k)$. Then, the stochastic gradient update of $\mathbf{W}$ yields

$$\mathbf{W}(k+1) = \mathbf{W}(k) + \mu_k \mathbf{W}(k)[\boldsymbol{\Lambda}^{-1}(k)\mathbf{y}(k)\mathbf{y}^T(k) - \mathbf{y}(k)\mathbf{y}^T(k)\boldsymbol{\Lambda}^{-1}(k)] \quad (4.65)$$

$$\boldsymbol{\Lambda}(k+1) = \boldsymbol{\Lambda}(k) + \mu'_k[\boldsymbol{\Lambda}^{-2}(k)\text{Diag}[\mathbf{y}(k)\mathbf{y}^T(k)] - \boldsymbol{\Lambda}^{-1}(k)] \quad (4.66)$$

where the stepsizes μ_k and μ'_k are possibly different. We note that, starting from an orthonormal matrix $\mathbf{W}(0)$, the sequence of estimates $\mathbf{W}(k)$ given by (4.65) is orthonormal up to the second-order term in μ_k only. To ensure in practice the convergence of this algorithm, is has been shown in [18] that it is necessary to orthonormalize $\mathbf{W}(k)$ quite often to compensate for the orthonormality drift in $O(\mu_k^2)$. Using continuous-time system theory and differential geometry [21], a modification of (4.65) has been proposed in [18]. It is clear that $\nabla_{\mathbf{W}} L$ is tangent to the curve defined by

$$\mathbf{W}(t) = \mathbf{W}(0)\exp[t(\boldsymbol{\Lambda}^{-1}\mathbf{y}(k)\mathbf{y}^T(k) - \mathbf{y}(k)\mathbf{y}^T(k)\boldsymbol{\Lambda}^{-1})]$$

for $t = 0$, where the matrix exponential is defined, for example, in [35, Chap. 11]. Furthermore, we note that this curve lies in the manifold of orthogonal matrices if $\mathbf{W}(0)$ is orthogonal because $\exp(\mathbf{A})$ is orthogonal if and only if $\mathbf{A}$ is skew-symmetric ($\mathbf{A}^T = -\mathbf{A}$) and matrix $\boldsymbol{\Lambda}^{-1}\mathbf{y}(k)\mathbf{y}^T(k) - \mathbf{y}(k)\mathbf{y}^T(k)\boldsymbol{\Lambda}^{-1}$ is clearly skew-symmetric. Moving on the curve $\mathbf{W}(t)$ from point $t = 0$ in the direction of increasing values of $\nabla_{\mathbf{W}} L$ amounts to letting t increase. Thus, a discretized version of the optimization of $L(\mathbf{W}, \boldsymbol{\Lambda})$ as a continuous function of $\mathbf{W}$ is given by the following update scheme

$$\mathbf{W}(k+1) = \mathbf{W}(k)\exp\{\mu_k[\boldsymbol{\Lambda}^{-1}(k)\mathbf{y}(k)\mathbf{y}^T(k) - \mathbf{y}(k)\mathbf{y}^T(k)\boldsymbol{\Lambda}^{-1}(k)]\} \quad (4.67)$$

and the coupled update equations (4.66) and (4.67) form the MALASE algorithm. As mentioned above the update factor $\exp\{\mu_k[\boldsymbol{\Lambda}^{-1}(k)\mathbf{y}(k)\mathbf{y}^T(k) - \mathbf{y}(k)\mathbf{y}^T(k)\boldsymbol{\Lambda}^{-1}(k)]\}$ is an orthogonal matrix. This ensures that the orthonormality property is preserved by the MALASE algorithm, provided that the algorithm is initialized with an orthogonal matrix $\mathbf{W}(0)$. However, it has been shown by the numerical experiments presented in [18], that it is not necessary to have $\mathbf{W}(0)$ orthogonal to ensure the convergence, since the MALASE algorithm steers $\mathbf{W}(k)$ towards the manifold of orthogonal matrices. The MALASE algorithm seems to involve high computational cost, due to the matrix exponential that applies in (4.67). However, since $\exp\{\mu_k[\mathbf{\Lambda}^{-1}(k)\mathbf{y}(k)\mathbf{y}^T(k) - \mathbf{y}(k)\mathbf{y}^T(k)\mathbf{\Lambda}^{-1}(k)]\}$ is the exponential of a sum of two rank-one matrices, the calculation of this matrix requires only $O(n^2)$ operations [18]. Originally, this algorithm which updates the EVD of the covariance matrix $\mathbf{C}_x(k)$ can be modified by a simple preprocessing to estimate the principal or minor r signal eigenvectors only, when the remaining $n - r$ eigenvectors are associated with a common eigenvalue $\sigma^2(k)$ (see Subsection 4.3.1). This algorithm, denoted MALASE(r) requires $O(nr)$ operations by iteration. Finally, note that a theoretical analysis of convergence has been presented in [18]. It is proved that in stationary environments, the stationary stable points of the algorithm (4.66), (4.67) correspond to the EVD of $\mathbf{C}_x$. Furthermore, the covariance of the asymptotic distribution of the estimated parameters is given for Gaussian independently distributed data $\mathbf{x}(k)$ using general results of Gaussian approximation (see Subsection 4.7.2).

4.6.5 Particular Case of Second-order Stationary Data

Finally, note that for $\mathbf{x}(k) = [x(k), x(k-1), \ldots, x(k-n+1)]^T$ comprising of time delayed versions of scalar valued second-order stationary data $x(k)$, the covariance matrix $\mathbf{C}_x(k) = \mathrm{E}[\mathbf{x}(k)\mathbf{x}^T(k)]$ is Toeplitz and consequently centro-symmetric. This property occurs in important applications—temporal covariance matrices obtained from a uniform sampling of a second-order stationary signals, and spatial covariance matrices issued from uncorrelated and band-limited sources observed on a centro-symmetric sensor array (e.g. on uniform linear arrays). This centro-symmetric structure of $\mathbf{C}_x$ allows us to use for real-valued data, the property[9] [14] that its EVD can be obtained from two orthonormal eigenbases of half-size real symmetric matrices. For example if n is even, $\mathbf{C}_x$ can be partitioned as follows

$$\mathbf{C}_x = \begin{bmatrix} \mathbf{C}_1 & \mathbf{C}_2^T \\ \mathbf{C}_2 & \mathbf{J}\mathbf{C}_1\mathbf{J} \end{bmatrix}$$

where $\mathbf{J}$ is a $n/2 \times n/2$ matrix with ones on its antidiagonal and zeroes elsewhere. Then, the n unit 2-norm eigenvectors $\mathbf{v}_i$ of $\mathbf{C}_x$ are given by $n/2$ symmetric and $n/2$ skew symmetric vectors $\mathbf{v}_i = \frac{1}{\sqrt{2}}\begin{bmatrix} \mathbf{u}_i \\ \varepsilon_i \mathbf{J}\mathbf{u}_i \end{bmatrix}$ where $\varepsilon_i = \pm 1$, respectively issued from the unit 2-norm eigenvectors $\mathbf{u}_i$ of $\mathbf{C}_1 + \varepsilon_i \mathbf{J}\mathbf{C}_2 = \frac{1}{2}\mathrm{E}\{[\mathbf{x}'(k) + \varepsilon_i \mathbf{J}\mathbf{x}''(k)][\mathbf{x}'(k) + \varepsilon_i \mathbf{J}\mathbf{x}''(k)]^T\}$ with $\mathbf{x}(k) = [\mathbf{x}'^T(k), \mathbf{x}''^T(k)]^T$. This property has been exploited [23, 26]

[9] Note that for Hermitian centro-symmetric covariance matrices, such property does not extend. But any eigenvector $\mathbf{v}_i$ satisfies the relation $[\mathbf{v}_i]_k = e^{i\phi_i}[\mathbf{v}_i^*]_{n-k}$, that can be used to reduce the computational cost by a factor of two.

to reduce the computational cost of the previously introduced eigenvectors adaptive algorithms. Furthermore, the conditioning of these two independent EVDs is improved with respect to the EVD of $\mathbf{C}_x$ since the difference between the two consecutive eigenvalues increases in general. Compared to the estimators that do not take the centro-symmetric structure into account, the performance ought to be improved. This has been proved in [26], using closed-form expressions of the asymptotic bias and covariance of eigenvectors power-based estimators with constant stepsize μ derived in [22] for independent Gaussian distributed data $\mathbf{x}(k)$. Finally, note that the deviation from orthonormality is reduced and the convergence speed is improved, yielding a better tradeoff between convergence speed and misadjustment.

4.7 CONVERGENCE AND PERFORMANCE ANALYSIS ISSUES

Several tools may be used to assess the convergence and the performance of the previously described algorithms. First of all, note that despite the simplicity of the LMS algorithm (see e.g. [36])

$$\mathbf{w}(k+1) = \mathbf{w}(k) + \mu\mathbf{x}(k)[y(k) - \mathbf{x}^T(k)\mathbf{w}(k)]$$

its convergence and associated analysis has been the subject of many contributions in the past three decades (see e.g. [68] and references therein). However, in-depth theoretical studies are still a matter of utmost interest. Consequently, due to their complexity with respect to the LMS algorithm, results about the convergence and performance analysis of subspaces or eigenvectors tracking will be much weaker.

To study the convergence of the algorithms introduced in the previous two sections from a theoretical point of view, the data $\mathbf{x}(k)$ will be supposed to be stationary and the stepsize μ_k will be considered as decreasing. In these conditions, according to the addressed problem, some questions arise. Does the sequence $\mathbf{W}(k)\mathbf{W}^T(k)$ converge almost surely to the signal $\mathbf{\Pi}_s$ or the noise projector $\mathbf{\Pi}_n$? And does the sequence $\mathbf{W}^T(k)\mathbf{W}(k)$ converge almost surely to $\mathbf{I}_r$ for the subspace tracking problem? Or does the sequence $\mathbf{W}(k)$ converge to the signal or the noise eigenvectors $[\pm\mathbf{u}_1, \ldots, \pm\mathbf{u}_r]$ or $[\pm\mathbf{u}_{n-r+1}, \ldots, \pm\mathbf{u}_n]$ for the eigenvectors tracking problems? These questions are very challenging, but using the stability of the associated ODE, a partial response will be given in Subsection 4.7.1.

Now, from a practical point of view, the stepsize sequence μ_k is reduced to a small constant μ to track signal or noise subspaces (or signal or noise eigenvectors) with possible nonstationary data $\mathbf{x}(k)$. Under these conditions, the previous sequences do not converge almost surely any longer even for stationary data $\mathbf{x}(k)$. Nevertheless, if for stationary data, these algorithms converge almost surely with a decreasing stepsize, their estimate $\boldsymbol{\theta}(k)$ ($\mathbf{W}(k)\mathbf{W}^T(k)$, $\mathbf{W}^T(k)\mathbf{W}(k)$ or $\mathbf{W}(k)$ according to the problem) will oscillate around their limit $\boldsymbol{\theta}_*$ ($\mathbf{\Pi}_s$ or $\mathbf{\Pi}_n$, $\mathbf{I}_r$, $[\pm\mathbf{u}_1, \ldots, \pm\mathbf{u}_r]$ or $[\pm\mathbf{u}_{n-r+1}, \ldots, \pm\mathbf{u}_n]$, according to the problem) with a constant small step size. In these later conditions, the performance of the algorithms will be assessed by the covariance matrix of the errors $[\boldsymbol{\theta}(k) - \boldsymbol{\theta}_*]$ using some results of Gaussian approximation recalled in Subsection 4.7.2.

Unfortunately, the study of the stability of the associated ODE and the derivation of the covariance of the errors are not always possible due to their complex forms. In these cases, the convergence and the performance of the algorithms for stationary data will be assessed by first-order analysis using coarse approximations. In practice, this analysis will be possible only for independent data $\mathbf{x}(k)$ and assuming the stepsize μ is sufficiently small to keep terms that are at most of the order of μ in the different used expansions. An example of such analysis has been used in [30, 75] to derive an approximate expression of the mean of the deviation from orthonormality $\mathrm{E}[\mathbf{W}^T(k)\mathbf{W}(k) - \mathbf{I}_r]$ for the estimate $\mathbf{W}(k)$ given by the FRANS algorithm (described in Subsection 4.5.1) that allows to explain the difference in behavior of this algorithm when estimating the noise and signal subspaces.

4.7.1 A Short Review of the ODE Method

The so-called ODE [13, 43] is a powerful tool to study the asymptotic behavior of the stochastic approximation algorithms of the general form[10]

$$\boldsymbol{\theta}(k+1) = \boldsymbol{\theta}(k) + \mu_k f[\boldsymbol{\theta}(k), \mathbf{x}(k)] + \mu_k^2 h[\boldsymbol{\theta}(k), \mathbf{x}(k)] \tag{4.68}$$

with $\mathbf{x}(k) = g[\boldsymbol{\xi}(k)]$, where $\boldsymbol{\xi}(k)$ is a Markov chain that does not depend on $\boldsymbol{\theta}, f(\boldsymbol{\theta}, \mathbf{x})$ and $h(\boldsymbol{\theta}, \mathbf{x})$ are regular enough functions, and where $(\mu_k)_{k \in \mathcal{N}}$ is a positive sequence of constants, converging to zero, and satisfying the assumption $\sum_k \mu_k = \infty$. Then, the convergence properties of the discrete time stochastic algorithm (4.68) is intimately connected to the stability properties of the deterministic ODE associated with (4.68), which is defined as the first-order ordinary differential equation

$$\frac{d\boldsymbol{\theta}(t)}{dt} = \bar{f}[\boldsymbol{\theta}(t)] \tag{4.69}$$

where the function $\bar{f}(\boldsymbol{\theta})$ is defined by

$$\bar{f}(\boldsymbol{\theta}) \stackrel{\text{def}}{=} \mathrm{E}\{f[\boldsymbol{\theta}, \mathbf{x}(k)]\} \tag{4.70}$$

where the expectation is taken only with respect to the data $\mathbf{x}(k)$ and $\boldsymbol{\theta}$ is assumed deterministic. We first recall in the following some definitions and results of stability theory of ODE (i.e., the asymptotic behavior of trajectories of the ODE) and then, we will specify its connection to the convergence of the stochastic algorithm (4.68). The stationary points of this ODE are the values $\boldsymbol{\theta}_*$ of $\boldsymbol{\theta}$ for which the driving term $\bar{f}(\boldsymbol{\theta})$

[10]The most common form of stochastic approximation algorithms corresponds to $h(.) = 0$. This residual perturbation term $\mu_k^2 h[\boldsymbol{\theta}(k), \mathbf{x}(k)]$ will be used to write the trajectories governed by the estimated projector $\mathbf{P}(k) = \mathbf{W}(k)\mathbf{W}^T(k)$.

vanishes; hence the term stationary ponts. This gives $\bar{f}(\boldsymbol{\theta}_*) = \mathbf{0}$, so that the motion of the trajectory ceases. A stationary point $\boldsymbol{\theta}_*$ of the ODE is said to be

- *stable* if for an arbitrary neighborhood of $\boldsymbol{\theta}_*$, the trajectory $\boldsymbol{\theta}(t)$ stays in this neighborhood for an initial condition $\boldsymbol{\theta}(0)$ in another neighborhood of $\boldsymbol{\theta}_*$;
- *locally asymptotically stable* if there exists a neighborhood of θ_* such that for all initial conditions $\boldsymbol{\theta}(0)$ in this neighborhood, the ODE (4.69) forces $\boldsymbol{\theta}(t) \to \boldsymbol{\theta}_*$ as $t \to \infty$;
- *globally asymptotically stable* if for all possible values of initial conditions $\boldsymbol{\theta}(0)$, the ODE (4.69) forces $\boldsymbol{\theta}(t) \to \boldsymbol{\theta}_*$ as $t \to \infty$;
- *unstable* if for all neighborhoods of $\boldsymbol{\theta}_*$, there exists some initial value $\boldsymbol{\theta}(0)$ in this neighborhood for which the ODE (4.69) do not force $\boldsymbol{\theta}(t)$ to converge to $\boldsymbol{\theta}_*$ as $t \to \infty$.

Assuming that the set of stationary points can be derived, two standard methods are used to test for stability. They are summarized in the following. The first consists in finding a Lyapunov function $L(\boldsymbol{\theta})$ for the differential equation (4.69), that is, a positive valued function that is decreasing along all trajectories. In this case, it is proved (see e.g. [12]) that the set of the stationary points $\boldsymbol{\theta}_*$ are asymptotically stable. This stability is local if this decrease occurs from an initial condition $\boldsymbol{\theta}(0)$ located in a neighborhood of the stationary points and global if the initial condition can be arbitrary. If $\boldsymbol{\theta}_*$ is a (locally or globally) stable stationary point, then such a Lyapunov function necessarily exists [12]. But for general nonlinear functions $\bar{f}(\boldsymbol{\theta})$, no general recipe exists for finding such a function. Instead, one must try many candidate Lyapunov functions in the hopes of uncovering one which works.

However, for specific functions $\bar{f}(\boldsymbol{\theta})$ which constitute negative gradient vectors of a positive scalar function $J(\boldsymbol{\theta})$

$$\bar{f}(\boldsymbol{\theta}) = -\nabla_{\boldsymbol{\theta}} J \quad \text{with } J > 0$$

then, all the trajectories of the ODE (4.69) converge to the set of the stationary points of the ODE (see Exercise 4.16). Consequently, the set of the stationary points is globally asymptotically stable for this ODE.

The second method consists in a local linearization of the ODE (4.69) about each stationary point $\boldsymbol{\theta}_*$ in which case a stationary point is locally asymptotically stable if and only if the locally linearized equation is asymptotically stable. Consequently the final conclusion amounts to an eigenvalue check of the matrix $\frac{d\bar{f}(\boldsymbol{\theta})}{d\boldsymbol{\theta}}_{|\boldsymbol{\theta}=\boldsymbol{\theta}_*}$. More precisely (see Exercise 4.17), if $\boldsymbol{\theta}_* \in \mathcal{R}^m$ is a stationary point of the ODE (4.69), and $\nu_1, \ldots, \nu_m$ are the eigenvalues of the $m \times m$ matrix $\frac{d\bar{f}(\boldsymbol{\theta})}{d\boldsymbol{\theta}}_{|\boldsymbol{\theta}=\boldsymbol{\theta}_*}$, then (see Exercise 4.17 or [12] for a formal proof)

- if all eigenvalues $\nu_1, \ldots, \nu_m$ have strictly negative real parts, $\boldsymbol{\theta}_*$ is a locally asymptotically stable point;
- if there exists ν_i among $\nu_1, \ldots, \nu_m$ such that $\Re(\nu_i) > 0$, $\boldsymbol{\theta}_*$ is an unstable point;

- if for all eigenvalues $\nu_1, \ldots, \nu_m$, $\Re(\nu_i) \leq 0$ and for at least one eigenvalue ν_{i_0} among $\nu_1, \ldots, \nu_m$, $\Re(\nu_{i_0}) = 0$, we cannot conclude.

Considering now the connection between the stability properties of the associated deterministic ODE (4.69) and the convergence properties of the discrete time stochastic algorithm (4.68), several results are available. First, the sequence $\boldsymbol{\theta}(k)$ generated by the algorithm (4.68) can only converge almost surely [13, 43] to a (locally or globally) asymptotically stable stationary point of the associated ODE (4.69). But deducing some convergence results about the stochastic algorithm (4.68) from the stability of the associated ODE is not trivial because a stochastic algorithm has much more complex asymptotic behavior than a given solution of its associated deterministic ODE. However under additional technical assumptions, it is proved [32] that if the ODE has a finite number of globally (up to a Lebesgue measure zero set of initial conditions) asymptotically stable stationary points $(\boldsymbol{\theta}_{*_i})_{i=1,\ldots,d}$ and if each trajectory of the ODE converges towards one of theses points, then the sequence $\boldsymbol{\theta}(k)$ generated by the algorithm (4.68) converges almost surely to one of these points. The conditions of the result are satisfied, in particular, if the mean field $\bar{f}(\boldsymbol{\theta})$ can be written as $\bar{f}(\boldsymbol{\theta}) = -\nabla_{\boldsymbol{\theta}} J$ where $\nabla_{\boldsymbol{\theta}} J$ is a positive valued function admitting a finite number of local minima. In this later case, this result has been extended for an infinite number of isolated minima in [33].

In adaptive processing, we do not wish for a decreasing stepsize sequence, since we would then lose the tracking capability of the algorithms. To be able to track the possible nonstationarity of the data $\mathbf{x}(k)$, the sequence of stepsize is reduced to a small constant parameter μ. In this case, the stochastic algorithm (4.68) does not converge almost surely even for stationary data and the rigorous results concerning the asymptotic behavior of (4.68) are less powerful. However, when the set of all stable points $(\boldsymbol{\theta}_{*_i})_{i=1,\ldots,d}$ of the associated ODE (4.69) is globally asymptotically stable (up to a zero measure set of initial conditions), the weak convergence approach developed by Kushner [44] suggests that for a sufficiently small μ, $\boldsymbol{\theta}(k)$ will oscillate around one of the limit points $\boldsymbol{\theta}_{*_i}$ of the decreasing stepsize stochastic algorithm. In particular, one should note that, when there exists more than one possible limit ($d \neq 1$), the algorithm may oscillate around one of them $\boldsymbol{\theta}_{*_i}$, and then move into a neighborhood of another equilibrium point $\boldsymbol{\theta}_{*_j}$. However, the probability of such events decreases to zero as $\mu \to 0$, so that their implication is marginal in most cases.

4.7.2 A Short Review of a General Gaussian Approximation Result

For constant stepsize algorithms and stationary data, we will use the following result proved in [13, Th. 2, p. 108] under a certain number of hypotheses. Consider the constant stepsize stochastic approximation algorithm (4.68). Suppose that $\boldsymbol{\theta}(k)$ converges almost surely to the unique globally asymptotically stable point $\boldsymbol{\theta}_*$ in the corresponding decreasing stepsize algorithm. Then, if $\boldsymbol{\theta}_\mu(k)$ denotes the value of $\boldsymbol{\theta}(k)$ associated with the algorithm of stepsize μ, we have when $\mu \to 0$ and $k \to \infty$ (where $\xrightarrow{\mathcal{L}}$ denotes the convergence in distribution and $\mathcal{N}(\mathbf{m}, \mathbf{C}_x)$, the Gaussian distribution of mean $\mathbf{m}$ and covariance $\mathbf{C}_x$)

$$\frac{1}{\sqrt{\mu}}\,[\boldsymbol{\theta}_\mu(k) - \boldsymbol{\theta}_*] \overset{\mathcal{L}}{\rightarrow} \mathcal{N}(\mathbf{0}, \mathbf{C}_\theta) \tag{4.71}$$

where $\mathbf{C}_\theta$ is the unique solution of the continuous-time Lyapunov equation

$$\mathbf{D}\mathbf{C}_\theta + \mathbf{C}_\theta\mathbf{D}^T + \mathbf{G} = \mathbf{O} \tag{4.72}$$

where $\mathbf{D}$ and $\mathbf{G}$ are, respectively, the derivative of the mean field $\bar{f}(\boldsymbol{\theta})$ and the following sum of covariances of the field $f[\boldsymbol{\theta}, \mathbf{x}(k)]$ of the algorithm (4.68)

$$\mathbf{D} \overset{\text{def}}{=} \frac{d\bar{f}(\boldsymbol{\theta})}{d\boldsymbol{\theta}}_{|\boldsymbol{\theta}=\boldsymbol{\theta}_*} \left([\mathbf{D}]_{i,j} \overset{\text{def}}{=} \frac{\partial \bar{f}_i(\boldsymbol{\theta})}{\partial \theta_j}\right) \tag{4.73}$$

$$\begin{aligned}\mathbf{G} &\overset{\text{def}}{=} \sum_{k=-\infty}^{\infty} \operatorname{Cov}\{f[\boldsymbol{\theta}_*, \mathbf{x}(k)], f[\boldsymbol{\theta}_*, \mathbf{x}(0)]\} \\ &= \sum_{k=-\infty}^{\infty} \operatorname{E}\{[f(\boldsymbol{\theta}_*, \mathbf{x}(k))][f(\boldsymbol{\theta}_*, \mathbf{x}(0))]^T\}.\end{aligned} \tag{4.74}$$

Note that all the eigenvalues of the derivative $\mathbf{D}$ of the mean field have strictly negative real parts since $\boldsymbol{\theta}_*$ is an asymptotically stable point of (4.69) and that for independent data $\mathbf{x}(k)$, $\mathbf{G}$ is simply the covariance of the field. Unless we have sufficient information about the data, which is often not the case, in practice we consider the simplifying hypothesis of independent identically Gaussian distributed data $\mathbf{x}(k)$.

It should be mentioned that the rigorous proof of this result (4.71) needs a very strong hypothesis on the algorithm (4.68), namely that $\boldsymbol{\theta}(k)$ converges almost surely to the unique globally asymptotically stable point $\boldsymbol{\theta}_*$ in the corresponding decreasing step size algorithm. However, the practical use of (4.71) in more general situations is usually justified by using a general diffusion approximation result formally [13, Th. 1, p. 104].

In practice, μ is small and fixed, but it is assumed that the asymptotic distribution of $\mu^{-1/2}(\boldsymbol{\theta}_\mu(k) - \boldsymbol{\theta}_*)$ when k tends to ∞ can still be approximated by a zero mean Gaussian distribution of covariance $\mathbf{C}_\theta$, and consequently that for large enough k, the distribution of the residual error $(\boldsymbol{\theta}_\mu(k) - \boldsymbol{\theta}_*)$ is a zero mean Gaussian distribution of covariance $\mu\mathbf{C}_\theta$ where $\mathbf{C}_\theta$ is solution of the Lyapunov equation (4.72). Note that the approximation $\operatorname{E}[(\boldsymbol{\theta}_\mu(k) - \boldsymbol{\theta}_*)(\boldsymbol{\theta}_\mu(k) - \boldsymbol{\theta}_*)^T] \approx \mu\mathbf{C}_\theta$ enables us to derive an expression of the asymptotic bias $\operatorname{E}[\boldsymbol{\theta}_\mu(k)] - \boldsymbol{\theta}_*$ from a perturbation analysis of the expectation of both sides of (4.68) when the field $f[\boldsymbol{\theta}(k), \mathbf{x}(k)]$ is linear in $\mathbf{x}(k)\mathbf{x}^T(k)$. An example of such a derivation is given in Subsection 4.7.3, [26] and Exercise 4.18.

Finally, let us recall that there is no relation between the asymptotic performance of the stochastic approximation algorithm (4.68) and its convergence rate. As is well known, the convergence rate depends on the transient behavior of the algorithm, for which no general result seems to be available. For this reason, different authors (e.g. [22, 26]) have resorted to simulations to compare the convergence speed of different algorithms whose associated stepsizes μ are chosen to provide the same value of the mean square error $\mathrm{E}\|(\boldsymbol{\theta}_\mu(k) - \boldsymbol{\theta}_*)\|_2 \approx \mu \mathrm{Tr}(\mathbf{C}_\theta)$.

4.7.3 Examples of Convergence and Performance Analysis

Using the previously described methods, two examples of convergence and performance analysis will be given. Oja's neuron algorithm as the simplest algorithm will allow us to present a comprehensive study of an eigenvector tracking algorithm. Then Oja's algorithm will be studied as an example of a subspace tracking algorithm.

Convergence and Performance Analysis of the Oja's Neuron Consider Oja's neuron algorithms (4.23) and (4.25) introduced in Section 4.4. The stationary points of their associated ODE

$$\frac{d\mathbf{w}(t)}{dt} = \mathbf{C}_x\mathbf{w}(t) - \mathbf{w}(t)[\mathbf{w}(t)^T\mathbf{C}_x\mathbf{w}(t)]\{\text{resp., } -\mathbf{C}_x\mathbf{w}(t) + \mathbf{w}(t)[\mathbf{w}(t)^T\mathbf{C}_x\mathbf{w}(t)]\} \quad (4.75)$$

are the roots of $\mathbf{C}_x\mathbf{w} = \mathbf{w}[\mathbf{w}^T\mathbf{C}_x\mathbf{w}]$ and thus are clearly given by $(\pm\,\mathbf{u}_k)_{k=1,\ldots,n}$. To study the stability of these stationarity points, consider the derivative $\mathbf{D}_\mathbf{w}$ of the mean field $\mathbf{C}_x\mathbf{w} - \mathbf{w}[\mathbf{w}^T\mathbf{C}_x\mathbf{w}]$ {resp., $-\mathbf{C}_x\mathbf{w} + \mathbf{w}[\mathbf{w}^T\mathbf{C}_x\mathbf{w}]$} at these points. Using a standard first-order perturbation, we obtain

$$\begin{aligned}\mathbf{D}_\mathbf{w}(\pm\,\mathbf{u}_k) &= \mathbf{C}_x - (\mathbf{w}^T\mathbf{C}_x\mathbf{w})\mathbf{I}_n - 2\mathbf{w}\mathbf{w}^T\mathbf{C}_{x|\mathbf{w}=\pm\mathbf{u}_k} \\ &\quad \times [\text{resp., } -\mathbf{C}_x + (\mathbf{w}^T\mathbf{C}_x\mathbf{w})\mathbf{I}_n + 2\mathbf{w}\mathbf{w}^T\mathbf{C}_{x|\mathbf{w}=\pm\mathbf{u}_k}].\end{aligned}$$

Because the eigenvalues of $\mathbf{D}_\mathbf{w}(\pm\,\mathbf{u}_k)$ are $-2\lambda_k, (\lambda_i - \lambda_k)_{i\neq k}$ [resp. $2\lambda_k$, $-(\lambda_i - \lambda_k)_{i\neq k}$], these eigenvalues are all real negative for $k = 1$ only, for the stochastic approximation algorithms (4.23), in contrast to the stochastic approximation algorithms (4.25) for which $\mathbf{D}_\mathbf{w}(\pm\,\mathbf{u}_k)$ has at least one nonnegative eigenvalue. Consequently only $\pm\,\mathbf{u}_1$ is locally asymptotically stable for the ODE associated with (4.23) and all the eigenvectors $(\pm\,\mathbf{u}_k)_{k=1,\ldots,n}$ are unstable for the ODE associated with (4.25) and thus only (4.23) (Oja's neuron for dominant eigenvector) can be retained.

Note that the coupled stochastic approximation algorithms (4.23) and (4.25) can be globally written as (4.68) as well. The associated ODE, given by

$$\frac{d}{dt}\begin{pmatrix}\mathbf{w}(t)\\ \lambda(t)\end{pmatrix} = \begin{pmatrix}\mathbf{C}_x\mathbf{w} - \mathbf{w}\mathbf{w}^T\mathbf{C}_x\mathbf{w}\\ \mathbf{w}^T\mathbf{C}_x\mathbf{w} - \lambda\end{pmatrix} \quad (4.76)$$

has the pairs $(\pm\mathbf{u}_k, \lambda_k)_{k=1,\dots n}$ as stationary points. The derivative $\mathbf{D}$ of the mean field at theses points is given by

$$\mathbf{D} = \begin{pmatrix} \mathbf{D}_\mathbf{w}(\pm\mathbf{u}_k) & \mathbf{0} \\ 2\mathbf{u}_k^T\mathbf{C}_x & -1 \end{pmatrix}$$

whose eigenvalues are $-2\lambda_k$, $(\lambda_i - \lambda_k)_{i\neq k}$ and -1. Consequently the pair $(\pm\mathbf{u}_1, \lambda_1)$ is the only locally asymptotically stable point for the associated ODE (4.76) as well.

More precisely, it is proved in [50] that if $\mathbf{w}(0)^T\mathbf{u}_1 > 0$ [resp. < 0], the solution $\mathbf{w}(t)$ of the ODE (4.69) tends exponentially to $\mathbf{u}_1$ [resp. $-\mathbf{u}_1$] as $t \to \infty$. The pair $(\pm\mathbf{u}_1, \lambda_1)$ is thus globally asymptotically stable for the associated ODE.

Furthermore, using the stochastic approximation theory and in particular [44, Th. 2.3.1], it is proved in [51] that Oja's neuron (4.23) with decreasing stepsize μ_k, converges almost surely to $+\mathbf{u}_1$ or $-\mathbf{u}_1$ as k tends to ∞.

We have now the conditions to apply the Gaussian approximation results of Subsection 4.7.2. To solve the Lyapunov equation, the derivative $\mathbf{D}$ of the mean field at the pair $(\pm\mathbf{u}_1, \lambda_1)$ is given by

$$\mathbf{D} = \begin{pmatrix} \mathbf{C}_x - \lambda_1\mathbf{I}_n - 2\lambda_1\mathbf{u}_1\mathbf{u}_1^T & \mathbf{0} \\ 2\lambda_1\mathbf{u}_1^T & -1 \end{pmatrix}.$$

In the case of independent Gaussian distributed data $\mathbf{x}(k)$, it has been proved [22, 26] that the covariance $\mathbf{G}$ (4.74) of the field is given by

$$\mathbf{G} = \begin{pmatrix} \mathbf{G}_\mathbf{w} & \mathbf{0} \\ \mathbf{0}^T & 2\lambda_1^2 \end{pmatrix}$$

with $\mathbf{G}_\mathbf{w} = \sum_{i=2}^n \lambda_1\lambda_i\mathbf{u}_i\mathbf{u}_i^T$. Solving the Lyapunov equation (4.72), the following asymptotic covariance $\mathbf{C}_\theta$ is obtained [22, 26]

$$\mathbf{C}_\theta = \begin{pmatrix} \mathbf{C}_\mathbf{w} & \mathbf{0} \\ \mathbf{0}^T & \lambda_1^2 \end{pmatrix}$$

with $\mathbf{C}_\mathbf{w} = \sum_{i=2}^n \frac{\lambda_1\lambda_i}{2(\lambda_1 - \lambda_i)}\mathbf{u}_i\mathbf{u}_i^T$. Consequently the estimates $[\mathbf{w}(k), \lambda(k)]$ of $(\pm\mathbf{u}_1, \lambda_1)$ given by (4.23) and (4.24) respectively, are asymptotically independent and Gaussian distributed with

$$\mathrm{E}(\|\mathbf{w}(k) - (\pm\mathbf{u}_1)\|^2) \sim \sum_{i=2}^n \frac{\mu\lambda_1\lambda_i}{2(\lambda_1 - \lambda_i)} \quad \text{and} \quad \mathrm{E}(\lambda(k) - \lambda_1)^2 \sim \mu\lambda_1^2.$$

We note that the behavior of the adaptive estimates $[\mathbf{w}(k), \lambda(k)]$ of $(\pm\mathbf{u}_1, \lambda_1)$ are similar to the behavior of their batch estimates. More precisely if $\mathbf{w}(k)$ and $\lambda(k)$ denote now the dominant eigenvector and the associated eigenvalue of the sample estimate

$\mathbf{C}(k) = \frac{1}{k}\sum_{i=1}^{k} \mathbf{x}(i)\mathbf{x}^T(i)$ of $\mathbf{C}_x$, a standard result [3, Th. 13.5.1, p. 541] gives

$$\sqrt{k}\,(\boldsymbol{\theta}(k) - \boldsymbol{\theta}_*) \xrightarrow{\mathcal{L}} \mathcal{N}(\mathbf{0}, \mathbf{C}_{\boldsymbol{\theta}}) \tag{4.77}$$

with $\mathbf{C}_{\boldsymbol{\theta}} = \begin{pmatrix} \mathbf{C}_{\mathbf{w}} & \mathbf{0} \\ \mathbf{0}^T & 2\lambda_1^2 \end{pmatrix}$ where $\mathbf{C}_{\mathbf{w}} = \sum_{i=2}^{n} \frac{\lambda_1\lambda_i}{(\lambda_1 - \lambda_i)^2}\mathbf{u}_i\mathbf{u}_i^T$. The estimates $\mathbf{w}(k)$ and $\lambda(k)$ are asymptotically uncorrelated and the estimation of the eigenvalue λ_1 is well conditioned in contrast to those of the eigenvector $\mathbf{u}_1$ whose conditioning may be very bad when λ_1 and λ_2 are very close.

Expressions of the asymptotic bias $\lim_{k\to\infty} \mathrm{E}[\boldsymbol{\theta}(k)] - \boldsymbol{\theta}_*$ can be derived from (4.71). A word of caution is nonetheless necessary because the convergence of $\mu^{-1/2}(\boldsymbol{\theta}_\mu(k) - \boldsymbol{\theta}_*)$ to a limiting Gaussian distribution with covariance matrix $\mathbf{C}_{\boldsymbol{\theta}}$ does not guarantee the convergence of its moments to those of the limiting Gaussian distribution. In batch estimation, both the first and the second moments of the limiting distribution of $\sqrt{k}(\boldsymbol{\theta}(k) - \boldsymbol{\theta}_*)$ are equal to the corresponding asymptotic moments for independent Gaussian distributed data $\mathbf{x}(k)$. In the following, we assume the convergence of the second-order moments allowing us to write

$$\mathrm{E}[(\boldsymbol{\theta}_\mu(k) - \boldsymbol{\theta}_*)]\,(\boldsymbol{\theta}_\mu(k) - \boldsymbol{\theta}_*)^T\,] = \mu\mathbf{C}_{\boldsymbol{\theta}} + o(\mu).$$

Let $\boldsymbol{\theta}_\mu(k) = \boldsymbol{\theta}_* + \delta\boldsymbol{\theta}_k$ with $\boldsymbol{\theta}_* = \begin{pmatrix} \mathbf{u}_1 \\ \lambda_1 \end{pmatrix}$. Provided the data $\mathbf{x}(k)$ are independent (which implies that $\mathbf{w}(k)$ and $\mathbf{x}(k)\mathbf{x}^T(k)$ are independent) and $\boldsymbol{\theta}_\mu(k)$ is stationary, taking the expectation of both sides of (4.23) and (4.24) gives[11]

$$\mathrm{E}[\mathbf{C}_x(\mathbf{u}_1 + \delta\mathbf{w}_k) - (\mathbf{u}_1 + \delta\mathbf{w}_k)(\mathbf{u}_1 + \delta\mathbf{w}_k)^T\mathbf{C}_x(\mathbf{u}_1 + \delta\mathbf{w}_k)] = \mathbf{0}$$
$$\mathrm{E}[(\mathbf{u}_1 + \delta\mathbf{w}_k)^T\mathbf{C}_x(\mathbf{u}_1 + \delta\mathbf{w}_k) - (\lambda_1 + \delta\lambda_k)] = 0.$$

Using a second-order expansion, we get after some algebraic manipulations

$$\begin{bmatrix} \mathbf{C}_x - \lambda_1\mathbf{I}_n - 2\lambda_1\mathbf{u}_1\mathbf{u}_1^T & \mathbf{0} \\ 2\lambda_1\mathbf{u}_1^T & -1 \end{bmatrix}\begin{bmatrix} \mathrm{E}(\delta\mathbf{w}_k) \\ \mathrm{E}(\delta\lambda_k) \end{bmatrix} + \mu\begin{bmatrix} -(2\lambda_1\mathbf{C}_{\mathbf{w}} + \mathrm{Tr}(\mathbf{C}_x\mathbf{C}_{\mathbf{w}})\mathbf{I}_n)\mathbf{u}_1 \\ \mathrm{Tr}(\mathbf{C}_x\mathbf{C}_{\mathbf{w}}) \end{bmatrix} = o(\mu).$$

Solving this equation in $\mathrm{E}(\delta\mathbf{w}_k)$ and $\mathrm{E}(\delta\lambda_k)$ using the expression of $\mathbf{C}_{\mathbf{w}}$, gives the following expressions of the asymptotic bias

$$\mathrm{E}[\mathbf{w}(k)] - \mathbf{u}_1 = -\mu\left(\sum_{i=2}^{n} \frac{\lambda_i^2}{4(\lambda_1 - \lambda_i)}\right)\mathbf{u}_1 + o(\mu) \quad \text{and} \quad \mathrm{E}[\lambda(k)] - \lambda_1 = o(\mu).$$

[11] We note that this derivation would not be possible for non-polynomial adaptations $f[\boldsymbol{\theta}(k), \mathbf{x}(k)]$.

We note that these asymptotic biases are similar to those obtained in batch estimation derived from a Taylor series expansion [77, p. 68] with expression (4.77) of $\mathbf{C}_\theta$.

$$\mathrm{E}[\mathbf{w}(k)] - \mathbf{u}_1 = -\frac{1}{k}\left(\sum_{i=2}^{n} \frac{\lambda_1\lambda_i}{2(\lambda_1-\lambda_i)^2}\right)\mathbf{u}_1 + o\left(\frac{1}{k}\right) \quad \text{and} \quad \mathrm{E}[\lambda(k)] - \lambda_1 = o\left(\frac{1}{k}\right).$$

Finally, we see that in adaptive and batch estimation, the square of these biases are an order of magnitude smaller that the variances in $O(\mu)$ or $O(\frac{1}{k})$.

This methodology has been applied to compare the theoretical asymptotic performance of several adaptive algorithms for minor and principal component analysis in [22, 26, 27]. For example, the asymptotic mean square error $\mathrm{E}(\|\mathbf{W}(k) - \mathbf{W}_*\|_{\mathrm{Fro}}^2)$ of the estimate $\mathbf{W}(k)$ given by the WSA algorithm (4.57) is shown in Figure 4.1, where the stepsize μ is chosen to provide the same value for $\mu\mathrm{Tr}(\mathbf{C}_\theta)$. We clearly see in this figure that the value $\beta_2/\beta_1 = 0.6$ optimizes the asymptotic mean square error/speed of convergence tradeoff.

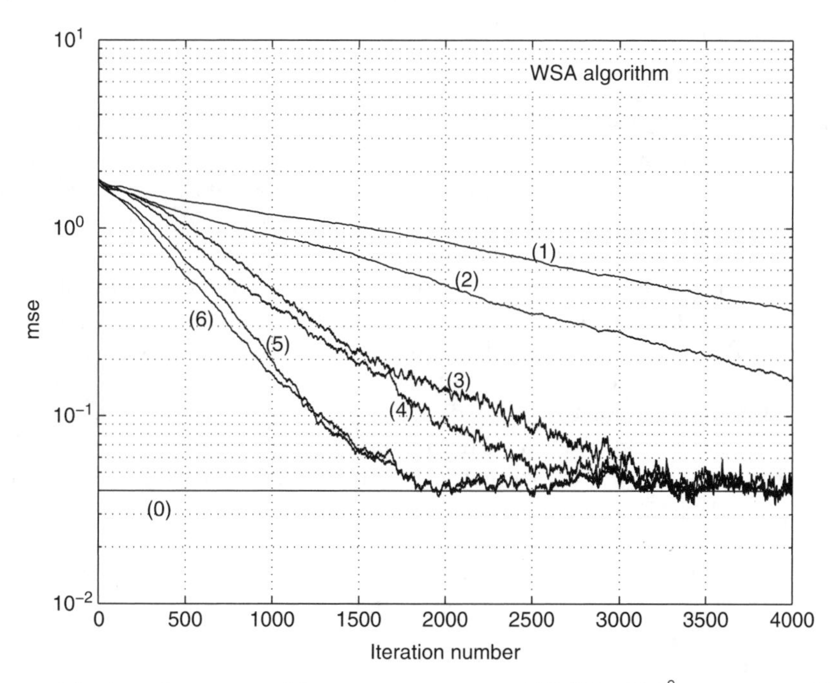

Figure 4.1 Learning curves of the mean square error $\mathrm{E}(\|\mathbf{W}(k) - \mathbf{W}_*\|_{\mathrm{Fro}}^2)$ averaging 100 independent runs for the WSA algorithm, for different values of parameter $\beta_2/\beta_1 = 0.96$ (1), 0.9 (2), 0.1 (3), 0.2 (4), 0.4 (5), and 0.6 (6) compared with $\mu\mathrm{Tr}(\mathbf{C}_\theta)$ (0) in the case $n = 4$, $r = 2$, $\mathbf{C}_x = \mathrm{Diag}(1.75, 1.5, 0.5, 0.25)$, where the entries of $\mathbf{W}(0)$ are chosen randomly uniformly in (0, 1).

Convergence and Performance Analysis of Oja's Algorithm Consider now Oja's algorithm (4.30) described in Subsection 4.5.1. A difficulty arises in the study of the behavior of $\mathbf{W}(k)$ because the set of orthonormal bases of the r-dominant subspace forms a *continuum* of attractors: the column vectors of $\mathbf{W}(k)$ do not tend in general to the eigenvectors $\mathbf{u}_1, \ldots, \mathbf{u}_r$, and we have no proof of convergence of $\mathbf{W}(k)$ to a particular orthonormal basis of their span. Thus, considering the asymptotic distribution of $\mathbf{W}(k)$ is meaningless. To solve this problem, in the same way as Williams [78] did when he studied the stability of the estimated projection matrix $\mathbf{P}(k) \stackrel{\text{def}}{=} \mathbf{W}(k)\mathbf{W}^T(k)$ in the dynamics induced by Oja's learning equation $\frac{d\mathbf{W}(t)}{dt} = [\mathbf{I}_n - \mathbf{W}(t)\mathbf{W}(t)^T]\mathbf{C}\mathbf{W}(t)$, viz

$$\frac{d\mathbf{P}(t)}{dt} = [\mathbf{I}_n - \mathbf{P}(t)]\mathbf{C}\mathbf{P}(t) + \mathbf{P}(t)\mathbf{C}[\mathbf{I}_n - \mathbf{P}(t)] \tag{4.78}$$

we consider the trajectory of the matrix $\mathbf{P}(k) \stackrel{\text{def}}{=} \mathbf{W}(k)\mathbf{W}^T(k)$ whose dynamics are governed by the stochastic equation

$$\mathbf{P}(k+1) = \mathbf{P}(k) + \mu_k f[\mathbf{P}(k), \mathbf{x}(k)\mathbf{x}^T(k)] + \mu_k^2 h[\mathbf{P}(k), \mathbf{x}(k)\mathbf{x}^T(k)] \tag{4.79}$$

with $f(\mathbf{P}, \mathbf{C}) \stackrel{\text{def}}{=} (\mathbf{I}_n - \mathbf{P})\mathbf{C}\mathbf{P} + \mathbf{P}\mathbf{C}(\mathbf{I}_n - \mathbf{P})$ and $h(\mathbf{P}, \mathbf{C}) \stackrel{\text{def}}{=} (\mathbf{I}_n - \mathbf{P})\mathbf{C}\mathbf{P}\mathbf{C}(\mathbf{I}_n - \mathbf{P})$. A remarkable feature of (4.79) is that the field f and the complementary term h depend only on $\mathbf{P}(k)$ and *not* on $\mathbf{W}(k)$. This fortunate circumstance makes it possible to study the evolution of $\mathbf{P}(k)$ without determining the evolution of the underlying matrix $\mathbf{W}(k)$. The characteristics of $\mathbf{P}(k)$ are indeed the most interesting since they completely characterize the estimated subspace. Since (4.78) has a unique global asymptotically stable point $\mathbf{P}_* = \mathbf{\Pi}_s$ [69], we can conjecture from the stochastic approximation theory [13, 44] that (4.79) converges almost surely to $\mathbf{P}_*$. And consequently the estimate $\mathbf{W}(k)$ given by (4.30) converges almost surely to the signal subspace in the meaning recalled in Subsection 4.2.4.

To evaluate the asymptotic distributions of the subspace projection matrix estimator given by (4.79), we must adapt the results of Subsection 4.7.2 because the parameter $\mathbf{P}(k)$ is here a $n \times n$ rank-r symmetric matrix. Furthermore, we note that some eigenvalues of the derivative of the mean field $\bar{f}(\mathbf{P}) = \mathrm{E}[f(\mathbf{P}, \mathbf{x}(k)\mathbf{x}^T(k))]$ are positive real. To overcome this difficulty, let us now consider the following parametrization of $\mathbf{P}(k)$ in a neighborhood of $\mathbf{P}_*$ introduced in [24, 25]. If $\{\theta_{ij}(\mathbf{P}) \mid 1 \leq i \leq j \leq n\}$ are the coordinates of $\mathbf{P} - \mathbf{P}_*$ in the orthonormal basis $(\mathbf{S}_{i,j})_{1 \leq i \leq j \leq n}$ defined by

$$\mathbf{S}_{i,j} = \begin{cases} \mathbf{u}_i\mathbf{u}_i^T & i = j \\ \dfrac{\mathbf{u}_i\mathbf{u}_j^T + \mathbf{u}_j\mathbf{u}_i^T}{\sqrt{2}} & i < j \end{cases}$$

with the inner product under consideration is $(\mathbf{A}, \mathbf{B}) \stackrel{\text{def}}{=} \mathrm{Tr}(\mathbf{A}^T\mathbf{B})$, then,

$$\mathbf{P} = \mathbf{P}_* + \sum_{1 \leq i,j \leq n} \theta_{ij}(\mathbf{P})\mathbf{S}_{i,j}$$

and $\theta_{ij}(\mathbf{P}) = \mathrm{Tr}\{\mathbf{S}_{ij}(\mathbf{P} - \mathbf{P}_*)\}$ for $1 \leq i \leq j \leq n$. The relevance of this basis is shown by the following relation proved in [24, 25]

$$\mathbf{P} = \mathbf{P}_* + \sum_{(i,j) \in P_s} \theta_{ij}(\mathbf{P})\,\mathbf{S}_{ij} + O(\|\mathbf{P} - \mathbf{P}_*\|^2_{\mathrm{Fro}}) \tag{4.80}$$

where $P_s \stackrel{\mathrm{def}}{=} \{(i,j) \mid 1 \leq i \leq j \leq n \text{ and } i \leq r\}$. There are $\frac{r}{2}(2n - r + 1)$ pairs in P_s and this is exactly the dimension of the manifold of the $n \times n$ rank-r symmetric matrices. This point, together with relation (4.80), shows that the matrix set $\{\mathbf{S}_{ij} \mid (i,j) \in P_s\}$ is in fact an orthonormal basis of the tangent plane to this manifold at point $\mathbf{P}_*$. In other words, a $n \times n$ rank-r symmetric matrix $\mathbf{P}$ lying less than ε away from $\mathbf{P}_*$ (i.e., $\|\mathbf{P} - \mathbf{P}_*\| < \varepsilon$) has negligible (of order ε^2) components in the direction of $\mathbf{S}_{ij}$ for $r < i \leq j \leq n$. It follows that, in a neighborhood of $\mathbf{P}_*$, the $n \times n$ rank-r symmetric matrices are uniquely determined by the $\frac{r}{2}(2n - r + 1) \times 1$ vector $\boldsymbol{\theta}(\mathbf{P})$ defined by: $\boldsymbol{\theta}(\mathbf{P}) \stackrel{\mathrm{def}}{=} \mathcal{S}^T \mathrm{vec}(\mathbf{P} - \mathbf{P}_*)$, where $\mathcal{S}$ denotes the following $n^2 \times \frac{r}{2}(2n - r + 1)$ matrix: $\mathcal{S} \stackrel{\mathrm{def}}{=} [\ldots, \mathrm{vec}(\mathbf{S}_{ij}), \ldots]$, $(i,j) \in P_s$. If $\mathcal{P}(\boldsymbol{\theta})$ denotes the unique (for $\|\boldsymbol{\theta}\|$ sufficiently small) $n \times n$ rank-r symmetric matrix such that $\mathcal{S}^T \mathrm{vec}(\mathcal{P}(\boldsymbol{\theta}) - \mathbf{P}_*) = \boldsymbol{\theta}$, the following one-to-one mapping is exhibited for sufficiently small $\|\boldsymbol{\theta}(k)\|$

$$\begin{aligned}\mathrm{vec}\{\mathcal{P}[\boldsymbol{\theta}(k)]\} = \mathrm{vec}(\mathbf{P}_*) + \mathcal{S}\boldsymbol{\theta}(k) + O(\|\boldsymbol{\theta}(k)\|^2) \longleftrightarrow \boldsymbol{\theta}(k) \\ = \mathcal{S}^T \mathrm{vec}[\mathbf{P}(k) - \mathbf{P}_*].\end{aligned} \tag{4.81}$$

We are now in a position to solve the Lyapunov equation in the new parameter $\boldsymbol{\theta}$. The stochastic equation governing the evolution of $\boldsymbol{\theta}(k)$ is obtained by applying the transformation $\mathbf{P}(k) \rightarrow \boldsymbol{\theta}(k) = \mathcal{S}^T \mathrm{vec}[\mathbf{P}(k) - \mathbf{P}_*]$ to the original equation (4.79), thereby giving

$$\boldsymbol{\theta}(k+1) = \boldsymbol{\theta}(k) + \mu_k \phi[\boldsymbol{\theta}(k), \mathbf{x}(k)] + \mu_k^2 \psi[\boldsymbol{\theta}(k), \mathbf{x}(k)] \tag{4.82}$$

where $\phi(\boldsymbol{\theta}, \mathbf{x}) \stackrel{\mathrm{def}}{=} \mathcal{S}^T \mathrm{vec}[f(\mathcal{P}(\boldsymbol{\theta}), \mathbf{x}\mathbf{x}^T)]$ and $\psi(\boldsymbol{\theta}, \mathbf{x}) \stackrel{\mathrm{def}}{=} \mathcal{S}^T \mathrm{vec}[h(\mathcal{P}(\boldsymbol{\theta}), \mathbf{x}\mathbf{x}^T)]$.

Solving now the Lyapunov equation associated with (4.82) after deriving the derivative of the mean field $\bar{\phi}(\boldsymbol{\theta})$ and the covariance of the field $\phi[\boldsymbol{\theta}(k), \mathbf{x}(k)]$ for independent Gaussian distributed data $\mathbf{x}(k)$, yields the covariance $\mathbf{C}_{\boldsymbol{\theta}}$ of the asymptotic distribution of $\boldsymbol{\theta}(k)$. Finally using mapping (4.81), the covariance $\mathbf{C}_P = \mathcal{S}\mathbf{C}_{\boldsymbol{\theta}}\mathcal{S}^T$ of the asymptotic distribution of $\mathbf{P}(k)$ is deduced [25]

$$\mathbf{C}_P = \sum_{1 \leq i \leq r < j \leq n} \frac{\lambda_i \lambda_j}{2(\lambda_i - \lambda_j)} (\mathbf{u}_i \otimes \mathbf{u}_j + \mathbf{u}_j \otimes \mathbf{u}_i)(\mathbf{u}_i \otimes \mathbf{u}_j + \mathbf{u}_j \otimes \mathbf{u}_i)^T. \tag{4.83}$$

To improve the learning speed and misadjustment tradeoff of Oja's algorithm (4.30), it has been proposed in [25] to use the recursive estimate (4.20) for $\mathbf{C}_x(k) = \mathrm{E}[\mathbf{x}(k)\mathbf{x}^T(k)]$. Thus the modified Oja's algorithm, called the smoothed Oja's algorithm, reads

$$\mathbf{C}(k+1) = \mathbf{C}(k) + \alpha\mu_k[\mathbf{x}(k)\mathbf{x}^T(k) - \mathbf{C}(k)] \tag{4.84}$$

$$\mathbf{W}(k+1) = \mathbf{W}(k) + \mu_k[\mathbf{I}_n - \mathbf{W}(k)\mathbf{W}^T(k)]\mathbf{C}(k)\mathbf{W}(k) \tag{4.85}$$

where α is introduced in order to normalize both algorithms because if the learning rate of (4.84) has no dimension, the learning rate of (4.85) must have the dimension of the inverse of the power of $\mathbf{x}(k)$. Furthermore α can take into account a better trade-off between the misadjustments and the learning speed. Note that the performance derivations may be extended to this smoothed Oja's algorithm by considering that the coupled stochastic approximation algorithms (4.84) and (4.85) can be globally written as (4.68) as well. Reusing now the parametrization $(\theta_{ij})_{1\leq i\leq j\leq n}$ because $\mathbf{C}(k)$ is symmetric as well, and following the same approach, we obtain now [25]

$$\mathbf{C}_P = \sum_{1\leq i\leq r<j\leq n} \frac{\alpha_{ij}\lambda_i\lambda_j}{2(\lambda_i - \lambda_j)}(\mathbf{u}_i \otimes \mathbf{u}_j + \mathbf{u}_j \otimes \mathbf{u}_i)(\mathbf{u}_i \otimes \mathbf{u}_j + \mathbf{u}_j \otimes \mathbf{u}_i)^T \tag{4.86}$$

with $\alpha_{ij} \stackrel{\text{def}}{=} \alpha/(\alpha + \lambda_i - \lambda_j) < 1$.

This methodology has been applied to compare the theoretical asymptotic performance of several minor and principal subspace adaptive algorithms in [24, 25]. For example, the asymptotic mean square error $\mathrm{E}(\|\mathbf{P}(k) - \mathbf{P}_*\|^2_{\mathrm{Fro}})$ of the estimate $\mathbf{P}(k)$ given by Oja's algorithm (4.30) and the smoothed Oja's algorithm (4.84) are shown in Figure 4.2, where the stepsize μ of the Oja's algorithm and the couple (μ, α) of the smoothed Oja's algorithm are chosen to provide the same value for $\mu\mathrm{Tr}(\mathbf{C}_P)$. We clearly see in this figure that the smoothed Oja's algorithm with $\alpha = 0.3$ provides faster convergence than the Oja's algorithm.

Regarding the issue of asymptotic bias, note that there is a real methodological problem to apply the methodology of the end of Subsection 4.7.3. The trouble stems from the fact that the matrix $\mathbf{P}(k) = \mathbf{W}(k)\mathbf{W}^T(k)$ does not belong to a linear vector space because it is constrained to have fixed rank $r < n$. The set of such matrices is not invariant under addition; it is actually a smooth submanifold of $\mathcal{R}^{n\times n}$. This is not a problem in the first-order asymptotic analysis because this approach amounts to approximating this manifold by its tangent plane at a point of interest. This tangent plane is linear indeed. In order to refine the analysis by developing a higher order theory, it becomes necessary to take into account the curvature of the manifold. This is a tricky business. As an example of these difficulties, one could show (under simple assumptions) that there exists no projection-valued estimators of a projection matrix that are unbiased at order $O(\mu)$; this can be geometrically pictured by representing the estimates as points on a curved manifold (here: the manifold of projection matrices).

Using a more involved expression of the covariance of the field (4.74), the previously described analysis can be extended to correlated data $\mathbf{x}(k)$. Expressions (4.83) and (4.86) extend provided that $\lambda_i\lambda_j$ is replaced by $\lambda_i\lambda_j + \lambda_{i,j}$ where $\lambda_{i,j}$ is

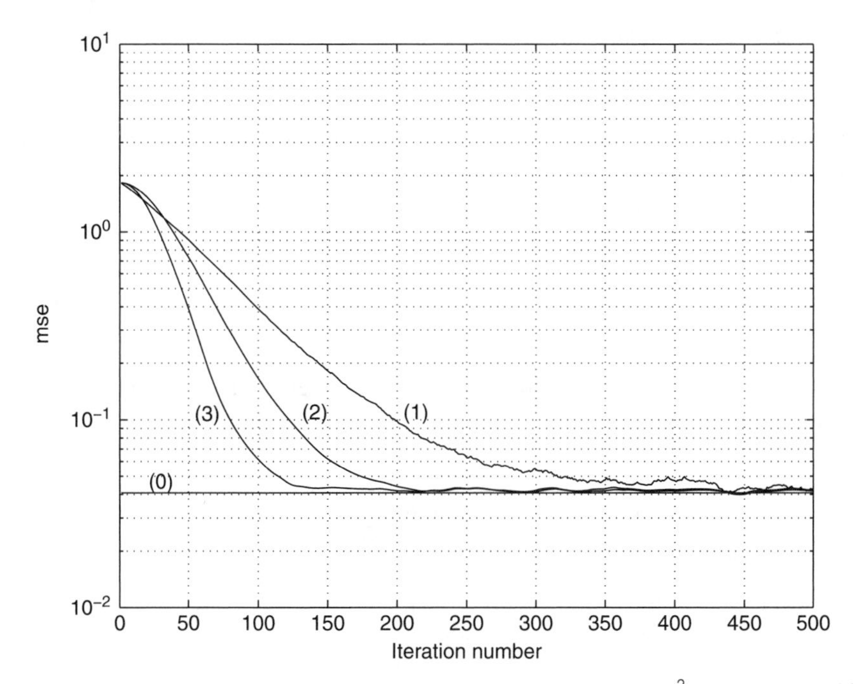

Figure 4.2 Learning curves of the mean square error $\mathrm{E}(\|\mathbf{P}(k) - \mathbf{P}_*\|^2_{\mathrm{Fro}})$ averaging 100 independent runs for the Oja's algorithm (1) and the smoothed Oja's algorithm with $\alpha = 1$ (2) and $\alpha = 0.3$ (3) compared with $\mu\mathrm{Tr}(\mathbf{C}_P)$ (0) in the same configuration $(\mathbf{C}_x, \mathbf{W}(0))$ as Figure 4.1.

defined in [25]. Note that when $\mathbf{x}(k) = (x_k, x_{k-1}, \ldots, x_{k-n+1})^T$ with x_k being an ARMA stationary process, the covariance of the field (4.74) and thus $\lambda_{i,j}$ can be expressed in closed form with the help of a finite sum [23].

The domain of learning rate μ for which the previously described asymptotic approach is valid and the performance criteria for which no analytical results could be derived from our first-order analysis, such as the speed of convergence and the deviation from orthonormality $d^2(\mu) \stackrel{\text{def}}{=} \|\mathbf{W}^T(k)\mathbf{W}(k) - \mathbf{I}_r\|^2_{\mathrm{Fro}}$ can be derived from numerical experiments only. In order to compare Oja's algorithm and the smoothed Oja's algorithm, the associated parameters μ and (α, μ) must be constrained to give the same value of $\mu\mathrm{Tr}(\mathbf{C}_P)$. In these conditions, it has been shown in [25] by numerical simulations that the smoothed Oja's algorithm provides faster convergence and a smaller deviation from orthonormality $d^2(\mu)$ than Oja's algorithm. More precisely, it has been shown that $d^2(\mu) \propto \mu^2$ [resp. $\propto \mu^4$] for Oja's [resp. the smoothed Oja's] algorithm. This result agrees with the presentation of Oja's algorithm given in Subsection 4.5.1 in which the term $O(\mu_k^2)$ was omitted from the orthonormalization of the columns of $\mathbf{W}(k)$.

Finally, using the theorem of continuity (e.g. [59, Th. 6.2a]), note that the behavior of any differentiable function of $\mathbf{P}(k)$ can be obtained. For example, in DOA tracking from the MUSIC algorithm[12] (see e.g. Subsection 4.8.1), the MUSIC estimates $[\theta_i(k)]_{i=1,\ldots,r}$ of the DOAs at time k can be determined as the r deepest minima of the localization function $\mathbf{a}^H(\theta)[\mathbf{I}_n - \mathbf{P}(k)]\mathbf{a}(\theta)$. Using the mapping $\mathbf{P}(k) \mapsto \boldsymbol{\theta}(k)$ where here $\boldsymbol{\theta}(k) \stackrel{\text{def}}{=} [\theta_1(k), \ldots, \theta_r(k)]^T$, the Gaussian asymptotic distribution of the estimate $\boldsymbol{\theta}(k)$ can be derived [24] and compared to the batch estimate. For example for a single source, it has been proved [24] that

$$\mathrm{Var}[\theta_1(k)] = \mu \frac{n\sigma_1^2}{2\alpha_1}\left(1 + \frac{\sigma_n^2}{n\sigma_1^2}\right)\frac{\sigma_n^2}{\sigma_1^2} + o(\mu)$$

where σ_1^2 is the source power and α_1 is a purely geometrical factor. Compared to the batch MUSIC estimate

$$\mathrm{Var}(\theta_1(k)) = \frac{1}{k}\frac{1}{\alpha_1}\left(1 + \frac{\sigma_n^2}{n\sigma_1^2}\right)\frac{\sigma_n^2}{\sigma_1^2} + o\left(\frac{1}{k}\right)$$

the variances are similar provided $\mu n\sigma_1^2$ is replaced by $\frac{2}{k}$. This suggests that the stepsize μ of the adaptive algorithm must be normalized by $n\sigma_1^2$.

4.8 ILLUSTRATIVE EXAMPLES

Fast estimation and tracking of the principal (or minor) subspace or components of a sequence of random vectors is a major tool for parameter and signal estimation in many signal processing communications and RADAR applications (see e.g. [11] and the references therein). We can cite, for example, the DOA tracking and the blind channel estimation including CDMA and OFDM communications as illustrations.

Going back to the common observation model (4.15) introduced in Subsection 4.3.1

$$\mathbf{x}(k) = \mathbf{A}(k)\mathbf{r}(k) + \mathbf{n}(k) \tag{4.87}$$

where $\mathbf{A}(k)$ is an $n \times r$ full column rank matrix with $r < n$, the different applications are issued from specific deterministic parametrizations $\mathbf{A}[\boldsymbol{\phi}(k)]$ of $\mathbf{A}(k)$ where $\boldsymbol{\phi}(k) \in \mathcal{R}^q$ is a slowly time-varying parameter compared to $\mathbf{r}(k)$. When this parametrization $\boldsymbol{\phi}(k) \mapsto \mathbf{A}[\boldsymbol{\phi}(k)]$ is nonlinear, $\boldsymbol{\phi}(k)$ is assumed identifiable from the signal subspace span$[\mathbf{A}(k)]$ or the noise subspace null$[\mathbf{A}^T(k)]$ which is its orthogonal complement,

[12]Naturally in this application, the data are complex-valued, but using the conjugate transpose operator instead of transpose, and a complex parametrization based on the orthonormal basis $(\mathbf{H}_{i,j})_{1 \le i,j \le n}$ where $\mathbf{H}_{i,j} = \mathbf{u}_i\mathbf{u}_i^H$ for $i = j$, $\frac{\mathbf{u}_i\mathbf{u}_j^H + \mathbf{u}_j\mathbf{u}_i^H}{\sqrt{2}}$ for $i < j$ and $\frac{\mathbf{u}_i\mathbf{u}_j^H - \mathbf{u}_j\mathbf{u}_i^H}{i\sqrt{2}}$ for $i > j$ instead of the orthonormal basis $(\mathbf{S}_{i,j})_{1 \le i \le j \le n}$, expressions (4.83) and (4.86) are still valid.

that is,

$$\text{span}[\mathbf{A}(\boldsymbol{\phi}(k))] = \text{span}[\mathbf{A}(\boldsymbol{\phi}'(k))] \Rightarrow \boldsymbol{\phi}'(k) = \boldsymbol{\phi}(k)$$

and when this parametrization $\boldsymbol{\phi}(k) \longmapsto \mathbf{A}[\boldsymbol{\phi}(k)]$ is linear, this identifiability is of course up to a multiplicative constant only.

4.8.1 Direction of Arrival Tracking

In the standard narrow-band array data model, $\mathbf{A}(\boldsymbol{\phi})$ is partitioned into r column vectors as $\mathbf{A}(\boldsymbol{\phi}) \stackrel{\text{def}}{=} [\mathbf{a}(\boldsymbol{\phi}_1), \ldots, \mathbf{a}(\boldsymbol{\phi}_r)]$, where $(\boldsymbol{\phi}_i)_{i=1,\ldots,r}$ denotes different parameters associated with the r sources (azimuth, elevation, polarization, ...). In this case, the parametrization is nonlinear. The simplest case corresponds to one parameter per source $(q = r)$ (e.g. for a uniform linear array $\mathbf{a}(\phi_i) = \left(1, e^{i2\pi\frac{d}{\lambda}\sin\phi_i}, \ldots, e^{i2\pi\frac{d(n-1)}{\lambda}\sin\phi_i}\right)^T$. For convenience and without loss of generality, we consider this case in the following. A simplistic idea to track the r DOAs would be to use an adaptive estimate $\widehat{\mathbf{\Pi}}_n(k)$ of the noise orthogonal projection matrix $\mathbf{\Pi}_n(k)$ given by $\mathbf{W}(k)\mathbf{W}^T(k)$ or $\mathbf{I}_n - \mathbf{W}(k)\mathbf{W}^T(k)$ where $\mathbf{W}(k)$ is, respectively, given by a minor or a dominant subspace adaptive algorithm introduced in Section 4.5,[13] and then to derive the estimated DOAs as the r minima of the cost function

$$\mathbf{a}^H(\phi)\,\widehat{\mathbf{\Pi}}_n\,(k)\mathbf{a}(\phi)$$

by a Newton–Raphson procedure

$$\phi_i(k+1) = \phi_i(k) - \frac{\Re[\,\mathbf{a}_i'^H[\phi(k)]\,\widehat{\mathbf{\Pi}}_n\,(k+1)\mathbf{a}_i^H(\phi(k=))]}{\Re[\,\mathbf{a}''^H(\phi_i(k))\,\widehat{\mathbf{\Pi}}_n\,(k+1)\mathbf{a}^H(\phi_i(k)) + \mathbf{a}'^H\,(\phi_i(k))\,\widehat{\mathbf{\Pi}}_n\,(k+1)\,\mathbf{a}'^H\,(\phi_i(k))]} \qquad i = 1, \ldots, r$$

where $\mathbf{a}_i' \stackrel{\text{def}}{=} \frac{d\mathbf{a}_i}{d\phi}$ and $\mathbf{a}_i'' \stackrel{\text{def}}{=} \frac{d^2\mathbf{a}_i}{d\phi^2}$. While this approach works for distant different DOAs, it breaks down when the DOAs of two or more sources are very close and particularly in scenarios involving targets with crossing trajectories. So the difficulty in DOA tracking is the association of the DOA estimated at different time points with the correct sources. To solve this difficulty, various algorithms for DOA tracking have been proposed in the literature (see e.g. [60] and the references therein). To maintain this correct association, a solution is to introduce the dynamic model governing the

[13]Of course, adapted to complex-valued data.

motion of the different sources

$$\boldsymbol{\phi}_i(k+1) \stackrel{\text{def}}{=} \begin{bmatrix} \phi_i(k+1) \\ \phi_i(k+1) \\ \phi_i(k+1) \end{bmatrix} = \begin{bmatrix} 1 & T & T^2/2 \\ 0 & 1 & T \\ 0 & 0 & 1 \end{bmatrix} \begin{bmatrix} \phi_i(k) \\ \phi_i'(k) \\ \phi_i''(k) \end{bmatrix} + \begin{bmatrix} n_{1,i}(k) \\ n_{2,i}(k) \\ n_{3,i}(k) \end{bmatrix}$$

where T denotes the sampling interval and $[n_{j,i}(k)]_{j=1,2,3}$ are random process noise terms that account for random perturbations about the constant acceleration trajectory. This enables us to predict the state (position, velocity, and acceleration) of each source in any interval of time using the estimated state in the previous interval. An efficient and computationally simple heuristic procedure has been proposed in [66]. It consists of four steps by iteration k. First, a prediction $\hat{\boldsymbol{\phi}}_i(k+1/k)$ of the state from the estimate $\hat{\boldsymbol{\phi}}_i(k/k)$ is obtained. Second, an update of the estimated noise projection matrix $\widehat{\mathbf{\Pi}}_n(k)$ given by a subspace tracking algorithm introduced in Section 4.5 is derived from the new snapshot $\mathbf{x}(k)$. Third, for each source i, an estimate $\hat{\boldsymbol{\phi}}_i(k+1)$ given by a Newton–Raphson step initialized by the predicted DOA $\hat{\boldsymbol{\phi}}_i(k+1/k)$ given by a Kalman filter of the first step whose measurement equation is given by

$$\hat{\boldsymbol{\phi}}_i(k) = [1, 0, 0] \begin{bmatrix} \phi_i(k) \\ \phi_i'(k) \\ \phi_i''(k) \end{bmatrix} + n_{4,i}(k)$$

where the observation $\hat{\boldsymbol{\phi}}_i(k)$ is the DOA estimated by the Newton–Raphson step at iteration $k-1$. Finally, the DOA $\hat{\boldsymbol{\phi}}_i(k+1/k)$ predicted by the Kalman filter is also used to smooth the DOA $\hat{\boldsymbol{\phi}}_i(k+1)$ estimated by the Newton–Raphson step, to give the new estimate $\hat{\boldsymbol{\phi}}_i(k+1/k+1)$ of the state whose its first component is used for tracking the r DOAs.

4.8.2 Blind Channel Estimation and Equalization

In communication applications, the matched filtering followed by symbol rate sampling or oversampling yields an n-vector data $\mathbf{x}(k)$ which satisfies the model (4.87), where $\mathbf{r}(k)$ contains different transmitted symbols b_k. Depending on the context, (SIMO channel or CDMA, OFDM, MC CDMA) with or without intersymbol interference, different parametrizations of $\mathbf{A}(k)$ arise which are generally linear in the unknown parameter $\boldsymbol{\phi}(k)$. The latter represents different coefficients of the impulse response of the channel that are assumed slowly time-varying compared to the symbol rate. In these applications, two problems arise. First, the updating of the estimated parameters $\boldsymbol{\phi}(k)$, that is, the adaptive identification of the channel can be useful to an optimal equalization based on an identified channel. Second, for particular models (4.87), a direct linear equalization $\mathbf{m}^T(k)\mathbf{x}(k)$ can be used from the adaptive estimation of the weight $\mathbf{m}(k)$. To illustrate subspace or component-based methods, two simple examples are given in the following.

For the two channel SIMO model, we assume that the two channels are of order m and that we stack the $m+1$ most recent samples of each channel to form the observed data $\mathbf{x}(k) = [\mathbf{x}_1(k), \mathbf{x}_2(k)]^T$. In this case we obtain the model (4.87) where $\mathbf{A}(k)$ is the following $2(m+1) \times (2m+1)$ Sylvester filtering matrix

$$\mathbf{A}(k) = \begin{pmatrix} \boldsymbol{\phi}_0(k) & \cdots & \cdots & \boldsymbol{\phi}_m(k) & & = \\ & \ddots & & & \ddots & \\ & & \boldsymbol{\phi}_0(k) & \cdots & \cdots & \boldsymbol{\phi}_m(k) \end{pmatrix}$$

and $\mathbf{r}(k) = (b_k, \ldots, b_{k-2m})^T$, with $\boldsymbol{\phi}_i(k) = [h_{i,1}(k), h_{i,2}(k)]^T$, $i = 0, \ldots m$ where $h_{i,j}$ represents the ith term of the impulse response of the j-th channel. These two channels do not share common zeros, guaranteeing their identifiability. In this specific two-channel case, the so called least square [80] and subspace [49] estimates of the impulse response $\boldsymbol{\phi}(k) = [\phi_0^T(k), \ldots, \phi_m^T(k)]^T$ defined up to a constant scale factor, coincide [81] and are given by $\boldsymbol{\phi}(k) = \mathbf{T}\mathbf{v}(k)$ with $\mathbf{v}(k)$ is the eigenvector associated with the unique smallest eigenvalue of $\mathbf{C}_x(k) = \mathrm{E}[\mathbf{x}(k)\mathbf{x}^T(k)]$ where $\mathbf{T}$ is the antisymmetric orthogonal matrix $\mathbf{I}_{m+1} \otimes \begin{bmatrix} 0 & 1 \\ -1 & 0 \end{bmatrix}$. Consequently an adaptive estimation of the slowly time-varying impulse response $\boldsymbol{\phi}(k)$ can be derived from the adaptive estimation of the eigenvector $\mathbf{v}(k)$. Note that in this example, the rank r of the signal subspace is given by $r = 2m+1$ whose order m of the channels that usually possess small leading and trailing terms is ill defined. For such channels it has been shown [46] that blind channel approximation algorithms should attempt to model only the significant part of the channel composed of the large impulse response terms because efforts toward modeling small leading and/or trailing terms lead to effective overmodeling, which is generically ill-conditioned and, thus, should be avoided. A detection procedure to detect the order of this significant part has been given in [45].

Consider now an asynchronous direct sequence CDMA system of r users without intersymbol interference. In this case, model (4.87) applies, where $\mathbf{A}(k)$ is given by

$$\mathbf{A}(k) = [a_1(k)\mathbf{s}_1, \ldots, a_r(k)\mathbf{s}_r]$$

where $a_i(k)$ and $\mathbf{s}_i$ are respectively the amplitude and the signature sequence of the i-th user and $\mathbf{r}(k) = (b_{k,1}, \ldots, b_{k,r})^T$ where $b_{k,i}$ is the symbol k of the i-th user. We assume that only the signature sequence of User 1, the user of interest, is known. Two linear multiuser detectors $\mathbf{m}^T(k)\mathbf{x}(k)$, namely, the decorrelation detector (i.e., that completely eliminates the multiple access interference caused by the other users) and the linear MMSE detector for estimating the symbol $b_{k,1}$, have been proposed in [76] in terms of the signal eigenvalues and eigenvectors. The scaled version of the respective weights $\mathbf{m}(k)$ of these detectors are given by

$$\mathbf{m}(k) = \mathbf{U}_s(k)[\boldsymbol{\Delta}(k) - \sigma_n^2(k)\mathbf{I}_r]^{-1}\mathbf{U}_s^T(k)\mathbf{s}_1$$

$$\mathbf{m}(k) = \mathbf{U}_s(k)\boldsymbol{\Delta}^{-1}(k)\mathbf{U}_s^T(k)\mathbf{s}_1$$

where $\mathbf{U}_s(k) = [\mathbf{v}_1(k), \ldots, \mathbf{v}_r(k)]$, $\mathbf{\Delta}(k) = \text{Diag}[\lambda_1(k), \ldots, \lambda_r(k)]$ and $\sigma_n^2(k) = \lambda_{r+1}(k)$ issued from the adaptive EVD of $\mathbf{C}_x(k) = \text{E}[\mathbf{x}(k)\mathbf{x}^T(k)]$ including the detection of the number r of user that can change by a rank tracking procedure (e.g. [72]).

4.9 CONCLUDING REMARKS

Although adaptive subspace and component-based algorithms were introduced in signal processing three decades ago, a rigorous convergence analysis has been derived only for the celebrated Oja's algorithm, whose Oja's neuron is a particular case, in stationary environments. In general all of these techniques are derived heuristically from standard iterative computational techniques issued from numerical methods of linear algebra. So a theoretical convergence and performance analysis of these algorithms is necessary, but seems very challenging. Furthermore, such analysis is not sufficient because these algorithms may present numerical instabilities due to rounding errors. Consequently, a comprehensive comparison of the different algorithms that have appeared in the literature from the performance (convergence speed, mean square error, distance to the orthonormality, tracking capabilities), computational complexity and numerical stability points of view—that are out the scope of this chapter—would be be very useful for practitioners.

The interest of the signal processing community in adaptive subspace and component-based schemes remains strong as is evident from the numerous articles and reports published in this area each year. But we note that these contributions mainly consist in the application of standard adaptive subspace and component-based algorithms in new applications and in refinements of well known subspace/component-based algorithms, principally to reduce their computational complexity and to numerically stabilize the minor subspace/component-based algorithms, whose literature is much more limited than the principal subspace and component-based algorithms.

4.10 PROBLEMS

4.1 Let λ_0 be a simple eigenvalue of a real symmetric $n \times n$ matrix $\mathbf{C}_0$, and let $\mathbf{u}_0$ be a unit 2-norm associated eigenvector, so that $\mathbf{C}\mathbf{u}_0 = \lambda_0\mathbf{u}_0$. Then a real-valued function $\lambda(.)$ and a vector function $\mathbf{u}(.)$ are defined for all $\mathbf{C}$ in some neighborhood (e.g. among the real symmetric matrices) of $\mathbf{C}_0$ such that

$$\lambda(\mathbf{C}_0) = \lambda_0, \ \mathbf{u}(\mathbf{C}_0) = \mathbf{u}_0, \ \text{and} \ \mathbf{C}\mathbf{u} = \lambda\mathbf{u} \quad \text{under the constraint } \|\mathbf{u}\|_2 = 1.$$

Using simple perturbations algebra manipulations, prove that the functions $\lambda(.)$ and $\mathbf{u}(.)$ are differentiable on some neighborhood of $\mathbf{C}_0$ and that the differentials at $\mathbf{C}_0$ are given by

$$\delta\lambda = \mathbf{u}_0^H(\delta\mathbf{C})\mathbf{u}_0 \quad \text{and} \quad \delta\mathbf{u} = -(\mathbf{C} - \lambda_0\mathbf{I}_n)^{\#}(\delta\mathbf{C})\mathbf{u}_0 \tag{4.88}$$

where # stands for the Moore Penrose inverse. Prove that if the constraint $\|\mathbf{u}\|_2 = 1$ is replaced by $\mathbf{u}_0^T\mathbf{u} = 1$, the differential $\delta\mathbf{u}$ given by (4.88) remains valid.

Now consider the same problem where $\mathbf{C}_0$ is a Hermitian matrix. To fix the perturbed eigenvector $\mathbf{u}$, the condition $\|\mathbf{u}\|^2 = 1$ is not sufficient. So suppose now that $\mathbf{u}_0^H\mathbf{u} = 1$. Note that in this case $\mathbf{u}$ no longer has unit 2-norm. Using the same approach as for the real symmetric case, prove that the functions $\lambda(.)$ and $\mathbf{u}(.)$ are differentiable on some neighborhood of $\mathbf{C}_0$ and that the differentials at $\mathbf{C}_0$ are now given by

$$\delta\lambda = \mathbf{u}_0^H(\delta\mathbf{C})\mathbf{u}_0 \quad \text{and} \quad \delta\mathbf{u} = -(\mathbf{C} - \lambda_0\mathbf{I}_n)^{\#}(\mathbf{I}_n - \mathbf{u}_0\mathbf{u}_0^H)(\delta\mathbf{C})\mathbf{u}_0. \tag{4.89}$$

In practice, different constraints are used to fix $\mathbf{u}$. For example, the SVD function of MATLAB forces all eigenvectors to be unit 2-norm with a real first element. Specify in this case the new expression of the differential $\delta\mathbf{u}$ given by (4.89). Finally, show that the differential $\delta\mathbf{u}$ given by (4.88) would be obtained with the condition $\mathbf{u}_0^H\delta\mathbf{u} = 0$, which is no longer derived from the constraint $\|\mathbf{u}\|_2 = 1$.

4.2 Consider an $n \times n$ real symmetric or complex Hermitian matrix $\mathbf{C}_0$ whose the r smallest eigenvalues are equal to σ^2 with $\lambda_{n-r} > \lambda_{n-r+1}$. Let $\mathbf{\Pi}_0$ the projection matrix onto the invariant subspace associated with σ^2. Then a matrix-valued function $\mathbf{\Pi}(.)$ is defined as the projection matrix onto the invariant subspace associated with the r smallest eigenvalues of $\mathbf{C}$ for all $\mathbf{C}$ in some neighborhood of $\mathbf{C}_0$ such that $\mathbf{\Pi}(\mathbf{C}_0) = \mathbf{\Pi}_0$. Using simple perturbations algebra manipulations, prove that the functions $\mathbf{\Pi}(.)$ is two times differentiable on some neighborhood of $\mathbf{C}_0$ and that the differentials at $\mathbf{C}_0$ are given by

$$\begin{aligned}\delta\mathbf{\Pi} = &-(\mathbf{\Pi}_0(\delta\mathbf{C})\mathbf{S}_0^{\#} + \mathbf{S}_0^{\#}(\delta\mathbf{C})\mathbf{\Pi}_0)\\ &+ \mathbf{S}_0^{\#}(\delta\mathbf{C})\mathbf{\Pi}_0(\delta\mathbf{C})\mathbf{S}_0^{\#} - \mathbf{\Pi}_0(\delta\mathbf{C})\mathbf{S}_0^{\#2}(\delta\mathbf{C})\mathbf{\Pi}_0 + \mathbf{S}_0^{\#}(\delta\mathbf{C})\mathbf{S}_0^{\#}(\delta\mathbf{C})\mathbf{\Pi}_0\\ &+ \mathbf{\Pi}_0(\delta\mathbf{C})\mathbf{S}_0^{\#}(\delta\mathbf{C})\mathbf{S}_0^{\#} - \mathbf{S}_0^{\#2}(\delta\mathbf{C})\mathbf{\Pi}_0(\delta\mathbf{C})\mathbf{\Pi}_0 - \mathbf{\Pi}_0(\delta\mathbf{C})\mathbf{\Pi}_0(\delta\mathbf{C})\mathbf{S}_0^{\#2}\end{aligned}$$

where $\mathbf{S}_0 \stackrel{\text{def}}{=} \mathbf{C}_0 - \sigma^2\mathbf{I}_n$.

4.3 Consider a Hermitian matrix $\mathbf{C}$ whose real and imaginary parts are denoted by $\mathbf{C}_r$ and $\mathbf{C}_i$ respectively. Prove that each eigenvalue eigenvector pair $(\lambda, \mathbf{u})$ of $\mathbf{C}$ is associated with the eigenvalue eigenvector pairs $\left[\lambda, \begin{pmatrix}\mathbf{u}_r\\ \mathbf{u}_i\end{pmatrix}\right]$ and $\left[\lambda, \begin{pmatrix}-\mathbf{u}_i\\ \mathbf{u}_r\end{pmatrix}\right]$ of the real symmetric matrix $\begin{bmatrix}\mathbf{C}_r & -\mathbf{C}_i\\ \mathbf{C}_i & \mathbf{C}_r\end{bmatrix}$ where $\mathbf{u}_r$ and $\mathbf{u}_i$ denote the real and imaginary parts of $\mathbf{u}$.

4.4 Consider what happens when the orthogonal iteration method (4.11) is applied with $r = n$ and under the assumption that all the eigenvalues of $\mathbf{C}$ are simple. The QR algorithm arises by considering how to compute the matrix

$\mathbf{T}_i \stackrel{\text{def}}{=} \mathbf{W}_i^T \mathbf{C} \mathbf{W}_i$ directly from this predecessor $\mathbf{T}_{i-1}$. Prove that the following iterations

$$\begin{aligned} \mathbf{T}_0 &= \mathbf{Q}_0^T \mathbf{C} \mathbf{Q}_0 \quad \text{where } \mathbf{Q}_0 \text{ is an arbitrary orthonormal matrix} \\ \text{for } i = 1, 2, \ldots \ \mathbf{T}_{i-1} &= \mathbf{Q}_i \mathbf{R}_i \quad \text{QR} \quad \text{factorization} \\ \mathbf{T}_i &= \mathbf{R}_i \mathbf{Q}_i \end{aligned}$$

produce a sequence $(\mathbf{T}_i, \mathbf{Q}_0\mathbf{Q}_i...\mathbf{Q}_i)$ that converges to $(\text{Diag}(\lambda_1, \ldots, \lambda_n), [\pm\mathbf{u}_1, \ldots, \pm\mathbf{u}_n])$.

4.5 Specify what happens to the convergence and the convergence speed, if the step $\mathbf{W}_i = \text{orthonorm}\{\mathbf{C}\mathbf{W}_{i-1}\}$ of the orthogonal iteration algorithm (4.11) is replaced by the following $\{\mathbf{W}_i = \text{orthonorm}\{(\mathbf{I}_n + \mu\mathbf{C})\mathbf{W}_{i-1}\}$. Same questions, for the step $\{\mathbf{W}_i = \text{orthonormalization of } \mathbf{C}^{-1}\ \mathbf{W}_{i-1}\}$, then $\{\mathbf{W}_i = \text{orthonormalization of } (\mathbf{I}_n - \mu\mathbf{C})\mathbf{W}_{i-1}\}$. Specify the conditions that must satisfy the eigenvalues of $\mathbf{C}$ and μ for these latter two steps. Examine the specific case $r = 1$.

4.6 Using the EVD of $\mathbf{C}$, prove that the solutions $\mathbf{W}$ of the maximizations and minimizations (4.7) are given by $\mathbf{W} = [\mathbf{u}_1, \ldots, \mathbf{u}_r]\,\mathbf{Q}$ and $\mathbf{W} = [\mathbf{u}_{n-r+1}, \ldots, \mathbf{u}_n]\,\mathbf{Q}$ respectively, where $\mathbf{Q}$ is an arbitrary $r \times r$ orthogonal matrix.

4.7 Consider the scalar function (4.14) $J(\mathbf{W}) \stackrel{\text{def}}{=} \text{E}(\|\mathbf{x} - \mathbf{W}\mathbf{W}^T\mathbf{x}\|^2)$ of $\mathbf{W} = [\mathbf{w}_1, \ldots, \mathbf{w}_r]$ with $\mathbf{C} \stackrel{\text{def}}{=} \text{E}(\mathbf{x}\mathbf{x}^T)$. Let $\nabla_{\mathbf{W}} = [\nabla_1, \ldots, \nabla_r]$ where $(\nabla_k)_{k=1,\ldots,r}$ is the gradient operator with respect to $\mathbf{w}_k$. Prove that

$$\nabla_{\mathbf{W}} J = 2(-2\mathbf{C} + \mathbf{C}\mathbf{W}\mathbf{W}^T + \mathbf{W}\mathbf{W}^T\mathbf{C})\mathbf{W}. \tag{4.90}$$

Then, prove that the stationary points of $J(\mathbf{W})$ are given by $\mathbf{W} = \mathbf{U}_r\mathbf{Q}$ where the r columns of $\mathbf{U}_r$ denote arbitrary r distinct unit-2 norm eigenvectors among $\mathbf{u}_1, \ldots, \mathbf{u}_n$ of $\mathbf{C}$ and where $\mathbf{Q}$ is an arbitrary $r \times r$ orthogonal matrix. Finally, prove that at each stationary point, $J(\mathbf{W})$ equals the sum of eigenvalues whose eigenvectors are not involved in $\mathbf{U}_r$.

Consider now the complex valued case where $J(\mathbf{W}) \stackrel{\text{def}}{=} \text{E}(\|\mathbf{x} - \mathbf{W}\mathbf{W}^T\mathbf{x}\|^2)$ with $\mathbf{C} \stackrel{\text{def}}{=} \text{E}(\mathbf{x}\mathbf{x}^H)$ and use the complex gradient operator (see e.g., [36]) defined by $\nabla_{\mathbf{W}} = \frac{1}{2}[\nabla_R + i\nabla_I]$ where ∇_R and ∇_I denote the gradient operators with respect to the real and imaginary parts. Show that $\nabla_{\mathbf{W}} J$ has the same form as the real gradient (4.90) except for a factor $1/2$ and changing the transpose operator by the conjugate transpose one. By noticing that $\nabla_{\mathbf{w}} J = \mathbf{O}$ is equivalent to $\nabla_R\, J = \nabla_I\, J = \mathbf{O}$, extend the previous results to the complex valued case.

4.8 With the notations of Exercise 4.7, suppose now that $\lambda_r > \lambda_{r+1}$ and consider first the real valued case. Show that the (i, j)th block $\nabla_i \nabla_j^T J$ of the block Hessian matrix $\mathbf{H}$ of $J(\mathbf{W})$ with respect to the nr-dimensional vector $[\mathbf{w}_1^T, \ldots, \mathbf{w}_r^T]^T$ is given by

$$\frac{1}{2}\nabla_i \nabla_j^T J = \delta_{ij}(-2\mathbf{C} + \mathbf{C}\mathbf{W}\mathbf{W}^T + \mathbf{W}\mathbf{W}^T\mathbf{C})$$
$$+ (\mathbf{w}_j^T\mathbf{C}\mathbf{w}_i)\mathbf{I}_n + (\mathbf{w}_j^T\mathbf{w}_i)\mathbf{C} + \mathbf{C}\mathbf{w}_j\mathbf{w}_i^T + \mathbf{w}_j\mathbf{w}_i^T\mathbf{C}.$$

After evaluating the EVD of the block Hessian matrix $\mathbf{H}$ at the stationary points $\mathbf{W} = \mathbf{U}_r\mathbf{Q}$, prove that $\mathbf{H}$ is nonnegative if $\mathbf{U}_r = [\mathbf{u}_1, \ldots, \mathbf{u}_r]$. Interpret in this case the zero eigenvalues of $\mathbf{H}$. Prove that when $\mathbf{U}_r$ contains an eigenvector different from $\mathbf{u}_1, \ldots, \mathbf{u}_r$, some eigenvalues of $\mathbf{H}$ are strictly negative. Deduce that all stationary points of $J(\mathbf{W})$ are saddle points except the points $\mathbf{W}$ whose associated matrix $\mathbf{U}_r$ contains the r dominant eigenvectors $\mathbf{u}_1, \ldots, \mathbf{u}_r$ of $\mathbf{C}$ which are global minima of the cost function (4.14).
Extend the previous results by considering the $2nr \times 2nr$ real Hessian matrix $\mathbf{H} = \nabla\nabla\, J$ with $\nabla \stackrel{\text{def}}{=} [\nabla_{R,1}^T, \ldots, \nabla_{R,r}^T, \nabla_{I,1}^T, \ldots, \nabla_{I,r}^T]^T$.

4.9 With the notations of the NP3 algorithm described in Subsection 4.5.1, write (4.40) in the form

$$\mathbf{G}(k+1) = \frac{1}{\beta}[\mathbf{G}^{-1/2}(k)(\mathbf{I}_n + \mathbf{a}\mathbf{b}^T + \mathbf{b}\mathbf{a}^T + \alpha\mathbf{a}\mathbf{a}^T)\mathbf{G}^{-T/2}(k)]^{-1/2}$$

with $\mathbf{a} \stackrel{\text{def}}{=} \frac{1}{\beta}\mathbf{G}(k)\mathbf{y}(k)$, $\mathbf{b} \stackrel{\text{def}}{=} \frac{1}{\beta}\mathbf{G}(k)\mathbf{z}(k)$ and $\alpha \stackrel{\text{def}}{=} \|\mathbf{x}(k)\|^2$. Then, using the EVD $\nu_1\mathbf{e}_1\mathbf{e}_1^T + \nu_2\mathbf{e}_2\mathbf{e}_2^T$ of the symmetric rank two matrix $\mathbf{a}\mathbf{b}^T + \mathbf{b}\mathbf{a}^T + \alpha\mathbf{a}\mathbf{a}^T$, prove equalities (4.41) and (4.42) where $\tau_i \stackrel{\text{def}}{=} 1 - 1/\sqrt{\nu_i + 1}$, $i = 1, 2$.

4.10 Consider the following stochastic approximation algorithm derived from Oja's algorithm (4.30) where the sign of the step size can be reversed and where the estimate $\mathbf{W}(k)$ is forced to be orthonormal at each time step

$$\mathbf{W}'(k+1) = \mathbf{W}(k) \pm \mu_k[\mathbf{I}_n - \mathbf{W}(k)\mathbf{W}^T(k)]\mathbf{x}(k)\mathbf{x}^T = (k)\mathbf{W}(k)$$
$$\mathbf{W}(k+1) = \mathbf{W}'(k+1)[\mathbf{W}'^T(k+1)\mathbf{W}'(k+1)]^{-1/2} \qquad (4.91)$$

where $[\mathbf{W}'^T(k+1)\mathbf{W}'(k+1)]^{-1/2}$ denotes the symmetric inverse square root of $\mathbf{W}'^T(k+1)\mathbf{W}'(k+1)$. To compute the later, use the updating equation of $\mathbf{W}'(k+1)$ and keeping in mind that $\mathbf{W}(k)$ is orthonormal, prove that $\mathbf{W}'^T(k+1)\mathbf{W}'(k+1) = \mathbf{I}_r \pm \mathbf{z}\mathbf{z}^T$ with $\mathbf{z} \stackrel{\text{def}}{=} \mu\|\mathbf{x}(k) - \mathbf{W}(k)\mathbf{y}(k)\|\mathbf{y}(k)$ where $\mathbf{y}(k) \stackrel{\text{def}}{=} \mathbf{W}^T(k)\mathbf{x}(k)$. Using identity (4.5.9), prove that $[\mathbf{W}'^T(k+1)\mathbf{W}'(k+1)]^{-1/2} = \mathbf{I}_r \pm \tau_k\mathbf{y}(k)\mathbf{y}^T(k)$ with $\tau_k \stackrel{\text{def}}{=} [1/\|\mathbf{y}(k)\|^2][(1/(1 + \mu^2\|\mathbf{x}(k) - \mathbf{W}(k)\mathbf{y}(k)\|^2\|\mathbf{y}(k)\|^2)^{1/2}) - 1]$. Finally, using the update equation of $\mathbf{W}(k+1)$, prove that algorithm (4.9.5) leads to $\mathbf{W}(k+1) = \mathbf{W}(k) \pm \mu_k \mathbf{p}(k)\mathbf{y}^T(k)$ with $\mathbf{p}(k) \stackrel{\text{def}}{=} \pm\tau_k/\mu_k\mathbf{W}(k)\mathbf{y}(k) + [1 + \tau_k\|\mathbf{y}(k)\|^2][\mathbf{x}(k) - \mathbf{W}(k)\mathbf{y}(k)]$.

Alternatively, prove that algorithm (4.9.5) leads to $\mathbf{W}(k+1) = \mathbf{H}(k)\mathbf{W}(k)$ where $\mathbf{H}(k)$ is the Householder transform given by $\mathbf{H}(k) = \mathbf{I}_n - 2\mathbf{u}(k)\mathbf{u}^T(k)$ where $\mathbf{u}(k) \stackrel{\text{def}}{=} \mathbf{p}(k)/\|\mathbf{p}(k)\|$.

4.11 Consider the scalar function (4.5.21) $J(\mathbf{W}) \stackrel{\text{def}}{=} \text{Tr}[\ln(\mathbf{W}^T\mathbf{C}\mathbf{W})] - \text{Tr}(\mathbf{W}^T\mathbf{W})$. Using the notations of Exercise 4.7, prove that

$$\nabla_{\mathbf{W}} J = 2[\mathbf{C}\mathbf{W}(\mathbf{W}^T\mathbf{C}\mathbf{W})^{-1} - \mathbf{W}]. \tag{4.92}$$

Then, prove that the stationary points of $J(\mathbf{W})$ are given by $\mathbf{W} = \mathbf{U}_r\,\mathbf{Q}$ where the r columns of $\mathbf{U}_r$ denotes arbitrary r distinct unit-2 norm eigenvectors among $\mathbf{u}_1, \ldots, \mathbf{u}_n$ of $\mathbf{C}$ and where $\mathbf{Q}$ is an arbitrary $r \times r$ orthogonal matrix. Finally, prove that at each stationary point, $J(\mathbf{W}) = \sum_{i=1}^{r} \ln(\lambda_{s_i}) - r$, where the r eigenvalues λ_{s_i} are associated with the eigenvectors involved in $\mathbf{U}_r$.

4.12 With the notations of Exercise 4.11 and using the matrix differential method [47, Chap. 6], prove that the Hessian matrix $\mathbf{H}$ of $J(\mathbf{W})$ with respect to the nr-dimensional vector $[\mathbf{w}_1^T, \ldots, \mathbf{w}_r^T]^T$ is given by

$$\begin{aligned}\frac{1}{2}\mathbf{H} = &-\mathbf{I}_{nr} - (\mathbf{W}^T\mathbf{C}\mathbf{W})^{-1} \otimes [\mathbf{C}\mathbf{W}(\mathbf{W}^T\mathbf{C}\mathbf{W})^{-1}\mathbf{W}^T\mathbf{C}] \\ &- \mathbf{K}_{rn}[\mathbf{C}\mathbf{W}(\mathbf{W}^T\mathbf{C}\mathbf{W})^{-1}] \otimes [(\mathbf{W}^T\mathbf{C}\mathbf{W})^{-1}\mathbf{W}^T\mathbf{C}] + (\mathbf{W}^T\mathbf{C}\mathbf{W})^{-1} \otimes \mathbf{C}\end{aligned}$$

where $\mathbf{K}_{rn}$ is the $nr \times rn$ commutation matrix [47, Chap. 2]. After evaluating this Hessian matrix $\mathbf{H}$ at the stationnary points $\mathbf{W} = \mathbf{U}_r\mathbf{Q}$ of $J(\mathbf{W})$ (4.46), substituting the EVD of $\mathbf{C}$ and deriving the EVD of $\mathbf{H}$, prove that when $\lambda_r > \lambda_{r+1}$, $\mathbf{H}$ is nonnegative if $\mathbf{U}_r = [\mathbf{u}_1, \ldots, \mathbf{u}_r]$. Interpret in this case the zero eigenvalues of $\mathbf{H}$. Prove that when $\mathbf{U}_r$ contains an eigenvector different from $\mathbf{u}_1, \ldots, \mathbf{u}_r$, some eigenvalues of $\mathbf{H}$ are strictly positive. Deduce that all stationary points of $J(\mathbf{W})$ are saddle points except the points $\mathbf{W}$ whose associated matrix $\mathbf{U}_r$ contains the r dominant eigenvectors $\mathbf{u}_1, \ldots, \mathbf{u}_r$ of $\mathbf{C}$ which are global maxima of the cost function (4.46).

4.13 Suppose the columns $[\mathbf{w}_1(k), \ldots, \mathbf{w}_r(k)]$ of the $n \times r$ matrix $\mathbf{W}(k)$ are orthonormal and let $\mathbf{W}'(k+1)$ be the matrix $\mathbf{W}(k) \pm \mu_k\mathbf{x}(k)\mathbf{x}^T(k)\mathbf{W}(k)$. If the matrix $\mathbf{S}(k+1)$ performs a Gram-Schmidt orthonormalization on the columns of $\mathbf{W}'(k+1)$, write this in explicit form for the columns of matrix $\mathbf{W}(k+1) = \mathbf{W}'(k+1)\mathbf{S}(k+1)$ as a power series expansion in μ_k and prove that

$$\begin{aligned}\mathbf{w}_i(k+1) = \mathbf{w}_i(k) + \mu_k\left[\mathbf{I}_n - \mathbf{w}_i(k)\mathbf{w}_i^T(k) - 2\sum_{j=1}^{i-1}\mathbf{w}_j(k)\mathbf{w}_j^T(k)\right] \\ \mathbf{x}(k)\mathbf{x}^T(k)\mathbf{w}_i(k) + O(\mu_k^2) \quad \text{for } i = 1, \ldots, r.\end{aligned}$$

Following the same approach with now $\mathbf{W}'(k+1) = \mathbf{W}(k) \pm \mathbf{x}(k)\mathbf{x}^T(k)\mathbf{W}(k)\mathbf{\Gamma}(k)$ where $\mathbf{\Gamma}(k) = \mu_k \mathrm{Diag}(1, \alpha_2, \ldots, \alpha_r)$, prove that

$$\mathbf{w}_i(k+1) = \mathbf{w}_i(k) + \alpha_i \mu_k \left[\mathbf{I}_n - \mathbf{w}_i(k)\mathbf{w}_i^T(k) - \sum_{j=1}^{i-1} \left(1 + \frac{\alpha_j}{\alpha_i}\right) \mathbf{w}_j(k)\mathbf{w}_j^T(k) \right]$$

$$\mathbf{x}(k)\mathbf{x}^T(k)\mathbf{w}_i(k) + O(\mu_k^2) \quad \text{for } i = 1, \ldots, r.$$

4.14 Specify the stationary points of the ODE associated with algorithm (4.58). Using the eigenvalues of the derivative of the mean field of this algorithm, prove that if $\lambda_{n-r+1} < 1$ and $\beta > \frac{\lambda_{n-r+1}}{\lambda_n} - 1$, the only asymptotically stable points of the associated ODE are the eigenvectors $\pm \mathbf{v}_{n-r+1}, \ldots, \pm \mathbf{v}_n$.

4.15 Prove that the set of the $n \times r$ orthogonal matrices $\mathbf{W}$ (denoted the Stiefel manifold $\mathbf{St}_{n,r}$) is given by the set of matrices of the form $e^{\mathbf{A}}\mathbf{W}$ where $\mathbf{W}$ is an arbitrary $n \times r$ fixed orthogonal matrix and $\mathbf{A}$ is a skew-symmetric matrix ($\mathbf{A}^T = -\mathbf{A}$).
Prove the following relation

$$J(\mathbf{W} + \delta\mathbf{W}) = J(\mathbf{W}) + \mathrm{Tr}[\delta\mathbf{A}^T(\mathbf{H}_2\mathbf{W}\mathbf{H}_1\mathbf{W}^T - \mathbf{W}\mathbf{H}_1\mathbf{W}^T\mathbf{H}_2)] + o(\delta\mathbf{W})$$

where $J(\mathbf{W}) = \mathrm{Tr}[\mathbf{W}\mathbf{H}_1\mathbf{W}^T\mathbf{H}_2]$ (where $\mathbf{H}_1$ and $\mathbf{H}_2$ are arbitrary $r \times r$ and $n \times n$ symmetric matrices) defined on the set of $n \times r$ orthogonal matrices. Then, give the differential dJ of the cost function $J(\mathbf{W})$ and deduce the gradient of $J(\mathbf{W})$ on this set of $n \times r$ orthogonal matrices

$$\nabla_{\mathbf{W}} J = [\mathbf{H}_2\mathbf{W}\mathbf{H}_1\mathbf{W}^T - \mathbf{W}\mathbf{H}_1\mathbf{W}^T\mathbf{H}_2]\mathbf{W}. \tag{4.93}$$

4.16 Prove that if $\bar{f}(\boldsymbol{\theta}) = -\nabla_{\boldsymbol{\theta}} J$, where $J(\boldsymbol{\theta})$ is a positive scalar function, $J[\boldsymbol{\theta}(t)]$ tends to a constant as t tends to ∞, and consequently all the trajectories of the ODE (4.69) converge to the set of the stationary points of the ODE.

4.17 Let $\boldsymbol{\theta}_*$ be a stationary point of the ODE (4.69). Consider a Taylor series expansion of $\bar{f}(\boldsymbol{\theta})$ about the point $\boldsymbol{\theta} = \boldsymbol{\theta}_*$

$$\bar{f}(\boldsymbol{\theta}) = \bar{f}(\boldsymbol{\theta}_*) + \frac{d\bar{f}(\boldsymbol{\theta})}{d\boldsymbol{\theta}}_{|\boldsymbol{\theta}=\boldsymbol{\theta}_*} (\boldsymbol{\theta} - \boldsymbol{\theta}_*) + O[(\boldsymbol{\theta} - \boldsymbol{\theta}_*)](\boldsymbol{\theta} - \boldsymbol{\theta}_*).$$

By admitting that the behavior of the trajectory $\boldsymbol{\theta}(t)$ of the ODE (4.69) in the neighborhood of $\boldsymbol{\theta}_*$ is identical to those of the associated linearized ODE $\frac{d\boldsymbol{\theta}(t)}{dt} = \mathbf{D}[\boldsymbol{\theta}(t) - \boldsymbol{\theta}_*]$ $\left(\text{with } \mathbf{D} \stackrel{\text{def}}{=} \frac{d\bar{f}(\boldsymbol{\theta})}{d\boldsymbol{\theta}}_{|\boldsymbol{\theta}=\boldsymbol{\theta}_*}\right)$ about the point $\boldsymbol{\theta}_*$, relate the stability of the stationary point $\boldsymbol{\theta}_*$ to the behavior of the eigenvalues of the matrix $\mathbf{D}$.

4.18 Consider the general stochastic approximation algorithm (4.68) in which the field $f[\boldsymbol{\theta}(k), \mathbf{x}(k)\mathbf{x}^T(k)]$ and the residual perturbation term $h[\boldsymbol{\theta}(k), \mathbf{x}(k)\mathbf{x}^T(k)]$ depend on the data $\mathbf{x}(k)$ through $\mathbf{x}(k)\,\mathbf{x}^T(k)$ and are linear in $\mathbf{x}(k)\mathbf{x}^T(k)$. The data $\mathbf{x}(k)$ are independent. The estimated parameter is here denoted $\boldsymbol{\theta}_\mu(k) \stackrel{\text{def}}{=} \boldsymbol{\theta}_* + \delta\boldsymbol{\theta}_k$. We suppose that the Gaussian approximation result (4.7.4) applies and that the convergence of the second-order moments allows us to write $\mathrm{E}[(\boldsymbol{\theta}_\mu(k) - \boldsymbol{\theta}_*)]\,[(\boldsymbol{\theta}_\mu(k) - \boldsymbol{\theta}_*)^T] = \mu\mathbf{C}_{\boldsymbol{\theta}} + o(\mu)$. Taking the expectation of both sides of (4.7.1), provided $\mu_k = \mu$ and $\boldsymbol{\theta}_\mu(k)$ stationary, gives that

$$\mathbf{0} = \mathrm{E}(f(\boldsymbol{\theta}_* + \delta\boldsymbol{\theta}_k, \mathbf{C}_x) = \frac{\partial f}{\partial \boldsymbol{\theta}}_{|\boldsymbol{\theta}=\boldsymbol{\theta}_*} \mathrm{E}(\delta\boldsymbol{\theta}_k) + \frac{\mu}{2}\frac{\partial^2 f}{\partial \boldsymbol{\theta}^2}_{|\boldsymbol{\theta}=\boldsymbol{\theta}_*} \mathrm{vec}(\mathbf{C}_{\boldsymbol{\theta}}) + o(\mu).$$

Deduce a general expression of the asymptotic bias $\lim_{k\to\infty} \mathrm{E}[\boldsymbol{\theta}(k)] - \boldsymbol{\theta}_*$.

REFERENCES

1. K. Abed Meraim, A. Chkeif, and Y. Hua, Fast orthogonal PAST algorithm, *IEEE Signal Process. letters*, 7, No. 3, pp. 60–62, (2000).
2. K. Abed Meraim, S. Attallah, A. Chkeif, and Y. Hua, Orthogonal Oja algorithm, *IEEE Signal Process. letters*, 7, No. 5, pp. 116–119, (2000).
3. T. W. Anderson, *An introduction to multivariate statistical analysis*, Second Edition, Wiley and Sons, 1984.
4. S. Attallah and K. Abed Meraim, Fast algorithms for subspace tracking, *IEEE Signal Process. letters*, 8, No. 7, pp. 203–206, (2001).
5. S. Attallah and K. Abed Meraim, Low cost adaptive algorithm for noise subspace estimation, *Electron. letters*, 48, No. 12, pp. 609–611, (2002).
6. S. Attallah, The generalized Rayleigh's quotient adaptive noise subspace algorithm: a Householder transformation-based implementation, *IEEE Trans. on circ. and syst. II*, 53, No. 81, pp. 3–7, (2006).
7. S. Attallah, J. H. Manton, and K. Abed Meraim, Convergence analysis of the NOJA algorithm using the ODE approach, *Signal Processing*, 86, pp. 3490–3495, (2006).
8. R. Badeau, B. David, and G. Richard, Fast approximated power iteration subspace tracking, *IEEE Trans. on Signal Process.*, 53, No. 8, pp. 2931–2941, (2005).
9. S. Bartelmaos, K. Abed Meraim and S. Attallah, "Fast algorithms for minor component analysis," *Proc. ICASSP*, Philadelphia, March 2005.
10. S. Bartelmaos, "Subspace tracking and mobile localization in UMTS," Unpublished doctoral thesis, *Telecom Paris and University Pierre et Marie Curie*, France, 2008.
11. S. Beau and S. Marcos, Range dependent clutter rejection using range recursive space time adaptive processing (STAP) algorithms, Submitted to *Signal Processing*, March 2008.
12. R. Bellman, *Stability theory of differential equations*, McGraw Hill, New York, 1953.
13. A. Benveniste, M. Métivier, and P. Priouret, *Adaptive algorithms and stochastic approximations*, Springer Verlag, New York, 1990.
14. A. Cantoni and P. Butler, Eigenvalues and eigenvectors of symmetric centrosymmetric matrices, *Linear Algebra and its Applications* 13, pp. 275–288, (1976).

15. T. Chen, Modified Oja's algorithms for principal subspace and minor subspace extraction, *Neural Processing letters*, 5, pp. 105–110, (1997).

16. T. Chen, Y. Hua, and W. Y. Yan, Global convergence of Oja's subspace algorithm for principal component extraction, *IEEE Trans. on Neural Networks*, 9, No. 1, pp. 58–67, (1998).

17. T. Chen and S. Amari, Unified stabilization approach to principal and minor components extraction algorithms, *Neural Networks*, 14, No. 10, pp. 1377–1387, (2001).

18. T. Chonavel, B. Champagne, and C. Riou, Fast adaptive eigenvalue decomposition: a maximum likelihood approach, *Signal Processing*, 83, pp. 307–324, (2003).

19. A. Cichocki and S. Amari, *Adaptive blind signal and image processing, learning algorithms and applications*, John Wiley and Sons, 2002.

20. P. Comon and G. H. Golub, Tracking a few extreme singular values and vectors in signal processing, *Proc. IEEE*, 78, No. 8, pp. 1327–1343, (1990).

21. J. Dehaene, "Continuous-time matrix algorithms, systolic algorithms and adaptive neural networks," Unpublished doctoral dissertation, Katholieke Univ. Leuven, Belgium, 1995.

22. J. P. Delmas and F. Alberge, Asymptotic performance analysis of subspace adaptive algorithms introduced in the neural network literature, *IEEE Trans. on Signal Process.*, 46, No. 1, pp. 170–182, (1998).

23. J. P. Delmas, Performance analysis of a Givens parameterized adaptive eigenspace algorithm, *Signal Processing*, 68, No. 1, pp. 87–105, (1998).

24. J. P. Delmas and J. F. Cardoso, Performance analysis of an adaptive algorithm for tracking dominant subspace, *IEEE Trans. on Signal Process.*, 46, No. 11, pp. 3045–3057, (1998).

25. J. P. Delmas and J. F. Cardoso, Asymptotic distributions associated to Oja's learning equation for Neural Networks, *IEEE Trans. on Neural Networks*, 9, No. 6, pp. 1246–1257, (1998).

26. J. P. Delmas, On eigenvalue decomposition estimators of centro-symmetric covariance matrices, *Signal Processing*, 78, No. 1, pp. 101–116, (1999).

27. J. P. Delmas and V. Gabillon, "Asymptotic performance analysis of PCA algorithms based on the weighted subspace criterion," in *Proc. ICASSP* Taipei, Taiwan, April 2009.

28. S. C. Douglas, S. Y. Kung, and S. Amari, A self-stabilized minor subspace rule, *IEEE Signal Process. letters*, 5, No. 12, pp. 328–330, (1998).

29. X. G. Doukopoulos, "Power techniques for blind channel estimation in wireless communications systems," Unpublished doctoral dissertation, IRISA-INRIA, University of Rennes, France, 2004.

30. X. G. Doukopoulos and G. V. Moustakides, "The fast data projection method for stable subspace tracking," *Proc. 13th European Signal Proc. Conf.*, Antalya, Turkey, September 2005.

31. X. G. Doukopoulos and G. V. Moustakides, Fast and Stable Subspace Tracking, *IEEE Trans. on Signal Process*, 56, No. 4, pp. 1452–1465, (2008).

32. J. C. Fort and G. Pagès, "Sur la convergence presque sure d'algorithmes stochastiques: le théoreme de Kushner-Clark theorem revisité," *Technical report*, University Paris 1, 1994. Preprint SAMOS.

33. J. C. Fort and G. Pagès, Convergence of stochastic algorithms: from the Kushner and Clark theorem to the Lyapunov functional method, *Advances in Applied Probability*, No. 28, pp. 1072–1094, (1996).
34. B. Friedlander and A. J. Weiss, On the second-order of the eigenvectors of sample covariance matrices, *IEEE Trans. on Signal Process.*, 46, No. 11, pp. 3136–3139, (1998).
35. G. H. Golub and C. F. Van Loan, *Matrix computations*, 3rd ed., the Johns Hopkins University Press, 1996.
36. S. Haykin, *Adaptive filter theory*, Englewoods Cliffs, NJ: Prentice Hall, 1991.
37. R. A. Horn and C. R. Johnson, *Matrix analysis*, Cambridge University Press, 1985.
38. Y. Hua, Y. Xiang, T. Chen, K. Abed Meraim, and Y. Miao, A new look at the power method for fast subspace tracking, *Digital Signal Process.*, 9, No. 2, pp. 297–314, (1999).
39. B. H. Juang, S. Y. Kung, and C. A. Kamm (Eds.), *Proc. IEEE Workshop on neural networks for signal processing*, Princeton, NJ, September 1991.
40. I. Karasalo, Estimating the covariance matrix by signal subspace averaging, *IEEE Trans. on on ASSP*, 34, No. 1, pp. 8–12, (1986).
41. J. Karhunen and J. Joutsensalo, Generalizations of principal componant analysis, optimizations problems, and neural networks, *Neural Networks*, 8, pp. 549–562, (1995).
42. S. Y. Kung and K. I. Diamantaras, Adaptive principal component extraction (APEX) and applications, *IEEE Trans. on ASSP*, 42, No. 5, pp. 1202–1217, (1994).
43. H. J. Kushner and D. S. Clark, *Stochastic approximation for constrained and unconstrained systems*, Applied math. Science, No. 26, Springer Verlag, New York, 1978.
44. H. J. Kushner, *Weak convergence methods and singular perturbed stochastic control and filtering problems*, Vol. 3 of Systems and Control: Foundations and applications, Birkhäuser, 1989.
45. A. P. Liavas, P. A. Regalia, and J. P. Delmas, Blind channel approximation: Effective channel order determination, *IEEE Transactions on Signal Process.*, 47, No. 12, pp. 3336–3344, (1999).
46. A. P. Liavas, P. A. Regalia, and J. P. Delmas, On the robustness of the linear prediction method for blind channel identification with respect to effective channel undermodeling/overmodeling, *IEEE Transactions on Signal Process.*, 48, No. 5, pp. 1477–1481, (2000).
47. J. R. Magnus and H. Neudecker, *Matrix differential calculus with applications in statistics and econometrics*, Wiley series in probability and statistics, 1999.
48. Y. Miao and Y. Hua, Fast subspace tracking and neural learning by a novel information criterion, *IEEE Trans. on Signal Process.*, 46, No. 7, pp. 1967–1979, (1998).
49. E. Moulines, P. Duhamel, J. F. Cardoso, and S. Mayrargue, Subspace methods for the blind identification of multichannel FIR filters, *IEEE Trans. Signal Process.*, 43, No. 2, pp. 516–525, (1995).
50. E. Oja, A simplified neuron model as a principal components analyzer, *J. Math. Biol.*, 15, pp. 267–273, (1982).
51. E. Oja and J. Karhunen, On stochastic approximation of the eigenvectors and eigenvalues of the expectation of a random matrix, *J. Math. anal. Applications*, 106, pp. 69–84, (1985).
52. E. Oja, *Subspace methods of pattern recognition*, Letchworth, England, Research Studies Press and John Wiley and Sons, 1983.
53. E. Oja, Principal components, minor components and linear neural networks, *Neural networks*, 5, pp. 927–935, (1992).

54. E. Oja, H. Ogawa, and J. Wangviwattana, Principal component analysis by homogeneous neural networks, Part I: The weighted subspace criterion, *IEICE Trans. Inform. and Syst.*, E75-D, pp. 366–375, (1992).

55. E. Oja, H. Ogawa, and J. Wangviwattana, Principal component analysis by homogeneous neural networks, Part II: Analysis and extensions of the learning algorithms, *IEICE Trans. Inform. Syst.*, E75-D, pp. 376–382, (1992).

56. N. Owsley, "Adaptive data orthogonalization," in *Proc. Conf. ICASSP*, pp. 109–112, 1978.

57. B. N. Parlett, *The symmetric eigenvalue problem*, Prentice Hall, Englewood Cliffs, N.J. 1980.

58. B. Picinbono, Second-order complex random vectors and normal distributions, *IEEE Trans. on Signal Process.*, 44, No. 10, pp. 2637–2640, (1996).

59. C. R. Rao, *Linear statistical inference and its applications*, New York, Wiley, 1973.

60. C. R. Rao, C. R. Sastry, and B. Zhou, Tracking the direction of arrival of moving targets, *IEEE Trans. on Signal Process.*, 472, No. 5, pp. 1133–1144, (1994).

61. P. A. Regalia, An adaptive unit norm filter with applications to signal analysis and Karhunen Loéve tranformations, *IEEE Trans. on Circuits and Systems*, 37, No. 5, pp. 646–649, (1990).

62. C. Riou, T. Chonavel, and P. Y. Cochet, "Adaptive subspace estimation – Application to moving sources localization and blind channel identification," in *Proc. ICASSP* Atlanta, GA, May 1996.

63. C. Riou, "Estimation adaptative de sous espaces et applications," Unpublished doctoral dissertation, ENSTB, University of Rennes, France, 1997.

64. H. Rutishauser, Computational aspects of F. L. Bauer's simultaneous iteration method, *Numer. Math*, 13, pp. 3–13, (1969).

65. H. Sakai and K. Shimizu, A new adaptive algorithm for minor component analysis, *Signal Processing*, 71, pp. 301–308, (1998).

66. J. Sanchez-Araujo and S. Marcos, A efficient PASTd algorithm implementation for multiple direction of arrival tracking, *IEEE Trans. on Signal Process.*, 47, No. 8, pp. 2321–2324, (1999).

67. T. D. Sanger, Optimal unsupervised learning in a single-layer linear feedforward network, *Neural Networks*, 2, pp. 459–473, (1989).

68. A. H. Sayed, *Fundamentals of adaptive filtering*, IEEE Press, Wiley-Interscience, 2003.

69. W. Y. Yan, U. Helmke, and J. B. Moore, Global analysis of Oja's flow for neural networks, *IEEE Trans. on Neural Networks*, 5, No. 5, pp. 674–683, (1994).

70. J. F. Yang and M. Kaveh, Adaptive eigensubspace algorithms for direction or frequency estimation and tracking, *IEEE Trans. on ASSP*, 36, No. 2, pp. 241–251, (1988).

71. B. Yang, Projection approximation subspace tracking, *IEEE Trans. on Signal Process.*, 43, No. 1, pp. 95–107, (1995).

72. B. Yang, An extension of the PASTd algorithm to both rank and subspace tracking, *IEEE Signal Process. Letters*, 2, No. 9, pp. 179–182, (1995).

73. B. Yang, Asymptotic convergence analysis of the projection approximation subspace tracking algorithms, *Signal Processing*, 50, pp. 123–136, (1996).

74. B. Yang and F. Gersemsky, "Asymptotic distribution of recursive subspace estimators," in *Proc. ICASSP* Atlanta, GA, May 1996.

75. L. Yang, S. Attallah, G. Mathew, and K. Abed-Meraim, Analysis of orthogonality error propagation for FRANS an HFRANS algorithms, *IEEE Trans. on Signal Process.*, 56, No. 9, pp. 4515–4521, (2008).

76. X. Wang and H. V. Poor, Blind multiuser detection, *IEEE Trans. on Inform. Theory*, 44, No. 2, pp. 677–689, (1998).

77. J. H. Wilkinson, *The algebraic eigenvalue problem*, New York: Oxford University Press, 1965.

78. R. Williams, "Feature discovery through error-correcting learning," *Technical Report 8501*, San Diego, CA: University of California, Institute of Cognitive Science, 1985.

79. L. Xu, E. Oja, and C. Suen, Modified Hebbian learning for curve and surface fitting, *Neural Networks*, 5, No. 3, pp. 441–457, (1992).

80. G. Xu, H. Liu, L. Tong, and T. Kailath, A least-squares approach to blind channel identification, *IEEE Trans. Signal Process.*, 43, pp. 2982–2993, (1995).

81. H. Zeng and L. Tong, Connections between the least squares and subspace approaches to blind channel estimation, *IEEE Trans. Signal Process.*, 44, pp. 1593–1596, (1996).

5

PARTICLE FILTERING

Petar M. Djurić and Mónica F. Bugallo
Stony Brook University, Stony Brook, NY

Many problems in adaptive filtering are nonlinear and non-Gaussian. Of the many methods that have been proposed in the literature for solving such problems, particle filtering has become one of the most popular. In this chapter we provide the basics of particle filtering and review its most important implementations. In Section 5.1 we give a brief introduction to the area, and in Section 5.2 we motivate the use of particle filtering by examples from several disciplines in science and engineering. We then proceed with an introduction of the underlying idea of particle filtering and present a detailed explanation of its essential steps (Section 5.3). In Section 5.4 we address two important issues of particle filtering. Subsequently, in Section 5.5 we focus on some implementations of particle filtering and compare their performance on synthesized data. We continue by explaining the problem of estimating constant parameters with particle filtering (Section 5.6). This topic deserves special attention because constant parameters lack dynamics, which for particle filtering creates some serious problems. We can improve the accuracy of particle filtering by a method known as Rao–Blackwellization. It is basically a combination of Kalman and particle filtering, and is discussed in Section 5.7. Prediction and smoothing with particle filtering are described in Sections 5.8 and 5.9, respectively. In the last two sections of the chapter, we discuss the problems of convergence of particle filters (Section 5.10) and computational issues and hardware implementation of the filters (Section 5.11). At the end of the chapter we present a collection of exercises which should help the reader in getting additional insights into particle filtering.

Adaptive Signal Processing: Next Generation Solutions. Edited by Tülay Adalı and Simon Haykin
Copyright © 2010 John Wiley & Sons, Inc.

5.1 INTRODUCTION

Some of the most important problems in signal processing require the sequential processing of data. The data describe a system that is mathematically represented by equations that model its evolution with time, where the evolution contains a random component. The system is defined by its state variables of which some are dynamic and some are constant. A typical assumption about the state of the system is that it is Markovian, which means that given the state at a previous time instant, the distribution of the state at the current time instant does not depend on any other older state.

The state is not directly observable. Instead, we acquire measurements which are functions of the current state of the system. The measurements are degraded by noise or some other random perturbation, and they are obtained sequentially in time. The main objective of sequential signal processing is the recursive estimation of the state of the system based on the available measurements. The methods that are developed for estimation of the state are usually called filters. We may also be interested in estimates of the state in the past or would like to predict its values at future time instants.

In general, we can categorize the studied systems as linear and nonlinear and as Gaussian and non-Gaussian. For any of them, the complete information about the state is in the probability density functions (PDFs) of the state. These PDFs can be grouped as filtering, predictive, and smoothing densities (for their precise definitions, see the next section). If a system is linear and Gaussian, we can obtain all the relevant densities exactly, that is, without approximations. A filter that produces exact solutions is known as the Kalman filter (KF), and it obtains them by using analytical expressions [3, 43, 57].

When the description of the system deviates from linearity and Gaussianity, the solutions often represent approximations and therefore are not optimal. In fact, for some systems, such solutions can become poor to the point of becoming misleading. The traditional filter that deals with nonlinear systems is the extended Kalman filter (EKF) [3, 35, 40, 57]. The EKF implements the approximation by linearization of the system with a Taylor series expansion around the latest estimate of the state followed by application of the Kalman's linear recursive algorithm.

The unscented Kalman filter (UKF) [42] and the Gaussian quadrature Kalman filter (QKF) [39] form a group of methods that also assume Gaussian PDFs in the system but do not require the evaluation and computation of Jacobian matrices. Instead, they implement the necessary integration for filtering with different methods. The UKF uses the unscented transform, which represents a Gaussian posterior of the state with a set of deterministically chosen samples that capture the first two moments of the Gaussian. These samples are propagated through the system, and the obtained values contain information about the first two moments of the new Gaussian. The QKF exploits the Gauss–Hermite quadrature integration rule [29, 59].

For problems where the Gaussian assumption about the posterior is invalid, one can use the Gaussian sum filter. This filter approximates the *a posteriori* PDF by a sum of Gaussians, where the update of each Gaussian can be carried out by a separate EKF [2, 70], UKF [41], or QKF [5].

There is another group of approaches for nonlinear/non-Gaussian filtering, which is based on a different paradigm. These methods exploit the idea of evaluating the required PDFs over deterministic grids [14, 46, 49, 71]. Namely, the state space is represented by a fixed set of nodes and values associated with them, which represent the computed PDFs at the nodes. The final estimates of the densities are obtained by piecewise linear (first-order spline) interpolations. These methods can work well for low-dimensional state spaces; otherwise, they become quickly computationally prohibitive.

The particle filtering methodology[1] is very similar in spirit as the one that exploits deterministic grids. However, there is a subtle but very important difference between them. Particle filtering uses random grids, which means that the location of the nodes vary with time in a random way. In other words, at one time instant the grid is composed of one set of nodes, and at the next time instant, this set of nodes is completely different. The nodes are called particles, and they have assigned weights, which can be interpreted as probability masses. The particles and the weights form a discrete random measure,[2] which is used to approximate the densities of interest. In the generation of particles, each particle has a "parent", and the parent its own parent and so on. Such sequence of particles is called a particle stream, and it represents one possible evolution of the state with time. Filters based on this methodology are called particle filters (PFs).

A main challenge in implementing PFs is to place the nodes of the grids (i.e., generate particles) in regions of the state space over which the densities carry significant probability masses and to avoid placing nodes in parts of the state space with negligible probability masses. To that end, one exploits concepts from statistics known as importance sampling [48] and sampling-importance-resampling [68], which are explained in the sequel. For the computation of the particle weights, one follows the Bayesian theory. The sequential processing amounts to recursive updating of the discrete random measure with the arrival of new observations, where the updates correspond to the generation of new particles and computation of their weights. Thus, one can view particle filtering as a method that approximates evolving densities by generating streams of particles and assigning weights to them.

The roots of particle filtering were established about 50 years ago with the methods for estimating the mean squared extension of a self-avoiding random walk on lattice spaces [33, 67]. One of the first applications of the methodology was on the simulation of chain polymers. The control community produced interesting work on sequential Monte Carlo integration methods in the 1960s and 1970s, for example [1, 34]. The popularity of particle filtering in the last 15 years was triggered by [31], where the sampling-importance resampling filter was presented and applied to tracking problems. The timing of [31] was perfect because it came during a period when computing power started to become widely available. Ever since, the amount of work on particle filtering has proliferated and many important advances have been made.

[1]In the literature, particle filtering is also known as sequential Monte Carlo methodology [26].
[2]The random measure is basically a probability mass function.

One driving force for the advancement of particle filtering is the ever increasing range of its applications. We list some of them in the next section. Here we only reiterate the advantage of PFs over other filters and why they have become popular. PFs can be applied to any state space model where the likelihood and the prior are computable up to proportionality constants. The accuracy of the method depends on how well we generate the particles and how many particles we use to represent the random measure. From a mathematical point of view, it is important to have some understanding of the theoretical properties of PFs, and in particular the convergence of their estimates to the true states of the system. Some important results on this theory will also be presented in this chapter. At last, we have to keep in mind the practical issues related to PFs. Namely, it is well known that particle filtering is computationally intensive, which is an issue because in sequential signal processing any algorithm has to complete the processing of the latest measurement by a certain deadline. However, particle filtering allows for significant parallelization of its operations, which may speed up its time of execution by orders of magnitude. We also address some basic computational issues on this subject in the chapter.

5.2 MOTIVATION FOR USE OF PARTICLE FILTERING

In a typical setup, the observed data are modeled by

$$\boldsymbol{y}(n) = g_2(\boldsymbol{x}(n), \boldsymbol{v}_2(n)) \tag{5.1}$$

where $n = 1, 2, \ldots$ is a time index, $\boldsymbol{y}(n)$ is a vector of observations, $\boldsymbol{x}(n)$ is the state (or signal) that needs to be estimated, $\boldsymbol{v}_2(n)$ is an observation noise vector, and $g_2(\cdot)$ is a known function, which in general may change with time. We assume that all of the vectors in the above equation conform properly with their dimensions.

It is assumed that the state $\boldsymbol{x}(n)$ varies according to

$$\boldsymbol{x}(n) = g_1(\boldsymbol{x}(n-1), \boldsymbol{v}_1(n)) \tag{5.2}$$

where $\boldsymbol{v}_1(n)$ is a state noise vector, and $g_1(\cdot)$ is a known function (which also might vary with time). The expression (5.1) is known as an observation equation and (5.2) as a state equation, and the two are referred to as the state–space model. The objective is to estimate the unobserved signal $\boldsymbol{x}(n)$ from the observations $\boldsymbol{y}(n)$.

■ EXAMPLE 5.1

Consider the tracking of an object based on bearings-only measurements. The object moves in a two-dimensional sensor field according to [32]

$$\boldsymbol{x}(n) = \boldsymbol{A}\boldsymbol{x}(n-1) + \boldsymbol{B}\boldsymbol{v}_1(n) \tag{5.3}$$

where the state vector $\boldsymbol{x}(n)$ is defined by the position and velocity of the object, or

$$\boldsymbol{x}(n) = [x_1(n) \quad x_2(n) \quad \dot{x}_1(n) \quad \dot{x}_2(n)]^\top$$

with $x_1(n)$ and $x_2(n)$ being the coordinates of the object at time instant n and $\dot{x}_1(n)$ and $\dot{x}_2(n)$ the components of the target velocity. The matrices $\boldsymbol{A}$ and $\boldsymbol{B}$ are defined by

$$\boldsymbol{A} = \begin{pmatrix} 1 & 0 & T_s & 0 \\ 0 & 1 & 0 & T_s \\ 0 & 0 & 1 & 0 \\ 0 & 0 & 0 & 1 \end{pmatrix} \quad \text{and} \quad \boldsymbol{B} = \begin{pmatrix} \frac{T_s^2}{2} & 0 \\ 0 & \frac{T_s^2}{2} \\ T_s & 0 \\ 0 & T_s \end{pmatrix}$$

where T_s is the sampling period. The state noise $v_1(n)$ models small accelerations of the object and is assumed to have a known PDF.

The bearings-only range measurements are obtained by J sensors placed at known locations in the sensor field. The sensors measure the angle of the target with respect to a reference axis. At time instant n, the j-th sensor, which is located at $(l_{1,j}, l_{2,j})$, gets the measurement

$$y_j(n) = \arctan\left(\frac{x_2(n) - l_{2,j}}{x_1(n) - l_{1,j}}\right) + v_{2,j}(n) \tag{5.4}$$

where the noise samples $v_{2,j}(n)$ have a known joint PDF, and are independent from the state noise $\boldsymbol{v}_1(n)$.

Given the measurements of J sensors, $\boldsymbol{y}(n) = [y_1(n) \; y_2(n) \; \cdots \; y_J(n)]^\top$, and the movement model of the object defined by (5.3), we want to track the object in time, that is, estimate its position and velocity. In the literature this problem is known as target tracking [66].

In this model the state equation is (5.3), and it is clear that it is represented by a linear function. The observation equation is given by the set of expressions (5.4), and they are all nonlinear. So, in this example, optimal filtering cannot be achieved even though the noises in the system are Gaussian. Instead, one has to resort to sub-optimal methods like extended Kalman filtering or particle filtering.

An alternative characterization of the state–space system can be given in terms of PDFs

$$f(\boldsymbol{x}(n) \,|\, \boldsymbol{x}(n-1)) \tag{5.5}$$

$$f(\boldsymbol{y}(n) \,|\, \boldsymbol{x}(n)) \tag{5.6}$$

where obviously (5.5) is derived from (5.2), and (5.6) is obtained from (5.1). The forms of the PDFs in (5.5) and (5.6) depend on the functions $g_1(\cdot)$ and $g_2(\cdot)$ as well as on the PDFs of $\boldsymbol{v}_1(n)$ and $\boldsymbol{v}_2(n)$.

The problem of estimation can have various twists. Here, we distinguish three different problems, which we refer to as filtering, prediction, and smoothing. In filtering, the goal is to obtain the a *posteriori* PDF of $\boldsymbol{x}(n)$ given all of the measurements from time instant one to n, which we express by $\mathbf{y}(1:n)$. This density is accordingly called filtering density and is denoted by $f(\boldsymbol{x}(n) \,|\, \boldsymbol{y}(1:n))$, $n \geq 1$. All the information about $\boldsymbol{x}(n)$ is in $f(\boldsymbol{x}(n) \mid \boldsymbol{y}(1:n))$, and for example, if one wants to find point estimates of $\boldsymbol{x}(n)$, such as the minimum mean square error (MMSE) estimate or the maximum a *posteriori* (MAP) estimate, one can obtain them from the filtering density.

In prediction, the goal is to find the predictive PDF $f(\boldsymbol{x}(n+k) \mid \boldsymbol{y}(1:n))$, where $k > 0$, $n > 0$. Again, all the information about a future value of the state given the measurements $\boldsymbol{y}(1:n)$ is in the predictive PDF and various point estimates can be obtained from it.

Finally, the problem of smoothing amounts to obtaining $f(\boldsymbol{x}(n) \mid \boldsymbol{y}(1:N))$, where $0 \leq n < N$, and where the density is called the smoothing PDF.[3] We can expect that with smoothing, in general, we get more accurate estimates of $\boldsymbol{x}(n)$. It is also clear that with smoothing, we have a delay in processing of the data. For example, we first acquire N measurements and then obtain the smoothing densities. We distinguish several types of smoothing problems, and we describe them later in the chapter.

The objective of sequential signal processing in the context presented here consists of tracking the PDFs of interest by exploiting recursive relationships, that is,

- $f(\boldsymbol{x}(n) \mid \boldsymbol{y}(1:n))$ from $f(\boldsymbol{x}(n-1) \mid \boldsymbol{y}(1:n-1))$ for the filtering problem, where $n \geq 1$;[4]
- $f(\boldsymbol{x}(n+k) \mid \boldsymbol{y}(1:n))$ from $f(\boldsymbol{x}(n+k-1) \mid \boldsymbol{y}(1:n))$ for the prediction problem where $k > 0$, $n \geq 0$; and
- $f(\boldsymbol{x}(n) \mid \boldsymbol{y}(1:N))$ from $f(\boldsymbol{x}(n+1) \mid \boldsymbol{y}(1:N))$ for the smoothing problem, where $0 \leq n < N$.

We reiterate that the complete information about the unknown values is in the respective densities of the unknowns. Many sequential methods provide only point estimates of these unknowns accompanied possibly with another metric that shows how variable the estimates are. By contrast, particle filtering has a much more ambitious aim than yielding point estimates. Its objective is to track in time the approximations of all the desired densities of the unknowns in the system. It is well known that when these densities are not Gaussian, there are not many available methods that can reach this goal. Thus, the motivation for using particle filtering is in its ability to estimate sequentially the densities of unknowns of non-Gaussian and/or nonlinear systems.

What are the forms of the recursions for filtering, prediction, and smoothing? Here we briefly explain only the recursion for obtaining the filtering PDF. The other two recursions are discussed in Sections 5.8 and 5.9. Suppose that at time $n-1$, we know the observations $\boldsymbol{y}(1:n-1)$ and the *a posteriori* PDF $f(\boldsymbol{x}(n-1) \mid \boldsymbol{y}(1:n-1))$. Once $\boldsymbol{y}(n)$ becomes available, we would like to update $f(\boldsymbol{x}(n-1) \mid \boldsymbol{y}(1:n-1))$ and modify it to $f(\boldsymbol{x}(n) \mid \boldsymbol{y}(1:n))$. To achieve this, we formally write

$$f(\boldsymbol{x}(n) \mid \boldsymbol{y}(1:n)) \propto f(\boldsymbol{y}(n) \mid \boldsymbol{x}(n)) f(\boldsymbol{x}(n) \mid \boldsymbol{y}(1:n-1)) \tag{5.7}$$

[3]In the expression for the smoothing PDF, we have included the zero as a possible value of the time index, that is, $n = 0$. The PDF of $\boldsymbol{x}(0)$, $f(\boldsymbol{x}(0))$, is the *a priori* PDF of the state before the first measurement is taken.

[4]When $n = 1$, $f(\boldsymbol{x}(n-1) \mid \boldsymbol{y}(1:n-1))$ is simply the *a priori* PDF of $\boldsymbol{x}(0)$, that is, $f(\boldsymbol{x}(0))$.

where $\propto$ signifies proportionality. The first factor on the right of the proportionality sign is the likelihood function of the unknown state, and the second factor is the predictive density of the state. For the predictive density we have

$$f(\boldsymbol{x}(n) \mid \boldsymbol{y}(1:n-1)) = \int f(\boldsymbol{x}(n) \mid \boldsymbol{x}(n-1)) f(\boldsymbol{x}(n-1) \mid \boldsymbol{y}(1:n-1)) \, d\boldsymbol{x}(n-1). \quad (5.8)$$

In writing (5.8) we used the property of the state that given $\boldsymbol{x}(n-1)$, $\boldsymbol{x}(n)$ does not depend on $\boldsymbol{y}(1:n-1)$. Now, the required recursive equation for the update of the filtering density is obtained readily by combining (5.7) and (5.8), that is, we formally have

$$\begin{aligned} f(\boldsymbol{x}(n) \mid \boldsymbol{y}(1:n)) \propto & f(\boldsymbol{y}(n) \mid \boldsymbol{x}(n)) \\ & \times \int f(\boldsymbol{x}(n) \mid \boldsymbol{x}(n-1)) f(\boldsymbol{x}(n-1) \mid \boldsymbol{y}(1:n-1)) \, d\boldsymbol{x}(n-1). \end{aligned}$$

Thus, on the left of the proportionality sign we have the filtering PDF at time instant n, and on the right under the integral, we see the filtering PDF at time instant $n-1$.

There are at least two problems in carrying out the above recursion, and they may make the recursive estimation of the filtering density very challenging. The first one is the solving of the integral in (5.8) and obtaining the predictive density $f(\boldsymbol{x}(n) \mid \boldsymbol{y}(1:n-1))$. In some cases it is possible to obtain the solution analytically, which considerably simplifies the recursive algorithm and makes it more accurate. The second problem is the combining of the likelihood and the predictive density in order to get the updated filtering density. These problems may mean that it is impossible to express the filtering PDF in a recursive form. For instance, in Example 5.1, we cannot obtain an analytical solution due to the nonlinearities in the observation equation. It will be seen in the sequel that the smoothing and the predictive densities suffer from analogous problems.

We can safely state that in many problems the recursive evaluation of the densities of the state–space model cannot be done analytically, and consequently we have to resort to numerical methods. As already discussed, an important class of systems which allows for exact analytical recursions is the one represented by linear state–space models with Gaussian noises. These recursions are known as Kalman filtering [3]. When analytical solutions cannot be obtained, particle filtering can be employed with elegance and with performance characterized by high accuracy.

Particle filtering has been used in many different disciplines. They include surveillance guidance, and obstacle avoidance systems, robotics, communications, speech processing, seismic signal processing, system engineering, computer vision, and econometrics. During the past decade, in practically all of these fields, the number of contributions has simply exploded. To give a perspective of how the subject has been vibrant with activities, we provide a few examples of developments in target tracking, positioning, and navigation.

In target tracking, there have been a large number of papers published that show the advantage of PFs over other filters in a variety of scenarios. For example, in [19], multiple targets were tracked by PFs using a combination of video and acoustic sensors while in [38], in a sonar application, the tracking is based on bearings-only measurements (as in Example 5.1). The tracking of the direction-of-arrival (DOA) of multiple moving targets using measurements of a passive sensor array was shown in [63]. PFs can also be used for simultaneous detection and tracking of multiple targets [62]. In [75], a moving acoustic source was tracked in a moderately reverberant room. A particle filtering-based method for the joint tracking of location and speaking activity of several speakers in a meeting room with a microphone array and multiple cameras was proposed in [64]. Tracking of multiaspect targets using image sequences with background clutter and where the target aspect changed were addressed in [13]. Mobility tracking in wireless communication networks based on received signal strength was explored in [60].

Positioning and navigation are related problems to target tracking [32]. In the former problem, an object estimates its own position, and in the latter, the object estimates more unknowns such as its velocity, attitude and heading, acceleration, and angular rates. For example, in [30] PFs were applied to navigation that uses a global positioning system whose measurements are distorted with multipath effects. In [45], PFs were used for maritime surface and underwater map-aided navigation. The surface navigation employed a radar sensor and a digital sea chart whereas the underwater navigation was based on a sonar sensor and a depth database. Land vehicle positioning by using synchronous and asynchronous sensor measurements was presented in [17]. Applications of positioning, navigation and tracking can, *inter alia*, be used for car collision avoidance in the car industry [32].

The above list is by no means exhaustive. The other areas where PFs are of interest are also rich with many interesting results and contributions. For more references on particle filtering applications, the reader should consult the review papers [6, 16, 23, 25] and the books [15, 26, 66].

5.3 THE BASIC IDEA

Consider the state of the system at time instant n, that is, $\boldsymbol{x}(n)$. Under the particle filtering framework, the *a posteriori* PDF of $\boldsymbol{x}(n)$, $f(\boldsymbol{x}(n) \mid \mathbf{y}(1:n))$, is approximated by a discrete random measure composed of M particles and weights

$$\chi(n) = \{\boldsymbol{x}^{(m)}(n), w^{(m)}(n)\}_{m=1}^{M} \tag{5.9}$$

where $\boldsymbol{x}^{(m)}(n)$ and $w^{(m)}(n)$ represent the m-th particle and weight, respectively. The particles are Monte Carlo samples of the system state, and the weights are nonnegative values that sum up to one and can be interpreted as probabilities of the particles. The previous measure allows for approximation of $f(\boldsymbol{x}(n) \mid \mathbf{y}(1:n))$ by

$$f(\boldsymbol{x}(n) \mid \mathbf{y}(1:n)) \approx \sum_{m=1}^{M} w^{(m)}(n)\delta(\boldsymbol{x}(n) - \boldsymbol{x}^{(m)}(n)) \tag{5.10}$$

where $\delta(\cdot)$ denotes the Dirac delta function. With this approximation, computations of expectations of functions of the random process $\boldsymbol{X}(n)$ simplify to summations, that is,

$$E(h(\boldsymbol{X}(n))) = \int h(\boldsymbol{x}(n))\, f(\boldsymbol{x}(n) \,|\, \boldsymbol{y}(1:n))\, d\boldsymbol{x}(n)$$

$$\Downarrow$$

$$E(h(\boldsymbol{X}(n))) \approx \sum_{m=1}^{M} w^{(m)}(n) h(\boldsymbol{x}^{(m)}(n))$$

where $E(\cdot)$ denotes expectation, and $h(\cdot)$ is an arbitrary function of $\boldsymbol{X}(n)$.

■ EXAMPLE 5.2

Assume that independent particles, $\boldsymbol{x}^{(m)}(n)$, can be drawn from $f(\boldsymbol{x}(n) \,|\, \boldsymbol{y}(1:n))$. In that case, all the particles have the same weights, $w^{(m)}(n) = 1/M$, and

$$E(h(\boldsymbol{X}(n))) = \int h(\boldsymbol{x}(n)) f(\boldsymbol{x}(n) \,|\, \boldsymbol{y}(1:n))\, d\boldsymbol{x}(n)$$

$$\Downarrow$$

$$\hat{E}(h(\boldsymbol{X}(n))) = \frac{1}{M} \sum_{m=1}^{M} h(\boldsymbol{x}^{(m)}(n)) \tag{5.11}$$

where $\hat{E}(\cdot)$ is an unbiased estimator of the conditional expectation $E(\cdot)$. Note that in this case we intrinsically approximate $f(\boldsymbol{x}(n) \,|\, \boldsymbol{y}(1:n))$ by the random measure

$$\chi(n) = \{\boldsymbol{x}^{(m)}(n),\ w^{(m)}(n) = 1/M\}_{m=1}^{M}.$$

If the variance $\sigma_h^2 < \infty$,[5] then the variance of $\hat{E}(\cdot)$ is given by

$$\sigma^2_{\hat{E}(h(\cdot))} = \frac{\sigma_h^2}{M}. \tag{5.12}$$

As $M \to \infty$, from the strong law of large numbers, we get that $\hat{E}(h(\boldsymbol{X}(n)))$ converges to $E(h(\boldsymbol{X}(n)))$ almost surely [27], that is

$$\hat{E}(h(\boldsymbol{X}(n))) \xrightarrow{a.s.} E(h(\boldsymbol{X}(n))) \tag{5.13}$$

and from the central limit theorem, we obtain that $\hat{E}(h(\boldsymbol{X}(n)))$ converges in distribution to a Gaussian distribution [27]

$$\hat{E}(h(\boldsymbol{X}(n))) \xrightarrow{d} \mathcal{N}\left(E(h(\boldsymbol{X}(n)),\ \frac{\sigma_h^2}{M} \right). \tag{5.14}$$

[5]The notation σ_h^2 symbolizes the variance of $h(\boldsymbol{X}(n))$.

The example shows that if we sample from $f(\boldsymbol{x}(n)\,|\,\boldsymbol{y}(1:n))$ a large number of particles M, we will be able to estimate $E(h(\boldsymbol{X}(n))$ with arbitrary accuracy. In practice, however, the problem is that we often cannot draw samples directly from the *a posteriori* PDF $f(\boldsymbol{x}(n)\,|\,\boldsymbol{y}(1:n))$. An attractive alternative is to use the concept of importance sampling [58]. The idea behind it is based on the use of another function for drawing particles. This function is called importance sampling function or proposal distribution, and we denote it by $\pi(\boldsymbol{x}(n))$.

When the particles are drawn from $\pi(\boldsymbol{x}(n))$, the estimate of $E(h(\boldsymbol{X}(n)))$ in (5.11) can be obtained either by

$$E(h(\boldsymbol{X}(n))) \approx \frac{1}{M}\sum_{m=1}^{M} w^{*(m)}(n)h(\boldsymbol{x}^{(m)}(n)) \tag{5.15}$$

or by

$$E(h(\boldsymbol{X}(n))) \approx \sum_{m=1}^{M} w^{(m)}(n)h(\boldsymbol{x}^{(m)}(n)) \tag{5.16}$$

where

$$w^{*(m)}(n) = \frac{f(\boldsymbol{x}^{(m)}(n)\,|\,\boldsymbol{y}(1:n))}{\pi(\boldsymbol{x}^{(m)}(n))} \tag{5.17}$$

and

$$w^{(m)}(n) = \frac{\widetilde{w}^{(m)}(n)}{\sum_{i=1}^{M}\widetilde{w}^{(i)}(n)} \tag{5.18}$$

where

$$\widetilde{w}^{(m)}(n) = c\, w^{*(m)}(n)$$

with c being some unknown constant. The symbols $w^{*(m)}(n)$ and $w^{(m)}(n)$ are known as true and normalized importance weights of the particles $\boldsymbol{x}^{(m)}(n)$, respectively. They are introduced to correct for the bias that arises due to sampling from a different function than the one that is being approximated, $f(\boldsymbol{x}(n)\,|\,\boldsymbol{y}(1:n))$. The estimate in (5.15) is unbiased whereas the one from (5.16) is with a small bias but often with a smaller mean-squared error than the one in (5.15) [55]. An advantage in using (5.16) over (5.15) is that we only need to know the ratio $f(\boldsymbol{x}(n)\,|\,\boldsymbol{y}(1:n))/\pi(\boldsymbol{x}(n))$ up to a multiplicative constant and not the exact ratio in order to compute the estimate of the expectation of $h(\boldsymbol{X}(n))$.

How is (5.18) obtained? Suppose that the true weight cannot be found and instead we can only compute it up to a proportionality constant, that is

$$\begin{aligned}\widetilde{w}^{(m)}(n) &= c\frac{f(\boldsymbol{x}^{(m)}(n)\,|\,\boldsymbol{y}(1:n))}{\pi(\boldsymbol{x}^{(m)}(n))}\\ &= c w^{*(m)}(n)\end{aligned} \tag{5.19}$$

where the constant c is unknown. Since we must have

$$1 = \int \frac{f(\boldsymbol{x}(n)\,|\,\boldsymbol{y}(1:n))}{\pi(\boldsymbol{x}(n))}\,\pi(\boldsymbol{x}(n))\,d\boldsymbol{x}(n)$$
$$\simeq \frac{1}{cM}\sum_{m=1}^{M}\widetilde{w}^{(m)}(n)$$

from where we can estimate c by

$$c \approx \frac{1}{M}\sum_{m=1}^{M}\widetilde{w}^{(m)}(n). \tag{5.20}$$

Now, by using (5.19), we can express (5.15) in terms of $\widetilde{w}^{(m)}(n)$. We have

$$E(h(\boldsymbol{X}(n))) \approx \frac{1}{M}\sum_{m=1}^{M} w^{*(m)}(n)h(\boldsymbol{x}^{(m)}(n))$$
$$= \frac{1}{cM}\sum_{m=1}^{M}\widetilde{w}^{(m)}(n)h(\boldsymbol{x}^{(m)}(n))$$
$$\approx \sum_{m=1}^{M}\frac{\widetilde{w}^{(m)}(n)}{\sum_{i=1}^{M}\widetilde{w}^{(i)}(n)}h(\boldsymbol{x}^{(m)}(n))$$
$$= \sum_{m=1}^{M} w^{(m)}(n)h(\boldsymbol{x}^{(m)}(n))$$

where $w^{(m)}(n)$ is the normalized weight from (5.18).

In summary, when we use a random measure to approximate a PDF, such as $f(\boldsymbol{x}(n)\,|\,\boldsymbol{y}(1:n))$, we can do it by

- drawing samples from a proposal distribution $\pi(\boldsymbol{x}(n))$, which needs to be known only up to a multiplicative constant, that is

$$\boldsymbol{x}^{(m)}(n) \sim \pi(\boldsymbol{x}(n)), \quad m = 1, 2, \ldots, M$$

- computing the weights of the particles $w^{(m)}(n)$ by (5.17) and (5.18).

The resulting random measure is of the form given by (5.9). As proposal distributions we choose ones that are easy to sample from and whose shape is close to the product of $h(\boldsymbol{x}(n))$ and $f(\boldsymbol{x}(n)\,|\,\boldsymbol{y}(1:n))$ [55]. A rule of thumb is that we would like to generate particles from regions of the support of $\boldsymbol{x}(n)$ where that product has large values.

In Figure 5.1 we see a posterior and a proposal distribution as well as particles and weights that form a random measure approximating the posterior. It is important to

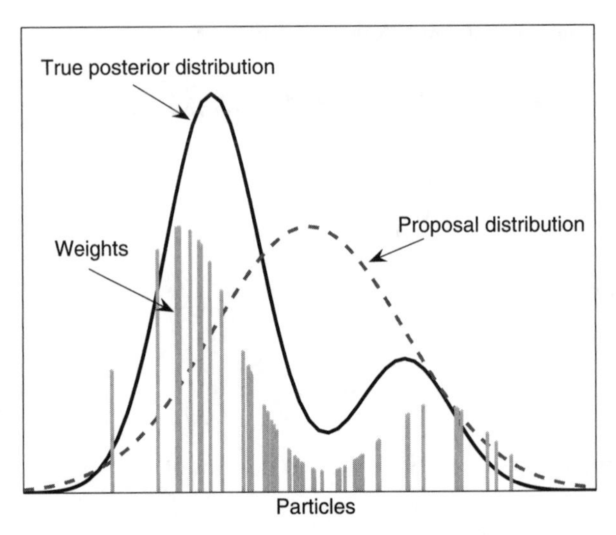

Figure 5.1 A posterior and a proposal distribution, and particles and weights that approximate the posterior.

note that the particle weights are *not* only functions of the posterior but also of the proposal distribution.

■ EXAMPLE 5.3

Suppose that we want to generate a random measure that approximates a Gaussian with mean one and variance one (target distribution). Let the proposal distribution be a standard Gaussian (that is, a Gaussian with mean zero and variance one). Figure 5.2 shows the target distribution, the drawn particles and their weights. From the figure, it may appear that there is a big discrepancy between the random measure and the Gaussian. However, when we compute the cumulative distribution functions (CDFs) of the Gaussian and the random measure, we can see that there is a good agreement between the two CDFs, as can be seen in Figure 5.3. Moreover, the figure also shows that there is better agreement between the two CDFs where there are more generated particles.

Since particle filtering exploits the concept of importance sampling heavily, we will address the choice of $\pi(\boldsymbol{x}(n))$ in more detail in Section 5.4. Now we turn our attention to the problem of recursive computation of the random measure $\chi(n)$.

How do we obtain $\chi(n)$ from $\chi(n-1)$? The random measures $\chi(n)$ and $\chi(n-1)$ approximate $f(\boldsymbol{x}(n)\,|\,\boldsymbol{y}(1:n))$ and $f(\boldsymbol{x}(n-1)\,|\,\boldsymbol{y}(1:n-1))$, respectively. We write (see (5.10))

$$f(\boldsymbol{x}(n-1)\,|\,\boldsymbol{y}(1:n-1)) \simeq \sum_{m=1}^{M} w^{(m)}(n-1)\delta(\boldsymbol{x}(n-1)-\boldsymbol{x}^{(m)}(n-1)).$$

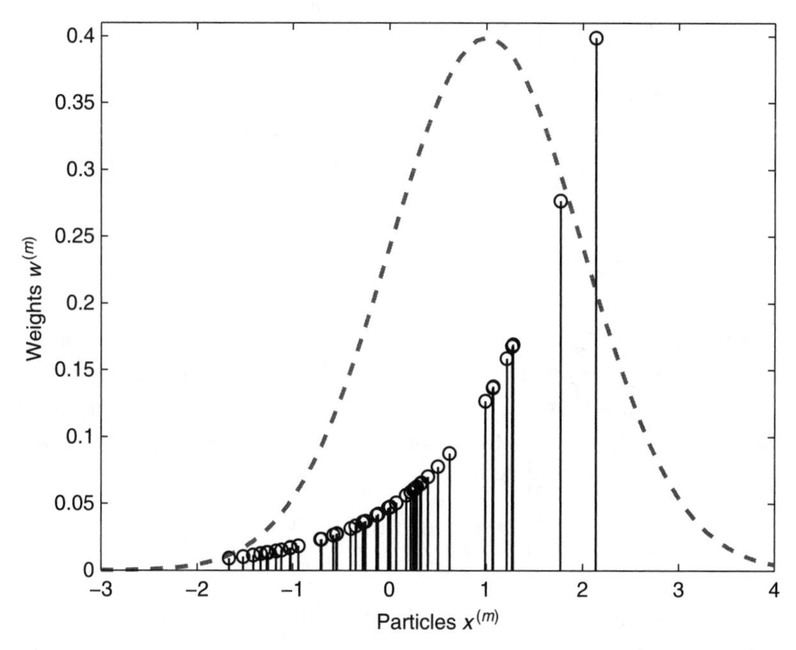

Figure 5.2 A target distribution $\mathcal{N}(1, 1)$, drawn particles from $\mathcal{N}(0, 1)$ and their weights.

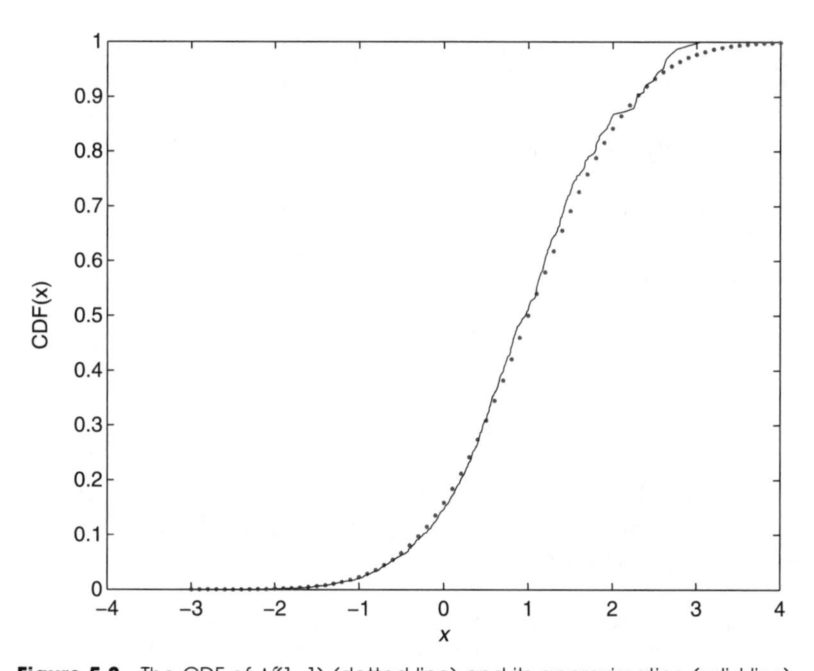

Figure 5.3 The CDF of $\mathcal{N}(1, 1)$ (dotted line) and its approximation (solid line).

For the filtering PDF $f(\boldsymbol{x}(n)\,|\,\boldsymbol{y}(1:n))$, we have

$$\begin{aligned} f(\boldsymbol{x}(n)\,|\,\boldsymbol{y}(1:n)) \propto & f(\boldsymbol{y}(n)\,|\,\boldsymbol{x}(n)) \\ & \times \int f(\boldsymbol{x}(n)\,|\,\boldsymbol{x}(n-1)) f(\boldsymbol{x}(n-1)\,|\,\boldsymbol{y}(1:n-1))\, d\boldsymbol{x}(n-1) \\ & \simeq f(\boldsymbol{y}(n)\,|\,\boldsymbol{x}(n)) \sum_{m=1}^{M} w^{(m)}(n-1) f(\boldsymbol{x}(n)\,|\,\boldsymbol{x}^{(m)}(n-1)). \end{aligned}$$

We want to represent the new filtering density with $\chi(n)$, and to that end we need to generate particles $\boldsymbol{x}(n)$ and compute their weights. If for particle generation we use the proposal distribution $\pi(\boldsymbol{x}(n)\,|\,\boldsymbol{x}^{(m)}(n-1), \boldsymbol{y}(1:n))$, the newly generated particle $\boldsymbol{x}^{(m)}(n)$ is appended to the particle stream $\boldsymbol{x}^{(m)}(1:n-1)$, and we compute the value of its weight according to

$$\tilde{w}^{(m)}(n) = w^{(m)}(n-1) \frac{f(\boldsymbol{y}(n)\,|\,\boldsymbol{x}^{(m)}(n)) f(\boldsymbol{x}^{(m)}(n)\,|\,\boldsymbol{x}^{(m)}(n-1))}{\pi(\boldsymbol{x}^{(m)}(n)\,|\,\boldsymbol{x}^{(m)}(n-1), \boldsymbol{y}(1:n))} \tag{5.21}$$

where $\tilde{w}^{(m)}(n)$ is a non-normalized weight of $\boldsymbol{x}^{(m)}(n)$.[6] We see that the weight of the m-th stream is obtained by updating its value at time instant $n-1$ with the factor

$$\frac{f(\boldsymbol{y}(n)\,|\,\boldsymbol{x}^{(m)}(n)) f(\boldsymbol{x}^{(m)}(n)\,|\,\boldsymbol{x}^{(m)}(n-1))}{\pi(\boldsymbol{x}^{(m)}(n)\,|\,\boldsymbol{x}^{(m)}(n-1), \boldsymbol{y}(1:n))}.$$

The so obtained weights are then normalized so that they sum up to one.

In summary, the PF implements two steps. One is the generation of particles for the next time instant and the other is the computation of the weights of these particles. Figure 5.4 depicts the graphical flow and interpretation of the resulting particle filtering algorithm, and Table 5.1 summarizes the mathematical expressions needed for carrying out the recursive update (from $\chi(n-1)$ to $\chi(n)$).

■ EXAMPLE 5.4

Consider the state-space model given by

$$\begin{aligned} x(n) &= x(n-1) + v_1(n) \\ y(n) &= x(n) + v_2(n) \end{aligned}$$

where $v_1(n)$ and $v_2(n)$ are independent standard Gaussian random variables.[7] A particular realization of such system is given in Figure 5.5 and the values of the first four time instants are shown in Table 5.2.

[6]The nonnormalized weight $\tilde{w}^{(m)}(n)$ should be distinguished from the true weight defined by (5.17).
[7]We use this example only to explain how the particle filtering algorithm proceeds. Note that this model is linear and Gaussian and that the data can optimally be processed by the Kalman filter.

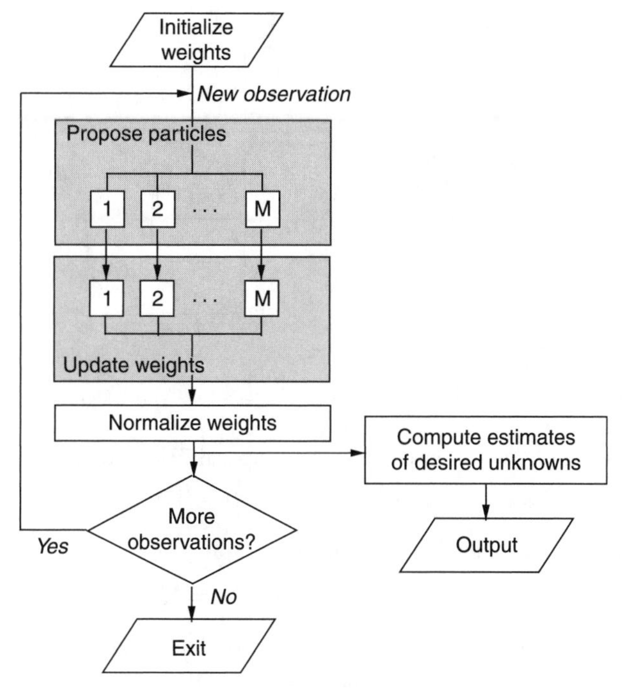

Figure 5.4 Flowchart of the sequential importance sampling algorithm.

Table 5.1 The sequential importance sampling algorithm

Particle generation

Basis

$$\pi(\boldsymbol{x}(0:n)\,|\,\boldsymbol{y}(1:n)) = \pi(\boldsymbol{x}(n)\,|\,\boldsymbol{x}(0:n-1),\boldsymbol{y}(1:n))\,\pi(\boldsymbol{x}(0:n-1)\,|\,\boldsymbol{y}(1:n-1))$$

$$\boldsymbol{x}^{(m)}(0:n-1) \sim \pi(\boldsymbol{x}(0:n-1)\,|\,\boldsymbol{y}(1:n-1))$$

$$w^{(m)}(n-1) \propto \frac{f(\boldsymbol{x}^{(m)}(0:n-1)\,|\,\boldsymbol{y}(n-1))}{\pi(\boldsymbol{x}^{(m)}(0:n-1)\,|\,\boldsymbol{y}(1:n-1))}$$

Augmentation of the trajectory $\boldsymbol{x}^{(m)}(0:n-1)$ with $\boldsymbol{x}^{(m)}(n)$

$$\boldsymbol{x}^{(m)}(n) \sim \pi(\boldsymbol{x}(n)\,|\,\boldsymbol{x}^{(m)}(0:n-1),\boldsymbol{y}(1:n)),\; m=1,\ldots,M$$

Weight update

$$w^{(m)}(n) \propto \frac{f(\boldsymbol{y}(n)\,|\,\boldsymbol{x}^{(m)}(n))\,f(\boldsymbol{x}^{(m)}(n)\,|\,\boldsymbol{x}^{(m)}(n-1))}{\pi(\boldsymbol{x}^{(m)}(n)\,|\,\boldsymbol{x}^{(m)}(0:n-1),\boldsymbol{y}(1:n))}w^{(m)}(n-1) \quad m=1,\ldots,M$$

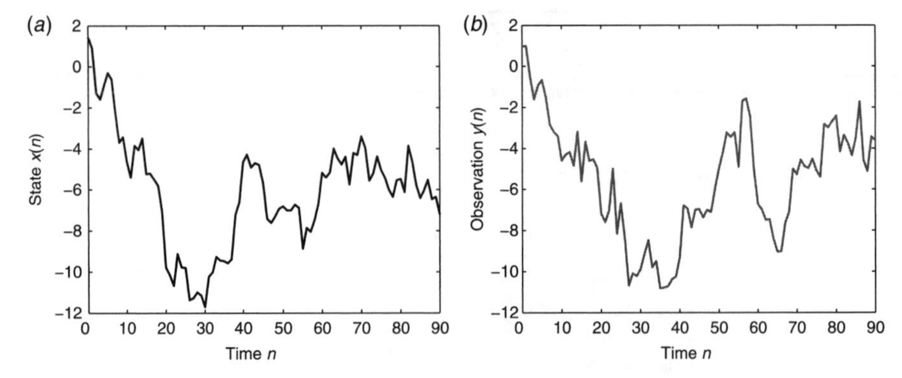

Figure 5.5 One realization of the state-space model of Example 5.4.

In this example, we choose to use the proposal distribution

$$\pi(x(n)) = f(x(n) \mid x(n-1)).$$

In other words, we generate the particles according to

$$x^{(m)} \sim f(x(n) \mid x^{(m)}(n-1)), \quad m = 1, 2, \ldots, M$$

that is, each particle stream has a separate proposal distribution. Since the proposal function is identical to the transition function $f(x(n) \mid x^{(m)}(n-1))$, from (5.21) we deduce that the update of the weights is according to

$$w^{(m)}(n) \propto f(y(n) \mid x^{(m)}(n))w^{(m)}(n-1).$$

The step by step execution of the sequential importance sampling algorithm for the first three time instants proceeds as follows.

Initialization $n = 0$

$$x^{(m)}(0) \sim \mathcal{N}(0, 1) \quad m = 1, 2, \ldots, M.$$

Note that all the weights are equal (Fig. 5.6).

Time instant $n = 1$

Generation of particles using the proposal distribution

$$x^{(m)}(1) \sim f(x(1) \mid x^{(m)}(0))$$
$$= \frac{1}{\sqrt{2\pi}}\exp\left(-\frac{(x(1) - x^{(m)}(0))^2}{2}\right).$$

Table 5.2 Values of the first four time instants for Example 5.4

	$n = 0$	$n = 1$	$n = 2$	$n = 3$
$x(n)$	1.44	0.89	−1.28	−1.59
$y(n)$	0.96	0.98	−0.51	−1.61

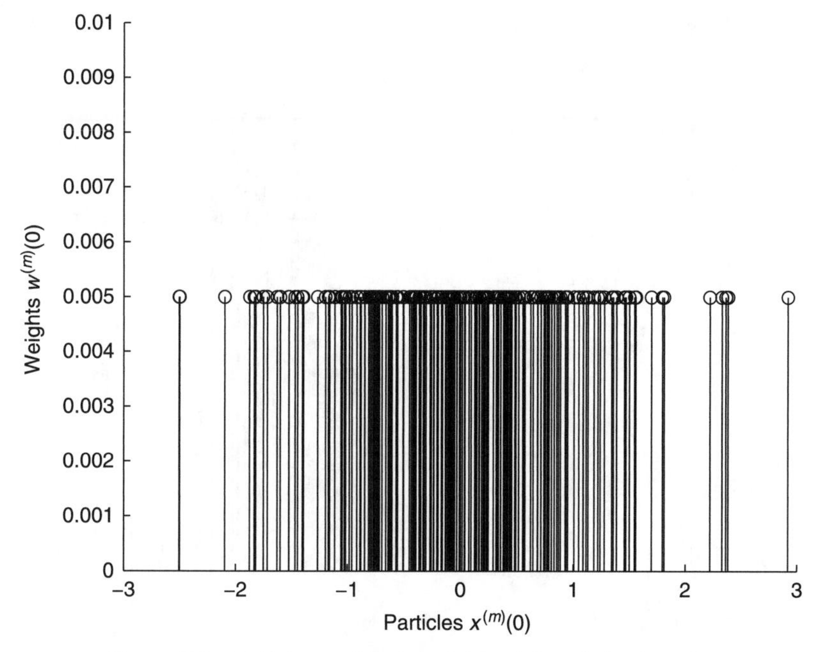

Figure 5.6 Particles and their weights at time instant $n = 0$.

Weight update

$$w^{(m)}(1) \propto \exp\left(-\frac{(y(1) - x^{(m)}(1))^2}{2}\right).$$

The particles and their weights are shown in Figure 5.7.

Time instant $n = 2$
Generation of particles using the proposal distribution

$$\begin{aligned}\mathbf{x}^{(m)}(2) &\sim f(x(2) \mid x^{(m)}(1)) \\ &= \frac{1}{\sqrt{2\pi}} \exp\left(-\frac{(x(2) - x^{(m)}(1))^2}{2}\right).\end{aligned}$$

Weight update

$$w^{(m)}(2) \propto w^{(m)}(1) \exp\left(-\frac{(y(2) - x^{(m)}(2))^2}{2}\right).$$

The particles and their weights are shown in Figure 5.8.

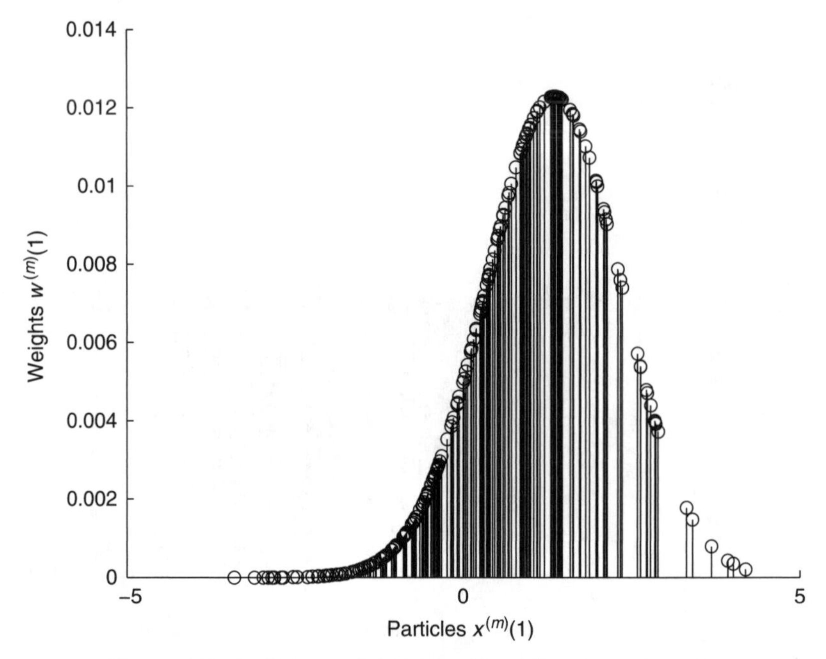

Figure 5.7 Particles and their weights at time instant $n = 1$.

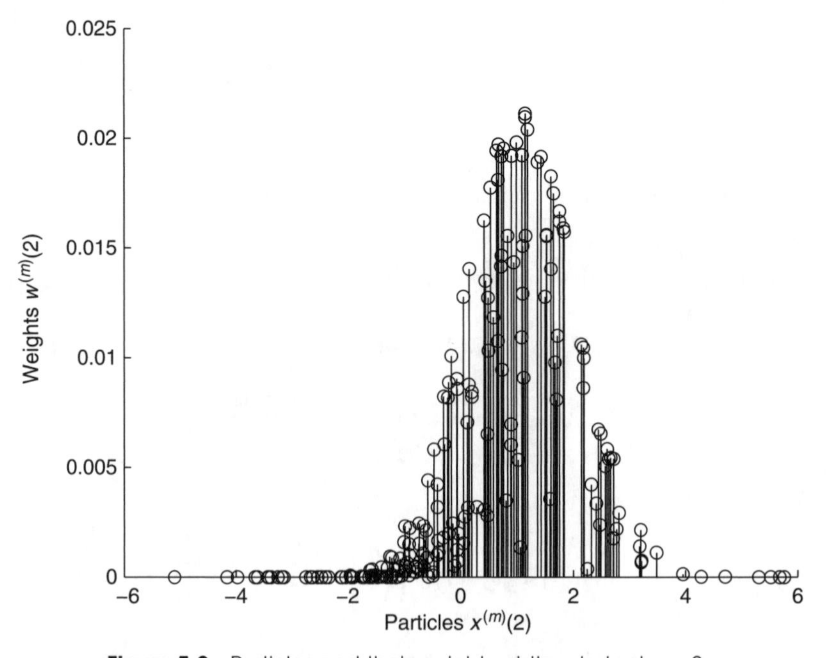

Figure 5.8 Particles and their weights at time instant $n = 2$.

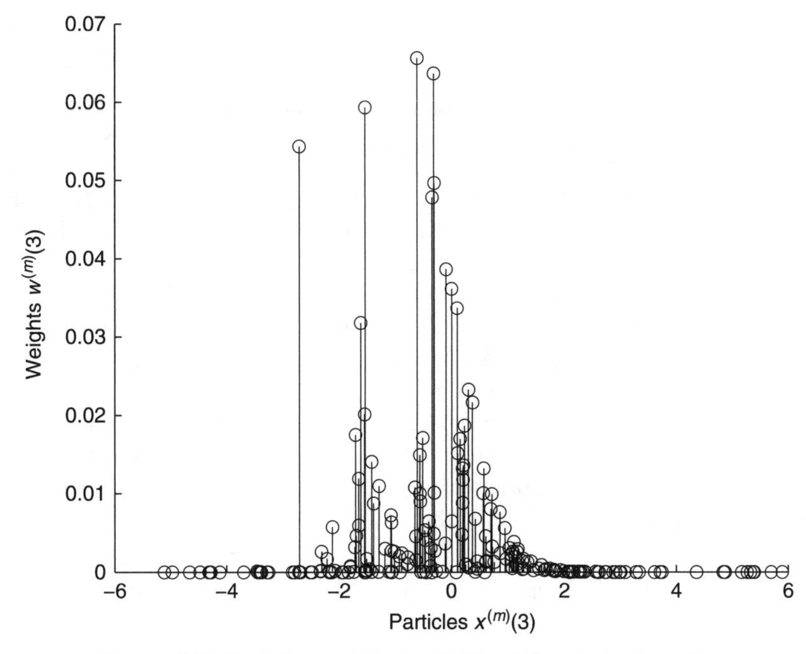

Figure 5.9 Particles and their weights at time instant $n = 3$.

Time instant $n = 3$

Generation of particles using the proposal distribution

$$x^{(m)}(3) \sim \frac{1}{\sqrt{2\pi}} \exp\left(-\frac{(x(3) - x^{(m)}(2))^2}{2}\right).$$

Weight update

$$w^{(m)}(3) \propto w^{(m)}(2) \exp\left(-\frac{(y(3) - x^{(m)}(3))^2}{2}\right).$$

The particles and their weights are shown in Figure 5.9.

The process continues in the same way as new observations become available. Note that the variance of the weights increases with time, which is an unwanted effect. Even at time instant $n = 3$, a few particles have large weights and the rest have small weights.

5.4 THE CHOICE OF PROPOSAL DISTRIBUTION AND RESAMPLING

In this section two important issues that affect the performance and implementation of particle filtering algorithms are discussed. One is the choice of proposal distributions and the other the concept of resampling, which turns out to be indispensable in the implementation of PFs.

5.4.1 Choice of Proposal Distribution

The proposal distribution (importance function) plays a crucial role in the performance of particle filtering. From a practical and intuitive point of view, it is desirable to use easy-to-sample proposal distributions that produce particles with a large enough variance in order to avoid exploration of the state space in too narrow regions and thereby contributing to losing the tracks of the state, but not too large to alleviate generation of too dispersed particles [55]. The support of the proposal distributions has to be the same as that of the targeted distribution.

As already pointed out, the approximation of the posterior distribution with the random measure obtained by the proposal distribution will improve if the proposal becomes very similar to the posterior. In fact, the optimal choice for the importance function is the posterior distribution, that is

$$\pi(\boldsymbol{x}(n)\,|\,\boldsymbol{x}^{(m)}(0:n-1), \boldsymbol{y}(1:n)) = f(\boldsymbol{x}(n)\,|\,\boldsymbol{x}^{(m)}(n-1), \boldsymbol{y}(n)) \tag{5.22}$$

which corresponds to the following weight calculation

$$w^{(m)}(n) \propto w^{(m)}(n-1) f(\boldsymbol{y}(n)\,|\,\boldsymbol{x}^{(m)}(n-1)).$$

This importance function minimizes the variance of the weights, $w^{(m)}(n)$, conditional on $\boldsymbol{x}^{(m)}(0:n-1)$ and $\boldsymbol{y}(1:n)$. However, the computation of the weights requires the integration of $\boldsymbol{x}(n)$, that is, solving

$$f(\boldsymbol{y}(n)\,|\,\boldsymbol{x}^{(m)}(n-1)) = \int f(\boldsymbol{y}(n)\,|\,\boldsymbol{x}(n)) f(\boldsymbol{x}(n)\,|\,\boldsymbol{x}^{(m)}(n-1)) d\boldsymbol{x}(n).$$

Thus, the implementation of the optimal importance function may be difficult for two reasons: First, direct sampling from the posterior (5.22) may not be easy, and second, the computation of the weights may require integration.

■ **EXAMPLE 5.5**

Consider the following decomposition of the posterior in (5.22)

$$f(\boldsymbol{x}(n)\,|\,\boldsymbol{x}^{(m)}(n-1), \boldsymbol{y}(n)) \propto f(\boldsymbol{y}(n)\,|\,\boldsymbol{x}(n)) f(\boldsymbol{x}(n)\,|\,\boldsymbol{x}^{(m)}(n-1)).$$

If the distributions on the right-hand side of the previous expressions are Gaussians, their product will also be Gaussian. Thus, the proposal is a Gaussian and sampling from it can readily be performed. It is worth pointing out that if the noises in the system are additive and Gaussian and the observation is a linear function of the state, we can sample from the optimal proposal distribution *even* if the function in the state equation is nonlinear.

A popular choice for the importance function is the prior

$$\pi(\boldsymbol{x}(n) \,|\, \boldsymbol{x}^{(m)}(0:n-1), \boldsymbol{y}(1:n)) = f(\boldsymbol{x}(n) \,|\, \boldsymbol{x}^{(m)}(n-1))$$

which yields importance weights proportional to the likelihood

$$w^{(m)}(n) \propto w^{(m)}(n-1) f(\boldsymbol{y}(n) \,|\, \boldsymbol{x}^{(m)}(n)).$$

The main advantage of this choice is the ease in the computation of the weights, which amounts to obtaining the likelihood function. However, the generation of the particles is implemented without the use of observations, and therefore not all of the available information is used to explore the state space. This may lead in some practical cases to poor estimation results. Strategies to improve this performance consist of the inclusion of a prediction step like that of the auxiliary PF [65] (see Subsection 5.5.2) or the use of a hybrid importance function if possible [37]. We refer to an importance function as a hybrid importance function if part of the state is proposed from a prior and the remaining state from the optimal importance function.

■ EXAMPLE 5.6

Consider a state space whose parameters can be divided in two groups $\boldsymbol{x}(n) = \{\boldsymbol{x}_1(n)\boldsymbol{x}_2(n)\}$ and where sampling can be carried out from $f(\boldsymbol{x}_2(n) \,|\, \boldsymbol{x}_1^{(m)}(n-1), \boldsymbol{x}_2^{(m)}(n-1))$ and $f(\boldsymbol{x}_1(n) \,|\, \boldsymbol{x}_2^{(m)}(n), \boldsymbol{x}_2^{(m)}(n-1), \boldsymbol{x}_1^{(m)}(n-1), \boldsymbol{y}(n))$. A hybrid proposal that combines the prior and the posterior importance functions is given by

$$\begin{aligned}\pi(\boldsymbol{x}(n) \,|\, \boldsymbol{x}^{(m)}(n-1), \boldsymbol{y}(n)) &= f(\boldsymbol{x}_2(n) \,|\, \boldsymbol{x}_1^{(m)}(n-1), \boldsymbol{x}_2^{(m)}(n-1)) \\ &\quad \times f(\boldsymbol{x}_1(n) \,|\, \boldsymbol{x}_2^{(m)}(n), \boldsymbol{x}_2^{(m)}(n-1), \boldsymbol{x}_1^{(m)}(n-1), \boldsymbol{y}(n))\end{aligned}$$

where $\boldsymbol{x}_2^{(m)}(n)$ is a sample from $f(\boldsymbol{x}_2(n) \,|\, \boldsymbol{x}_1^{(m)}(n-1), \boldsymbol{x}_2^{(m)}(n-1))$. The update of the weights can readily be obtained from the general expression given by (5.21).

5.4.2 Resampling

In particle filtering the discrete random measure degenerates quickly and only few particles are assigned meaningful weights. This degradation leads to a deteriorated functioning of particle filtering. Figure 5.10 illustrates three consecutive time instants of the operation of a PF which does not use resampling. Particles are represented by circles and their weights are reflected by the corresponding diameters. Initially, all the particles have the same weights, that is, all the diameters are equal, and at each time step the particles are propagated and assigned weights (note that the true posterior at each time instant is depicted in the figure.) As time evolves, all the particles except

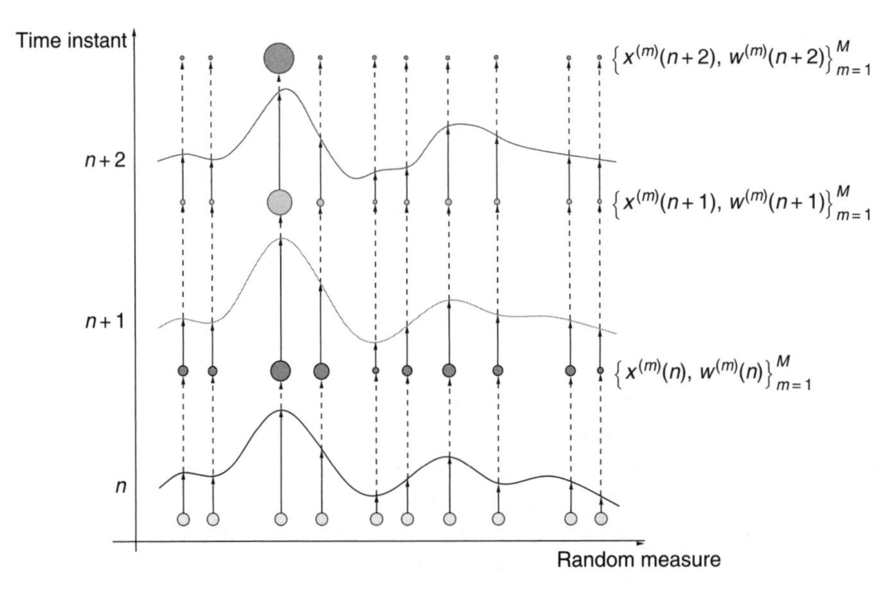

Figure 5.10 A pictorial description of particle filtering without resampling.

for very few are assigned negligible weights. In the last step there is only one particle with significant weight.

The mechanism of resampling as well as the use of good importance function can reduce this degeneracy. A measure of this degeneracy is the effective particle size defined by [54]

$$M_{\text{eff}} = \frac{M}{1 + \text{Var}(w^{*(m)}(n))}$$

where $w^{*(m)}(n) = \dfrac{f(\boldsymbol{x}(n))}{\pi(\boldsymbol{x}(n))}$ is the true particle weight. This metric can be estimated as

$$\hat{M}_{\text{eff}} = \frac{1}{\sum_{m=1}^{M}(w^{(m)}(n))^2}$$

with $w^{(m)}(n)$ being the normalized weight corresponding to the m-th particle at time instant n. If the effective particle size is below a predefined threshold, resampling is carried out. Clearly, when all the particles have the same weights, the variance of the weights is zero and the particle size is equal to the number of particles, M. The other extreme occurs when all the particles except one have negligible weights, and the particle size is equal to one.

Resampling eliminates particles with small weights and replicates particles with large weights. In general, if the random measure at time instant n is $\chi(n)$, it proceeds as follows.

1. Draw M particles, $\boldsymbol{x}^{(k_m)}(n)$, from the distribution given by $\chi(n)$, where the k_ms are the indexes of the drawn particles.
2. Let $\boldsymbol{x}^{(m)}(n) = \boldsymbol{x}^{(k_m)}(n)$, and assign equal weights $\frac{1}{M}$ to the particles.

The generation of the M resampled particles can be implemented in different ways. One possible scheme is illustrated in Figure 5.11 with $M = 5$ particles. There, the left column of circles depicts particles before resampling and the diameters of the circles are proportional to the weights of the particles. The dashed lines from the particles on the left to the middle line, which is equally divided in five areas, illustrate the mapping of the weights. The right column of circles are the particles after resampling. In general, the large particles are replicated and the small particles are removed. For example, the particle with the largest weight is replicated three times and the next two particles in size are preserved. The remaining two particles have small weights, and they are removed. After resampling all the circles have equal diameters, that is, all the weights are set to $\frac{1}{M}$.

Figure 5.12 illustrates the random measures and the actual probability distributions of interest in two consecutive time steps of the particle filtering algorithm that performs resampling after each cycle. In the figure, the solid curves represent the

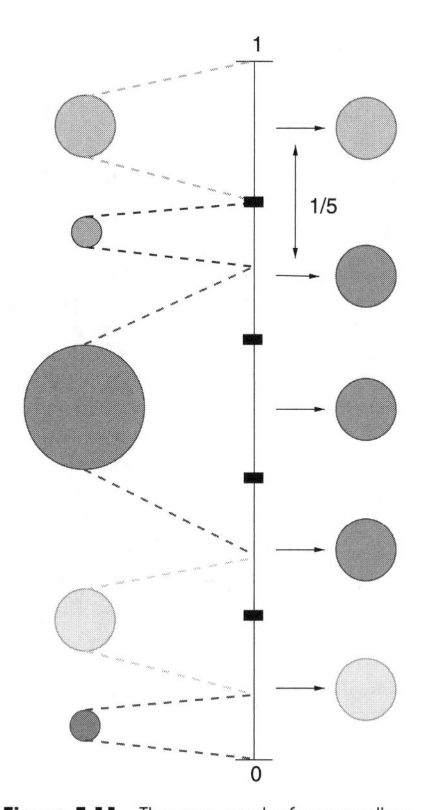

Figure 5.11 The concept of resampling.

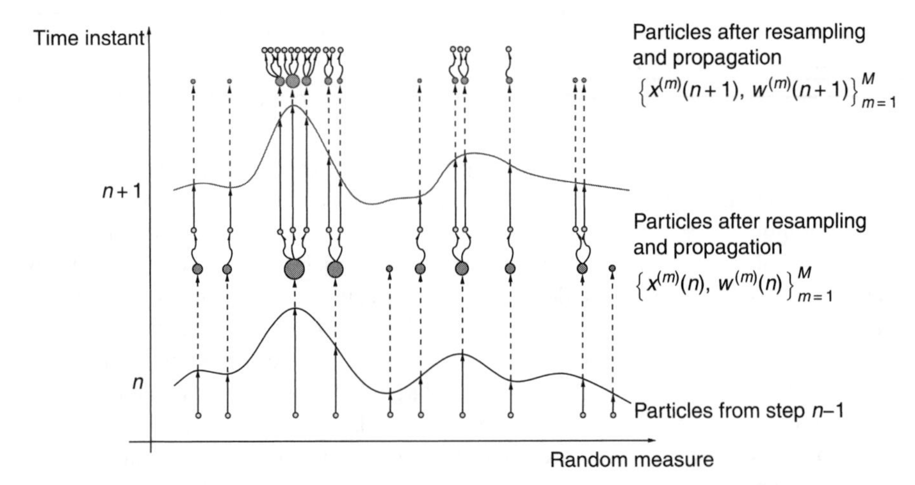

Figure 5.12 A pictorial description of particle filtering with resampling.

distributions of interest, which are approximated by the discrete measures. The sizes of the circles reflect the weights that are assigned to the particles after being generated from the proposal distribution. The resampling step eliminates the particles with the smallest weights and replicates those with large weights. The new set of particles have equal weights and are propagated in the next time step.

A summary of a particle filtering algorithm that includes resampling is displayed in the flowchart of Figure 5.13. At time instant n, a new set of particles is generated, and their weights are computed. Thereby the random measure $\chi(n)$ is generated and

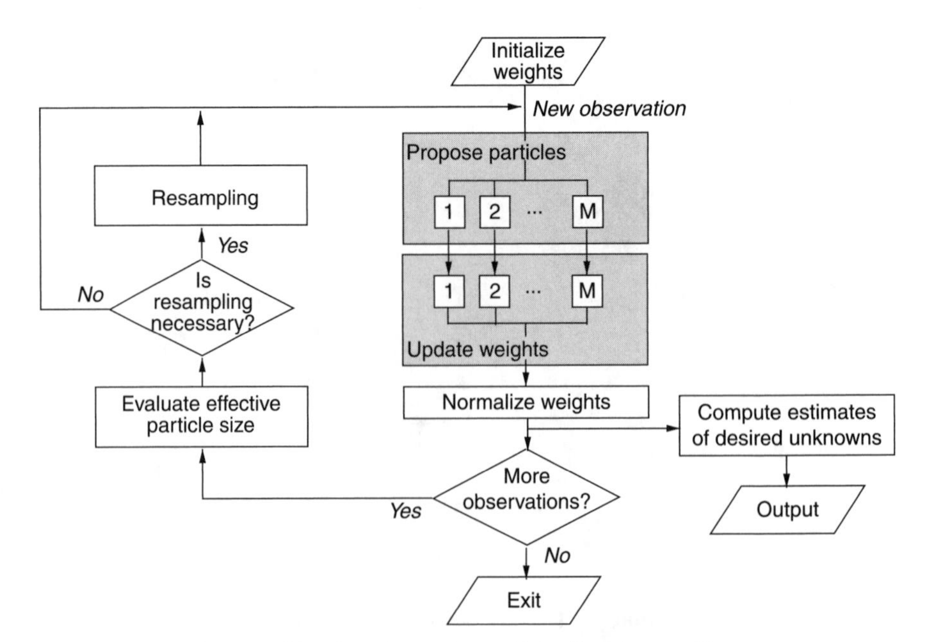

Figure 5.13 Flowchart of a particle filtering algorithm.

can be used for the estimation of the desired unknowns. Before the next step $n+1$, the effective particle size is estimated and resampling is carried out if necessary.

It is important to note that resampling becomes a major obstacle for efficient implementation of particle filtering algorithms in parallel very large scale integration (VLSI) hardware devices, because it creates full data dependencies among processing units [11]. Although some methods have been recently proposed [12], parallelization of resampling algorithms remains an open area of research.

5.5 SOME PARTICLE FILTERING METHODS

In this section we present three different particle filtering methods. They are also known as sampling-importance-resampling (SIR), auxiliary particle filtering (APF), and Gaussian particle filtering (GPF). The common feature of each of these methods is that at time instant n, they are represented by a discrete random measure given by $\chi(n) = \{\boldsymbol{x}^{(m)}(n), w^{(m)}(n)\}_{m=1}^{M}$, where, as before, $\boldsymbol{x}^{(m)}(n)$ is the m-th particle of the state vector at time instant n, $w^{(m)}(n)$ is the weight of that particle, and M is the number of particles. For each of these filters, we show how this random measure is obtained from $\chi(n-1)$ by using the observation vector $\boldsymbol{y}(n)$.

5.5.1 SIR particle filtering

The SIR method is the simplest of all of the particle filtering methods. It was proposed in [31] and was named bootstrap filter. Namely, the SIR method employs the prior density for drawing particles, which implies that the weights are only proportional to the likelihood of the drawn particles.[8] We now explain this in more detail.

Recall that if the particles are generated from a density function $\pi(\boldsymbol{x}(n))$, and the weights of the particles in the previous time step were $w^{(m)}(n-1)$, upon the reception of the measurement $\boldsymbol{y}(n)$, the weights are updated by

$$\tilde{w}^{(m)}(n) = w^{(m)}(n-1)\frac{f(\boldsymbol{y}(n)\,|\,\boldsymbol{x}^{(m)}(n))f(\boldsymbol{x}^{(m)}(n)\,|\,\boldsymbol{x}^{(m)}(n-1))}{\pi(\boldsymbol{x}^{(m)}(n)\,|\,\boldsymbol{x}^{(m)}(n-1), \boldsymbol{y}(1:n))}$$

where $\tilde{w}^{(m)}(n)$ denotes a nonnormalized weight. If the proposal distribution is equal to the prior, that is

$$\pi(\boldsymbol{x}(n)\,|\,\boldsymbol{x}^{(m)}(n-1), \boldsymbol{y}(1:n)) = f(\boldsymbol{x}(n)\,|\,\boldsymbol{x}^{(m)}(n-1))$$

the computation of the weights simplifies to

$$\tilde{w}^{(m)}(n) = w^{(m)}(n-1)f(\boldsymbol{y}(n)\,|\,\boldsymbol{x}^{(m)}(n)).$$

[8]Here we assume that the weights from the previous time instant are all equal due to resampling.

Furthermore, if the weights from the previous time step were all equal (because of resampling), the previous update equation becomes even simpler

$$\tilde{w}^{(m)}(n) = f(\boldsymbol{y}(n) \mid \boldsymbol{x}^{(m)}(n))$$

that is, the weight of the particle $\boldsymbol{x}^{(m)}(n)$ is only proportional to the likelihood of that particle.

Next, we explain in more detail the steps of the SIR method. In the first step we draw candidate particles $\boldsymbol{x}_c^{(m)}(n)$ according to

$$\boldsymbol{x}_c^{(m)}(n) \sim f(\boldsymbol{x}(n) \mid \boldsymbol{x}^{(m)}(n-1)), \ m = 1, 2, \ldots M.$$

In the second step, we compute the non-normalized weights of these particles by

$$\tilde{w}^{(m)}(n) = f(\boldsymbol{y}(n) \mid \boldsymbol{x}_c^{(m)}(n)), \ m = 1, 2, \ \ldots M$$

and then normalize them so that they sum up to one, that is, we use

$$w^{(m)}(n) = \frac{\tilde{w}^{(m)}(n)}{\sum_{j=1}^{M} \tilde{w}^{(j)}(n)}.$$

With the completion of this step, we have the random measure $\{\boldsymbol{x}_c^{(m)}(n), w^{(m)}(n)\}_{m=1}^{M}$, which should be used for computations of desired estimates.

In the third and final step we perform resampling of the particles $\boldsymbol{x}_c^{(m)}(n)$ according to the multinomial probability mass function defined by the weights $w^{(m)}(n)$. Namely, we draw indices k_m, for $m = 1, 2, \ldots, M$, where $Pr(k_m = i) = w^{(i)}(n)$. Once the indexes are drawn, we set

$$\boldsymbol{x}^{(m)}(n) = \boldsymbol{x}_c^{(k_m)}(n).$$

The weights of the particles $\boldsymbol{x}^{(m)}(n)$ are set to $w^{(m)}(n) = 1/M$. The whole SIR procedure is summarized in Table 5.3.

If resampling was not implemented as a third step, we have

$$\boldsymbol{x}^{(m)}(n) = \boldsymbol{x}_c^{(m)}(n)$$
$$\tilde{w}^{(m)}(n) = w^{(m)}(n-1) f(\boldsymbol{y}(n) \mid \boldsymbol{x}_c^{(m)}(n))$$

for $m = 1, 2, \ldots M$, and then we normalize the weights. After normalization, we would typically perform a test to decide if resampling is needed. If it was necessary, we would implement it as described; if not, we would proceed with the next time step. Recall that with resampling some particles are removed and the ones that are preserved may be replicated. We reiterate that, in general, resampling does not have to be implemented at every time step, but here we adopt to do so. If resampling is not performed at the end

Table 5.3 The SIR algorithm

```
Initialization
For m = 1, ..., M
  Sample x^(m)(0) ~ f(x(0))
  w^(m)(0) = 1/M
Recursions
For n = 1, 2, ...
  For m = 1, 2, ..., M
  Proposal of candidate particles
    Sample x_c^(m)(n) ~ f(x(n) | x^(m)(n-1))
  Computation of weights
    Evaluate the weights, w~^(m)(n) = f(y(n) | x_c^(m)(n))
    Normalize the weights, w^(m)(n) = w~^(m)(n) / sum_{j=1}^{M} w~^(j)(n)
  Resampling
    Sample k_m, where Pr(k_m = i) = w^(i)(n)
    Set x^(m)(n) = x_c^(k_m)(n) and
    w^(m)(n) = 1/M
```

of the recursion, the generated particles in the first step represent the support of the random measure used for the next time instant.

A clear advantage of the SIR method is that it is very simple for implementation. Its disadvantage is that in drawing the particles $\mathbf{x}^{(m)}(n)$ for exploring the state space, we do not use the observation $\mathbf{y}(n)$. In other words, once the particles are generated, the only thing we can do to steer the particles towards the region of the state space with large probability density is by resampling. If all of the generated particles are already far away from such regions, resampling will not help.

5.5.2 Auxiliary Particle Filtering

The APF attempts to improve the ability of the PF in exploring the state space by using the latest measurements. We know that with particle filtering, we approximate the filtering density $f(\boldsymbol{x}(n) \mid \boldsymbol{y}(1:n))$ by the mixture density

$$f(\boldsymbol{y}(n) \mid \boldsymbol{x}(n)) \sum_{m=1}^{M} w^{(m)}(n-1) f(\boldsymbol{x}(n) \mid \boldsymbol{x}^{(m)}(n-1)). \tag{5.23}$$

The underlying idea behind APF is to propose samples $\boldsymbol{x}^{(m)}(n)$ from this density. In order to do so, we introduce an auxiliary variable, which is an index variable. We denote it by k and we index it by m, so that we write it as k_m. We draw it from the set $\{1, 2, \ldots, M\}$,[9] and it denotes the particle stream which we want to update. Thus, if we draw $k_m = 5$, we work with the 5th stream, if we have $k_m = 11$, it is the 11th stream and so on.

First we describe the basic APF method. This method makes easy the problem of drawing $\boldsymbol{x}(n)$ from (5.23) by using estimates of $\boldsymbol{x}(n)$ for each stream of particles. If we denote the estimates by $\hat{\boldsymbol{x}}^{(m)}(n)$, we modify (5.23) and create a proposal distribution given by

$$\sum_{m=1}^{M} w^{(m)}(n-1) f(\boldsymbol{y}(n) \mid \hat{\boldsymbol{x}}^{(m)}(n)) f(\boldsymbol{x}(n) \mid \boldsymbol{x}^{(m)}(n-1)). \tag{5.24}$$

An estimate of $\boldsymbol{x}^{(m)}(n)$ can be any value of $\boldsymbol{x}(n)$ that is a good representative, which means that it should be a value that can easily be computed and has high likelihood. For example, if the state equation is

$$\boldsymbol{x}(n) = g_1(\boldsymbol{x}(n)) + \boldsymbol{v}_1(n)$$

and the noise vector $\boldsymbol{v}_1(n)$ is zero mean, an estimate of $\boldsymbol{x}^{(m)}(n)$ could be

$$\hat{\boldsymbol{x}}^{(m)}(n) = g_1(\boldsymbol{x}^{(m)}(n-1)).$$

With the estimates $\hat{\boldsymbol{x}}^{(m)}(n)$ and the new form of the proposal distribution (5.24), it is much easier to propose new particles $\boldsymbol{x}^{(m)}(n)$. The idea has a subtle point: we use (5.24) as a joint distribution of the auxiliary variable and the state. The implications is that first we draw the auxiliary variable (index) k_m from a multinomial distribution, where $Pr(k_m = i) \propto w^{(i)}(n-1)\, f(\boldsymbol{y}(n) \mid \hat{\boldsymbol{x}}^{(i)}(n))$.[10] The drawn index, say i, identifies the distribution from which we draw $\boldsymbol{x}^{(m)}(n)$, $f(\boldsymbol{x}(n) \mid \boldsymbol{x}^{(k_m=i)}(n-1))$, and so we proceed by drawing a particle from $f(\boldsymbol{x}(n) \mid \boldsymbol{x}^{(k_m=i)}(n-1))$. Once the particles are drawn, as with SIR, the last step is the computation of the weights.

Before we derive the formula for the update of the weights, we express the proposal distribution in a form that will make the derivation easy. First, we rewrite

[9] In fact, the number of drawn auxiliary variables can be different from M [65].
[10] This is basically the same procedure applied for resampling.

the proposal as

$$\pi(\boldsymbol{x}(n), k_m \mid \boldsymbol{y}(1:n)) = f(\boldsymbol{x}(n) \mid k_m, \boldsymbol{y}(1:n)) f(k_m \mid \boldsymbol{y}(1:n)).$$

The first factor on the right is

$$f(\boldsymbol{x}(n) \mid k_m, \boldsymbol{y}(1:n)) = f(\boldsymbol{x}(n) \mid \boldsymbol{x}^{(k_m)}(n-1))$$

because according to (5.24), given k_m, $\boldsymbol{y}(n)$ does not affect the density of $\boldsymbol{x}(n)$ and furthermore, given k_m, the density of $\boldsymbol{x}(n)$ is not a function of $\boldsymbol{y}(1:n-1)$ either, and instead it is simply the prior. This follows from the Markovian nature of the state variable. Recall that k_m is an index that points to the k_m-th stream and all its particles $\boldsymbol{x}^{(k_m)}(0:n-1)$. Thus, once, k_m is known, so is $\boldsymbol{x}^{(k_m)}(n-1)$, implying that all of the measurements $\boldsymbol{y}(1:n-1)$ become irrelevant in expressing the density of $\boldsymbol{x}(n)$.

For the second factor we can write

$$f(k_m \mid \boldsymbol{y}(1:n)) \propto w^{(k_m)}(n-1) f(\boldsymbol{y}(n) \mid \hat{\boldsymbol{x}}^{(k_m)}(n))$$

where the weight $w^{(k_m)}(n-1)$ is a function of $\boldsymbol{y}(1:n-1)$ and, clearly, $f(\boldsymbol{y}(n) \mid \hat{\boldsymbol{x}}^{(k_m)}(n))$ is a function of $\boldsymbol{y}(n)$. It is obvious that $f(k_m \mid \boldsymbol{y}(1:n))$ represents the probability of drawing k_m.

Therefore, for the update of the weight, we have the following

$$\begin{aligned}
\widetilde{w}^{(m)}(n) &= w^{(k_m)}(n-1) \frac{f(\boldsymbol{y}(n) \mid \boldsymbol{x}^{(m)}(n)) f(\boldsymbol{x}^{(m)}(n) \mid \boldsymbol{x}^{(k_m)}(n-1))}{\pi(\boldsymbol{x}^{(m)}(n), k_m \mid \boldsymbol{y}(1:n))} \\
&= w^{(k_m)}(n-1) \frac{f(\boldsymbol{y}(n) \mid \boldsymbol{x}^{(m)}(n)) f(\boldsymbol{x}^{(m)}(n) \mid \boldsymbol{x}^{(k_m)}(n-1))}{w^{(k_m)}(n-1) f(\boldsymbol{y}(n) \mid \hat{\boldsymbol{x}}^{(k_m)}(n)) f(\boldsymbol{x}^{(m)}(n) \mid \boldsymbol{x}^{(k_m)}(n-1))} \\
&= \frac{f(\boldsymbol{y}(n) \mid \boldsymbol{x}^{(m)}(n))}{f(\boldsymbol{y}(n) \mid \hat{\boldsymbol{x}}^{(k_m)}(n))}.
\end{aligned}$$

In summary, the procedure is rather straightforward. Given the particles and their weights at time instant $n-1$, $\boldsymbol{x}^{(m)}(n-1)$ and $w^{(m)}(n-1)$, respectively, first we compute estimates $\hat{\boldsymbol{x}}^{(m)}(n)$. For these estimates we evaluate the weights by

$$\hat{w}^{(m)}(n) \propto w^{(m)}(n-1) f(\boldsymbol{y}(n) \mid \hat{\boldsymbol{x}}^{(m)}(n)). \tag{5.25}$$

Next we draw the indexes of the particle streams that we will continue to append. The indexes are drawn from the multinomial distribution with parameters $\hat{w}^{(m)}(n)$. With this, we effectively perform resampling. After this step, we draw the particles of $\boldsymbol{x}(n)$ according to

$$\boldsymbol{x}^{(m)}(n) \sim f(\boldsymbol{x}(n) \mid \boldsymbol{x}^{(k_m)}(n-1)), \quad m = 1, 2, \ldots, M.$$

The last step amounts to computing the weights of these particles by

$$w^{(m)}(n) \propto \frac{f(\boldsymbol{y}(n) \mid \boldsymbol{x}^{(m)}(n))}{f(\boldsymbol{y}(n) \mid \hat{\boldsymbol{x}}^{(k_m)}(n))}. \tag{5.26}$$

The algorithm is summarized in Table 5.4.

So, what do we gain by computing estimates of $\boldsymbol{x}(n)$ and implementing the drawing of particles by auxiliary variables? By using the estimates of $\boldsymbol{x}^{(m)}(n)$, we look ahead to how good the particle streams may be. Rather than resampling from samples obtained from the prior, we first resample by using the latest measurement and then propagate from the surviving streams. Thereby, at the end of the recursion instead of having particles propagated without the use of $\boldsymbol{y}(n)$, we have particles moved in directions preferred by $\boldsymbol{y}(n)$. With SIR, the data $\boldsymbol{y}(n)$ affect the direction of particle propagation later than they do with APF.

Table 5.4 Auxiliary particle filter

Initialization

For $m = 1, 2, \ldots, M$

  Sample $\boldsymbol{x}(0)^{(m)} \sim f(\boldsymbol{x}(0))$

$$w^{(m)}(0) = \frac{1}{M}$$

Recursions

For $n = 1, 2, \ldots$

  For $m = 1, 2, \ldots, M$

  Estimation of next particles

    Compute $\hat{\boldsymbol{x}}^{(m)}(n) = E(\boldsymbol{x}(n) \mid \boldsymbol{x}^{(m)}(n-1))$

  Sample the indexes k_m of the streams that survive

    Sample $k_m = i$ with probability

    $w^{(i)}(n-1) f(\boldsymbol{y}(n) \mid \hat{\boldsymbol{x}}^{(i)}(n))$

  Sample the new particles for time instant n

    $\boldsymbol{x}^{(m)}(n) \sim f(\boldsymbol{x}(n) \mid \boldsymbol{x}^{(k_m)}(n-1))$

  Computation of weights

    Evaluate the weights $\tilde{w}^{(m)}(n) = \dfrac{f(\boldsymbol{y}(n) \mid \boldsymbol{x}^{(m)}(n))}{f(\boldsymbol{y}(n) \mid \hat{\boldsymbol{x}}^{(k_m)}(n))}$

    Normalize the weights $w^{(m)}(n) = \dfrac{\tilde{w}^{(m)}(n)}{\sum_{j=1}^{M} \tilde{w}^{(j)}(n)}$

It is important to note that one cannot guarantee that the basic APF method will perform better than the SIR algorithm, the help of the latest observation notwithstanding. The reason for this is that the new particles are still generated by the prior only.

Now we describe a more general version of the APF method. As with the basic APF, one preselects the streams that are propagated. Instead of a preselection based on the weights (5.25), we use weights that we denote by $w_a^{(m)}(n)$. After the k_m is selected, for propagation, we use the proposal distribution $\pi(\boldsymbol{x}(n)\,|\,\boldsymbol{x}^{(k_m)}(n-1), \boldsymbol{y}(n))$, that is

$$\boldsymbol{x}^{(m)}(n) \sim \pi(\boldsymbol{x}(n)\,|\,\boldsymbol{x}^{(k_m)}(n-1), \boldsymbol{y}(n)). \tag{5.27}$$

The new weights are then computed according to

$$w^{(m)}(n) \propto \frac{w^{(k_m)}(n-1)}{w_a^{(k_m)}(n)} \frac{f(\boldsymbol{y}(n)\,|\,\boldsymbol{x}^{(m)}(n)) f(\boldsymbol{x}^{(m)}(n)\,|\,\boldsymbol{x}^{(k_m)}(n-1))}{\pi(\boldsymbol{x}^{(m)}(n)\,|\,\boldsymbol{x}^{(k_m)}(n-1), \boldsymbol{y}(n))}. \tag{5.28}$$

Note that for the basic APF we have

$$w_a^{(k_m)}(n) \propto w^{(k_m)}(n-1) f(\boldsymbol{y}(n)\,|\,\hat{\boldsymbol{x}}^{(k_m)}(n))$$

and

$$\pi(\boldsymbol{x}^{(m)}(n)\,|\,\boldsymbol{x}^{(k_m)}(n-1), \boldsymbol{y}(n)) = f(\boldsymbol{x}^{(m)}(n)\,|\,\boldsymbol{x}^{(k_m)}(n-1)).$$

As with the standard PFs, the performance of the APF depends strongly on the proposal distribution $\pi(\boldsymbol{x}(n)\,|\,\boldsymbol{x}^{(k_m)}(n-1), \boldsymbol{y}(n))$. However, its performance also depends on the choice of the weights $w_a^{(m)}(n)$.

5.5.3 Gaussian Particle Filtering

With Gaussian particle filtering, we approximate the posterior and the predictive densities of the states with Gaussians [50] as is done by the EKF [3]. However, unlike with the EKF, we do not linearize any of the nonlinearities in the system. The method is very simple to implement, and its advantage over other particle filtering methods is that it does not require resampling. Another advantage is that the estimation of constant parameters is not a problem as is the case with other methods (see Section 5.6). In brief, the treatment of dynamic and constant states is identical. On the other hand, its disadvantage is that if the approximated densities are not Gaussians, the estimates may be inaccurate and the filter may diverge as any other method that uses Gaussian approximations. In that case, an alternative could be the use of Gaussian sum particle filtering where the densities in the system are approximated by mixture Gaussians [51].

The method proceeds as follows. Let the random measure at time instant $n-1$ be given by $\chi(n-1) = \{\boldsymbol{x}^{(m)}(n-1), 1/M\}_{m=1}^{M}$. In the first step, we draw samples $\boldsymbol{x}_c^{(m)}(n)$ from the prior, that is

$$\boldsymbol{x}_c^{(m)}(n) \sim f(\boldsymbol{x}(n) \,|\, \boldsymbol{x}^{(m)}(n-1)).$$

For the obtained samples, we compute their weights according to

$$\widetilde{w}^{(m)}(n) = f(\boldsymbol{y}(n) \,|\, \boldsymbol{x}_c^{(m)}(n))$$

and then we normalize them. Now, the assumption is that the particles $\boldsymbol{x}_c^{(m)}(n)$ and their weights $w^{(m)}(n)$ approximate a normal distribution whose moments are estimated by

$$\boldsymbol{\mu}(n) = \sum_{m=1}^{M} w^{(m)}(n)\boldsymbol{x}_c^{(m)}(n)$$

$$\boldsymbol{\Sigma}(n) = \sum_{m=1}^{M} w^{(m)}(n)(\boldsymbol{x}_c^{(m)}(n) - \boldsymbol{\mu}(n))(\boldsymbol{x}_c^{(m)}(n) - \boldsymbol{\mu}(n))^{\top}.$$

Finally, the particles that are used for propagation at the next time instant $n+1$ are generated by

$$\boldsymbol{x}^{(m)}(n) \sim \mathcal{N}(\boldsymbol{\mu}(n), \boldsymbol{\Sigma}(n))$$

and they are all assigned the same weights. The method is summarized by Table 5.5.

5.5.4 Comparison of the Methods

In this Subsection, we show the performances of the presented methods (SIR, APF, and GPF) on simulated data.

■ **EXAMPLE 5.7**

The data are considered to be generated according to a stochastic volatility model, which is defined by [6, 16, 31, 47]

$$x(n) = \alpha x(n-1) + v_1(n) \tag{5.29}$$

$$y(n) = \beta v_2(n) e^{x(n)/2} \tag{5.30}$$

where the unknown state variable $x(n)$ is called log-volatility, $v_1(n) \sim \mathcal{N}(0, \sigma^2)$, $v_2(n) \sim \mathcal{N}(0, 1)$, and α and β are parameters known as persistence in volatility shocks and modal volatility, respectively. The Gaussian processes $v_1(n)$ and $v_2(n)$ are assumed independent. This model is very much researched in the econometrics literature. Given the observations $y(n)$ and the model parameters, we want to estimate the unobserved state $x(n)$.

Table 5.5 The GPF algorithm

Initialization

For $m = 1, 2, \ldots, M$

Sample $\mathbf{x}_c^{(m)}(0) \sim f(\mathbf{x}(0))$

$$w^{(m)}(0) = \frac{1}{M}$$

Recursions

For $n = 1, 2, \ldots$

Drawing particles from the predictive density

$$\mathbf{x}_c^{(m)}(n) \sim f(\mathbf{x}(n) \mid \mathbf{x}^{(m)}(n-1)), \quad m = 1, 2, \ldots, M$$

Computation of the weights

$$\tilde{w}^{(m)}(n) = f(\mathbf{y}(n) \mid \mathbf{x}_c^{(m)}(n))$$

$$w^{(m)}(n) = \frac{\tilde{w}^{(m)}(n)}{\sum_{j=1}^{M} \tilde{w}^{(j)}(n)}$$

Computation of the moments of the filtering density

$$\boldsymbol{\mu}(n) = \sum_{m=1}^{M} w^{(m)}(n)\, \mathbf{x}_c^{(m)}(n)$$

$$\boldsymbol{\Sigma}(n) = \sum_{m=1}^{M} w^{(m)}(n)\, (\mathbf{x}_c^{(m)}(n) - \boldsymbol{\mu}(n))\, (\mathbf{x}_c^{(m)}(n) - \boldsymbol{\mu}(n))^{\top}$$

Drawing particles for propagation at the next time instant

$$\mathbf{x}^{(m)}(n) \sim \mathcal{N}(\boldsymbol{\mu}(n), \boldsymbol{\Sigma}(n)), \quad m = 1, 2, \ldots, M.$$

In Figure 5.14, we can see one realization of the observation $y(n)$ generated with model parameters $\alpha = 0.9702$, $\beta = 0.5992$ and $\sigma^2 = 0.178$. The figure also shows the hidden process $x(n)$ and its MSE estimate obtained by SIR with $M = 2000$ particles. The estimates obtained by the other two filters are not shown because they basically yielded the same estimates.

In Figure 5.15 we can see the computed MSEs per sample in 50 different realizations when the number of particles was $M = 500$. The results suggest that the three methods have comparable performance.[11]

[11]The performances of the APF and GPF may be improved if the proposal distributions provide better particles. For an application of an APF with improved performance, see [15].

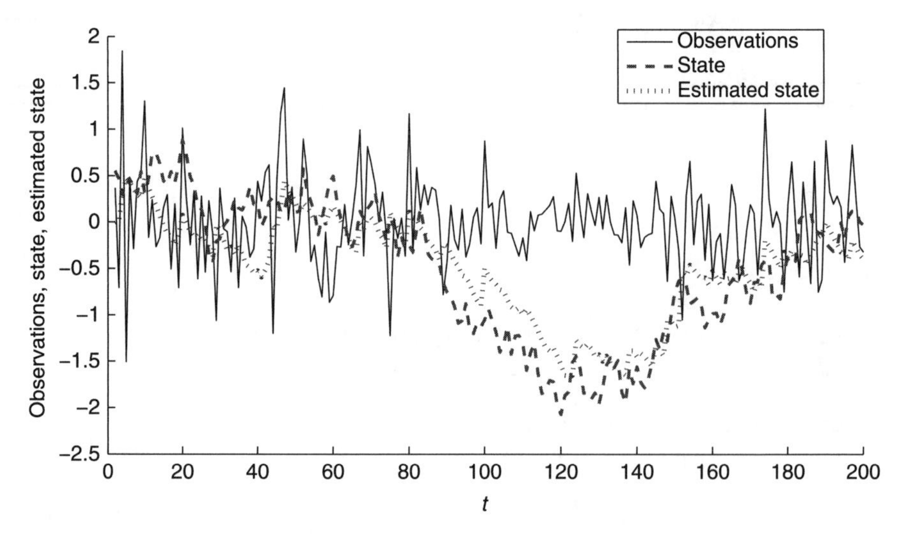

Figure 5.14 Stochastic volatility model: One realization of the observations, the state and its estimate.

One can expect that with the number of particles, the performance of the PFs improves. Indeed this is the case. In Figure 5.16, we can see the MSE per sample of the three filters as a function of the used number of particles. The number of generated realizations in the sample with a fixed number of particles was 1000. The results also show that for large enough number of particles the methods yield indistinguishable results.

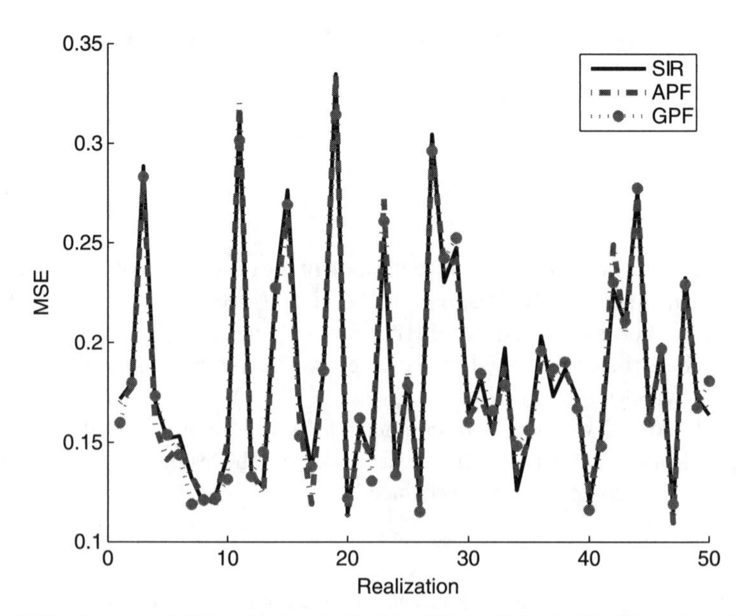

Figure 5.15 Average MSEs (obtained with $M = 500$ particles in 50 different realizations.

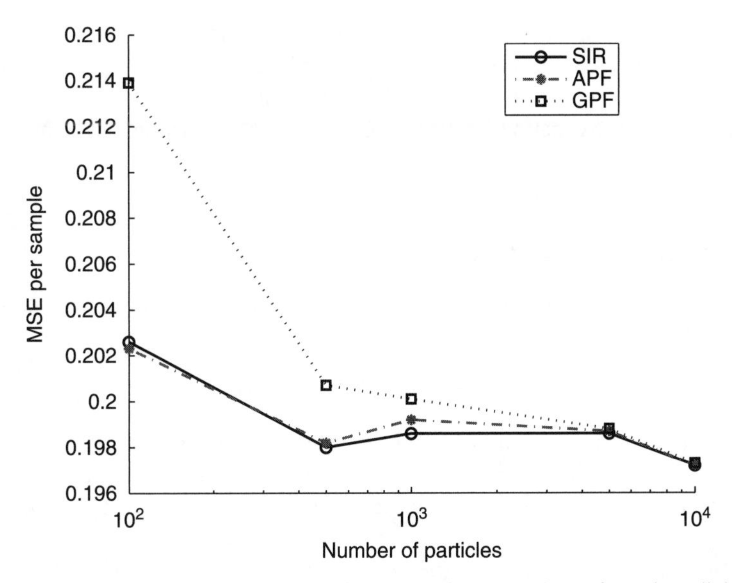

Figure 5.16 MSEs per sample as a function of the number of used particles.

5.6 HANDLING CONSTANT PARAMETERS

The particle filtering methodology was originally devised for the estimation of dynamic signals rather than static parameters. The most efficient way to address the problem is to integrate out the unknown parameters when possible, either analytically (see next section) [20] or by Monte Carlo procedures [72]. The former methods, however, depend on the feasibility of integration that is, on the mathematical model of the system. Generic solutions, useful for any model, are scant and limited in performance. A common feature of most of the approaches is that they introduce artificial evolution of the fixed parameters and thereby treat them in a similar way as the dynamic states of the model [31, 56]. Some methods insert the use of Markov chain Monte Carlo sampling to preserve diversity of the particles [10, 28]. A recent work [24] introduces a special class of PFs called density-assisted PFs that approximate the filtering density with a predefined parametric density by generalizing the concepts of Gaussian PFs and Gaussian sum PFs. These new filters can cope with constant parameters more naturally than previously proposed methods. In this section the problem of handling static parameters by PFs is reviewed under a kernel-based auxiliary PF method [56] and the density-assisted particle filtering technique [24].

We can reformulate the state-space model introduced in Section 1.2 to explicitly incorporate fixed parameters as

$$\begin{aligned} \boldsymbol{x}(n) &= g_1(\boldsymbol{x}(n-1),\, \boldsymbol{\theta},\, \boldsymbol{v}_1(n)) \\ \boldsymbol{y}(n) &= g_2(\boldsymbol{x}(n),\, \boldsymbol{\theta},\, \boldsymbol{v}_2(n)) \end{aligned}$$

where all the symbols have the same meaning as before and $\boldsymbol{\theta}$ is a vector of fixed parameters. Based on the observations $\boldsymbol{y}(n)$ and the assumptions, the objective is to estimate $\boldsymbol{x}(n)$ and $\boldsymbol{\theta}$ recursively. In the particle filtering context, this amounts to obtaining the approximation of $f(\boldsymbol{x}(n), \boldsymbol{\theta} \mid \boldsymbol{y}(1:n))$ by updating the approximation for $f(\boldsymbol{x}(n-1), \boldsymbol{\theta} \mid \boldsymbol{y}(1:n-1))$.

5.6.1 Kernel-based Auxiliary Particle Filter

The inclusion of fixed parameters in the model implies extending the random measure to the form

$$\chi(n) = \{\boldsymbol{x}^{(m)}(n), \boldsymbol{\theta}^{(m)}(n), w^{(m)}(n)\}_{m=1}^{M}$$

where the index n in the samples of $\boldsymbol{\theta}$ indicates the approximation of the posterior at time n and *not* time-variation of the parameter vector. The random measure approximates the density of interest $f(\boldsymbol{x}(n), \boldsymbol{\theta} \mid \boldsymbol{y}(n))$, which can be decomposed as

$$f(\boldsymbol{x}(n), \boldsymbol{\theta} \mid \boldsymbol{y}(1:n)) \propto f(\boldsymbol{y}(n) \mid \boldsymbol{x}(n), \boldsymbol{\theta}) f(\boldsymbol{x}(n) \mid \boldsymbol{\theta}, \boldsymbol{y}(1:n-1)) f(\boldsymbol{\theta} \mid \boldsymbol{y}(1:n-1)).$$

From the previous expression it is clear that there is a need for approximation of the density $f(\boldsymbol{\theta} \mid \boldsymbol{y}(1:n-1))$. In [56], a density of the form

$$f(\boldsymbol{\theta} \mid \boldsymbol{y}(1:n-1)) \approx \sum_{m=1}^{M} w^{(m)}(n) \mathcal{N}(\boldsymbol{\theta} \mid \bar{\boldsymbol{\theta}}^{(m)}(n), h^2 \boldsymbol{\Sigma}_{\boldsymbol{\theta}}(n))$$

is used, and it represents a mixture of Gaussian distributions, where the mixands $\mathcal{N}(\boldsymbol{\theta} \mid \bar{\boldsymbol{\theta}}^{(m)}(n), h^2\boldsymbol{\Sigma}_{\boldsymbol{\theta}}(n))$ are weighted by the particle weights $w^{(m)}(n)$. The parameters of the mixands are obtained using the previous time instant particles and weights, that is

$$\bar{\boldsymbol{\theta}}^{(m)}(n) = \sqrt{1-h^2}\,\boldsymbol{\theta}^{(m)}(n-1) + (1 - \sqrt{1-h^2}) \sum_{i=1}^{M} w^{(i)}(n-1)\boldsymbol{\theta}^{(i)}(n-1)$$

$$\boldsymbol{\Sigma}_{\boldsymbol{\theta}}(n) = \sum_{m=1}^{M} w^{(m)}(n)(\boldsymbol{\theta}^{(m)}(n) - \bar{\boldsymbol{\theta}}(n))(\boldsymbol{\theta}^{(m)}(n) - \bar{\boldsymbol{\theta}}(n))^{\top}$$

where h^2 is a smoothing parameter computed by $h^2 = 1 - \left(\frac{3\gamma - 1}{2\gamma}\right)^2$, and $\gamma \in (0,1)$ represents a discount factor, typically around 0.95–0.99.

We now describe the implementation of this idea in the context of auxiliary particle filtering. Assume that at time instant $n-1$, we have the random measure $\chi(n-1) = \{\boldsymbol{x}^{(m)}(n-1), \boldsymbol{\theta}^{(m)}(n-1), w^{(m)}(n-1)\}$. Then, we proceed as follows.

1. Estimate the next particle by $\hat{\boldsymbol{x}}^{(m)}(n) = E(\boldsymbol{x}(n) \mid \boldsymbol{x}^{(m)}(n-1), \boldsymbol{\theta}^{(m)}(n-1))$. This step is identical to the APF.

2. Sample the indexes k_m of the streams that survive, where $k_m = i$ with probability $w_a^{(i)} \propto w^{(i)}(n-1)f(y(n)\,|\,\hat{\boldsymbol{x}}^{(i)}(n),\ \bar{\boldsymbol{\theta}}^{(i)}(n-1))$. Here we basically resample so that the most promising streams are kept in the random measure and the less promising are removed.
3. Draw particles of the fixed parameter vector according to

$$\boldsymbol{\theta}^{(m)}(n) \sim \mathcal{N}(\bar{\boldsymbol{\theta}}^{(k_m)}(n-1),\ h^2\boldsymbol{\Sigma}_{\boldsymbol{\theta}}(n-1))$$

where the means and the covariance of the mixands are obtained at the end of the cycle of computations for time instant $n-1$. This step takes care of the vector of constant parameters.
4. Draw particles of the state according to

$$\boldsymbol{x}^{(m)}(n) \sim f(\boldsymbol{x}(n)\,|\,\boldsymbol{x}^{(k_m)}(n-1),\ \boldsymbol{\theta}^{(m)}(n)).$$

With this step we complete the drawing of new particles.
5. Update and normalize the weights, where

$$w^{(m)}(n) \propto \frac{f(y(n)\,|\,\boldsymbol{x}^{(m)}(n),\ \boldsymbol{\theta}^{(m)}(n))}{f(y(n)\,|\,\hat{\boldsymbol{x}}^{(k_m)}(n),\ \bar{\boldsymbol{\theta}}^{(k_m)}(n-1))}.$$

With this step we associate weights to the particles.
6. Compute the parameters of the kernels used in constructing the mixture Gaussian according to

$$\bar{\boldsymbol{\theta}}^{(m)}(n) = \sqrt{1-h^2}\,\boldsymbol{\theta}^{(m)}(n) + (1-\sqrt{1-h^2})\sum_{i=1}^{M} w^{(i)}(n)\boldsymbol{\theta}^{(i)}(n)$$

$$\boldsymbol{\Sigma}_{\boldsymbol{\theta}}(n) = \sum_{m=1}^{M} w^{(m)}(n)(\boldsymbol{\theta}^{(m)}(n) - \bar{\boldsymbol{\theta}}(n))(\boldsymbol{\theta}^{(m)}(n) - \bar{\boldsymbol{\theta}}(n))^{\top}.$$

■ EXAMPLE 5.8

We go back to the problem described in Example 5.7 where we want to estimate the hidden stochastic volatility. Suppose that in addition to the unknown $\boldsymbol{x}(n)$ we also do not know the parameters $\boldsymbol{\theta} = [\alpha\ \beta\ \sigma^2]^{\top}$. If we were to apply the kernel-based APF, we would first generate particles of $\boldsymbol{x}^{(m)}(0)$, and $\boldsymbol{\theta}^{(m)}(0)$ from the priors and implement the method shown in Table 5.5. As a first step in the recursion, we estimate $\hat{\boldsymbol{x}}^{(m)}(n),\ m = 1, 2, \cdots, M$ by

$$\hat{\boldsymbol{x}}^{(m)}(n) = \alpha^{(m)}(n-1)\boldsymbol{x}^{(m)}(n-1).$$

Then we obtain the weights $w_a^{(i)} \propto w^{(i)}(n-1)f(y(n)\,|\,\hat{\boldsymbol{x}}^{(i)}(n),\ \bar{\boldsymbol{\theta}}^{(i)}(n-1))$, where

$$f(y(n)\,|\,\hat{\boldsymbol{x}}^{(i)}(n),\ \bar{\boldsymbol{\theta}}^{(i)}(n-1)) = \mathcal{N}(0,\ \bar{\beta}^{2^{(i)}}(n-1)e^{\hat{x}^{(i)}(n)})$$

and sample the indexes k_m according the computed weights. We follow with generating the new particles of the constant parameters

$$\boldsymbol{\theta}^{(m)}(n) \sim \mathcal{N}(\bar{\boldsymbol{\theta}}^{(k_m)}(n-1), h^2\boldsymbol{\Sigma}_{\boldsymbol{\theta}}(n-1))$$

and the unknown states

$$\boldsymbol{x}^{(m)}(n) \sim \mathcal{N}(\alpha^{(k_m)}(n)\boldsymbol{x}^{(k_m)}(n-1), \sigma^{2^{(k_m)}}(n)).$$

The weights are computed by

$$w^{(m)}(n) \propto \frac{\mathcal{N}(0, \beta^{2^{(m)}}(n)e^{\boldsymbol{x}^{(m)}(n)})}{\mathcal{N}(0, \bar{\beta}^{2^{(k_m)}}(n-1)e^{\hat{\boldsymbol{x}}^{(k_m)}(n)})}.$$

Finally the kernel parameters $\bar{\boldsymbol{\theta}}^{(m)}(n)$ and $\boldsymbol{\Sigma}_{\boldsymbol{\theta}}(n)$ are found as shown in Table 5.5.

■ EXAMPLE 5.9

Consider as in Example 5.1 the problem of tracking one target using two static sensors which collected bearings-only measurements. We extend the problem to the case where the measurements are biased. Then, the mathematical formulation of the problem is given by

$$\begin{aligned}\boldsymbol{x}(n) &= \boldsymbol{A}\boldsymbol{x}(n-1) + \boldsymbol{B}\boldsymbol{v}_1(n)\\ \boldsymbol{y}(n) &= g_2(\boldsymbol{x}(n)) + \boldsymbol{\theta} + \boldsymbol{v}_2(n)\end{aligned}$$

where all the parameters have the same meaning as before and $\boldsymbol{\theta} = [b_1\ b_2]^\top$ represents a vector of unknown constant biases.

In step 1, we generate $\hat{\boldsymbol{x}}^{(m)}(n)$ by

$$\hat{\boldsymbol{x}}^{(m)}(n) = \boldsymbol{A}\boldsymbol{x}^{(m)}(n).$$

The resampling is carried out by using weights that are proportional to $w^{(i)}(n-1)\mathcal{N}(g_2(\hat{\boldsymbol{x}}^{(i)}(n)) + \bar{\boldsymbol{\theta}}^{(i)}(n-1), \boldsymbol{\Sigma}_v)$, where the Gaussian is computed at $\boldsymbol{y}(n)$. The remaining steps can readily be deduced from the scheme described by Table 5.5.

5.6.2 Density-assisted Particle Filter

As seen in the previous section, the GPFs approximate the predictive and filtering densities by Gaussian densities whose parameters are estimated from the particles and their weights. Similarly, the Gaussian sum PFs [51] approximate these densities with mixtures of Gaussians. The approximating densities can be other than Gaussians or mixtures of Gaussian, and therefore, we refer to the general class of filters of this type as density-assisted PFs (DAPFs) [24], which are a generalization of the Gaussian and Gaussian sum PFs. The main advantages of DAPFs is that they do not necessarily use resampling in the sense carried out by standard PFs and they do not share the limitation regarding the estimation of constant model parameters. Here we explain how we can use DAPFs when we have constant parameters in the model.

Let $\chi(n-1) = \{\boldsymbol{x}^{(m)}(n-1), \boldsymbol{\theta}^{(m)}(n-1), w^{(m)}(n-1)\}_{m=1}^{M}$ be the random measure at time instant $n-1$, $\boldsymbol{x}^{(m)}(n-1)$ and $\boldsymbol{\theta}^{(m)}(n-1)$ the particles of $\boldsymbol{x}(n-1)$ and $\boldsymbol{\theta}$, respectively, and $w^{(m)}(n-1)$ their associated weights. If we denote the approximating density of $f(\boldsymbol{x}(n-1), \boldsymbol{\theta} \,|\, \boldsymbol{y}(1:n-1))$ by $\pi(\boldsymbol{\phi}(n-1))$, where $\boldsymbol{\phi}(n-1)$ are the parameters of the approximating density, the steps of the density assisted particle filtering are the following.

1. Draw particles according to

$$\{\boldsymbol{x}^{(m)}(n-1), \boldsymbol{\theta}^{(m)}(n-1)\} \sim \pi(\boldsymbol{\phi}(n-1)).$$

2. Draw particles according to

$$\boldsymbol{x}^{(m)}(n) \sim f(\boldsymbol{x}(n) \,|\, \boldsymbol{x}^{(m)}(n-1), \boldsymbol{\theta}^{(m)}(n-1)).$$

3. Set $\boldsymbol{\theta}^{(m)}(n) = \boldsymbol{\theta}^{(m)}(n-1)$.
4. Update and normalize the weights

$$w^{(m)}(n) \propto f(\boldsymbol{y}(n) \,|\, \boldsymbol{x}^{(m)}(n), \boldsymbol{\theta}^{(m)}(n)).$$

5. Estimate the parameters, $\phi(n)$, of the density from

$$\chi(n) = \{\boldsymbol{x}^{(m)}(n), \boldsymbol{\theta}^{(m)}(n), w^{(m)}(n)\}_{m=1}^{M}.$$

The problem of standard PFs regarding constant parameters is avoided in the first step by drawing particles of the constants from an approximation of the posterior. It is important to note that one can combine standard PFs and DAPFs, in that we apply the standard PFs for the dynamic variables and DAPFs for the constant parameters. We show this in the next example.

■ EXAMPLE 5.10

Again we have the problem from Example 5.9, where we track a target in a two-dimensional plane. We assume that the marginal posterior density of the bias vector $\boldsymbol{\theta}$ is a Gaussian density. Suppose that at time instant $n-1$, we have the random measure $\chi(n-1) = \{\boldsymbol{x}^{(m)}(n-1), \boldsymbol{\theta}^{(m)}(n-1), w^{(m)}(n-1)\}_{m=1}^{M}$. From the random measure, we can compute the parameters of the approximating

Gaussian of the bias vector by

$$\boldsymbol{\mu}(n-1) = \sum_{m=1}^{M} w^{(m)}(n-1)\boldsymbol{\theta}^{(m)}(n-1)$$

$$\boldsymbol{\Sigma}(n-1) = \sum_{m=1}^{M} w^{(m)}(n-1)(\boldsymbol{\theta}^{(m)}(n-1) - \boldsymbol{\mu}(n-1))(\boldsymbol{\theta}^{(m)}(n-1) - \boldsymbol{\mu}(n-1))^{\top}.$$

Then we draw the particles of $\boldsymbol{\theta}$ for the next time step, that is, $\boldsymbol{\theta}^{(m)}(n) \sim \mathcal{N}(\boldsymbol{\mu}(n-1), \boldsymbol{\Sigma}(n-1))$. Once we have the particles of the biases, we proceed by using the favorite PF. For example, if it is the APF, first we project the dynamic particles as in the previous example, that is

$$\hat{\boldsymbol{x}}^{(m)}(n) = \hat{\boldsymbol{A}}\boldsymbol{x}^{(m)}(n).$$

This is followed by resampling of the streams whose weights are given by $w^{(k_m)}(n-1)\mathcal{N}(g_2(\hat{\boldsymbol{x}}^{(k_m)}(n)) + \boldsymbol{\theta}^{(k_m)}(n), \boldsymbol{\Sigma}_v)$, and where the Gaussian is computed at $\boldsymbol{y}(n)$. Once the indexes of the streams for propagation are known, we draw the particles of the dynamic variables, $\boldsymbol{x}^{(m)}(n)$. Next, the new weights of the streams are computed by

$$w^{(m)}(n) \propto \frac{f(\boldsymbol{y}(n) \mid \boldsymbol{x}^{(m)}(n), \boldsymbol{\theta}^{(k_m)}(n))}{f(\boldsymbol{y}(n) \mid \hat{\boldsymbol{x}}^{(k_m)}(n), \boldsymbol{\theta}^{(k_m)}(n))}.$$

5.7 RAO-BLACKWELLIZATION

In many practical problems, the considered dynamic nonlinear system may have some states that are conditionally linear given the nonlinear states of the system. When the applied methodology is particle filtering, this conditional linearity can be exploited using the concept of Rao–Blackwellization [20, 25]. Rao–Blackwellization is a statistical procedure that is used for reducing variance of estimates obtained by Monte Carlo sampling methods [18], and by employing it, we can have improved filtering of the unknown states.

The main idea consists of tracking the linear states differently from the nonlinear states by treating the linear parameters as nuisance parameters and marginalizing them out of the estimation problem. This strategy allows for more accurate estimates of the unknowns because the dimension of the space that is explored with particles is reduced and therefore it is much better searched. At every time instant the particles of the nonlinear states are propagated randomly, and once they are known, the problem is linear in the rest of the states. Therefore, one can find their 'optimal' values by employing Kalman filtering and associate them with the sampled nonlinear states. Some recent applications of Rao–Blackwellized PFs include tracking of maneuvering targets in clutter [61] and joint target tracking using kinematic radar information [4]. In [44], a computational complexity analysis of Rao–Blackwellized PFs is provided.

For the scenario of a nonlinear system with conditionally linear states, we write the state space model as

$$\begin{aligned} \boldsymbol{x}_n(n) &= g_{1,n}(\boldsymbol{x}_n(n-1)) + \boldsymbol{A}_{1,n}(\boldsymbol{x}_n(n-1))\boldsymbol{x}_l(n-1) + \boldsymbol{v}_{1,n}(n) \\ \boldsymbol{x}_l(n) &= g_{1,l}(\boldsymbol{x}_n(n-1)) + \boldsymbol{A}_{1,l}(\boldsymbol{x}_n(n-1))\boldsymbol{x}_l(n-1) + \boldsymbol{v}_{1,l}(n) \\ \boldsymbol{y}_t &= g_2(\boldsymbol{x}_n(n)) + \boldsymbol{A}_2(\boldsymbol{x}_n(n))\boldsymbol{x}_l(n) + \boldsymbol{v}_2(n) \end{aligned}$$

where the system state, $\boldsymbol{x}(n)$, includes nonlinear and conditionally linear components, that is, $\boldsymbol{x}^\top(n) = [\boldsymbol{x}_n^\top(n)\boldsymbol{x}_l^\top(n)]$, with $\boldsymbol{x}_n(n)$ and $\boldsymbol{x}_l(n)$ being the nonlinear and conditionally linear states, respectively; $\boldsymbol{v}_{1,n}(n)$ and $\boldsymbol{v}_{1,l}(n)$ are state noise vectors at time instant n which are assumed to be Gaussian; $g_{1,n}(\cdot)$ and $g_{1,l}(\cdot)$ are nonlinear state transition functions; $\boldsymbol{A}_{1,n}$ and $\boldsymbol{A}_{1,l}$ are matrices whose entries may be functions of the nonlinear states; $g_2(\cdot)$ is a nonlinear measurement function of the nonlinear states; $\boldsymbol{A}_2$ is another matrix whose entries may be functions of the nonlinear states; and $\boldsymbol{v}_2(n)$ is observation noise vector at time n which is also assumed to be Gaussian and independent from the state noises.

Suppose that at time instant $n-1$, the random measure composed of M streams is given by

$$\chi(n-1) = \{\boldsymbol{x}_n^{(m)}(n-1),\, w^{(m)}(n-1)\}_{m=1}^{M}$$

and that the linear state $\boldsymbol{x}_l(n-1)$ in the m-th stream is Gaussian distributed, that is

$$\boldsymbol{x}_l(n-1) \sim \mathcal{N}(\hat{\boldsymbol{x}}_l^{(m)}(n-1),\, \hat{\boldsymbol{C}}_{\boldsymbol{x}_l}^{(m)}(n-1))$$

where $\hat{\boldsymbol{x}}_l^{(m)}(n-1)$ is the estimate of $\boldsymbol{x}_l$ in the m-th stream at time instant $n-1$ and $\hat{\boldsymbol{C}}_{\boldsymbol{x}_l}^{(m)}(n-1)$ is the covariance matrix of that estimate. The scheme runs as follows.

1. Generate the nonlinear state particles according to the marginalized prior PDF

$$\boldsymbol{x}_n^{(m)}(n) \sim f(\boldsymbol{x}_n(n) \,|\, \boldsymbol{x}_n^{(m)}(n-1), \boldsymbol{y}(1:n-1))$$

where

$$\begin{aligned} f(\boldsymbol{x}_n(n) \,|\, \boldsymbol{x}_n^{(m)}(n-1), \boldsymbol{y}(1:n-1)) &= \int f(\boldsymbol{x}_n(n) \,|\, \boldsymbol{x}_n^{(m)}(n-1), \boldsymbol{x}_l^{(m)}(n-1)) \\ &\quad \times f(\boldsymbol{x}_l^{(m)}(n-1) \,|\, \boldsymbol{x}_n^{(m)}(n-1), \boldsymbol{y}(1:n-1))\, d\boldsymbol{x}_l^{(m)}(n-1). \end{aligned}$$

If the distributions inside the integral are Gaussians, $f(\boldsymbol{x}_n(n) \mid \boldsymbol{x}_n^{(m)}(n-1), \boldsymbol{y}(1:n-1))$ is also a Gaussian distribution. It can be shown that

$$f(\boldsymbol{x}_n(n) \mid \boldsymbol{x}_n^{(m)}(n-1), \boldsymbol{y}(1:n-1)) = \mathcal{N}(\boldsymbol{\mu}_{\boldsymbol{x}_n}^{(m)}(n), \boldsymbol{\Sigma}_{\boldsymbol{x}_n}^{(m)}(n))$$

where

$$\boldsymbol{\mu}_{\boldsymbol{x}_n}^{(m)}(n) = g_{1,n}(\boldsymbol{x}_n^{(m)}(n-1)) + \boldsymbol{A}_{1,n}^{(m)}(n-1)\hat{\boldsymbol{x}}_l^{(m)}(n-1) \tag{5.31}$$

$$\boldsymbol{\Sigma}_{\boldsymbol{x}_n}^{(m)}(n) = \boldsymbol{A}_{1,n}^{(m)}(n-1)\hat{\boldsymbol{C}}_{\boldsymbol{x}_l}^{(m)}(n-1)\boldsymbol{A}_{1,n}^{(m)\top}(n-1) + \boldsymbol{C}_{v1,n} \tag{5.32}$$

where we have dropped in the notation that the matrix $\boldsymbol{A}_{1,n}^{(m)}(n-1)$ may be a function of $\boldsymbol{x}_n^{(m)}(n-1)$. The symbol $\hat{\boldsymbol{x}}_l^{(m)}(n-1)$ is the estimate of the linear state at time instant $n-1$ from the m-th stream, and $\hat{\boldsymbol{C}}_{\boldsymbol{x}_l}^{(m)}(n-1)$ is the covariance matrix of that estimate.

2. Update the linear states with quasi measurements. We define the quasi measurements by

$$\begin{aligned} z^{(m)}(n) &= \boldsymbol{x}_n^{(m)}(n) - g_{1,n}(\boldsymbol{x}_n^{(m)}(n-1)) \\ &= \boldsymbol{A}_{1,n}^{(m)}(n-1)\hat{\boldsymbol{x}}_l^{(m)}(n-1) + \boldsymbol{v}_{1,n}(n) \end{aligned}$$

and use them to improve the estimate $\hat{\boldsymbol{x}}_l^{(m)}(n-1)$. The new estimate $\hat{\hat{\boldsymbol{x}}}_l^{(m)}(n-1)$ is obtained by applying the measurement step of the KF, that is,

$$\hat{\hat{\boldsymbol{x}}}_l^{(m)}(n-1) = \hat{\boldsymbol{x}}_l^{(m)}(n-1) + \boldsymbol{L}^{(m)}(n-1)(z^{(m)}(n) - \boldsymbol{A}_{1,n}^{(m)}(n-1)\hat{\boldsymbol{x}}_l^{(m)}(n-1))$$

$$\boldsymbol{L}^{(m)}(n-1) = \hat{\boldsymbol{C}}_{\boldsymbol{x}_l}^{(m)}(n-1) \times (\boldsymbol{A}_{1,n}^{(m)}(n-1)\boldsymbol{A}_{1,n}^{(m)}(n-1)\hat{\boldsymbol{C}}_{\boldsymbol{x}_l}^{(m)}(n-1)\boldsymbol{A}_{1,n}^{(m)\top}(n-1) + \boldsymbol{C}_{\boldsymbol{v}_{1,n}})^{-1}$$

$$\hat{\hat{\boldsymbol{C}}}_{\boldsymbol{x}_l}^{(m)}(n-1) = (\boldsymbol{I} - \boldsymbol{L}^{(m)}(n-1)\boldsymbol{A}_{1,n}^{(m)}(n-1))\hat{\boldsymbol{C}}_{\boldsymbol{x}_l}^{(m)}(n-1).$$

3. Perform time update of the linear states according to

$$\tilde{\boldsymbol{x}}_l^{(m)}(n) = g_{1,l}(\boldsymbol{x}_n^{(m)}(n-1)) + \boldsymbol{A}_{1,l}^{(m)}(n-1)\hat{\hat{\boldsymbol{x}}}_l^{(m)}(n-1)$$

$$\tilde{\boldsymbol{C}}_{\boldsymbol{x}_l}^{(m)}(n) = \boldsymbol{A}_{1,l}^{(m)}(n-1)\hat{\hat{\boldsymbol{C}}}_{\boldsymbol{x}_l}^{(m)}(n-1)\boldsymbol{A}_{1,l}^{(m)\top}(n-1) + \boldsymbol{C}_{\boldsymbol{v}_{1,l}}.$$

4. Compute the weights by

$$w^{(m)}(n) \propto w^{(m)}(n-1) f(\boldsymbol{y}(n) \mid \boldsymbol{x}_n^{(m)}(0:n), \boldsymbol{y}(1:n-1))$$

where

$$f(\boldsymbol{y}(n) \mid \boldsymbol{x}_n^{(m)}(0:n), \boldsymbol{y}(1:n-1)) = \int f(\boldsymbol{y}(n) \mid \boldsymbol{x}_n^{(m)}(n), \boldsymbol{x}_l^{(m)}(n)) \times f(\boldsymbol{x}_l^{(m)}(n) \mid \boldsymbol{x}_n^{(m)}(0:n), \boldsymbol{y}(1:n-1)) d\boldsymbol{x}_l^{(m)}(n).$$

Again, if the two densities of $\boldsymbol{x}_l^{(m)}(n)$ inside the integral are Gaussian densities, the integral can be solved analytically. We obtain

$$f(\boldsymbol{y}(n) \mid \boldsymbol{x}_n^{(m)}(0:n), \boldsymbol{y}(1:n-1)) = \mathcal{N}(\boldsymbol{\mu}_{\boldsymbol{y}}^{(m)}(n), \boldsymbol{\Sigma}_{\boldsymbol{y}}^{(m)}(n))$$

where

$$\boldsymbol{\mu}_{\boldsymbol{y}}^{(m)}(n) = g_2(\boldsymbol{x}_n^{(m)}(n)) + \boldsymbol{A}_2^{(m)}(n)\tilde{\boldsymbol{x}}_l^{(m)}(n)$$

$$\boldsymbol{\Sigma}_{\boldsymbol{y}}^{(m)}(n) = \boldsymbol{A}_2^{(m)}(n)\tilde{\boldsymbol{C}}_{\boldsymbol{x}_l}^{(m)}(n)\boldsymbol{A}_2^{(m)^\top}(n) + \boldsymbol{C}_{\boldsymbol{v}_2}.$$

5. Finally we carry out the measurement update of the linear states

$$\hat{\boldsymbol{x}}_l^{(m)}(n) = \tilde{\boldsymbol{x}}_l^{(m)}(n) + \boldsymbol{K}^{(m)}(n)(\boldsymbol{y}(n) - g_2(\boldsymbol{x}_n^{(m)}(n)) - \boldsymbol{A}_2^{(m)}(n)\tilde{\boldsymbol{x}}_l^{(m)}(n))$$

$$\boldsymbol{K}^{(m)}(n) = \tilde{\boldsymbol{C}}_{\boldsymbol{x}_l}^{(m)}(n)\boldsymbol{A}_2^{(m)}(n)(\boldsymbol{A}_2^{(m)}(n)\tilde{\boldsymbol{C}}_{\boldsymbol{x}_l}^{(m)}(n)\boldsymbol{A}_2^{(m)^\top}(n) + \boldsymbol{C}_{\boldsymbol{v}_2})^{-1}$$

$$\hat{\boldsymbol{C}}_{\boldsymbol{x}_l}^{(m)}(n) = (\boldsymbol{I} - \boldsymbol{K}^{(m)}(n)\boldsymbol{A}_2^{(m)}(n))\tilde{\boldsymbol{C}}_{\boldsymbol{x}_l}^{(m)}(n).$$

In general, the Rao–Blackwellization amounts to the use of a bank of KFs, one for each particle stream. So, with M particle streams, we will have M KFs.

■ EXAMPLE 5.11

The problem of target tracking using biased measurements presented in Example 5.9 can be addressed using a Rao–Blackwellized scheme. Here we outline the steps of the scheme. First, we assume that at time instant $n-1$ we have the random measure $\chi(n-1) = \{\boldsymbol{x}^{(m)}(n-1), w^{(m)}(n-1)\}_{m=1}^M$ and the estimates of the bias in each particle stream $\hat{\boldsymbol{\theta}}^{(m)}(n-1)$ and the covariances of the estimates $\hat{\boldsymbol{C}}_{\boldsymbol{\theta}}^{(m)}(n-1)$.

At time instant n, we first generate the particles of the dynamic variables

$$\boldsymbol{x}^{(m)}(n) \sim \mathcal{N}(\boldsymbol{A}\boldsymbol{x}^{(m)}(n-1), \boldsymbol{B}\boldsymbol{C}_{\boldsymbol{v}_1}\boldsymbol{B}^\top).$$

The second and the third steps are skipped because $\boldsymbol{\theta}$ only appears in the measurement equation. The computation of the weights proceeds by

$$w^{(m)}(n) \propto w^{(m)}(n-1)\,\mathcal{N}(\boldsymbol{\mu}_y^{(m)}(n), \boldsymbol{\Sigma}_y^{(m)}(n))$$

where

$$\boldsymbol{\mu}_y^{(m)}(n) = g_2(\boldsymbol{x}^{(m)}(n)) + \hat{\boldsymbol{\theta}}^{(m)}(n)$$
$$\boldsymbol{\Sigma}_y^{(m)}(n) = \hat{\boldsymbol{C}}_{\boldsymbol{\theta}}^{(m)}(n-1) + \boldsymbol{C}_{v_2}.$$

In the last step we update the estimates of the biases and their covariances by

$$\hat{\boldsymbol{\theta}}^{(m)}(n) = \hat{\boldsymbol{\theta}}^{(m)}(n-1) + \boldsymbol{K}^{(m)}(n)(\boldsymbol{y}(n) - g_2(\boldsymbol{x}^{(m)}(n)) - \hat{\boldsymbol{\theta}}^{(m)}(n-1))$$
$$\boldsymbol{K}^{(m)}(n) = \hat{\boldsymbol{C}}_{\boldsymbol{\theta}}^{(m)}(n-1)(\hat{\boldsymbol{C}}_{\boldsymbol{\theta}}^{(m)}(n-1) + \boldsymbol{C}_{v_2})^{-1}$$
$$\hat{\boldsymbol{C}}_{\boldsymbol{\theta}}^{(m)}(n) = (\boldsymbol{I} - \boldsymbol{K}^{(m)}(n))\hat{\boldsymbol{C}}_{\boldsymbol{\theta}}^{(m)}(n).$$

5.8 PREDICTION

Recall that the prediction problem revolves around the estimation of the predictive density $f(\boldsymbol{x}(n+k)\ \boldsymbol{y}(1:n))$, where $k > 0$. The prediction of the state is important in many applications. One of them is for model selection, where from a set of models $\mathcal{M}_l$, $l = 1, 2, \ldots, L$, one has to choose the best model according to a given criterion, for example the one based on MAP [9]. Then, one has to work with predictive probability distributions of the form $f(\boldsymbol{y}(n+1)\ \boldsymbol{y}(n), \mathcal{M}_l)$, where

$$f(\boldsymbol{y}(n+1)\,|\,\boldsymbol{y}(1:n), \mathcal{M}_l) = \int f(\boldsymbol{y}(n+1)\,|\,\boldsymbol{x}(n+1), \mathcal{M}_l) \times f(\boldsymbol{x}(n+1)\,|\,\boldsymbol{y}(1:n), \mathcal{M}_l)\, d\boldsymbol{x}(n+1)$$

where the second factor of the integrand is the predictive density of the state. The above integral often cannot be solved analytically, which is why we have interest in the prediction problem.

First we address the case $k = 1$, that is, the approximation of the predictive density $f(\boldsymbol{x}(n+1)\,|\,\boldsymbol{y}(1:n))$. Theoretically, we can obtain it from the filtering PDF $f(\boldsymbol{x}(n+1)\,|\,\boldsymbol{y}(1:n))$ by using the following integral

$$f(\boldsymbol{x}(n+1)\,|\,\boldsymbol{y}(1:n)) = \int f(\boldsymbol{x}(n+1)\,|\,\boldsymbol{x}(n)) f(\boldsymbol{x}(n)\,|\,\boldsymbol{y}(1:n))\, d\boldsymbol{x}(n). \tag{5.33}$$

Note that we already saw this expression in (5.7). Recall also that the PF provides an estimate of the filtering density in the form

$$f(\boldsymbol{x}(n) \,|\, \boldsymbol{y}(1:n)) \simeq \sum_{m=1}^{M} w^{(m)}(n)\ \delta(\boldsymbol{x}(n) - \boldsymbol{x}^{(m)}(n)) \tag{5.34}$$

which when used in (5.33) yields an estimate of the predictive density in the form of a mixture density given by

$$f(\boldsymbol{x}(n+1) \,|\, \boldsymbol{y}(1:n)) \simeq \sum_{m=1}^{M} w^{(m)}(n)\, f(\boldsymbol{x}(n+1) \,|\, \boldsymbol{x}^{(m)}(n)). \tag{5.35}$$

If we express this PDF by using a discrete random measure, then a natural way of expressing this measure is by $\{\boldsymbol{x}^{(m)}(n+1),\, w^{(m)}(n)\}_{m=1}^{M}$, where

$$\boldsymbol{x}^{(m)}(n+1) \sim f(\boldsymbol{x}(n+1) \,|\, \boldsymbol{x}^{(m)}(n))$$

and

$$f(\boldsymbol{x}(n+1) \,|\, \boldsymbol{y}(1:n)) \simeq \sum_{m=1}^{M} w^{(m)}(n)\ \delta(\boldsymbol{x}(n+1) - \boldsymbol{x}^{(m)}(n+1)).$$

In summary, we obtain an estimate of the predictive PDF $f(\boldsymbol{x}(n+1) \,|\, \boldsymbol{y}(1:n))$ from the random measure $\chi(n) = \{\boldsymbol{x}^{(m)}(n),\, w^{(m)}(n)\}_{m=1}^{M}$ by simply (a) drawing particles $\boldsymbol{x}^{(m)}(n+1)$ from $f(\boldsymbol{x}(n+1) \,|\, \boldsymbol{x}^{(m)}(n))$, for $m = 1, 2, \ldots, M$ and (b) assigning the weights $w^{(m)}(n)$ to the drawn particles $\boldsymbol{x}^{(m)}(n+1)$.[12]

Next we do the case $k = 2$. The strategy for creating the random measure that approximates $f(\boldsymbol{x}(n+2) \,|\, \boldsymbol{y}(1:n))$ is analogous. Now

$$\begin{aligned} f(\boldsymbol{x}(n+2) \,|\, \boldsymbol{y}(1:n)) = \int\!\!\int & f(\boldsymbol{x}(n+2) \,|\, \boldsymbol{x}(n+1))\, f(\boldsymbol{x}(n+1) \,|\, \boldsymbol{x}(n)) \\ & \times f(\boldsymbol{x}(n) \,|\, \boldsymbol{y}(1:n))\, d\boldsymbol{x}(n)\, d\boldsymbol{x}(n+1). \end{aligned} \tag{5.36}$$

Here we have two integrals, where the first involves the integration of $\boldsymbol{x}(n)$ carried out in the same way as above. We reiterate that we can approximate $f(\boldsymbol{x}(n) \,|\, \boldsymbol{y}(1:n))$ as in

[12] Another approximation would be by drawing the particles from the mixture in (5.35), that is, by first sampling the mixand and then drawing the particle from it. In that case all the drawn particles have the same weight.

(5.34), and after the integration we obtain

$$\begin{aligned}f(\boldsymbol{x}(n+2)\,|\,\boldsymbol{y}(1:n)) &= \int f(\boldsymbol{x}(n+2)\,|\,\boldsymbol{x}(n+1))f(\boldsymbol{x}(n+1)\,|\,\boldsymbol{y}(1:n))\,d\boldsymbol{x}(n+1)\\ &\simeq \int f(\boldsymbol{x}(n+2)\,|\,\boldsymbol{x}(n+1))\sum_{m=1}^{M} w^{(m)}(n)\delta(\boldsymbol{x}(n+1)\\ &\qquad - \boldsymbol{x}^{(m)}(n+1))\,d\boldsymbol{x}\,(n+1)\\ &= \sum_{m=1}^{M} w^{(m)}(n) f(\boldsymbol{x}(n+2)\,|\,\boldsymbol{x}^{(m)}(n+1)).\end{aligned}$$

Further, we approximate this PDF by

$$f(\boldsymbol{x}(n+2)\,|\,\boldsymbol{y}(1:n)) \simeq \sum_{m=1}^{M} w^{(m)}(n)\delta(\boldsymbol{x}(n+2) - \boldsymbol{x}^{(m)}(n+2))$$

where as before we obtain $\boldsymbol{x}^{(m)}(n+2)$ by drawing it from $f(\boldsymbol{x}(n+2)\,|\,\boldsymbol{x}^{(m)}(n+1))$.

Thus, the approximation of the predictive density $f(\boldsymbol{x}(n+2)\,|\,\boldsymbol{y}(1:n))$ is obtained by

1. drawing particles $\boldsymbol{x}^{(m)}(n+1)$ from $f(\boldsymbol{x}(n+1)\,|\,\boldsymbol{x}^{(m)}(n))$
2. generating $\boldsymbol{x}^{(m)}(n+2)$ from $f(\boldsymbol{x}(n+2)\,|\,\boldsymbol{x}^{(m)}\,(n+1))$; and
3. associating with the samples $\boldsymbol{x}^{(m)}(n+2)$ the weights $w^{(m)}(n)$ and thereby forming the random measure $\{\boldsymbol{x}^{(m)}(n+2), w^{(m)}(n)\}_{m=1}^{M}$.

At this point, it is not difficult to generalize the procedure for obtaining the approximation of $f(\boldsymbol{x}(n+k)\,|\,\boldsymbol{y}(1:n))$ for any $k > 0$. We propagate the particle streams from time instant n to time instant $n+k$ by using the transition PDFs $f(\boldsymbol{x}(n+i)\,|\,\boldsymbol{x}(n+i-1))$, $i = 1, 2, \ldots, k$, and associating with the last set of generated particles $\boldsymbol{x}^{(m)}(n+k)$ the weights $w^{(m)}(n)$ to obtain the random measure $\{\boldsymbol{x}^{(m)}(n+k), w^{(m)}(n)\}_{m=1}^{M}$.

5.9 SMOOTHING

In Section 5.2, we pointed out that sometimes we may prefer to estimate $\boldsymbol{x}(n)$ based on data $\boldsymbol{y}(1:n+k)$, where $k > 0$. All the information about $\boldsymbol{x}(n)$ in that case is in the smoothing PDF $f(\boldsymbol{x}(n)\,|\,\boldsymbol{y}(n+k))$. With particle filtering, we approximate this density with a random measure in a similar way as we approximated filtering PDFs.

When we deal with the problem of smoothing, we discriminate between two types of smoothing. One of them is called fixed-lag smoothing and the interest there is in the PDF $f(\boldsymbol{x}(n)\,|\,\boldsymbol{y}(1:n+k))$, where k is a fixed lag. The other is called fixed interval smoothing, where the objective is to find the PDF $f(\boldsymbol{x}(n)\,|\,\boldsymbol{y}(1:N))$ for any n where $0 \leq n < N$.

A naive approach to these two problems would be to conduct particle filtering in the usual way and store all the random measures, and for obtaining approximations of the smoothing densities use the set of particles at time instant n and the weights of these streams at time instants $n+k$ or N, respectively. In other words, we use

$$f(\boldsymbol{x}(n)\,|\,\boldsymbol{y}(1:n+k)) \simeq \sum_{m=1}^{M} w^{(m)}(n+k)\delta(\boldsymbol{x}(n)-\boldsymbol{x}^{(m)}(n))$$

and

$$f(\boldsymbol{x}(n)\,|\,\boldsymbol{y}(1:N)) \simeq \sum_{m=1}^{M} w^{(m)}(N)\delta(\boldsymbol{x}(n)-\boldsymbol{x}^{(m)}(n)).$$

These approximations may be good if k or $N-n$ are small numbers. For even moderate values of k or $N-n$, the approximations may become quite inaccurate, especially when resampling is implemented at each time instant. Namely, with resampling, we deplete the number of surviving streams k or $N-n$ lags later, which may make the representation at time instant n rather poor.

In the sequel, we show two approaches for improved smoothing. They are both based on backward smooth recursions, where the first recursion uses only the set of weights generated during the forward filtering pass [52] and the second one generates an additional set of weights during the backward filtering pass [25]. Here we address the fixed-interval smoothing problem because modifications for fixed-lag smoothing are straightforward.

First we write the joint a posteriori PDF of the states as follows

$$f(\boldsymbol{x}(0:N)\,|\,\boldsymbol{y}(1:N)) = f(\boldsymbol{x}(N)\,|\,\boldsymbol{y}(1:N)) \prod_{n=0}^{N-1} f(\boldsymbol{x}(n)\,|\,\boldsymbol{x}(n+1), \boldsymbol{y}(1:N)).$$

We note that

$$\begin{aligned} f(\boldsymbol{x}(n)\,|\,\boldsymbol{x}(n+1), \boldsymbol{y}(1:N)) &= f(\boldsymbol{x}(n)\,|\,\boldsymbol{x}(n+1), \boldsymbol{y}(1:n)) \\ &\propto f(\boldsymbol{x}(n+1)\,|\,\boldsymbol{x}(n)) f(\boldsymbol{x}(n)\,|\,\boldsymbol{y}(1:n)). \end{aligned}$$

From the above two expressions, we obtain the recursion that takes us from the smoothing PDF $f(\boldsymbol{x}(n+1:N)\,|\,\boldsymbol{y}(1:N))$ to $f(\boldsymbol{x}(n:N)\,|\,\boldsymbol{y}(1:N))$, that is,

$$\begin{aligned} f(\boldsymbol{x}(n:N)\,|\,\boldsymbol{y}(1:N)) &= f(\boldsymbol{x}(n+1:N)\,|\,\boldsymbol{y}(1:N)) f(\boldsymbol{x}(n)\,|\,\boldsymbol{x}(n+1), \boldsymbol{y}(1:N)) \\ &= f(\boldsymbol{x}(n+1:N)\,|\,\boldsymbol{y}(1:N)) f(\boldsymbol{x}(n)\,|\,\boldsymbol{x}(n+1), \boldsymbol{y}(1:n)) \\ &\propto f(\boldsymbol{x}(n+1:N)\,|\,\boldsymbol{y}(1:N)) f(\boldsymbol{x}(n+1)\,|\,\boldsymbol{x}(n)) f(\boldsymbol{x}(n)\,|\,\boldsymbol{y}(1:n)). \end{aligned} \tag{5.37}$$

If we approximate $f(\boldsymbol{x}(n+1:N)\,|\,\boldsymbol{y}(1:N))$ by the smoothing random measure $\chi_s(n+1) = \{\boldsymbol{x}^{(m)}(n+1), w_s^{(m)}(n+1)\}_{m=1}^{M}$, where the weights $w_s^{(m)}(n+1)$ are the smoothing weights of the particles, then the problem is in obtaining the smoothing weights from the above recursion.

From (5.37), it is clear that we have to run the PF in the usual way from time instant 0 to time instant N. The random measure at time instant N, $\chi(N)$ is identical to the smoothing random measure $\chi_s(N)$, which means that $w_s^{(m)}(N) = w^{(m)}(N)$. We obtain the random measure $\chi_s(N-1)$ from $\chi_s(N)$ and $\chi(N-1)$ by exploiting (5.37) as follows. First we draw a particle from the random measure $\chi_s(N)$. Let this particle be $\boldsymbol{x}^{(k)}(N)$. Then the nonnormalized smoothing weights are computed by

$$\tilde{w}_s^{(m)}(N-1) = w^{(m)}(N-1) f(\boldsymbol{x}^{(k)}(N) \,|\, \boldsymbol{x}^{(m)}(N-1)), \;\; m = 1, 2, \ldots, M.$$

Subsequently the weights $\tilde{w}_s^{(m)}(N-1)$ are normalized to $w_s^{(m)}(N-1)$, and the smoothing random measure $\chi_s(N-1) = \{\boldsymbol{x}^{(m)}(N-1), w_s^{(m)}(N-1)\}_{m=1}^{M}$ is formed. With this computation completed, we proceed with finding $\chi_s(N-2)$. We repeat the above steps by first drawing $\boldsymbol{x}(N-1)$ from $\chi_s(N-1)$, and then computing the weights $w_s(N-2)$ and so on. The method is summarized in Table 5.6. The procedure can be repeated to obtain independent realizations of the smoothing random measures.

Table 5.6 Smoothing algorithm I

Forward particle filtering

Run particle filtering in the forward direction and store all the random measures

$\chi(n) = \{\boldsymbol{x}^{(m)}(n), w^{(m)}(n)\}_{m=1}^{M}$, $n = 1, 2, \ldots, N$

Backward recursions

Set the smoothing weights

$w_s^{(m)}(N) = w^{(m)}(N)$ and

$\chi_s(N) = \{\boldsymbol{x}^{(m)}(N), w_s^{(m)}(N)\}_{m=1}^{M}$

For $n = N-1, \ldots, 1, 0$

Drawing a particle from $\chi_s(n+1)$

Draw $\boldsymbol{x}(n+1)$ from $\chi_s(n+1)$ and let the drawn particle be $\boldsymbol{x}^{(k)}(n+1)$

Computation of the smoothing weights $w_s(n)$

Compute the smoothing non-normalized weights of $\boldsymbol{x}^{(m)}(n)$ by

$\tilde{w}_s^{(m)}(n) = w^{(m)}(n) f(\boldsymbol{x}^{(k)}(n+1) \,|\, \boldsymbol{x}^{(m)}(n))$, $m = 1, 2, \ldots, M$

Normalize the smoothing weights by computing

$$w_s^{(m)}(n) = \frac{\tilde{w}_s^{(m)}(n)}{\sum_{j=1}^{M} \tilde{w}_s^{(j)}(n)}$$

Construction of the smoothing random measure $\chi_s(n)$

Set $\chi_s(n) = \{\boldsymbol{x}^{(m)}(n), w_s^{(m)}(n)\}_{m=1}^{M}$

The second algorithm is based on an identity from [46], which connects the smoothing density $f(\boldsymbol{x}(n)\,|\,\boldsymbol{y}(1:N))$ with $f(\boldsymbol{x}(n+1)\,|\,\boldsymbol{y}\ (1:N))$ through an integral and which is given by

$$f(\boldsymbol{x}(n)\,|\,\boldsymbol{y}(1:N)) = f(\boldsymbol{x}(n)\,|\,\boldsymbol{y}(1:n)) \times \int \frac{f(\boldsymbol{x}(n+1)\,|\,\boldsymbol{x}(n)) f(\boldsymbol{x}(n+1)\,|\,\boldsymbol{y}(1:N))}{f(\boldsymbol{x}(n+1)\,|\,\boldsymbol{y}(1:n))}\, d\boldsymbol{x}(n+1). \tag{5.38}$$

The identity can readily be proved by noting that

$$f(\boldsymbol{x}(n)\,|\,\boldsymbol{y}(1:N)) = \int f(\boldsymbol{x}(n), \boldsymbol{x}(n+1)\,|\,\boldsymbol{y}(1:N))\, d\boldsymbol{x}(n+1) \tag{5.39}$$

and

$$\begin{aligned} f(\boldsymbol{x}(n), \boldsymbol{x}(n+1)\,|\,\boldsymbol{y}(1:N)) &= f(\boldsymbol{x}(n+1)\,|\,\boldsymbol{y}(1:N)) f(\boldsymbol{x}(n)\,|\,\boldsymbol{x}(n+1), \boldsymbol{y}(1:n)) \\ &= f(\boldsymbol{x}(n+1)\,|\,\boldsymbol{y}(1:N)) \\ &\quad \times \frac{f(\boldsymbol{x}(n+1)\,|\,\boldsymbol{x}(n)) f(\boldsymbol{x}(n)\,|\,\boldsymbol{y}(1:n))}{f(\boldsymbol{x}(n+1)\,|\,\boldsymbol{y}(1:n))}. \end{aligned} \tag{5.40}$$

We use the following approximations

$$f(\boldsymbol{x}(n)\,|\,\boldsymbol{y}(1:N)) \simeq \sum_{m=1}^{M} w_s^{(m)}(n) \delta(\boldsymbol{x}(n) - \boldsymbol{x}^{(m)}(n))$$

$$f(\boldsymbol{x}(n)\,|\,\boldsymbol{y}(1:n)) \simeq \sum_{m=1}^{M} w^{(m)}(n) \delta(\boldsymbol{x}(n) - \boldsymbol{x}^{(m)}(n))$$

and

$$\begin{aligned} f(\boldsymbol{x}(n+1)\,|\,\boldsymbol{y}(1:n)) &= \int f(\boldsymbol{x}(n+1)\,|\,\boldsymbol{x}(n)) f(\boldsymbol{x}(n)\,|\,\boldsymbol{y}(1:n))\, d\boldsymbol{x}(n) \\ &\approx \sum_{m=1}^{M} w^{(m)}(n) f(\boldsymbol{x}(n+1)|\boldsymbol{x}^{(m)}(n)) \end{aligned}$$

to obtain

$$w_s^{(m)}(n) = \sum_{j=1}^{M} w_s^{(j)}(n+1) \frac{w^{(m)}(n) f(\boldsymbol{x}^{(j)}(n+1)\,|\,\boldsymbol{x}^{(m)}(n))}{\sum_{l=1}^{M} w^{(l)}(n) f(\boldsymbol{x}^{(j)}(n+1)\,|\,\boldsymbol{x}^{(l)}(n))}.$$

The algorithm is summarized in Table 5.7. The second algorithm is more complex than the first.

Table 5.7 Smoothing algorithm II

Forward PF

Run PF in the forward direction and store all the random measures

$\chi(n) = \{\mathbf{x}^{(m)}(n), w^{(m)}(n)\}_{m=1}^{M}, \quad n = 1, 2, \cdots, N$

Backward recursions

Set the smoothing weights

$w_s^{(m)}(N) = w^{(m)}(N)$ and

$\chi_s(N) = \{\mathbf{x}^{(m)}(N), w_s^{(m)}(N)\}_{m=1}^{M}$

For $n=N-1, \ldots, 1, 0$

Computation of the smoothing weights $w_s(n)$

Compute the smoothing weights of $\mathbf{x}^{(m)}(n)$ by

$$w_s^{(m)}(n) = \sum_{j=1}^{M} w_s^{(j)}(n+1) \frac{w^{(m)}(n) f(\mathbf{x}^{(j)}(n+1) \mid \mathbf{x}^{(m)}(n))}{\sum_{l=1}^{M} w^{(l)}(n) f(\mathbf{x}^{(j)}(n+1) \mid \mathbf{x}^{(l)}(n))}$$

Construction of the smoothing random measure $\chi_s(n)$

Set $\chi_s(n) = \{\mathbf{x}^{(m)}(n), w_s^{(m)}(n)\}_{m=1}^{M}$

5.10 CONVERGENCE ISSUES

Whenever we develop algorithms for estimation, we study their performance and often express it in some probabilistic/statistical terms. For example, when we deal with parameter estimation, a benchmark for performance comparison of unbiased estimators is the Cramér–Rao bound. In particle filtering, where we basically rely on the Bayesian methodology, we use the concept of posterior Cramér–Rao bounds (PCRBs) and in obtaining them we combine the information about the unknowns extracted from observed data and prior information [74]. The PCRBs represent the MSEs that can be achieved by an estimator. The computation of the PCRB in the context of sequential signal processing is presented in [73]. Thus, often we find that particle filtering algorithms are compared with PCRBs of the estimated states, where the PCRBs themselves are computed by using Monte Carlo integrations. The difference between the MSEs and the PCRBs provides us with quantitative information about how far the algorithm is from optimal performance. An important issue that has not been successfully resolved yet is the required number of particles that are needed for achieving a desired performance. A related problem is the curse of dimensionality, that is, that for satisfactory accuracy the number of particles most likely will go up steeply as the dimension of the state increases [22].

An important type of performance issue is related to the convergence of particle filtering methods. For example, one would like to know if the methods converge, and if they do, in what sense and with what rate. We know that particle filtering is based on approximations of PDFs by discrete random measures and we expect that the approximations deteriorate as the dimension of the state space increases but hope that as the number of particles $M \rightarrow \infty$, the approximation of the density will improve. We also know that the approximations are based on particle streams and that they are dependent because of the necessity of resampling. Therefore classic limit theorems for studying convergence cannot be applied, which makes the analysis more difficult.

The study of the convergence of particle filtering is outside the scope of this chapter. Here we only state some results without any proofs. For detailed technical details, the reader is referred to [21]. Here we summarize three results: (a) almost sure (weak) convergence of the empirical distributions to the true ones, (b) the set of sufficient conditions that ensure asymptotic convergence of the mean square error to zero, and (c) conditions for uniform convergence in time.

First we define the concept of weak convergence in the context of particle filtering. We say that a sequence of random measures

$$\begin{aligned} \chi^M(n) &= \{\boldsymbol{x}^{(m)}(n), w^{(m)}(n)\}_{m=1}^M \\ &= \sum_{m=1}^M w^{(m)}(n)\delta(\boldsymbol{x}(n) - \boldsymbol{x}^{(m)}(n)) \end{aligned}$$

approximating the *a posteriori* PDF $f(\boldsymbol{x}(n)|\ \boldsymbol{y}(1:n))$ converges weekly to $f(\boldsymbol{x}(n)\,|$ $\boldsymbol{y}(1:\mathrm{n}))$ if for any continuous bounded function $h(\boldsymbol{x}(n))$, we have

$$\lim_{M\rightarrow\infty} \int h(\boldsymbol{x}(n))\chi^M(n)\,d\boldsymbol{x}(n) = \int h(\boldsymbol{x}(n))f(\boldsymbol{x}(n)\,|\,\boldsymbol{y}(1:n))\,d\boldsymbol{x}(n).$$

Now we quote a theorem from [21].

Theorem 1 *If $f(\boldsymbol{x}(n)\,|\,\boldsymbol{x}(n-1))$ is Feller and the likelihood function $f(\boldsymbol{y}(n)\,|\,\boldsymbol{x}(n))$ is bounded, continuous and strictly positive, then*

$$\lim_{M\rightarrow\infty} \chi(n)^M = f(\boldsymbol{x}(n)\,|\,\boldsymbol{y}(1:n))$$

almost surely.

The next result is about the convergence of the MSE. We define the error by

$$e^M(n) = \int h(\boldsymbol{x}(n))\chi^M(n)\,d\boldsymbol{x}(n) - \int h(\boldsymbol{x}(n))f(\boldsymbol{x}(n)\,|\,\boldsymbol{y}(1:n))\,d\boldsymbol{x}(n).$$

Theorem 2 *If the likelihood function $f(\boldsymbol{y}(n)\,|\,\boldsymbol{x}(n))$ is bounded, for all $n \geq 1$ there exists a constant $c(n)$ independent of M such that for any bounded function $h(\boldsymbol{x}(n))$*

$$E(e^2(n)) \leq c(n)\frac{\|h(\boldsymbol{x}(n))\|^2}{M} \tag{5.41}$$

where $E(\cdot)$ is an expectation operator, $h(\boldsymbol{x}(n))$ has the same meaning as before and $\|h(\boldsymbol{x}(n))\|$ denotes the supremum norm, that is,

$$\|h(\boldsymbol{x}(n))\| = \sup_{\boldsymbol{x}(n)} |\, h(\boldsymbol{x}(n))\,|.$$

For a given precision on the MSE in (5.41), the number of particles depends on $c(n)$, which itself can depend on the dimension of the state space. The theorem holds for bounded functions of $\boldsymbol{x}(n)$, which means that it does not hold for $h(\boldsymbol{x}(n)) = \boldsymbol{x}(n)$. In other words, the result does not hold for the minimum MSE estimate of the state. In summary, the last theorem ensures that

$$\lim_{M\to\infty} \int h(\boldsymbol{x}(n))\chi^M(n)\, d\boldsymbol{x}(n) \to \int h(\boldsymbol{x}(n)) f(\boldsymbol{x}(n)\,|\,\boldsymbol{y}(1:n))\, d\boldsymbol{x}(n)$$

in the mean square sense and that the rate of convergence of the approximation error $E(e^2(n))$ is in $1/M$.

Clearly, the upper bound on the right hand side of the inequality in (5.41) depends on $c(n)$, and it can be shown that it increases with n. This implies that to have a certain precision, one has to increase the number of particles with n. To avoid the increase of $c(n)$ with time, we must have some mixing assumptions about the dynamic model that will allow for exponential forgetting of the error with time [21].

Before we proceed, we define a mixing kernel $r(\boldsymbol{x}(n)\,|\,\boldsymbol{x}(n-1))$ by

$$r(\boldsymbol{x}(n)\,|\,\boldsymbol{x}(n-1)) = f(\boldsymbol{y}(n)\,|\,\boldsymbol{x}(n)) f(\boldsymbol{x}(n)\,|\,\boldsymbol{x}(n-1)).$$

We now make the assumption that the mixing kernel is weakly dependent on $\boldsymbol{x}(n-1)$, and we express it by

$$\varepsilon\lambda(\boldsymbol{x}(n)) \le r(\boldsymbol{x}(n)\,|\,\boldsymbol{x}(n-1)) \le \varepsilon^{-1}\lambda(\boldsymbol{x}(n))$$

where $\lambda(\boldsymbol{x}(n))$ is some PDF. If in addition for

$$\rho(n) = \frac{\sup_x f(\boldsymbol{y}(n)\,|\,\boldsymbol{x}(n))}{\inf_\phi \int f(\boldsymbol{y}(n)\,|\,\boldsymbol{x}(n)) \int \phi(\boldsymbol{x}(n-1)) f(\boldsymbol{x}(n)\,|\,\boldsymbol{x}(n-1))\, d\boldsymbol{x}(n-1)\, d\boldsymbol{x}(n)}$$

where $\phi(\boldsymbol{x}(n-1))$ is some PDF, we assume that

$$\rho(n) < \rho < \infty$$

we can show that the following theorem holds true:

Theorem 3 *For any bounded function $h(\boldsymbol{x}(n))$ and for all n, there exists a constant $c(\varepsilon)$ which is independent of M such that*

$$E(e^2(n)) \le c(\varepsilon)\frac{\|h(\boldsymbol{x}(n))\|^2}{M}. \tag{5.42}$$

In summary, if the optimal filter is quickly mixing, which means that the filter has a short memory, then uniform convergence of the PF is ensured. If the state vector contains a constant parameter, the dynamic model is not ergodic and uniform convergence cannot be achieved.

The material in this section is based on [21]. An alternative approach to studying the theoretical properties of PFs can be found in [53].

5.11 COMPUTATIONAL ISSUES AND HARDWARE IMPLEMENTATION

The particle filtering methodology is computationally very intensive. However, many of the required computations can be performed in parallel, which may produce significant savings in computing time. Particle filtering is intended for applications where real time processing is required and where deadlines for processing the incoming samples are imposed. Therefore, it is important to study the computational requirements of the methodology and the possibilities for its hardware implementations.

It is clear that particle filtering algorithms coded for a personal computer must often be modified so that when they are implemented in hardware they can have higher speeds. On the other hand, implementation in hardware usually requires fixed point arithmetics which can affect the performance of PFs. Even implementations on digital signal processors (DSPs) that are highly pipelined and support some concurrency may not be befitting for high-speed real-time particle filtering. In working on modifications of algorithms so that they are more suitable for hardware implementation, one has to address issues that include the reduction of computational complexity, high throughput designs, scalability of parallel implementation, and reducing memory requirements. Good general approaches to resolving them are based on joint algorithmic and architectural designs. Some designs that are tractable for VLSI implementations are discussed in [69].

One of the first efforts on development of hardware architectures for supporting high speed particle filtering is presented in [11]. The achieved speed improvements on a field-programmable gate array (FPGA) platform over implementations on (at that time) state-of-the-art DSPs was about 50 times. An FPGA prototype of a PF was presented in [8].

We recall that the three basic operations of particle filtering are particle generation, weight computations, and resampling. Of the three steps, the most computationally intensive are the particle generation and weight computation. Resampling is not computationally intensive but poses a different set of problems.

The speed of particle filtering on an algorithmic level can be increased by reducing the number of operations, using operational concurrency between the particle generation and weight computation. A designer is often faced with various problems including ways of implementing the required mathematical operations and choosing the level of pipelining for the hardware blocks. It has been found that for pipelined hardware implementation which is not parallel, the maximum achievable speed is proportional to $1/(2MT_{clk})$, where, as before, M represents the number of particles and T_{clk} is the clock period [11].

In [11], a hardware designed with processing elements (PEs) and a central unit (CU) was proposed. Each PE processes fractions of the particles (meaning particle propagation and their weight computations) whereas the CU controls the PEs and performs resampling. The resampling step represents a bottleneck in the parallel implementation of the filter, and it introduces other disadvantages. It increases the sampling period of the filter, that is, it slows down its operation, and it increases the memory requirements of the filter. The resampling also entails intensive data exchange within the filter, where large sets of data have to be exchanged through the interconnection network. In brief, besides being a bottleneck, the resampling step increases the complexity of the CU.

As pointed out, there are many implementation issues with particle filtering. One of them is the development of resampling schemes which are better for hardware implementation. For example, schemes that require less operations, less memory and less memory access are desirable. It is also preferable to have a resampling scheme with a deterministic processing time. Designs of algorithms that allow for overlap of the resampling step with the remaining steps are particularly desirable. Some generic architectures for sampling and resampling are presented in [7]. GPFs do not require resampling in the usual sense and they are better for parallel implementation. The data exchange that is required for them is deterministic and is much lower than for standard PFs, and they do not require the storage of particles between sampling instants. Some of the implementations of GPFs proposed in [11] achieve speeds twice that of standard PFs. The computation requirements of the GPFs are, however, higher in that they need more random number generators and more multipliers.

Other issues include the use of fixed and floating point arithmetic and scalability. The area and speed of the filter design depend on various factors including the number of used particles and the levels of implemented parallelism.

Reconfigurable architectures for PFs have also been explored [8]. Since PFs can be applied to many different problems, the idea is to develop architectures that can be used for different problems and thereby allow for flexibility and generality. Namely, designs on adaptable platforms can be configured for various types of PFs by modifying minimal sets of control parameters. We can view this effort as the development of programmable particle filtering hardware. One idea is to be able to select a PF from a set of possibilities where all of the available filters have maximum resource sharing in that the processing blocks are maximally reused, and the interconnection and interface between them is carried out by distributed buffers, controllers, and multiplexers [8]. An example of a reconfigurable PF for real-time bearings-only tracking is presented in [36]. Reconfigurability can include the type of used PF, the dimension of the state spaces, and the number of employed particles.

5.12 ACKNOWLEDGMENTS

The work of the authors on particle filtering has been supported by the National Science Foundation (Awards CCR-0220011 and CCF-0515246) and the Office of Naval Research (Award N00014-09-1-1154).

5.13 EXERCISES

5.1 Consider the following dynamic system

$$x(n) = x(n-1) + v_1(n)$$
$$y(n) = x(n) + v_2(n)$$

where $\boldsymbol{x}(n)$ and $y(n)$ are scalars and $v_1(n)$ and $v_2(n)$ are independent Gaussian noises of parameters μ_1 and σ_1^2, and μ_2 and σ_2^2, respectively. This system evolves for $n = 0{:}500\, s$.

(a) If the prior proposal distribution is chosen for generation of particles, derive the weight update expression.

(b) If the posterior proposal distribution is chosen for generation of particles, derive the weight update expression.

(c) Program a particle filter in MATLAB based on Table 5.3 for the previous system and using as proposal distribution the prior distribution.

(d) Extend the previous program to plot the evolution of the weights. Justify the degeneracy of the weights as time evolves.

(e) Implement a particle filter that uses the posterior proposal distribution and compare its performance with the previous one in terms of mean square error (MSE). The MSE is defined as

$$\text{MSE}(n) = \frac{1}{J}\sum_{j=1}^{J} (\hat{x}^j(n) - x^j(n))^2$$

where J denotes the number of simulation runs, and $x(n)$ and $\hat{x}^j(n)$ are the true and estimated (in the j-th run) values of the state at time instant n, respectively.

5.2 For the model from the previous example, implement the SIR PF with two possible resampling schemes: one that performs the resampling at each time step, and another one where the resampling is carried out depending on the effective particle size measure. Compare the performances of the filters for different numbers of particles M and different thresholds of effective particle sizes.

5.3 A model for univariate nonstationary growth can be written as

$$x(n) = \alpha x(n-1) + \beta \frac{x(n-1)}{1 + x^2(n-1)} + \gamma \cos(1.2(n-1)) + v_1(n)$$

$$y(n) = \frac{x^2(n)}{20} + v_2(n),\ n = 1, \ldots, N$$

where $\boldsymbol{x}(0) = 0.1$, $\alpha = 0.5$, $\beta = 25$, $\gamma = 8$, $N = 500$, $v_1(n) \sim N(0, 1)$ and $v_2(n) \sim \mathcal{N}(0, 1)$. The term in the process equation that is independent of $x(n)$

but varies with time can be interpreted as time-varying noise. The likelihood $f(y(n) \mid x(n))$ has bimodal nature due to the form of the measurement equation. Implement the SIR PF, APF, and GPF and compare their performances in terms of MSE.

5.4 Consider the recursive estimation of a target trajectory in a two-dimensional plane. The dynamic state space model is given by

$$\boldsymbol{x}(n) = \boldsymbol{x}(n-1) + T(\boldsymbol{s}(n) + \boldsymbol{v}_1(n))$$

$$y_i(n) = 10\log_{10}\left(\frac{P_0}{|\boldsymbol{x}(n) - \boldsymbol{r}_i|^{\theta}}\right) + v_{2,i}(n)$$

where $\boldsymbol{x}(n) \in \Re^2$ is the vehicle position at time n, $T = 5\,s$ is the observation period, $\boldsymbol{s}(n) \in \Re^2$ is the measured vehicle speed at time n, $\boldsymbol{v}_1(n) \sim \mathcal{N}(0, \boldsymbol{I}_2)$ is a 2×1 vector of a Gaussian error in the measurement of the speed, $y_i(n)$ is the measurement of the signal power arriving from the i-th sensor, $\boldsymbol{r}_i \in \Re^2$ is the (known) position of the i-th sensor, P_0 is the transmitted power which attenuates exponentially with the distance according to the unknown parameter $\theta \in [1, \theta_{max})$, and $v_{2,i}(n) \sim \mathcal{N}(0, 5)$ is a Gaussian error in the measurement of $y_i(n)$. All the noise processes are statistically independent. Implement and compare two different particle filters that estimate jointly $\boldsymbol{x}(0:n)$ and θ from the sequence of observations $\boldsymbol{y}(1:n)$, where $\boldsymbol{y}(n) = [y_1(n)\; y_2(n)\; y_3(n)]^\top$. Assume that the prior distribution of the state is $\boldsymbol{x}(0) \sim f(\boldsymbol{x}(0)) = \mathcal{N}(0, 10\boldsymbol{I}_2)$, $\theta_{max} = 6$ and the prior distribution of the fixed parameter is uniform, that is, $f(\theta) \sim \mathcal{U}(1, 6)$.

5.5 The state-space model for detection of digital signals in flat fading channels can be represented as

$$\boldsymbol{h}(n) = \boldsymbol{D}\boldsymbol{h}(n-1) + \boldsymbol{g}\, v_1(n)$$
$$y(n) = \boldsymbol{g}^\top \boldsymbol{h}(n) s(n) + v_2(n)$$

where

$$\boldsymbol{h}(n) = [h(n)\; h(n-1)]^\top$$

is a vector of complex Rayleigh fading coefficient whose temporal correlation is modeled by an AR(2) process with known parameters a_1 and a_2, that is

$$\boldsymbol{D} = \begin{bmatrix} -a_1 & -a_2 \\ 1 & 0 \end{bmatrix}$$

where $a_1 = -1.993$ and $a_2 = 0.996$, and $\boldsymbol{g} = [\mathbf{1}\,\mathbf{0}]^\top$, $v_1(n)$ is a complex Gaussian noise with zero mean and unit variance, $s(n)$ is an M-ary transmitted

signal, $v_2(n)$ is a complex Gaussian noise with zero mean and variance σ^2. At any time instant n, the unknowns of the problem are $s(n)$ and $\boldsymbol{h}(n)$. (Note that this is also a nonlinear problem but where $\boldsymbol{h}(n)$ is a continuous variable and $s(n)$ is a discrete variable.)
Implement a marginalized PF that detects the transmitted signal $s(n)$ (note that given $s(n)$, the system is linear in $\boldsymbol{h}(n)$). Compare the obtained PF with a one that solves the problem without marginalizing out the conditionally linear parameter.

5.6 Consider the model

$$x(n) = \alpha + \beta x(n-1) + v_1(n)$$

$$y(n) = e^{x(n)/2} v_2(n)$$

where $x(n)$ is the state of the system, $v_1 \sim \mathcal{N}(0, \sigma^2)$, $y(n)$ is an observation and $v_2(n) \sim \mathcal{N}(0, 1)$. The parameters α and β and the variance σ^2 are all unknown. The objective is to track the dynamic state of the system $x(n)$. Develop a PF for tracking the unknowns in the system. (Hint: For modeling the prior of the variance σ^2, use the inverse Gamma PDF.)

5.7 A frequent problem in signal processing is the tracking of the instantaneous frequency of a sinusoid. In other words, let $x(n)$ represent the unknown frequency of a sinusoid modeled by a random walk

$$x(n) = x(n-1) + v_1(n)$$

where $v_1(n) \sim N(0, \sigma_{v_1}^2)$ and where the noise variance $\sigma_{v_1}^2$ is known and small. We observe $y(n)$ given by

$$y(n) = a_1 \cos(2\pi x(n)n) + a_2 \sin(2\pi x(n)n) + v_2(n)$$

with the amplitudes a_1 and a_2 being unknown, and where $v_2(n) \sim N(0, \sigma_{v_2}^2)$, where $\sigma_{v_2}^2$ is assumed known. Develop a PF that applies Rao–Blackwellization (the amplitudes a_1 and a_2 are conditionally linear parameters). Extend the PF so that it can track the signal frequency when the noise variances in the system are unknown.

REFERENCES

1. H. Akashi and H. Kumamoto, Construction of discrete-time nonlinear filter by Monte Carlo methods with variance reducing techniques, *Systems and Control*, 19, 211, (1975).
2. D. L. Alspach and H. W. Sorenson, Nonlinear Bayesian estimation using Gaussian sum approximation, *IEEE Trans. on Automatic Control*, 17, 439, (1972).

3. B. D. O. Anderson and J. B. Moore, *Optimal filtering*, Mineola, NY: Dover Publications, Inc., 1979.
4. D. Angelova and L. Mihaylova, Joint target tracking and classification with particle filtering and mixture Kalman filtering using kinematic radar information, *Digital Signal Processing*, 16, 180, (2006).
5. I. Arasaratnam, S. Haykin, and R. J. Elliott, Discrete-time nonlinear filtering algorithms using Gauss-Hermite quadrature, *Proc. of the IEEE*, 95, 953, 2007.
6. M. S. Arulampalam, S. Maskell, N. Gordon, and T. Clapp, A tutorial on particle filters for online nonlinear/non-Gaussian Bayesian tracking, *IEEE Trans. on Signal Processing*, 50, 174, (2002).
7. A. Athalye, Miodrag Bolić, S. Hong, and P. M. Djurić, Generic hardware architectures for sampling and resampling in particle filters, *EURASIP Journal on Applied Signal Processing*, 2005, 2888 (2005).
8. A. Athalye, "Design and Implementation of Reconfigurable Hardware for Real-Time Particle Filtering," Unpublished doctoral dissertation, Stony Brook University, (2007).
9. J. M. Bernardo and A. F. M. Smith, *Bayesian Theory*, John Wiley & Sons, New York, (1994).
10. C. Berzuini, N. G. Best, W. R. Gilks, and C. Larizza, Dynamic conditional independence models and Markov chain Monte Carlo methods, *Journal of the American Statistical Association*, 92, 1403, (1997).
11. M. Bolić, "Architectures for Efficient Implementation of Particle Filters", Unpublished doctoral dissertation, Stony Brook University, (2004).
12. M. Bolić, P. M. Djurić, and S. Hong, Resampling algorithms and architectures for distributed particle filters, *IEEE Trans. Signal Processing*, 53, 2442, (2005).
13. M. G. S. Bruno, Bayesian methods for multiaspect target tracking in image sequences, *IEEE Trans. on Signal Processing*, 52, 1848, (2004).
14. R. S. Bucy, Bayes theorem and digital realization for nonlinear filters, *Journal of Astronautical Sciences*, 80, 73, (1969).
15. O. Cappé, E. Moulines, and T. Ryden, *Inference in Hidden Markov Models*, Springer, New York, (2005).
16. O. Cappé, S. J. Godsill, and E. Moulines, An overview of existing methods and recent advances in sequential Monte Carlo, *Proc. of the IEEE*, 95, 899, (2007).
17. F. Caron, M. Davy, E. Duflos, and P. Vanheeghe, Particle filtering for multisensor data fusion with switching observation models: Application to land vehicle positioning, *IEEE Trans. on Signal Processing*, 55, 2703, (2007).
18. G. Casella and C. P. Robert, Rao-Blackwellization of sampling schemes, *Biometrika*, 84, 81, (1996).
19. V. Cevher, A. C. Sankaranarayanan, J. H. McClellan, and R. Chellappa, Target tracking using a joint acoustic video system, *IEEE Trans. on Multimedia*, 9, 715 (2007).
20. R. Chen and J. S. Liu, Mixture Kalman filters, *Journal of the Royal Statistical Society*, 62, 493, 2000.
21. D. Crisan and A. Doucet, A survey of convergence results on particle filtering methods for practitioners, *IEEE Trans. on Signal Processing*, 50, 736, (2002).
22. F. Daum and J. Huang, "Curse of dimensionality and particle filters," *Proc. of IEEE Aerospace Conference*, IV, 1979, (2003).

23. P. M. Djurić, J. H. Kotecha, J. Zhang, Y. Huang, T. Ghirmai, M. F. Bugallo, and J. Míguez, Particle filtering, *IEEE Signal Processing Magazine*, 20, 19, (2003).
24. P. M. Djurić, M. F. Bugallo, and J. Míguez, "Density assisted particle filters for state and parameter estimation", *Proc. of the IEEE International Conference on Acoustics, Speech, and Signal Processing*, Montreal, Canada, (2004).
25. A. Doucet, S. J. Godsill, and C. Andrieu, On sequential Monte Carlo sampling methods for Bayesian filtering, *Statistics and Computing*, 10, 197, (2000).
26. A. Doucet, N. de Freitas, and N. Gordon, *Sequential Monte Carlo Methods in Practice*, Springer, New York, (2001).
27. A. Doucet, N. de Freitas, and N. Gordon, "An introduction to sequential Monte Carlo methods", In A. Doucet, N. de Freitas, and N. Gordon, Eds., *Sequential Monte Carlo Methods in Practice*, Springer, New York, (2001).
28. P. Fearnhead, Markov chain Monte Carlo, sufficient statistics and particle filters, *Journal of Computational and Graphical Statistics*, 11, 848, (2002).
29. S. Frühwirth-Schnatter, Applied state space modeling of non-Gaussian time series using integration-based Kalman filtering, *Statistics and Computing*, 4, 259, (1994).
30. A. Giremus, J.-Y. Tourneret, and V. Calmettes, A particle filtering approach for joint detection/estimation of multipath effects on GPS measurements, *IEEE Trans. on Signal Processing*, 55, 1275 (2007).
31. N. J. Gordon, D. J. Salmond, and A. F. M. Smith, Novel approach to nonlinear/non-Gaussian Bayesian state estimation, *IEE Proc.-F*, 140, 107, (1993).
32. F. Gustaffson, F. Gunnarsson, N. Bergman, U. Forssel, J. Jansson, R. Karlsson, and P.-J. Nordlund, Particle filtering for positioning, navigation, and tracking, *IEEE Trans. on Signal Processing*, 50, 425, (2002).
33. J. M. Hammersley and K. W. Morton, Poor man's Monte Carlo, *Journal of the Royal Statistical Society, Series B*, 16, 23, (1954).
34. J. E. Handshin, Monte Carlo techniques for prediction and filtering of nonlinear stochastic processes, *Automatica*, 6, 555, (1970).
35. A. C. Harvey, *Forecasting, Structural Time Series Models and the Kalman Filter*, Cambridge: Cambridge University Press, (1989).
36. S. Hong, J. Lee, A. Athalye, P. M. Djurić, and W. D. Cho, Design methodology for domain-specific particle filter realization, *IEEE Trans. on Circuits and Systems – I: Regular Papers*, 54, 1987, (2007).
37. Y. Huang and P. M. Djurić, A blind particle filtering detector of signals transmitted over flat fading channels, *IEEE Trans. on Signal Processing*, 52, 1891, (2004).
38. C. Hue, J.-P. Le Cadre and P. Perez, Sequential Monte Carlo methods for multiple target tracking and data fusion, *IEEE Trans. on Signal Processing*, 50, 309, (2002).
39. K. Ito and K. Xiong, Gaussian filters for nonlinear filtering problems, *IEEE Trans. on Automatic Control*, 45, 910, (2000).
40. A. H. Jazwinski, *Stochastic Processes and Filtering Theory*, New York: Academic Press, (1970).
41. S. J. Julier and J. K. Uhlmann, "A New extension of the Kalman filter to nonlinear systems", *Proc. of AeroSense: The 11th International Symposium on Aerospace/Defence Sensing, Simulation and Controls* (1997).

42. S. Julier, J. Uhlmann, and H. F. Durante-White, A new method for nonlinear transformation of means and covariances in filters and estimators, *IEEE Trans. on Automatic Control*, 45, 477 (2000).
43. R. E. Kalman, A new approach to linear filtering and prediction Pproblems, *Trans. of the ASME-Journal of Basic Engineering*, 82, 35, (1960).
44. R. Karlsson and F. Gustafsson, Complexity analysis of the marginalized particle filter, *IEEE Trans. on Signal Processing*, 53, 4408, (2005).
45. R. Karlsson and F. Gustafsson, Bayesian surface and underwater navigation, *IEEE Trans. on Signal Processing*, 54, 4204, (2006).
46. G. Kitagawa, Non-Gaussian state-space modeling of nonstationary time series, *Journal of the American Statistical Association*, 82, 1032, (1987).
47. G. Kitagawa, Monte Carlo filter and smoother for non-Gaussian nonlinear state space models, *Journal of Computational and Graphical Statistics*, 1, 25, (1996).
48. T. Kloek and H. K. van Dijk, Bayesian estimates of system equation parameters: An application of integration by Monte Carlo, *Econometrica*, 46, 1, (1978).
49. S. C. Kramer and H. W. Sorenson, Recursive Bayesian estimation using piece-wise constant approximations, *Automatica*, 24, 789, (1988).
50. J. Kotecha and P. M. Djurić, Gaussian particle filtering, *IEEE Trans. on Signal Processing*, 51, 2592, (2003).
51. J. Kotecha and P. M. Djurić, Gaussian sum particle filtering, *IEEE Trans. on Signal Processing*, 51, 2602, (2003).
52. H. R. Künsch, "State space and hidden Markov models", in O. E. Barndorff-Nielsen, D. R. Cox, and C. Klueppelberg, Eds., *In Complex Stochastic Systems*, 109, Boca Raton: CRC Publishers, (2001).
53. H. R. Künsch, Recursive Monte Carlo filters: Algorithms and theoretical analysis, *The Annals of Statistics*, 33, 1983, (2005).
54. J. S. Liu and R. Chen, Sequential Monte Carlo methods for dynamical systems, *Journal of the American Statistical Association*, 93, 1032, 1998.
55. J. S. Liu, *Monte Carlo Strategies in Scientific Computing*, Springer, New York, 2001.
56. J. Liu and M. West, "Combined parameter and state estimation in simulation-based filtering," in A. Doucet, N. de Freitas, and N. Gordon, Eds. *Sequential Monte Carlo Methods in Practice*, 197, Springer, New York, (2001).
57. L. Ljung and T. Södereström, *Theory and Practice of Recursive Identification*, Cambridge, MA: The MIT Press, (1987).
58. A. Marshall, "The use of multi-stage sampling schemes in Monte Carlo computations", in M. Meyer, Ed., *Symposium on Monte Carlo Methods*, Wiley, pp. 123–140, (1956).
59. C. J. Masreliez, Approximate non-Gaussian filtering with linear state and observation relations, *IEEE Trans. on Automatic Control*, 20, 107, (1975).
60. L. Mihaylova, D. Angelova, S. Honary, D. R. Bull, C.N. Canagarajah, and B. Ristic, Mobility tracking in cellular networks using particle filtering, *IEEE Trans. on Wireless Communications*, 6, 3589 (2007).
61. M. R. Morelande and S. Challa, Manoeuvering target tracking in clutter using particle filters, *IEEE Trans. on Aerospace and Electronic Systems*, 41, 252, (2005).

62. M. R. Morelande, C. M. Kreucher, and K. Kastella, A Bayesian approach to multiple target detection and tracking, *IEEE Trans. on Acoustics, Speech and Signal Processing*, 55, 1589 (2007).

63. M. Orton and W. Fitzgerald, A Bayesian approach to tracking multiple targets using sensor arrays and particle filters, *IEEE Trans. on Signal Processing*, 50, 216, (2002).

64. D. Gatica-Perez, G. Lathoud, J.-M. Odobez, and I. McCowan, Audiovisual probabilistic tracking of multiple speakers in meetings, *IEEE Trans. on Audio, Speech, and Language Processing*, 15, 601, (2007).

65. M. Pitt and N. Shepard, Filtering via simulation: auxiliary particle filters, *Journal of the American Statistical Association*, 94, 590, 1999.

66. B. Ristić, S. Arulampalam, and N. Gordon, *Beyond the Kalman Filter*, Boston, MA: Artech House (2004).

67. M. N. Rosenbluth and A. W. Rosenbluth, Monte Carlo calculation of the average extension of molecular chains, *Journal of Chemical Physics*, 23, 356, (1956).

68. D. B. Rubin, Comment on 'The calculation of posterior distributions by data augmentation', by M. A Tanner and W. H. Wong, *Journal of the American Statistical Association*, 82, 543, (1987).

69. A. C. Sankaranarayanan, R. Chellappa, and A. Srivastava, "Algorithmic and architectural design methodology for particle filters in hardware," *ICCD '05: Proc. of the 2005 International Conference on Computer Design*, (2005).

70. H. W. Sorenson and D. L. Alspach, Nonlinear Bayesian estimation using Gaussian sums, *Automatica*, 7, 465 (1971).

71. H. W. Sorenson, "Recursive estimation for nonlinear dynamic systems", in J. C. Spall, Ed. *Bayesian Analysis of Time Series and Dynamic Models*, New York: Dekker, (1988).

72. G. Storvik, Particle filters for state-space models with presence of unknown static parameters, *IEEE Trans. on Signal Processing*, 50, 281, (2002).

73. P. Tichavský and C. H. Muravchik and A. Nehorai, Posterior Cramér-Rao bounds for discrete-time nonlinear filtering, *IEEE Trans. on Signal Processing*, 46, 1386, (1998).

74. H. L. Van Trees, *Detection, Estimation, and Modulation Theory*, John Wiley & Sons, New York, (1968).

75. D. B. Ward, E. A. Lehmann, and R. C. Williamson, Particle filtering algorithms for tracking an acoustic source in a reverberant environment, *IEEE Trans. Speech Audio Processing*, 11, 826, (2003).

6

NONLINEAR SEQUENTIAL STATE ESTIMATION FOR SOLVING PATTERN-CLASSIFICATION PROBLEMS

Simon Haykin and Ienkaran Arasaratnam
McMaster University, Canada

6.1 INTRODUCTION

Sequential state estimation has established itself as one of the essential elements of signal processing and control theory. Typically, we think of its use being confined to dynamic systems, where we are given a set of observables and the requirement is to estimate the hidden state of the system on which the observables are dependant. However, when the issue of interest is that of pattern-classification (recognition), we usually do not think of sequential estimation as a relevant tool for solving such problems. Perhaps, this oversight may be attributed to the fact that pattern-classification machines are usually viewed as static rather than dynamic systems. In this chapter, we take a different view: Specifically, we look to nonlinear sequential state estimation as a tool for solving pattern-classification problems, where the problem is solved through an iterative supervised learning process. In so doing, we demonstrate that this approach to solving pattern-classification problems offers several computational advantages compared to traditional methods, particularly when the problem of interest is difficult or large-scale.

Adaptive Signal Processing: Next Generation Solutions. Edited by Tülay Adalı and Simon Haykin
Copyright © 2010 John Wiley & Sons, Inc.

The chapter is structured as follows: Section 6.2 briefly discusses the back-propagation (BP) algorithm and support-vector machine (SVM) learning as two commonly used procedures for pattern classification; these two approaches are used later in the chapter as a framework for experimental comparison with the sequential state estimation approach for pattern classification. Section 6.3 describes how the idea of nonlinear sequential state estimation can be used as supervised training of a multilayer perceptron. With the material of Section 6.3 at hand, the stage is set for specializing the well known extended Kalman filter (EKF) for the supervised training of a multilayer perceptron in Section 6.4. Section 6.5 describes a difficult pattern-classification task, which is used as a basis for comparing the EKF algorithm with the BP and SVM algorithms. The chapter concludes with a summary and discussion in Section 6.6.

6.2 BACK-PROPAGATION AND SUPPORT VECTOR MACHINE-LEARNING ALGORITHMS: REVIEW

In this section, we review two commonly used pattern-classification algorithms, namely, the back-propagation algorithm for the supervised training of multilayer perceptrons and the support vector machine for radial basis function (RBF) networks. The reason for including this review in the chapter is essentially to describe the two frameworks against which a pattern-classifier trained with the extended Kalman filter will be compared later on in the chapter.

6.2.1 Back-Propagation Learning

Basically, the back-propagation (BP) algorithm is an error-correction algorithm, providing a generalization of the ubiquitous least-mean squares (LMS) algorithm. As such, the BP algorithm has many of the attributes associated with the LMS algorithm, when it is implemented in the stochastic (supervised) mode of learning:

- simplicity of implementation;
- computational complexity, linear in the number of weights; and
- robustness.

Most importantly, the BP algorithm is not an algorithm intended for the optimal supervised training of a multilayer perceptron. Rather, it is an algorithm for computing the partial derivatives of a cost function with respect to a prescribed set of adjustable parameters [1].

Speaking of the multilayer perceptron (MLP), Figure 6.1 depicts the architectural graph of this neural network, which has two hidden layers and an output layer consisting of three computational units. Figure 6.2 depicts a portion of the MLP, in which two kinds of signals are identified in the network.

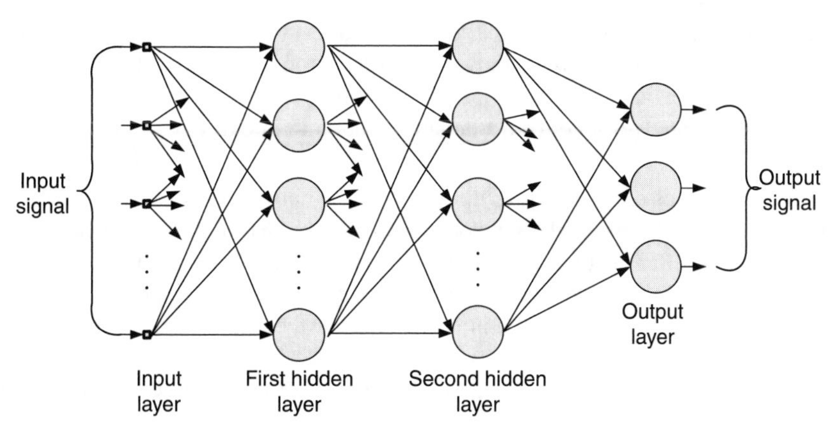

Figure 6.1 Architectural graph of a multilayer perceptron with two hidden layers.

1. *Function Signals*. A functional signal is an input signal (stimulus) that comes in at the input end of the network, and propagates forward neuron by neuron through the network, and emerges at the output end of the network as an output signal. We refer to such a signal as a functional signal for two reasons: First, it is presumed to perform a useful function at the output of the network. Second, at each neuron of the network through which a function signal passes, the signal is calculated as a function of the inputs and the associated weights applied to that neuron. The function signal is also referred to as the input signal.

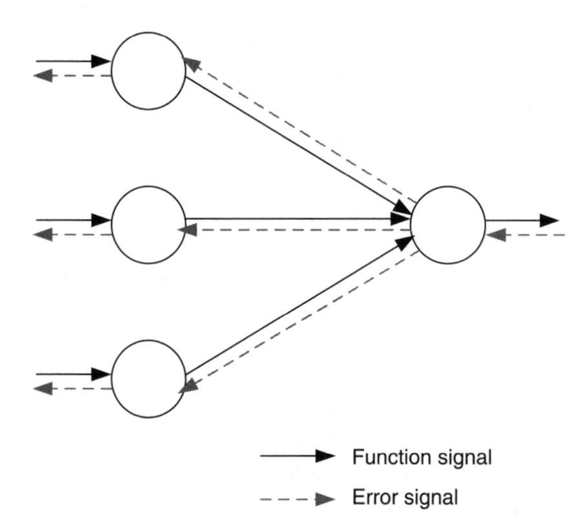

Figure 6.2 Illustration of the direction of the two basic signal flows in a MLP: forward propagation of function signals and back-propagation of error signals.

2. *Error Signals*. An error signal originates at the output neuron of the network and propagates backward (layer by layer) through the network. We refer to the signal as an error signal because its computation by every neuron of the network involves an error-dependant function in one form or another.

The output neurons constitute the output layer of the network. The remaining neurons constitute hidden layers of the network. Thus, the hidden units are not part of the output or input of the network—hence their designation as hidden. The first hidden layer is fed from the input layer made up of sensory units (source nodes); the resulting outputs of the first hidden layer are in turn applied to the next hidden layer; and so on for the rest of the network.

Each hidden or output neuron of a MLP is designed to perform two computations:

1. the computation of the function signal appearing at the output of each neuron, which is expressed as a continuous nonlinear function of the input signal and synaptic weights associated with that neuron;
2. the computation of an estimate of the gradient vector (i.e., the gradients of the error surface with respect to the weights connected to the inputs of a neuron), which is needed for the backward pass thorough the network.

The hidden neurons act as feature detectors; as such, they play a critical role in the operation of the MLP. As the learning process progresses across the MLP, the hidden neurons begin to gradually discover the salient features that characterize the training data. They do so by performing a nonlinear transformation on the input data into a new space called the feature space. In this new space, the classes of interest in a pattern-classification task, for example, may be more easily separated from each other than could be the case in the original input data space. Indeed, it is the formation of this feature space through supervised learning that distinguishes the MLP from Rosenblatt's perceptron.

As already mentioned, the BP algorithm is an error-correction learning algorithm, which proceeds in two phases.

1. In the *forward phase*, the synaptic weights of the network are fixed, and the input signal is propagated through the network, layer by layer, until it reaches the output. Thus, in this phase, changes are confined to activation potentials and outputs of the neurons in the network.
2. In the *backward phase*, an error signal is produced by comparing the output of the network with a desired response. The resulting error signal is propagated through the network, again layer by layer, but this time the propagation is performed in the backward direction. In this second phase, successive adjustments are made to the synaptic weights of the network. Calculation of the adjustments of the output layer is straightforward, but it is more challenging for the hidden layers.

6.2.2 Support Vector Machine

The support vector machine (SVM) is basically a binary classification machine with some highly elegant properties. It was pioneered by Vapnik and the first description was presented in [3]. Detailed descriptions of the machine are presented in [1, 2].

Unlike the BP algorithm, supervised training of the SVM looks to optimization theory for its derivation. To develop insights into SVM theory, formulation of the learning algorithm begins simply with the case of a simple pair of linearly separable patterns. In such a setting, the objective is to maximize the margin of separation between the two classes, given a set of training data made up of input vector-desired response pairs. Appealing to statistical learning theory, we find that the maximization of the margin of separation is equivalent to minimizing the Euclidean norm of the adjustable parameter vector. Building on this important finding, the optimization design of a linearly separable pattern-classifier is formulated as a constrained convex optimization problem, expressed in words as: Minimize sequentially the adjustable parameter vector of a linearly separable pattern-classifier, subject to the constraint that the parameter vector satisfies the supervisory training data. Hence, in mathematical terms, the objective function is expressed as a Lagrange function where the Lagrange multipliers constitute a set of auxiliary nonnegative variables equal in number to the training sample size. There are three important points that emerge from differentiating the Lagrangian function with respect to the adjustable parameters, including a bias term.

1. The Lagrange multipliers are divided into two sets: one zero, and the other nonzero.
2. The equality constraints are satisfied only by those Lagrange multipliers that are non-zero in accordance with the Karush–Kuhn–Turker (KKT) conditions of convex optimization theory.
3. The supervisory data points for which the KKT conditions are satisfied are called support vectors, which, in effect, define those data points that are the most difficult to classify. In light of this third point, the training data set is said to be sparse in the sense that only a fraction of them define the boundaries of the margin of separation.

Given such a constrained convex optimization problem, it is possible to construct another problem called the dual problem. Naturally, this second problem has the same optimal value as the primal problem, that is, the original optimization problem, but with an important difference: The optimal solution to the primal problem is expressed in terms of free network parameters, whereas the Lagrange multipliers provide the optimal solution to the dual problem. This statement is of profound practical importance as it teaches us that the decision boundary for the linearly separable pattern-classification problem can be computed without having to find the optimal value of the parameter vector. To be more specific, the objective function to be optimized in the dual problem takes the following form

$$Q(\alpha) = \sum_{i=1}^{N} \alpha_i - \frac{1}{2}\sum_{i=1}^{N}\sum_{j=1}^{N} \alpha_i\alpha_j d_i d_j \mathbf{x}_i^T \mathbf{x}_j \tag{6.1}$$

subject to the constraints

$$\sum_{i=1}^{N} \alpha_i d_i = 0$$

$$\alpha_i \geq 0 \quad \text{for } i = 1, 2, \ldots N$$

where α_i in (6.1) are the Lagrange multipliers, and the training data sample is denoted by $\{\mathbf{x}_i, d_i\}_{i=1}^{N}$ with $\mathbf{x}_i$ denoting an input vector, d_i denoting the corresponding desired response, and N denoting the sample size.

Careful examination of (6.1) reveals essential ingredients of SVM learning, namely, the requirement to solve a quadratic programming problem, which can become a difficult proposition to execute in computational terms, particularly when the sample size N is large.

What we have addressed thus far is the optimal solution to a linearly separable pattern-classification problem. To extend the SVM learning theory to the more difficult case of nonseparable pattern-classification problems, a set of variables called slack variables are introduced into the convex optimization problem; these new variables measure the deviation of a point from the ideal condition of perfect pattern separability. It turns out that in mathematical terms, the impact of slack variables is merely to modify the second constraint in the dual optimization problem described in (6.1) as follows

$$0 \leq \alpha_i \leq C \quad \text{for } i = 1, 2, \ldots N$$

where C is a user-specified positive parameter. Except for this modification, formulation of the dual optimization problem for nonseparable pattern-classification problems remains intact. Moreover, the support vectors are defined in the same way as before. The new parameter C controls the tradeoff between the complexity of the machine and the number of nonseparable patterns; it may therefore be viewed as the reciprocal of a parameter commonly referred to the regularization parameter. When parameter C is assigned a large value, the implication is that there is high confidence in the practical quality of the supervisory training data. Conversely, when C is assigned a small value, the training data sample is considered to be noisy, suggesting that less emphasis should be placed on its use and more attention be given to the adjustable parameter vector.

Thus far in the discussion, we have focused attention on an algorithmic formulation of the SVM. Clearly, we need a network structure for its implementation. In this context, the radial-basis function (RBF) provides a popular choice. Figure 6.3 depicts a structure of a RBF network, in which the nonlinear hidden layer of size K (smaller than the sample size N) consists of Gaussian units; nonlinearity of the hidden layer is needed to account for nonseparability of the input patterns. Each Gaussian unit is

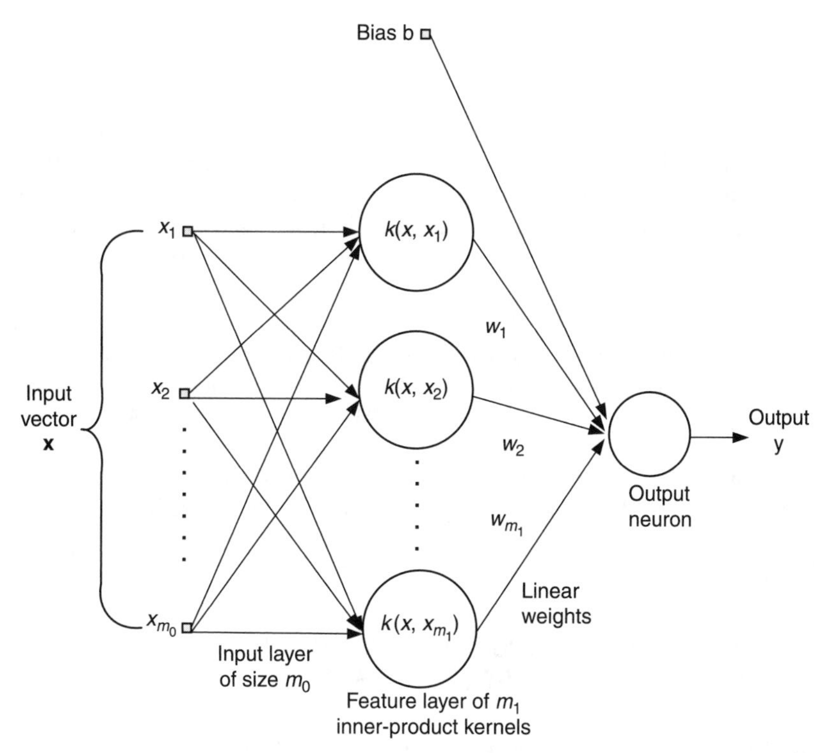

Figure 6.3 Architecture of support vector machine, using a radial basis function network.

defined in terms of its m_0-dimensional center, where m_0 is the dimensionality of the input data space. To complete the specifications of the hidden layer, we need to define also the effective width of each Gaussian function acting as a receptive field. To simplify the network design, it is common practice to assign a common width to all the Gaussian units. As for the output layer, it is linear with m_1 adjustable parameters $w_1, w_2, \ldots w_{m1}$, and an adjustable bias.

To propose an application of the SVM algorithm to the RBF structure of Figure 6.3, we keep the following points in mind.

1. The support vectors picked out of the supervisory training data by the SVM algorithm define the centers of the Gaussian units in the hidden layer.
2. The outputs of the Gaussian units, produced in response to the data applied to the source (input) nodes of the RBF network, define the inputs applied to the adjustable parameters. In effect, $\mathbf{x}_i$ in (6.1) is now replaced by the vector

$$\mathbf{\Phi}(\mathbf{x}_i) = [\Phi(\mathbf{x}_i, \bar{\mathbf{x}}_1), \Phi(\mathbf{x}_i, \bar{\mathbf{x}}_2), \ldots \Phi(\mathbf{x}_i, \bar{\mathbf{x}}_k)]^T \tag{6.2}$$

where $\{\bar{\mathbf{x}}_j\}_{j=1}^K$ denotes the set of support vectors computed by the SVM algorithm. Thus, using (6.2) in (6.1), we see that the inner product $\mathbf{x}_i^T \mathbf{x}_j$ in (6.1) is replaced by the new inner product

$$k(\mathbf{x}_i, \mathbf{x}_j) = \mathbf{\Phi}^T(\mathbf{x}_i)\,\mathbf{\Phi}(\mathbf{x}_j) \tag{6.3}$$

which is called the Mercer kernel. This terminology follows from the fact that the functions assigned to the hidden units of the RBF network must satisfy the celebrated Mercer's theorem, which is indeed satisfied by the choice of Gaussian functions. Equation (6.3) provides the basis for the kernel trick, which states the following.

> Insofar as pattern-classification in the output space of the RBF network is concerned, specification of the Mercer kernel $k(\mathbf{x}_i, \mathbf{x}_j)$ is sufficient, the implication of which is that there is no need to explicitly compute the adjustable parameters of the output layer.

For this reason, a network structure exemplified by the RBF network, is referred to as a kernel machine and the SVM learning algorithm used to train it is referred to as a kernel method.

6.3 SUPERVISED TRAINING FRAMEWORK OF MLPs USING NONLINEAR SEQUENTIAL STATE ESTIMATION

To describe how a nonlinear sequential state estimator can be used to train a MLP in a supervised manner, consider a MLP with s synaptic weights and p output nodes. With n denoting a time-step in the supervised training of the MLP, let the vector $\mathbf{w}_n$ denote the entire set of synaptic weights in the MLP computed at time step n. For example, we may construct $\mathbf{w}_n$ by stacking the weights associated with neuron 1 in the first hidden layer on top of each other, followed by those of neuron 2, carrying on in this manner until we have accounted for all the neurons in the first hidden layer; then we do the same for the second and any other hidden layer in the MLP, until all the weights in the MLP have been accounted for in the vector $\mathbf{w}_n$ in the orderly fashion just described.

With sequential state estimation in mind, the state-space model of the MLP under training is defined by the following pair of models (see Fig. 6.4).

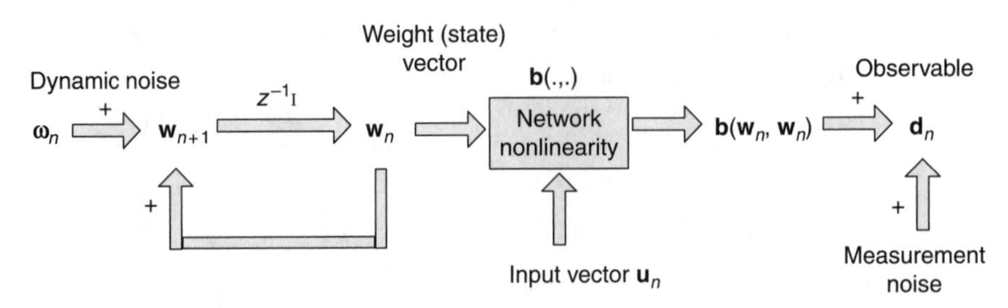

Figure 6.4 Nonlinear state-space model depicting the underlying dynamics of a MLP undergoing supervised training.

1. *System (state) model*, which is described by the first-order autoregressive equation

$$\mathbf{w}_{n+1} = \mathbf{w}_n + \boldsymbol{\omega}_n. \tag{6.4}$$

The dynamic noise $\boldsymbol{\omega}_n$ is white Gaussian noise of zero mean and covariance matrix $\mathbf{Q}_n$ which is purposely included in the system model to anneal the supervised training of the MLP over time. In the early stages of the training session, $\mathbf{Q}_n$ is large in order to encourage the supervised-learning algorithm to escape local minima, then it is gradually reduced to some finite but small value.

2. *Measurement model*, which is described by the equation

$$\mathbf{d}_n = \mathbf{b}(\mathbf{w}_n, \mathbf{u}_n) + \mathbf{v}_n \tag{6.5}$$

where the new entities are defined as follows:

- $\mathbf{d}_n$ is the vector denoting observables
- $\mathbf{u}_n$ is the vector denoting the input signal applied to the network
- $\mathbf{v}_n$ is the vector denoting the measurement noise process of zero mean and diagonal covariance matrix $\mathbf{R}_n$. The source of this noise is attributed to the way in which $\mathbf{d}_n$ is actually obtained.

The vector-valued measurement function $\mathbf{b}$ (.,.) in (6.5) accounts for the overall nonlinearity of the MLP from the input to the output layer; it is the only source of nonlinearity in the state–space model of the MLP. Here, the notion of state refers to an externally adjustable state, which manifests itself in adjustments applied applied to the MLP's weights through supervised training—hence the inclusion of the weight vector $\mathbf{w}_n$ in the state-space model described by both (6.4) and (6.5).

6.4 THE EXTENDED KALMAN FILTER

Given the training sample $\{\mathbf{u}_n, \mathbf{d}_n\}_{n=1}^{N}$, the issue of interest is how to undertake the supervised training of the MLP by means of a sequential state estimator. Since the MLP is nonlinear by virtue of the nonlinear measurement model of (6.5), the sequential state estimator would have to be correspondingly nonlinear. With this requirement in mind, we begin the discussion by considering how the extended Kalman filter (EKF) can be used to fulfil this rule.

For the purpose of our present discussion, the relevant equations of the EKF algorithm summarized in Table 6.1 are the following two, using the terminology of the state–space model of (6.4) and (6.5).

1. The *innovations process*, defined by

$$\boldsymbol{\alpha}_n = \mathbf{d}_n - \mathbf{b}(\hat{\mathbf{w}}_{n|n-1}, \mathbf{u}_n) \tag{6.6}$$

where the desired response $\mathbf{d}_n$ plays the role of the observable for the EKF.

Table 6.1 Summary of the EKF algorithm for supervised training of the MLP

Training sample: $\{\mathbf{u}_n, \mathbf{d}_n\}$, $n = 1, 2, \ldots N$
where $\mathbf{u}_n$ is the input vector applied to the MLP and $\mathbf{d}_n$ is the corresponding desired response.
MLP and Kalman filter: Parameters and variables

$\mathbf{b}(.,.)$	:	vector-valued measurement function
$\mathbf{B}$	:	linearized measurement matrix
$\mathbf{w}_n$	:	weight vector at time step n
$\hat{\mathbf{w}}_{n\|n-1}$	:	predicted estimate of the weight vector
$\hat{\mathbf{w}}_{n\|n}$	:	filtered estimate of the weight vector
$\mathbf{y}_n$	:	output vector of the MLP produced in response to $\mathbf{u}_n$
$\mathbf{Q}_n$	:	covariance matrix of dynamic noise $\boldsymbol{\omega}_n$
$\mathbf{R}_n$	:	covariance matrix of measurement noise $\mathbf{v}_n$
$\mathbf{G}_n$	:	Kalman gain
$\mathbf{P}_{n\|n-1}$	:	prediction error covariance matrix
$\mathbf{P}_{n\|n}$	:	filtering error covariance matrix

Initialization: The initial weights $\hat{\mathbf{w}}_{1|0}$ are drawn from a uniform distribution with zero mean and variance equal to the reciprocal of the number of synaptic connections feeding into a node (fan-in). The associated covariance of the initial weight estimate is fixed at $\delta\mathbf{I}$, where δ can be 'some' multiples of 10, and $\mathbf{I}$ is the identity matrix.
Recursive computation: For $n = 1, 2, \ldots$ compute the following:

$$
\begin{aligned}
\mathbf{G}_n &= \mathbf{P}_{n|n-1}\mathbf{B}_n^T[\mathbf{B}_n\mathbf{P}_{n|n-1}\mathbf{B}_n^T + \mathbf{R}_n]^{-1} \\
\boldsymbol{\alpha}_n &= \mathbf{d}_n - \mathbf{b}_n(\hat{\mathbf{w}}_{n|n-1}, \mathbf{u}_n) \\
\hat{\mathbf{w}}_{n|n} &= \hat{\mathbf{w}}_{n|n-1} + \mathbf{G}_n\boldsymbol{\alpha}_n \\
\mathbf{P}_{n|n} &= \mathbf{P}_{n|n-1} - \mathbf{G}_n\mathbf{B}_n\mathbf{P}_{n|n-1} \\
\hat{\mathbf{w}}_{n+1|n} &= \hat{\mathbf{w}}_{n|n} \\
\mathbf{P}_{n+1|n} &= \mathbf{P}_{n|n} + \mathbf{Q}_n
\end{aligned}
$$

2. The *weight (state) update*, defined by

$$\hat{\mathbf{w}}_{n|n} = \hat{\mathbf{w}}_{n|n-1} + \mathbf{G}_n\boldsymbol{\alpha}_n \tag{6.7}$$

where $\hat{\mathbf{w}}_{n|n-1}$ is the predicted (old) state estimate of the MLP's weight vector $\mathbf{w}$ at time n, given the desired response up to and including time $(n - 1)$, and $\hat{\mathbf{w}}_{n|n}$ is the filtered (updated) estimate of $\mathbf{w}$ on the receipt of observable $\mathbf{d}_n$. The matrix $\mathbf{G}_n$ is the *Kalman gain*, which is an integral part of the EKF algorithm.

Examining the underlying operation of the MLP, we find that the term $\mathbf{b}(\hat{\mathbf{w}}_{n|n-1}, \mathbf{u}_n)$ is the actual output vector $\mathbf{y}_n$ produced by the MLP with its old weight vector $\hat{\mathbf{w}}_{n|n-1}$ in response to the input vector $\mathbf{u}_n$. We may therefore rewrite the combination of (6.6) and (6.7) as a single equation:

$$\hat{\mathbf{w}}_{n|n} = \hat{\mathbf{w}}_{n|n-1} + \mathbf{G}_n(\mathbf{d}_n - \mathbf{y}_n). \tag{6.8}$$

On the basis of this insightful equation, we may now depict the supervised training of the MLP as the combination of two mutually coupled components that form a closed-loop feedback system, as shown in Figure 6.5.

1. The top part of the figure depicts the supervised learning process as viewed partly from the network's perspective. With the weight vector set at its old (predicted) value $\hat{\mathbf{w}}_{n|n-1}$, the MLP computes the actual weight vector $\mathbf{y}_n$ as the predicted estimate of the observable—namely $\hat{\mathbf{d}}_{n|n-1}$.
2. The bottom part of the figure depicts the EKF in its role as the facilitator of the training process. Supplied with $\hat{\mathbf{d}}_{n|n-1} = \mathbf{y}_n$, the EKF updates the old estimate of the weight vector by operating on the current desired response $\mathbf{d}_n$. The filtered estimate of the weight vector, namely, $\hat{\mathbf{w}}_{n|n}$, is thus computed in accordance with (6.8). The EKF supplies $\hat{\mathbf{w}}_{n|n}$ to the MLP via a bank of unit-time delays.

With the transition matrix being equal to the identity matrix, as evidenced by (6.4), we may set $\hat{\mathbf{w}}_{n+1|n}$ equal to $\hat{\mathbf{w}}_{n|n}$ for the next iteration. This equality permits the supervised training cycle to be repeated until the training session is completed.

Note that in the supervised training framework of Figure 6.5, the training sample $(\mathbf{u}_n, \mathbf{d}_n)$ is split between the MLP and the EKF: The input vector $\mathbf{u}_n$ is applied to the MLP as the excitation, and the desired response $\mathbf{d}_n$ is applied to the EKF as the

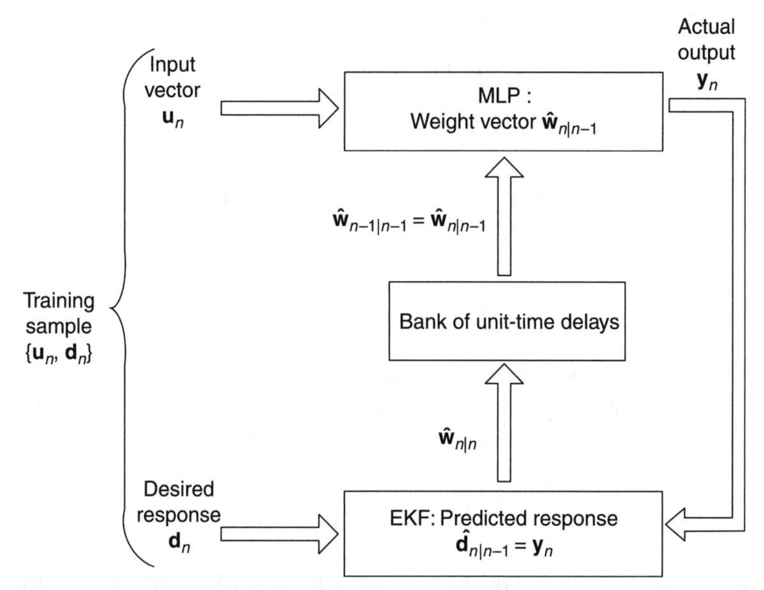

Figure 6.5 Closed-loop feedback system embodying the MLP and the EKF: 1. The MLP, with weight vector $\hat{\mathbf{w}}_{n|n-1}$, operates on the input vector $\mathbf{u}_n$ to produce the output vector $\mathbf{y}_n$; 2. The EKF, supplied with the prediction $\hat{\mathbf{d}}_{n|n-1} = \mathbf{y}_n$, operates on the current desired response $\mathbf{d}_n$ to produce the filtered weight estimate $\hat{\mathbf{w}}_{n|n}$ thereby preparing the closed-loop feedback system for the next iteration.

observable, which is dependant on the hidden weight (state) vector $\mathbf{w}_n$. From the predictor-corrector property of the Kalman filter and its variants and extensions, we find that examination of the block diagram of Figure 6.5 leads us to make the following insightful statement: The multilayer perceptron, undergoing training, performs the role of the predictor, and the EKF, providing the supervision, performs the role of the corrector. Thus, whereas in traditional applications of the Kalman filter for sequential state estimation, the roles of the predictor and corrector are embodied in the Kalman filter itself, in supervised training applications these two roles are split between the MLP and the EKF. Such a split of responsibilities in supervised learning is in perfect accord with the way in which the input and the desired response of the training sample are split in Figure 6.5.

6.4.1 The EKF Algorithm

For us to be able to apply the EKF algorithm as the facilitator of the supervised learning task, we have to linearize the measurement equation (6.5) by retaining first-order terms in the Taylor series expansion of the nonlinear part of the equation. With $\mathbf{b}(\mathbf{w}_n, \mathbf{u}_n)$ as the only source of nonlinearity, we may approximate (6.5) as

$$\mathbf{d}_n = \mathbf{B}_n \mathbf{w}_n + \mathbf{v}_n \tag{6.9}$$

where $\mathbf{B}_n$ is the p-by-s measurement matrix of the linearized model. The linearization process involves computing the partial derivatives of the p outputs of the MLP with respect to its s weights, obtaining the required matrix

$$\mathbf{B} = \begin{pmatrix} \dfrac{\partial b_1}{\partial w_1} & \dfrac{\partial b_1}{\partial w_2} & \cdots & \dfrac{\partial b_1}{\partial w_s} \\ \dfrac{\partial b_2}{\partial w_1} & \dfrac{\partial b_2}{\partial w_2} & \cdots & \dfrac{\partial b_2}{\partial w_s} \\ \vdots & \vdots & \vdots & \vdots \\ \dfrac{\partial b_p}{\partial w_1} & \dfrac{\partial b_p}{\partial w_2} & \cdots & \dfrac{\partial b_p}{\partial w_s} \end{pmatrix} \tag{6.10}$$

where b_i, $i = 1, 2, \ldots p$, in (6.10) denotes the i-th element of the vectorial function $\mathbf{b}(.,.)$, and the partial derivatives on the right-hand side of (6.10) are evaluated at $\mathbf{w}_n = \hat{\mathbf{w}}_{n|n-1}$. Recognizing that the dimensionality of the weight vector $\mathbf{w}$ is s, it follows that the matrix product $\mathbf{Bw}$ is a p-by-1 vector, which is in agreement with the dimensionality of the observable $\mathbf{d}$.

6.5 EXPERIMENTAL COMPARISON OF THE EXTENDED KALMAN FILTERING ALGORITHM WITH THE BACK-PROPAGATION AND SUPPORT VECTOR MACHINE LEARNING ALGORITHMS

In this section, we consider a binary class classification problem as shown in Figure 6.6(a). It consists of three concentric circles of radii 0.3, 0.8, and 1. It is a

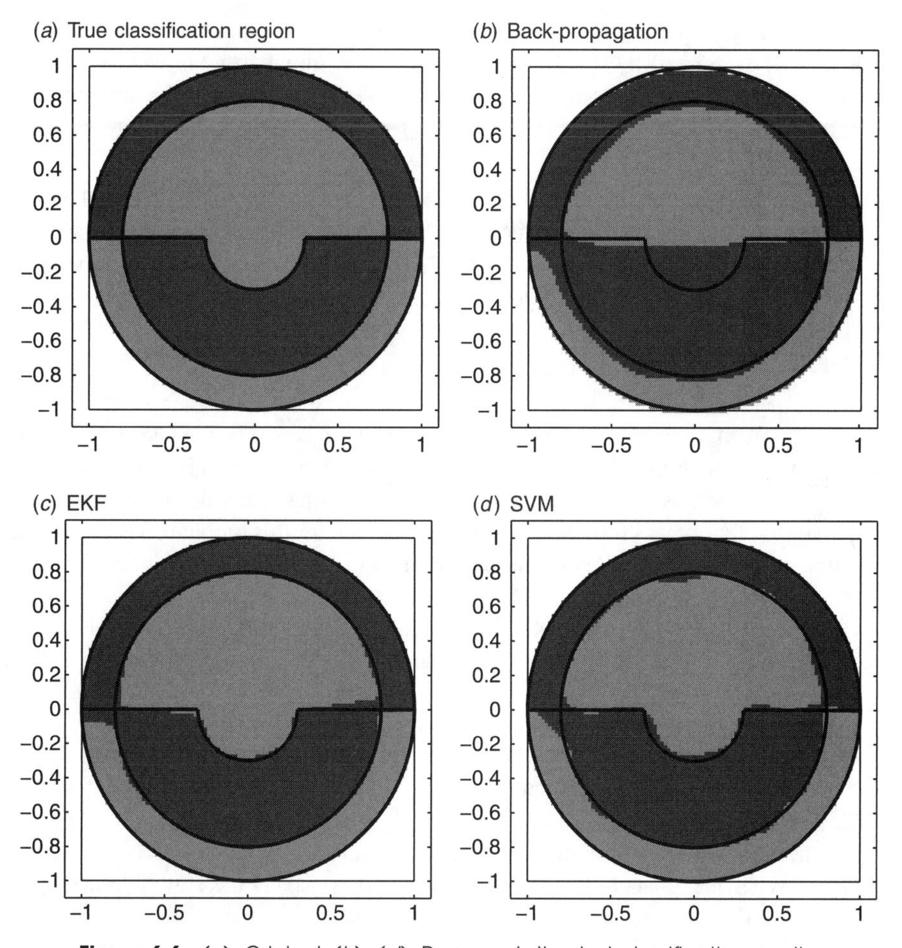

Figure 6.6 (*a*): Original; (*b*)–(*d*): Representative test classification results.

challenging problem because the two regions are disjoint and non-convex. For the experiment, we used the following supervised training algorithms

- MLP trained by the BP;
- MLP trained by the EKF; and
- RBF trained by the SVM.

The structure of the MLP was chosen to have two inputs, one output, and two hidden layers, with five neurons each. Hence, the MLP has a total of 51 adjustable parameters or weights (bias included). All neurons use the hyperbolic tangent function

$$\varphi(v) = \tanh(v) = \frac{\exp(v) - \exp(-v)}{\exp(v) + \exp(-v)}.$$

The learning-rate parameter of the back-propagation algorithm was fixed at 0.01. For the MLPs trained by the BP and the EKF, the initial weights were set up as described in Section 6.4. For the EKF, two more covariance matrices were additionally required.

1. The covariance matrix of dynamic noise $\mathbf{Q}_n$ was annealed such that $\mathbf{Q}_n = (\frac{1}{\lambda} - 1)\mathbf{P}_{n|n}$, where $\mathbf{P}_{n|n}$ is the error covariance associated with the weight estimate at time instant n, and $\lambda \in (0, 1)$ is the forgetting factor as defined in a recursive least-squares algorithm; this approximately assigns exponentially decaying weights to past observables; λ was fixed at 0.9995.
2. The variance of measurement noise R_n was fixed at unity.

For the SVM, the kernel in he RBF network was chosen to be the Gaussian radial-basis function. A kernel of width $\sigma = 0.2$ was found to be a good choice for this problem. This value was chosen based on the accuracy of test classification results. The soft margin of unity was found to be more appropriate for this problem. The quadratic programming code, available as an in-built routine in the MATLAB optimization toolbox, was used for training the SVM.

For the purpose of training 1000 data points were randomly picked from the considered region. In the MLP-based classification, each training epoch contained 100 examples randomly drawn from the training data set. For the purpose of testing, the test data were prepared as follows: A grid of 100 × 100 data points were chosen from the square region [−1, 1] × [−1, 1] [see Fig. 6.6(*a*)] and then the grid points falling outside the unit circle were discarded. In so doing, we were able to obtain 7825 test points.

At the end of every 10-training epoch interval, the MLP was presented with the test data set. We made 50 such independent experiment trials and the ensemble-averaged correct classification rate was computed. In contrast, for the SVM, this test grid was presented at the end of the training session. Figures 6.7(*a*) and 6.7(*b*) show the results

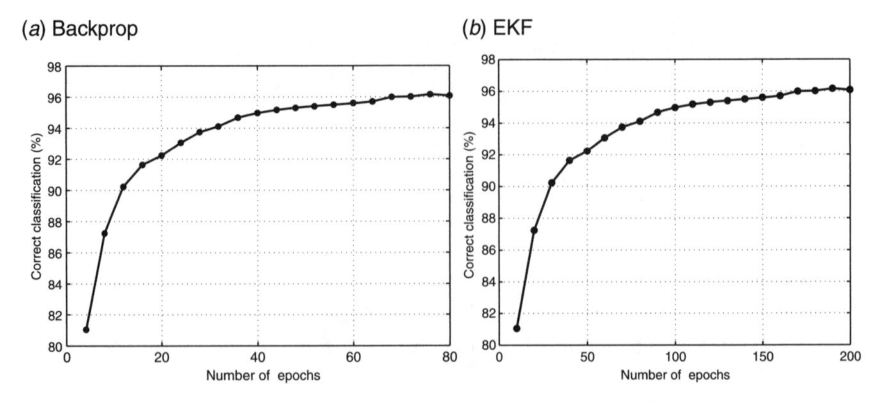

Figure 6.7 Ensemble-averaged (over 50 runs) correct classification rate versus number of epochs.

Table 6.2 Comparison of classification accuracy and computational times averaged over 50 independent experiments

Algorithm	Test accuracy (%)	Time (s)
BP	90	63
EKF	96	8
SVM	96	182

of the BP-trained MLP and the EKF-trained MLP, respectively. At the end of 5000 training epochs, we obtained the following results: the BP-trained MLP achieves nearly 90 percent classification accuracy; whereas, at the end of 200 training epochs, the EKF-trained MLP achieves nearly 96 percent classification accuracy. Figures 6.6(*b*)–6.6(*d*) depict representative test classification results of the employed algorithms.

We also computed the computational times of each algorithm averaged over 50 independent trials when executing on an Intel Pentium Core Duo processor of cycle speed 2.1 GHz. The summarized results are presented in Table 6.2. To execute 5000 epochs of the BP, roughly a minute was required, whereas, the EKF required only 8 seconds for completing its 200 training epochs. As expected, the SVM was found to take more time (nearly 3 minutes) to complete its training session.

Based on the computational time-accuracy tradeoff, we may say that the EKF-trained MLP is the best choice for solving the difficult binary pattern-classification problem illustrated by Figure 6.6(*a*). Such a tradeoff may equally apply to other difficult pattern classification problems, and possibly large-scale ones.

6.6 CONCLUDING REMARKS

In the neural networks and learning machines literature, the solution to pattern-classification problems is commonly tackled, for example, by using a multilayer perceptron (MLP) trained with the back-propagation algorithm or a radial-basis function (RBF) network using the support-vector machine (SVM) learning algorithm [1]. In this chapter, we have discussed another way of tackling the pattern-classification problem, namely, a multilayer perceptron trained using the extended Kalman filter (EKF). The rationale for this unusual approach is that the supervised learning process, involving the training of a multilayer perceptron may be viewed as a state-estimation problem, in which case the weight vector characterizing the multilayer perceptron is viewed as a hidden state. An attractive feature of using the EKF for estimating the unknown weight vector is the improved utilization of the information contained in the training sample by propagating the covariance matrix of the filtered estimation error, hence the accelerated rate of learning. This point is borne out by the shorter time taken to complete the learning process, compared to both the BP and SVM algorithms. Moreover, the experimental results presented in Section 6.5 show that the EKF approach to the supervised training of a multilayer perceptron achieves a pattern-classification accuracy that matches the results obtainable with the SVM.

To conclude, we may say that the supervised training of a multilayer perceptron, using the EKF as the tool to oversee the supervision, deserves serious consideration when the requirement is to solve difficult, and possibly large-scale pattern-classification problems.

6.7 PROBLEMS

6.1 You are given a set of supervisory data samples of large enough size. The data are to be used for training of a learning machine, which is required to solve a difficult pattern-classification exemplified by that of Figure 6.6(*a*).
Describe the attributes of a learning machine that is required to tackle such pattern-classification problem.

6.2 Discuss the pros and cons of the extended Kalman filtering algorithm used to train a multilayer perceptron for solving the pattern-classification problem of Figure 6.6(*a*) compared to the back-propagation algorithm.

6.3 Repeat the experiment described in Section 6.5 to verify your conclusions in Problem 6.2.

6.4 Discuss the pros and cons of the extended Kalman filtering algorithm used to train a multilayer perceptron for solving the pattern-classification problem of Figure 6.6(*a*) compared to a radial-basis function network trained using the support vector machine (SVM) learning algorithm.

6.5 Repeat the experiment described in Section 6.5 to verify your conclusions in Problem 6.4.

6.6 The learning capability of the human brain extends beyond pattern-classification. To probe further, consider two people who love listening to music. One is in their mid-thirties, and the other is in their mid-sixties. Suppose a piece of music is being played on the radio some time in 2010: If the piece of music goes back to the 1970s, it is the second person who very much enjoys it. On the other hand, when the piece of music played on the same radio goes back to the 1990s, it is the first person who very much enjoys it.
Explain the main reason that is responsible for this difference.

REFERENCES

1. S. Haykin, *Neural networks and learning machines*, 3rd ed., Prentice Hall, NJ, 2009.
2. B. Schölkolf and A. J. Smola, *Learning with kernels: Support vector machines, regularization, optimization, and beyond*, MA: MIT Press, 2003.
3. V. N. Vapnik, *Statistical learning theory*, NY: Wiley, 1998.

7

BANDWIDTH EXTENSION OF TELEPHONY SPEECH

Bernd Iser and Gerhard Schmidt
Harman/Becker Automotive Systems, Ulm, Germany

In this chapter an introduction on bandwidth extension of telephony speech is given. Why current telephone networks apply a limiting bandpass, what kind of bandpass is used, and what can be done to (re)increase the bandwidth on the receiver side without changing the transmission system is presented. Therefore, several approaches—most of them based on the source-filter model for speech generation—are discussed. The task of bandwidth extension algorithms that make use of this model can be divided into two subtasks: an excitation signal extension part and a wideband envelope estimation part. Different methods for accomplishing these tasks, like nonlinear processing, the use of signal and noise generators, or modulation approaches on the one hand and codebook approaches, linear mapping schemes or neural networks on the other, are presented.

7.1 INTRODUCTION

Speech is the most natural and convenient way of human communication. This is the reason for the big success of the telephone system since its invention in the nineteenth century [1]. At that time people didn't think of high quality speech but nowadays this has changed. Often customers are not satisfied with the quality of service provided by the telephone system especially when compared to other audio sources, such as radio or compact disc. The degradation of speech quality using analog telephone systems is

Adaptive Signal Processing: Next Generation Solutions. Edited by Tülay Adalı and Simon Haykin
Copyright © 2010 John Wiley & Sons, Inc.

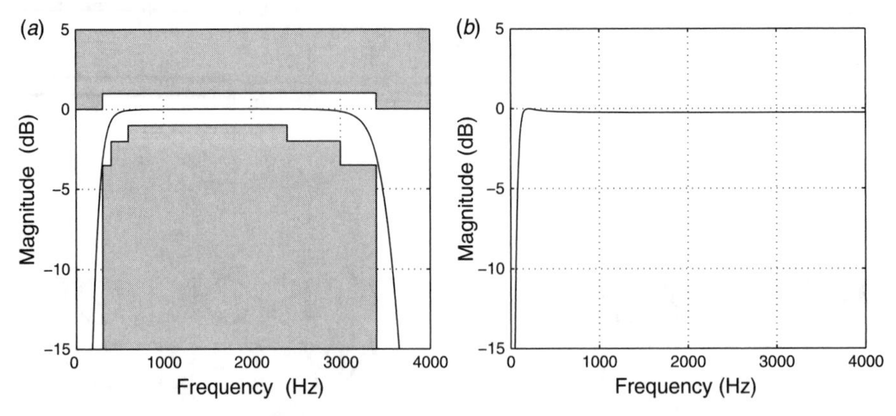

Figure 7.1 (*a*) Analog telephone bandpass according to ITU recommendation G.151 (16) (used in analog phone connections) and (*b*) highpass in GSM enhanced full rate codec according to (10) (used in the GSM cell phone network).

caused by the introduction of band limiting filters within amplifiers used to keep a certain signal level in long local loops [22]. These filters have a passband from approximately 300 Hz to 3400 Hz [see Fig. 7.1*a*] and were applied to reduce crosstalk between different channels.

As we can see in the time-frequency analysis of a speech signal transmitted over a telephone network [depicted in Fig. 7.3*b*] the application of such a bandpass attenuates large speech portions. Digital networks, such as Integrated Service Digital Network (ISDN) and Global System for Mobile Communication (GSM), are able to transmit speech in higher quality since signal components below 300 Hz as well as components between 3.4 kHz and 4 kHz can be transmitted [see Fig. 7.1*b*]. However, this is only true if the entire call (in terms of its routing) remains in those

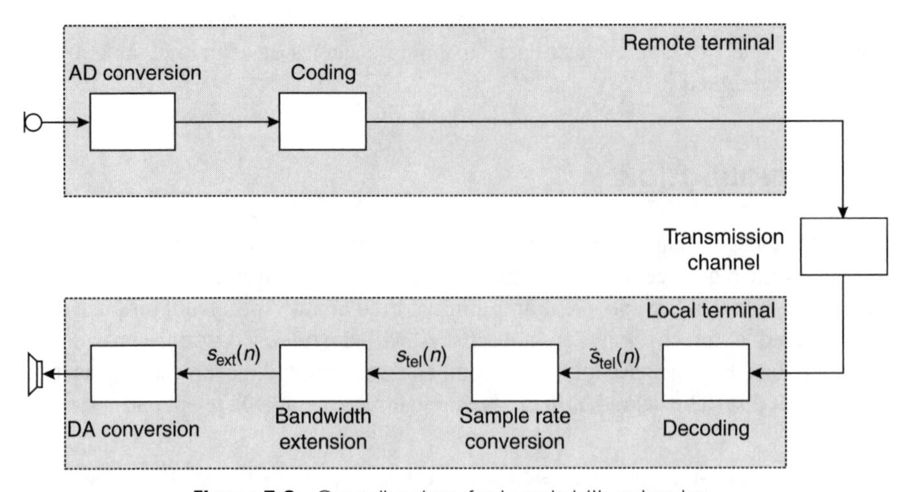

Figure 7.2 Overall system for bandwidth extension.

networks—when leaving into an analog telephone network the speech signal is once again bandlimited. Furthermore, we still have a band limitation of 4 kHz.

Thus, great efforts have been made to increase the quality of telephone speech signals in recent years. Wideband codecs are able to increase the bandwidth up to 7 kHz or even higher at only moderate complexity [17]. Nevertheless, applying these codecs would mean modifying the current networks. Another possibility is to increase the bandwidth after transmission by means of bandwidth extension. The basic idea of these enhancements is to estimate the speech signal components above 3400 Hz and below 300 Hz and to complement the signal in the new frequency bands with this estimate. In this case the telephone networks are left untouched. Figure 7.2

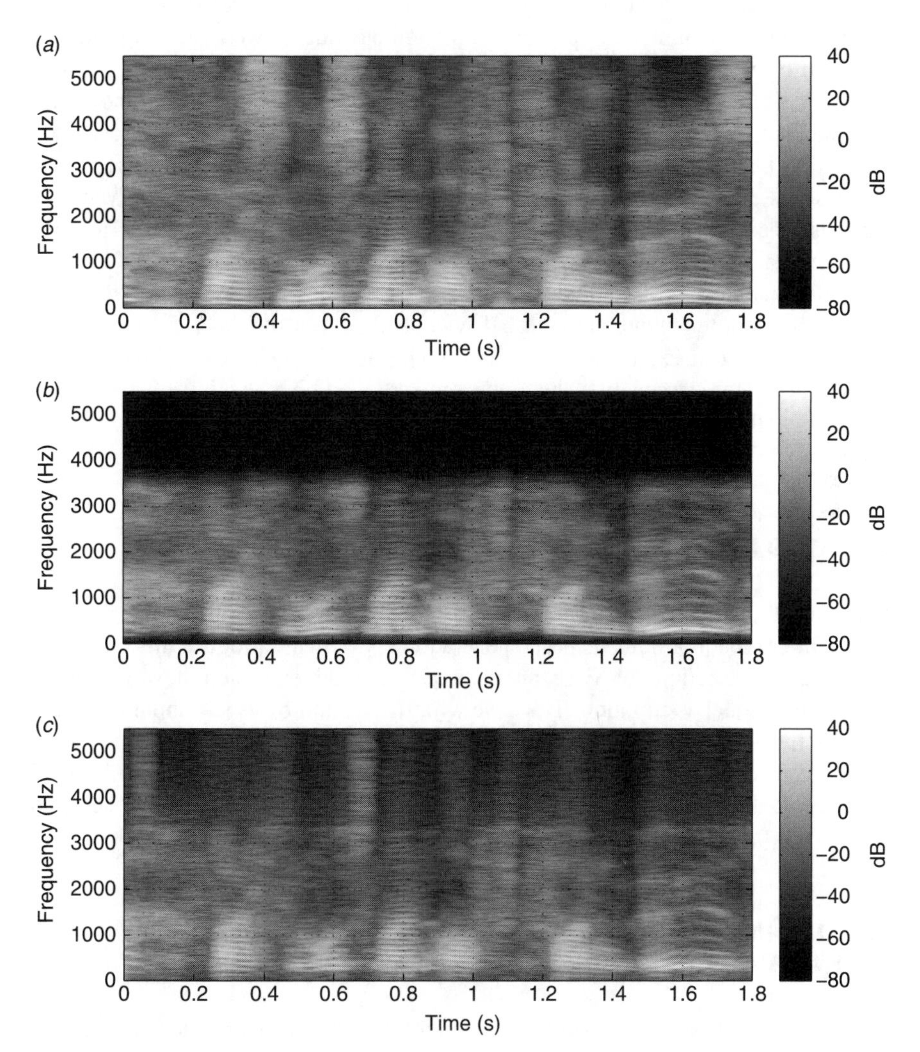

Figure 7.3 Spectrograms of (*a*) Wideband speech, (*b*) Bandlimited speech, and (*c*) Reconstructed speech.

shows the basic structure of a telephone connection with a bandwidth extension (BWE) system included in the receiving path of a telephone connection. The incoming signal $\tilde{s}_{\text{tel}}(n)$ at the local terminal is first upsampled from 8 kHz to the desired sampling rate, for example, 11 or 16 kHz. The resulting signal $s_{\text{tel}}(n)$ has now the desired rate but contains no signal components above 4 kHz (assuming anti-imaging filters of sufficient quality within the upsampling process). In the bandwidth extension unit the signal is analyzed within the telephone band and the missing frequency components are generated and added to the original input signal. The resulting signal is denoted as $s_{\text{ext}}(n)$.

Additionally, three time-frequency analyses are presented in Figure 7.3. The first analysis depicts a wideband speech signal $s(n)$ as it would be recorded close to the mouth of the communication partner on the remote side. If we assume not to have any kind of errors or distortions on the transmission a bandlimited signal $s_{\text{tel}}(n)$ as depicted in the center diagram would be received at the local side. The truncation of the frequency range is clearly visible. Without any additional processing the local communication partner would be listening to this signal. If bandwidth extension is applied a signal $s_{\text{ext}}(n)$ as depicted in part (*c*) of Figure 7.3 would be reconstructed. Even if the signal is not exactly the same as the original one, it sounds more natural and, as a variety of listening tests indicate, the speech quality in general is increased as well [14].

Additionally, we have to note that bandwidth extension is not limited to the bandwidth of current telephone networks. If wideband codecs such as [17] will be used BWE schemes can extend the frequency range above 7 kHz. Early investigations show that in these scenarios the improvement of speech quality is even better. However, in this chapter we will focus on extending the bandwidth of 4 kHz telephone speech.

7.2 ORGANIZATION OF THE CHAPTER

Since most of the recent schemes for bandwidth extension utilize the so-called source-filter model of human speech generation [7] we will introduce this model and a few basics in Section 7.4. Early attempts of bandwidth extension, however, do not rely on this model assumption. Thus, we will give a short overview about nonmodel based schemes first in the next section. The main part of this contribution is about model-based extension schemes, which are discussed in Section 7.5. Finally, an outlook on the evaluation of bandwidth extension systems for telephony speech is presented in Section 7.6.

7.3 NONMODEL-BASED ALGORITHMS FOR BANDWIDTH EXTENSION

The first approaches for bandwidth extension that have been documented neither make use of any particular model for the speech generation process nor do they make use of any *a priori* knowledge concerning speech properties. These so-called nonmodel-based algorithms are the most simple methods for increasing the bandwidth. As reported in [20] the first experiments were conducted in 1933 using a nonlinear (quadratic) characteristic. Later on the British Broadcasting Corporation (BBC) was interested in increasing the bandwidth during telephone contributions [6]. Mainly these are two different nonmodel based algorithms. In the following we will discuss these (historical) methods briefly as some of these ideas will be touched upon again in later sections.

7.3.1 Oversampling with Imaging

This method makes use of the spectral components that occur when upsampling (inserting zeros) a signal either without any anti-imaging filter at all or with a filter $H_{\mathrm{AI}}(e^{j\Omega})$ showing only some slow decay above half of the desired sampling frequency (see Fig. 7.4). This upsampling process results in a spectrum that is mirrored at half of the original sampling frequency. This method profits from the noise-like nature of the excitation signal concerning unvoiced utterances. Unvoiced utterances result in a spectrum which has most of its energy in the higher frequency regions and therefore exactly these portions are mirrored. The temporal behavior of unvoiced components below the cutoff frequency correlates strongly with those above the cut-off frequency which is preserved by applying this mirroring.

A drawback of this method concerning telephony speech is that the lower frequency part of the spectrum is not extended at all. Furthermore if a signal that is bandlimited according to the frequency response of an analog telephone system with 8 kHz sampling rate is upsampled a spectral gap is produced around 4 kHz. A rather costly alternative would be to first downsample the signal to a sampling rate that is equal to twice that of the maximum signal bandwidth (e.g. 2×3600 Hz) before upsampling once again. An example is depicted in Figure 7.5. The upper plot shows a time-frequency analysis of the incoming bandlimited telephone signal $\tilde{s}_{\mathrm{tel}}(n)$. After inserting zeros in between neighboring samples of the input signal (upsampling with $r = 2$) and filtering with an antiimaging filter that has its 3 dB cutoff frequency at about 7 kHz (instead of 4 kHz) a wideband signal $s_{\mathrm{ext}}(n)$ as depicted in the lower diagram of Figure 7.5 is generated.

The results crucially depend on the effective bandwidth of the original signal and the upsampling ratio. For increasing the bandwidth of signals that are bandlimited to 8 kHz up to 12 kHz, for example, this method works surprisingly well, but in the case of telephony speech this method produces poor results [5, 8].

7.3.2 Application of Nonlinear Characteristics

The application of nonlinear characteristics to periodic signals produces harmonics as we will see in more detail in Section 7.5.1. This can be exploited for increasing the

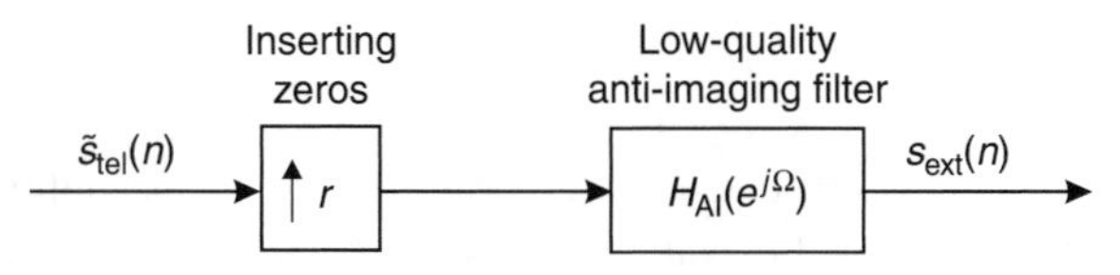

Figure 7.4 Bandwidth extension based on oversampling and imaging.

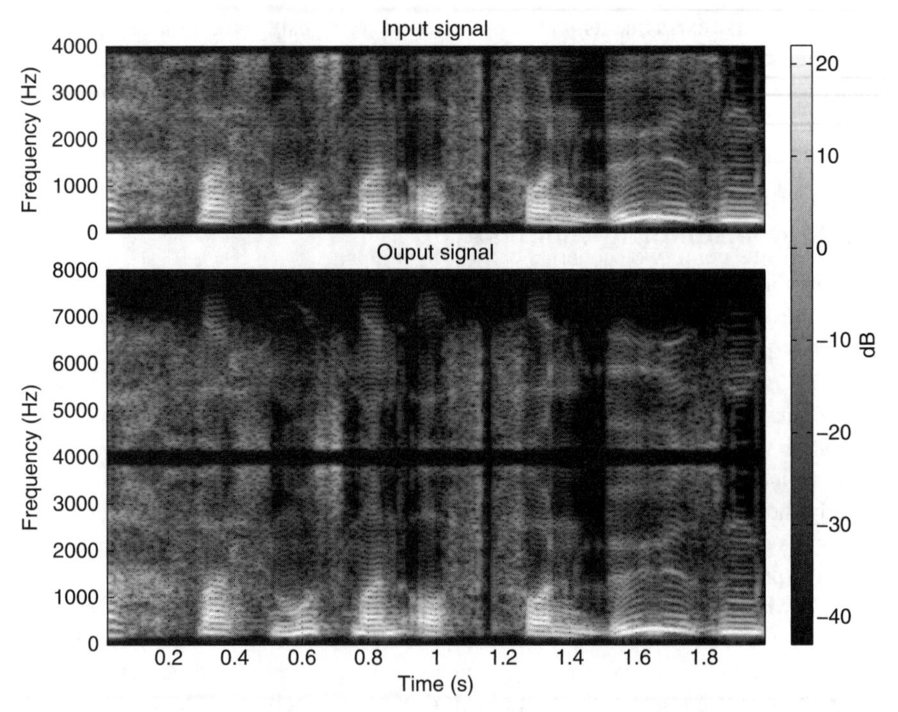

Figure 7.5 Bandwidth extension using upsampling and low-quality antiimaging filters. A time-frequency analysis of the input signal $\tilde{s}_{\rm tel}(n)$ is depicted in the upper diagram. After upsamping with an upsampling ratio of $r = 2$ and filtering with an anti-imaging filter having a cut-off frequency of about 7 kHz an output signal $\tilde{s}_{\rm tel}(n)$ as depicted in the lower diagram is generated.

bandwidth similar to the method described above. The components generated out of the telephone band are usually attenuated by an empirically determined filter. An advantage of this method compared to that presented above is that concerning telephony speech the lower frequency part is extended as well. Also no spectral gap occurs within the higher frequency part. A drawback might be the aliasing that can occur depending on the effective bandwidth, the sampling rate, and the nonlinear characteristic that has been applied. In Section 7.5.1 a small selection of nonlinear characteristics and their properties is presented.

The results produced by this approach are similar to the results of the method described in Section 7.3.1. Concerning telephony speech the results achieved are not satisfying [6, 31].

7.4 BASICS

The basic idea of bandwidth extension algorithms is to extract information on the missing components out of the available narrowband signal $s_{\rm tel}(n)$. In order to find information that is suitable for this task most of the algorithms employ the so-called source-filter model of speech generation [7, 11, 33].

7.4.1 Source-Filter Model

This model is motivated by the anatomical analysis of the human speech apparatus (see Fig. 7.6). A flow of air coming from the lungs is pressed through the vocal cords. At this point two scenarios can be distinguished.

- In the first scenario the vocal cords are loose causing a turbulent (noise-like) air flow.
- In the second scenario the vocal cords are tense and closed. The pressure of the air coming from the lungs increases until it causes the vocal cords to open. Now the pressure decreases rapidly and the vocal cords close once again. This scenario results in a periodic signal.

The signal that could be observed directly behind the vocal cords is called an excitation signal. This excitation signal has the property of being spectrally flat. After passing the vocal cords the air flow goes through several cavities such as the pharynx cavity, the nasal cavity, and the mouth cavity. In all these cavities the air flow undergoes frequency dependent reflections and resonances depending on the geometry of the cavity.

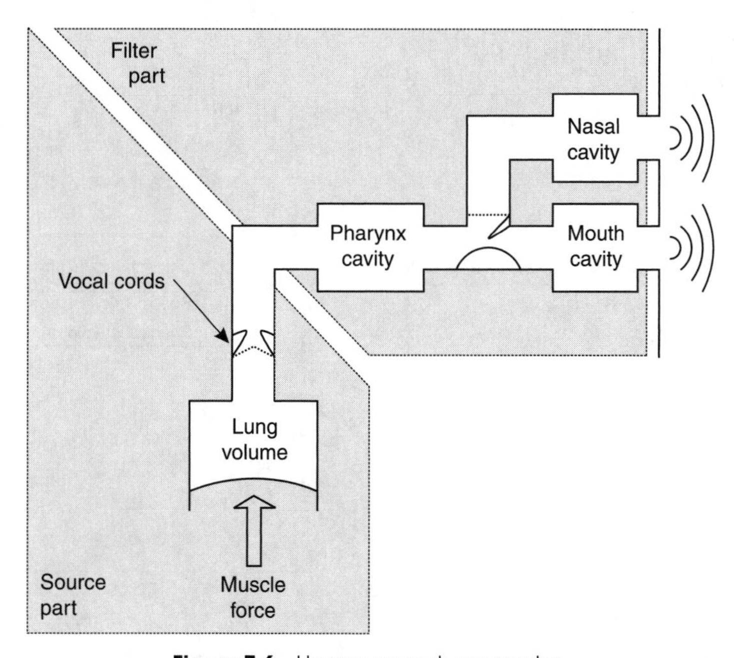

Figure 7.6 Human speech apparatus.

The source-filter model tries to rebuild the two scenarios that are responsible for the generation of the excitation signal by using two different signal generators (see Fig. 7.7):

- a noise generator for rebuilding unvoiced (noise-like) utterances, and
- a pulse train generator for rebuilding voiced (periodic) utterances.

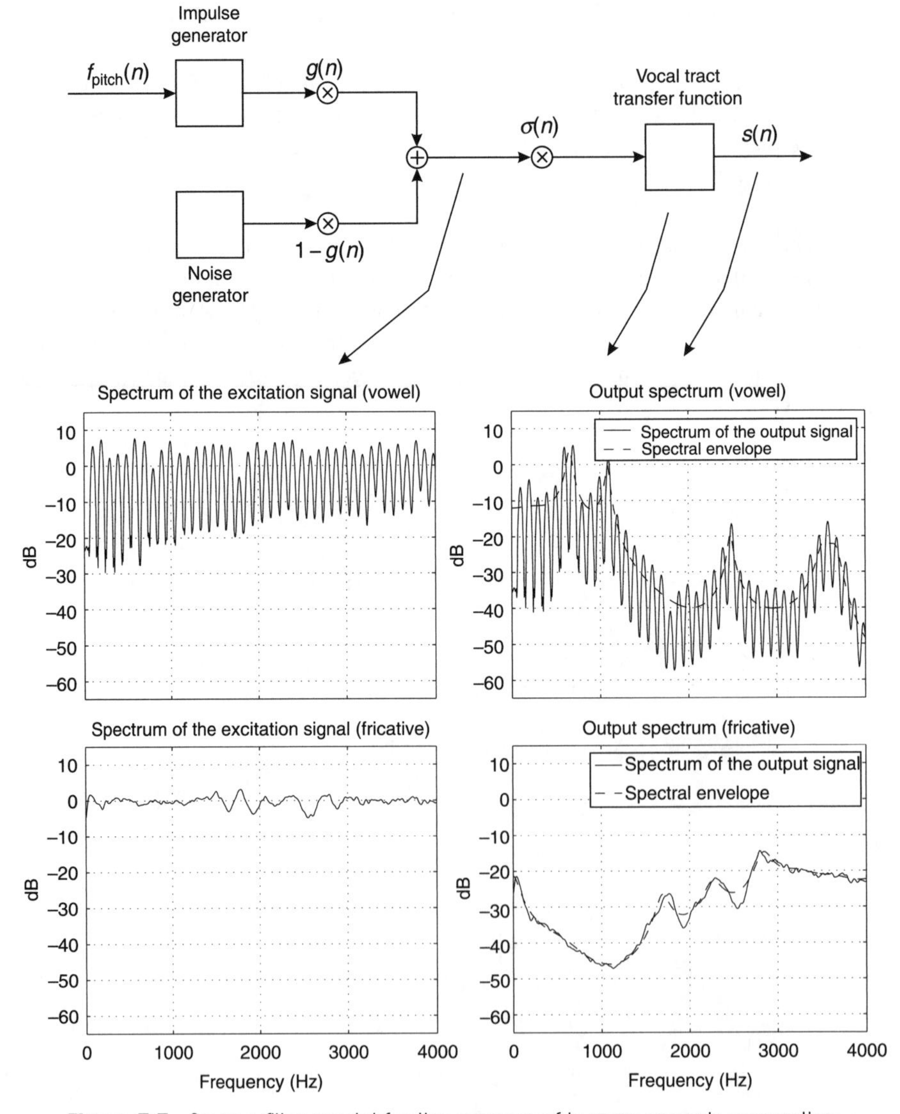

Figure 7.7 Source-filter model for the process of human speech generation.

The inverse of the pulse duration is called pitch frequency denoted in Figure 7.7 as $f_{\text{pitch}}(n)$. These signal generators can be accessed either in a binary manner

$$g(n) \in \{0, 1\} \tag{7.1}$$

or continuously

$$0 \leq g(n) \leq 1. \tag{7.2}$$

For rebuilding the influence of the cavities a low-order all-pole filter with the frequency response

$$H(e^{j\Omega}, n) = \frac{\sigma(n)}{A(e^{j\Omega}, n)} = \frac{\sigma(n)}{1 - \sum_{i=1}^{N_{\text{pre}}} \tilde{a}_i(n)\, e^{-j\Omega i}} \tag{7.3}$$

is employed.[1] The order N_{pre} of the all-pole model is chosen usually in the range of 10 to 20. Since the excitation signal is a spectrally flat signal, the transfer function of this all-pole model represents the spectral envelope of the speech signal. The parameters $\tilde{a}_i(n)$ of the all-pole model can be computed by solving the so-called Yule–Walker equation system

$$\underbrace{\begin{bmatrix} r_{ss,0}(n) & r_{ss,1}(n) & \cdots & r_{ss,N_{\text{pre}}-1}(n) \\ r_{ss,1}(n) & r_{ss,0}(n) & \cdots & r_{ss,N_{\text{pre}}-2}(n) \\ \vdots & \vdots & \ddots & \vdots \\ r_{ss,N_{\text{pre}}-1}(n) & r_{ss,N_{\text{pre}}-2}(n) & \cdots & r_{ss,0}(0) \end{bmatrix}}_{\mathbf{R}_{ss}(n)} \underbrace{\begin{bmatrix} \tilde{a}_1(n) \\ \tilde{a}_2(n) \\ \vdots \\ \tilde{a}_{N_{\text{pre}}}(n) \end{bmatrix}}_{\tilde{\mathbf{a}}(n)} = \underbrace{\begin{bmatrix} r_{ss,1}(n) \\ r_{ss,2}(n) \\ \vdots \\ r_{ss,N_{\text{pre}}}(n) \end{bmatrix}}_{\mathbf{r}_{ss}(n)}. \tag{7.4}$$

The coefficients $r_{ss,i}(n)$ represent the short-term autocorrelation at lag i estimated around the time index n. Finally, the gain parameter $\sigma(n)$ in (7.3) is computed as the square root of the output power of a predictor error filter with coefficients $\tilde{a}_i(n)$

$$\begin{aligned} \sigma(n) &= \sqrt{r_{ss,0}(n) - \sum_{i=1}^{N_{\text{pre}}} \tilde{a}_i(n)\, r_{ss,i}(n)} \\ &= \sqrt{r_{ss,0}(n) - \tilde{\mathbf{a}}^{\mathrm{T}}(n)\mathbf{r}_{ss}(n)}. \end{aligned} \tag{7.5}$$

Due to the special character of the matrix $\mathbf{R}_{ss}(n)$ Eqs. (7.4) and (7.5) can be solved in an order-recursive manner by using, for example, the Levinson–Durbin recursion

[1]We have used the tilde notation for the coefficients $\tilde{a}_i(n)$ in order to avoid a conflict with the definitions of standard transformations such as Fourier or z-transform. The Fourier transform is defined as $A(e^{j\Omega}, n) = \sum_{i=-\infty}^{\infty} a_i(n)e^{-j\Omega i}$. By comparing the coefficients we get $a_0(n) = 1$, $a_i(n) = -\tilde{a}_i(n)$ for $1 \leq i \leq N_{\text{pre}}$, and $a_i(n) = 0$ else.

[9, 26]. Since speech can be assumed to have a stationary character only for short periods of time, the parameters of the model need to be estimated periodically, that is, every 5 to 10 ms. However, by utilizing about 10 coefficients $a_i(n)$ and the gain $\sigma(n)$ one is able to estimate the spectral envelope of a speech signal in a reliable manner. In Figure 7.7 the estimated envelope of the sound /š/ as present in the word *shut* as well as the envelope of the sound /æ/ as present in the word *bat* are depicted.[2] By comparing the estimated envelopes with the true short-term spectra (depicted in the right two diagrams) the reliability becomes evident.

7.4.2 Parametric Representations of the Spectral Envelope

Due to their compact representation of short-term spectral envelopes the prediction parameters $a_i(n)$ [or $\tilde{a}_i(n)$] play a major role in speech coding and bandwidth extension. However, if cost functions that take the human auditory perception into account are applied the prediction parameters are often transformed into so-called cepstral coefficients [29]

$$c_i(n) = \frac{1}{2\pi} \int_{-\pi}^{\pi} \ln\{H(e^{j\Omega}, n)\} e^{j\Omega i} \, d\Omega. \tag{7.6}$$

By applying the natural logarithm on the complex filter spectrum we have to use its complex type which is defined as [3]

$$\ln\{z\} = \ln|z| + j \arg\{z\}, \tag{7.7}$$

where arg{z} denotes the angle of z within the complex plain. As we will see later on, the resulting cepstral coefficients $c_i(n)$ are real due to the special symmetry properties of $H(e^{j\Omega}, n)$ as well as due to the fact that $H(e^{j\Omega}, n)$ represents a stable all-pole infinite impulse response (IIR) filter. Applying a Fourier transform on both sides of (7.6) leads us to

$$\sum_{i=-\infty}^{\infty} c_i(n) e^{-j\Omega i} = \ln\{H(e^{j\Omega}, n)\}. \tag{7.8}$$

The cepstral coefficients used in this chapter are so-called linear predictive cepstral coefficients. As indicated by the name these coefficients are computed on the basis of linear predictive coefficients. By exchanging the filter spectrum $H(e^{j\Omega}, n)$ by its representation using an all-pole model we obtain

$$\begin{aligned} \sum_{i=-\infty}^{\infty} c_i(n) z^{-i} \bigg|_{z=e^{j\Omega}} &= \ln\left\{\frac{\sigma(n)}{A(z, n)}\right\}\bigg|_{z=e^{j\Omega}} \\ &= \ln\{\sigma(n)\} - \ln\{A(z, n)\}\big|_{z=e^{j\Omega}}. \end{aligned} \tag{7.9}$$

[2]The notation used for the phonetic symbols is according to [33].

Let us now have a closer look at the last term $\ln\{A(z, n)\}$ in (7.9)

$$\begin{aligned}\ln\{A(z, n)\} &= \ln\left\{1 - \sum_{i=1}^{N_{\text{pre}}} \tilde{a}_i(n) z^{-i}\right\} \\ &= \ln\left\{\sum_{i=0}^{N_{\text{pre}}} a_i(n) z^{-i}\right\}.\end{aligned} \tag{7.10}$$

By representing this expression as a product of zeros with modified coefficients $b_i(n)$ we can write

$$\begin{aligned}\ln\left\{\sum_{i=0}^{N_{\text{pre}}} a_i(n) z^{-i}\right\} &= \ln\left\{\prod_{i=0}^{N_{\text{pre}}} [1 - b_i(n) z^{-1}]\right\} \\ &= \sum_{i=0}^{N_{\text{pre}}} \ln\{1 - b_i(n) z^{-1}\}.\end{aligned} \tag{7.11}$$

By exploiting the following series expansion [3]

$$\ln\{1 - b\,z^{-1}\} = -\sum_{k=1}^{\infty} \frac{b^k}{k} z^{-k}, \quad \text{for } |z| > |b|, \tag{7.12}$$

that holds for factors that converge within the unit circle, which is the case here since $A(z, n)$ is analytic inside the unit circle [29], we can further rewrite (7.11) [and thus also (7.10)]

$$\begin{aligned}\ln\{(A(z, n)\} &= -\sum_{i=0}^{N_{\text{pre}}} \sum_{k=1}^{\infty} \frac{b_i^k(n)}{k} z^{-k} \\ &= -\sum_{k=1}^{\infty} \sum_{i=0}^{N_{\text{pre}}} \frac{b_i^k(n)}{k} z^{-k}.\end{aligned} \tag{7.13}$$

If we now compare (7.9) with (7.13) we observe that the two sums do not have equal limits. This means that the $c_i(n)$ are equal to zero for $i < 0$. For $i > 0$ we can set the $c_i(n)$ equal to the inner sum in (7.13). For $i = 0$ we have $c_0(n) = \ln\{\sigma(n)\}$ from (7.9). In conclusion, we can state

$$c_i(n) = \begin{cases} \displaystyle\sum_{m=0}^{N_{\text{pre}}} \frac{b_m^i(n)}{i}, & \text{for } i > 0, \\ \ln\{\sigma(n)\}, & \text{for } i = 0, \\ 0, & \text{for } i < 0. \end{cases} \tag{7.14}$$

This leads to the assertion

$$\ln\left\{\frac{\sigma(n)}{A(z,n)}\right\} = \ln\{\sigma(n)\} + \sum_{i=1}^{\infty} c_i(n) z^{-i}. \tag{7.15}$$

If the N_{pre} predictor coefficients $a_i(n)$ [or $\bar{a}_i(n)$] are known we can derive a simple recursive computation of the linear predictive cepstral coefficients $c_i(n)$ by differentiating both sides of (7.15) with respect to z and equating the coefficients of alike powers of z

$$\begin{aligned} -\frac{d}{dz}\left[\ln\left\{1 - \sum_{i=1}^{N_{\text{pre}}} a_i(n) z^{-i}\right\}\right] &= \frac{d}{dz}\left[\sum_{i=1}^{\infty} c_i(n) z^{-i}\right] \\ \sum_{i=1}^{N_{\text{pre}}} i a_i(n) z^{-i-1}\left[-1 + \sum_{i=1}^{N_{\text{pre}}} a_i(n) z^{-i}\right]^{-1} &= -\sum_{i=1}^{\infty} i c_i(n) z^{-i-1}. \end{aligned} \tag{7.16}$$

Multiplying both sides of (7.16) with $[-1 + \sum_{i=1}^{N_{\text{pre}}} a_i(n) z^{-i}]$ leads to

$$\sum_{i=1}^{N_{\text{pre}}} i a_i(n) z^{-i-1} = \sum_{i=1}^{\infty} i c_i(n) z^{-i-1} - \sum_{k=1}^{\infty} \sum_{i=1}^{N_{\text{pre}}} k c_k(n) a_i(n) z^{-k-i-1}. \tag{7.17}$$

If we now consider the equation above for equal powers of z, we find that starting from the left, the first two terms only contribute a single term each up to $z^{-N_{\text{pre}}-1}$. We will label the order with i. The last term in contrast produces an amount of terms that depends on i. Setting all terms that belong to the same power of z equal results in

$$i a_i(n) = i c_i(n) - \sum_{k=1}^{i-1} k c_k(n) a_{i-k}(n), \quad \text{for } i \in \{1, \ldots, N_{\text{pre}}\}. \tag{7.18}$$

Solving this equation for $c_i(n)$ results in

$$c_i(n) = a_i(n) + \frac{1}{i} \sum_{k=1}^{i-1} k c_k(n) a_{i-k}(n), \quad \text{for } i \in \{1, \ldots, N_{\text{pre}}\}. \tag{7.19}$$

For $i > N_{\text{pre}}$ the first term on the right side still needs to be considered in (7.17) whereas the term on the left side does not contribute any more to powers of z larger than $N_{\text{pre}} - 1$ and can therefore be omitted. Therefore, we can solve (7.17) for $i > N_{\text{pre}}$ as

$$c_i(n) = \frac{1}{i}\sum_{k=1}^{i-1} k c_k(n) a_{i-k}(n), \qquad \text{for } i > N_{\text{pre}}. \tag{7.20}$$

By summarizing all of the results we can formulate a recursive computation of cepstral coefficients from linear predictive coefficients as

$$c_i(n) = \begin{cases} 0, & \text{for } i < 0, \\ \ln\{\sigma(n)\}, & \text{for } i = 0, \\ a_i(n) + \dfrac{1}{i}\displaystyle\sum_{k=1}^{i-1} k c_k(n) a_{i-k}(n), & \text{for } 1 \leq i \leq N_{\text{pre}}, \\ \dfrac{1}{i}\displaystyle\sum_{k=1}^{i-1} k c_k(n) a_{i-k}(n), & \text{for } i > N_{\text{pre}}. \end{cases} \tag{7.21}$$

This means that the average gain of the filter is represented by the coefficient $c_0(n)$ while its shape is described by the coefficients $c_1(n)$, $c_2(n)$, etc. In most cases the series $c_i(n)$ fades towards zero rather quickly. Thus, it is possible to approximate all coefficients above a certain index as zero. We will compute in the following only the coefficients from $i = 0$ to $i = N_{\text{cep}} - 1$, with

$$N_{\text{cep}} = \frac{3}{2} N_{\text{pre}} \tag{7.22}$$

and assume all other coefficients to be zero. We have elaborated the relation between cepstral and predictor coefficients since we will use a cost function based on cepstral coefficients in most of the succeeding bandwidth extension schemes—either directly or within a parameter training stage.

Furthermore, the model according to (7.3) represents an IIR filter. Thus, after any modification of the coefficients $a_i(n)$ [or $\tilde{a}_i(n)$, respectively] it needs to be checked whether stability is still ensured. These tests can be avoided if a transformation into so-called line spectral frequencies (LSFs) [19] is applied. These parameters are the angles of the zeros of the polynomials $P(z)$ and $Q(z)$ in the z-domain

$$P(z, n) = A(z, n) + z^{-(N_{\text{pre}}+1)} A(z^{-1}, n), \tag{7.23}$$

$$Q(z, n) = A(z, n) - z^{-(N_{\text{pre}}+1)} A(z^{-1}, n). \tag{7.24}$$

where $P(z)$ is a mirror polynomial and $Q(z)$ an antimirror polynomial. The mirror property is characterized by

$$P(z, n) = z^{-(N_{\text{pre}}-1)} P(z^{-1}, n), \tag{7.25}$$

$$Q(z, n) = z^{-(N_{\text{pre}}-1)} Q(z^{-1}, n). \tag{7.26}$$

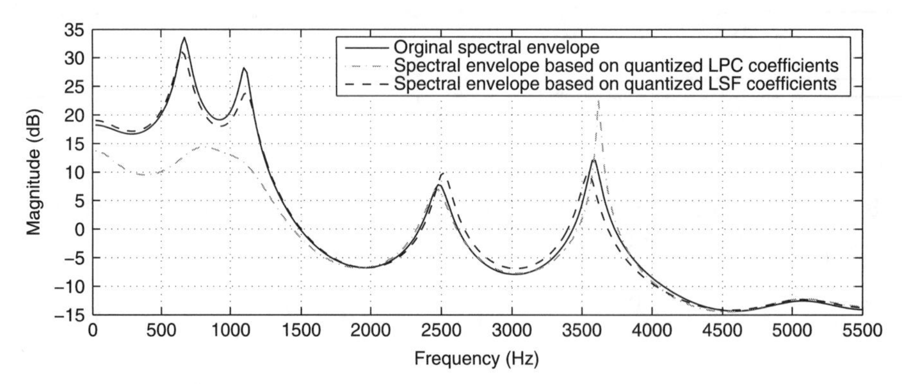

Figure 7.8 Original spectral envelope and two quantized versions (6 bits).

As shown in [34] $A(z, n)$ is guaranteed to be a minimum phase filter [and therefore the synthesis filter $H(z, n) = \sigma(n)/A(z, n)$ is guaranteed to be a stable filter] if the zeros of $P(z, n) = 0$ and $Q(z, n) = 0$ lie on the unit circle and if they are increasing monotonously and alternating. In fact $P(z, n)$ has a real zero at $z = -1$ and $Q(z, n)$ at $z = 1$.

In Figure 7.8 two versions of an original spectral envelope (also depicted) quantized with 6 bits are depicted. Representations using LPC coefficients as well as LSF coefficients have been used. It is obvious that LSF coefficients are more robust against quantization than LPC coefficients [35].[3] However, the benefit concerning robustness gained by the usage of line spectral frequencies does not always justify the increased computational complexity that results from the search for the zeroes.

7.4.3 Distance Measures

Distance measures play an important role in speech quality assessment, speech coding, and—as we will see in the next sections—also in training and search procedures in bandwidth extension schemes. Plenty of different distance measures for all kinds of applications exist [36]. Since we are dealing with bandwidth extension and one major task is the extension of the spectral envelope, we will focus here on distance measures that are appropriate to evaluate distances between parametric representations of spectral envelopes. Most of the spectral distance measures are L_p-norm-based measures

$$d_p(H(e^{j\Omega}, n), \widehat{H}(e^{j\Omega}, n)) = \left[\frac{1}{2\pi} \int_{-\pi}^{\pi} |H(e^{j\Omega}, n) - \widehat{H}(e^{j\Omega}, n)|^p \, d\Omega \right]^{\frac{1}{p}}. \tag{7.27}$$

[3]The 6 bit quantization (see Fig. 7.8) leads in the LPC case to large overestimation errors around 3600 Hz (fourth formand) as well as to large underestimation errors below 1300 Hz (first and second formand). The quantized LSF envelope version exhibits a much smaller error.

The most common choices for p are $p = 1, p = 2$, and $p = \infty$ resulting in the so-called city block distance, Euclidean distance and Minkowski distance. Since the representation of the magnitude spectrum in a logarithmic manner is very popular, another well known distance measure has emerged from this kind of representation. Furthermore, this distance measure can be computed directly on the parametric representation of the spectral envelope. It is the so-called cepstral distance, which is defined as

$$d_{\text{cep}}(c_i(n),\ \hat{c}_i(n)) = \sum_{i=-\infty}^{\infty} (c_i(n) - \hat{c}_i(n))^2, \tag{7.28}$$

where the $c_i(n)$ and $\hat{c}_i(n)$ denote linear predictive cepstral coefficients as described before. An interesting property of this definition can be shown using Parsevals theorem [33] [compare (7.6)]

$$d_{\text{cep}}(c_i(n), \hat{c}_i(n)) = \frac{1}{2\pi} \int_{-\pi}^{\pi} \left| \ln\{H(e^{j\Omega}, n)\} - \ln\{\widehat{H}(e^{j\Omega}, n)\} \right|^2 d\Omega. \tag{7.29}$$

For the tasks required in bandwidth extension schemes it is necessary to compare only the shape of two spectral envelopes but not the gain. For this reason, we let the summing index start at $i = 1$ in (7.28). Furthermore, we assume—as explained before—that the cepstral series does not show significant values for large indices. Thus, we modify the original definition (7.28) for our purposes to

$$\tilde{d}_{\text{cep}}(c_i(n), \hat{c}_i(n)) = \sum_{i=1}^{N_{\text{cep}}} (c_i(n) - \hat{c}_i(n))^2. \tag{7.30}$$

This represents an estimate for the average squared logarithmic difference in the frequency domain of two power normalized envelopes

$$\begin{aligned} \tilde{d}_{\text{cep}}(c_i(n), \hat{c}_i(n)) &\approx \frac{1}{2\pi} \int_{-\pi}^{\pi} \left| \ln\left\{ \frac{H(e^{j\Omega}, n)}{\sigma(n)} \right\} - \ln\left\{ \frac{\widehat{H}(e^{j\Omega}, n)}{\hat{\sigma}(n)} \right\} \right|^2 d\Omega \\ &= \frac{1}{2\pi} \int_{-\pi}^{\pi} |\ln\{A(e^{j\Omega}, n)\} - \ln\{\widehat{A}(e^{j\Omega}, n)\}|^2 \, d\Omega. \end{aligned} \tag{7.31}$$

For the last modification in (7.31) it has been used that $\ln\{1/A(e^{j\Omega})\} = -\ln\{A(e^{j\Omega})\}$. This distance measure will be utilized for most of the schemes presented in the following sections of this chapter.

7.5 MODEL-BASED ALGORITHMS FOR BANDWIDTH EXTENSION

In contrast to the methods described before the following algorithms make use of the already discussed source-filter model and thereby of *a priori* knowledge. As already stated in Section 7.4 the task of bandwidth extension following the source-filter model can be divided into two subtasks:

- the generation of a broadband excitation signal, and
- the estimation of the broadband spectral envelope.

In the next two sections we will discuss possibilities to accomplish these tasks. In Figure 7.9 a basic structure of a model-based bandwidth extension scheme is depicted.

We assume an input signal $s_{\mathrm{tel}}(n)$ that has already been upsampled to the desired sample rate but with a conventional sampling rate conversion (meaning that we do not have significant signal energy above 4 kHz). To generate the estimated broadband excitation signal $\hat{e}_{\mathrm{bb}}(n)$ often the spectral envelope of the input signal is first removed by means of a predictor error filter

$$e_{\mathrm{nb}}(n) = s_{\mathrm{tel}}(n) + \sum_{i=1}^{N_{\mathrm{pre,nb}}} a_{\mathrm{nb},i}(n) s_{\mathrm{tel}}(n-i). \tag{7.32}$$

The coefficients $a_{nb,i}(n)$ are computed on a block basis every 5 to 10 ms using the methods described in Section 7.4.1. The index “nb” indicates that telephone or narrow band quantities are addressed, while “bb” denotes extended or broadband

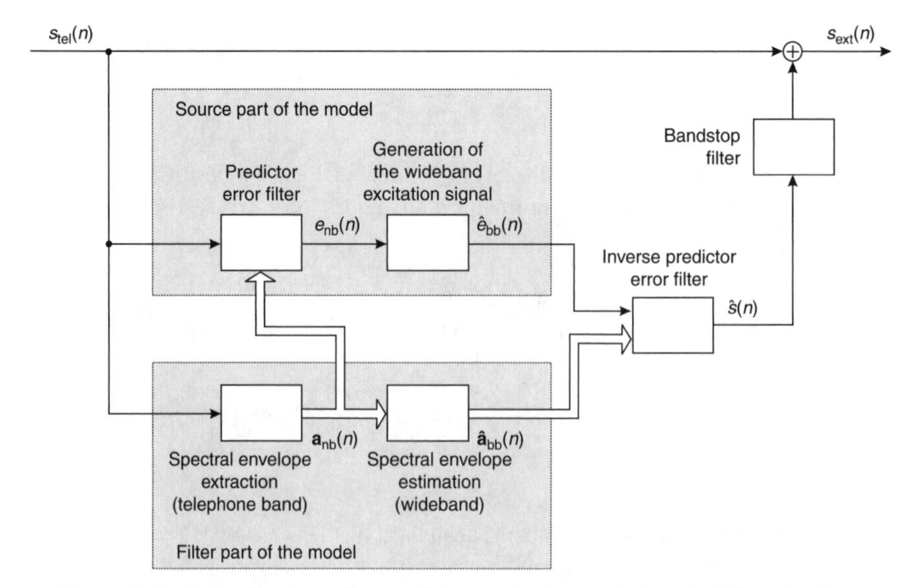

Figure 7.9 Basic structure of a model-based scheme for bandwidth extension.

signals and parameters. The spectrally whitened input signal $e_{\text{nb}}(n)$ is used for the estimation of the broadband excitation signal $\hat{e}_{\text{bb}}(n)$ while the extracted spectral envelope, described by its predictor coefficient vector

$$\mathbf{a}_{\text{nb}}(n) = [a_{\text{nb},1}(n), a_{\text{nb},2}(n), \ldots, a_{\text{nb},N_{\text{pre,nb}}}(n)]^T \tag{7.33}$$

is utilized for estimating the broadband spectral envelope. We will describe the latter term also by a vector containing the coefficients of a prediction filter

$$\mathbf{a}_{\text{bb}}(n) = [\hat{a}_{\text{bb},1}(n), \hat{a}_{\text{bb},2}(n), \ldots, \hat{a}_{\text{bb},N_{\text{pre,bb}}}(n)]^T. \tag{7.34}$$

Usually, this vector consists of more coefficients than its narrowband counterpart

$$N_{\text{pre,bb}} \geq N_{\text{pre,bb}}. \tag{7.35}$$

Finally, both model parts are combined using an inverse predictor error filter with coefficients $\hat{a}_{\text{bb},i}(n)$ that is excited with the estimated broadband excitation signal $\hat{e}_{\text{bb}}(n)$

$$\hat{s}(n) = \hat{e}_{\text{bb}}(n) - \sum_{i=1}^{N_{\text{pre,bb}}} \hat{s}(n-i)\hat{a}_{\text{bb},i}(n). \tag{7.36}$$

In some implementations a power adjustment of this signal is necessary. Since we want to reconstruct the signal only in those frequency ranges that are not transmitted over the telephone line a bandstop filter is applied. Only those frequencies that are not transmitted should pass the filter (e.g. frequencies below 200 Hz and above 3.8 kHz). Finally the reconstructed and upsampled telephone signal is added (see Fig. 7.9).

7.5.1 Generation of the Excitation Signal

For the generation of the broadband excitation signal mainly three classes of approaches exist. These classes include modulation techniques, nonlinear processing, and the application of function generators. We will describe all of them briefly in the next sections.

Modulation Techniques Modulation technique is a term that implies the processing of the excitation signal in the time domain by performing a multiplication with a modulation function

$$\hat{e}_{\text{bb}}(n) = e_{\text{nb}}(n) \cdot 2\cos(\Omega_0 n). \tag{7.37}$$

This multiplication with a cosine function in the time domain corresponds to the convolution with two dirac impulses in the frequency domain

$$\begin{aligned}\widehat{E}_{\rm bb}(e^{j\Omega}) &= E_{\rm nb}(e^{j\Omega}) * [\delta(\Omega - \Omega_0) + \delta(\Omega + \Omega_0)] \\ &= E_{\rm nb}(e^{j(\Omega-\Omega_0)}) + E_{\rm nb}(e^{j(\Omega+\Omega_0)}). \end{aligned} \tag{7.38}$$

In Figure 7.10 an example for a modulation approach is depicted. Part (*a*) shows a time-frequency analysis of an input signal consisting of several harmonic sine terms with a basic fundamental frequency that slowly increases over time. If we multiply

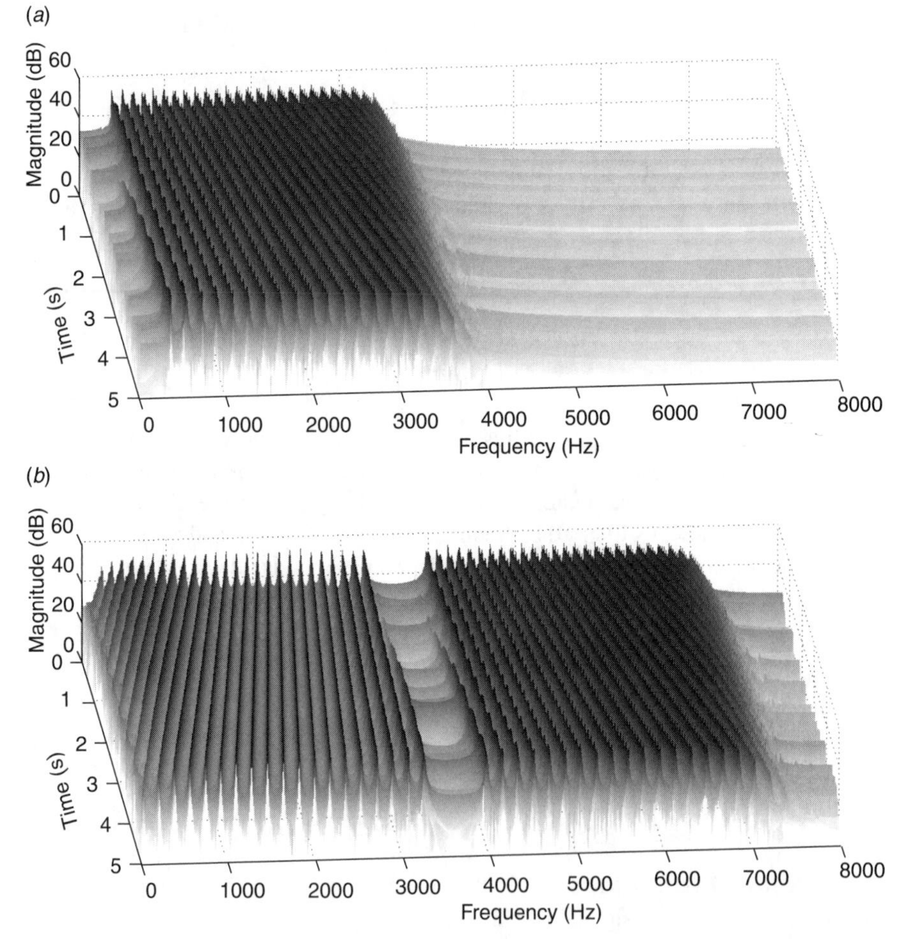

Figure 7.10 The effect of a cosine modulation. The spectral gap around 3800 Hz in the extended signal $\hat{e}_{\rm bb}(n)$ appears due to missing low frequency components within the input signal $e_{\rm nb}(n)$. (*a*) Harmonic (bandlimited) signal $e_{\rm nb}(n)$ (*b*) Harmonic (extended) signal $\hat{e}_{\rm bb}(n)$.

this signal with a 3800 Hz modulation function a signal, as depicted in part (*b*) of Figure 7.10, is generated.

Depending on the modulation frequency sometimes the spectrum resulting from the negative shift and that of the positive overlap. For that reason bandpass filtering might be necessary before applying the modulation function. Additionally the output signal might contain signal components within the telephone band. This is the case in the example that we have depicted in Figure 7.10. In those cases a filter also can be applied—this time the filter is applied to the output $\hat{e}_{\text{bb}}(n)$—that lets only those frequencies that are desired pass. Additionally the original signal, respectively its whitened version $e_{\text{nb}}(n)$, can be used within the telephone band.

The reason for performing such a shift in frequency can easily be understood when we consider the processing of voiced utterances. The periodic signal that is typical for voiced utterances can be extended by modulation techniques. By incorporating a pitch detection one could even perform an adaptive shift and thereby keep the pitch structure even at the transition regions from the telephone band to the extension area. As seen in Section 7.4 the excitation signal of unvoiced utterances is noise-like and therefore we do not have a structure we have to take care of. This means that we also can apply this excitation extension method for unvoiced utterances without having to worry about continuing the signal structure along the frequency axis.

Nonlinear Processing One major problem of the above discussed modulation techniques is the pitch detection if the algorithm is designed (pitch-) adaptive. Especially in the low frequency part bothersome artifacts occur if, concerning voiced utterances, the harmonics of the pitch frequency are misplaced. This means that the performance of the algorithm crucially depends on the performance of the pitch detection.

Another possibility to extend the excitation signal is the application of nonlinear characteristics. Nonlinearities have the property that they produce harmonics when applied to a periodic signal. This once again takes the case of voiced utterances into account. There exists a variety of nonlinear characteristics which all have different properties. A quadratic characteristic on one hand produces only even harmonics. A cubic characteristic on the other hand produces only odd harmonics. The effect of the application of a nonlinear characteristic can be explained best for the quadratic characteristic. The application of a quadratic characteristic in the time domain corresponds to the convolution of the signal with itself in the frequency domain

$$\hat{e}_{\text{bb}}(n) = e_{\text{nb}}^2(n) \circ\!\!-\!\!\bullet\; E_{\text{nb}}(e^{j\Omega}) * E_{\text{nb}}(e^{j\Omega}) = \widehat{E}_{\text{bb}}(e^{j\Omega}). \tag{7.39}$$

If we assume a line spectrum in the case of voiced sounds the effect becomes clear. Every time the lines match during the shift within the convolution, the resulting signal will have a strong component. In contrast to the above presented method where we had the convolution with a dirac at the arbitrary (or determined by a pitch estimation algorithm) frequency Ω_0 we convolve the signal with dirac impulses at proper harmonics of Ω_0. The effect of other nonlinear characteristics can be explained by their representation as a power series. Some possible characteristics are as follows.

- Half-way rectification: $f(x) = \begin{cases} x, & \text{if } x > 0, \\ 0 & \text{else.} \end{cases}$
- Full-way rectification: $f(x) = |x|$.
- Saturation characteristic: $f(x) = \begin{cases} a + (x - a)\,b, & \text{if } x > a, \\ a - (x + a)\,b, & \text{if } x < a, \\ x, & \text{else.} \end{cases}$
- Quadratic characteristic: $f(x) = x^2$.
- Cubic characteristic: $f(x) = x^3$.
- Tanh characteristic: $f(x) = \tanh(\mu x)$.

Where a, b and μ are arbitrary positive parameters. Another property is that they (depending on the effective bandwidth, the sampling rate and the kind of characteristic) produce components out of the Nyquist border. Therefore, the signal has to be upsampled before applying the nonlinear characteristic and filtered by a lowpass with the cutoff frequency corresponding to half of the desired sampling rate to avoid aliasing completely before downsampling. A second property is that these nonlinear characteristics produce strong components around 0 Hz, which have to be removed. After the application of a nonlinear characteristic the extended excitation signal might have undergone an additional coloration. This can be taken into account by applying a prediction error filter (whitening filter). A disadvantage of nonlinear characteristics is that in the case of harmonic noise (e.g. engine noise in the car environment) the harmonics of this noise signal are complemented as well. Furthermore, the processed signals of quadratic and cubic characteristics show a wide dynamic range and a power normalization is required. Figure 7.11 shows the effect of nonlinear characteristics on the example of a half way rectification without any post processing (no predictor error filter).

Function Generators The last class of algorithms for excitation signal extension are so-called function generators. The simplest form of a function generator is a sine generator. Similar to the use of adaptive modulation techniques this method needs a pitch estimation. The sine generators work in the time domain. The parameters (amplitude, and frequency) of the sine generators are obtained by employing the estimated broadband spectral envelope and the estimate for the fundamental frequency and its harmonics respectively. The advantage of the use of sine generators lies in the discrimination between actual values for amplitude and frequency and the desired values for amplitude and frequency. The sine generators are designed to change their actual values within a maximum allowed change of the value towards the desired value. This prevents artifacts due to a step of amplitude or frequency from one frame to another. Another advantage of these sine generators is that artifacts due to a step in phase of the low frequency components do not appear due to the time

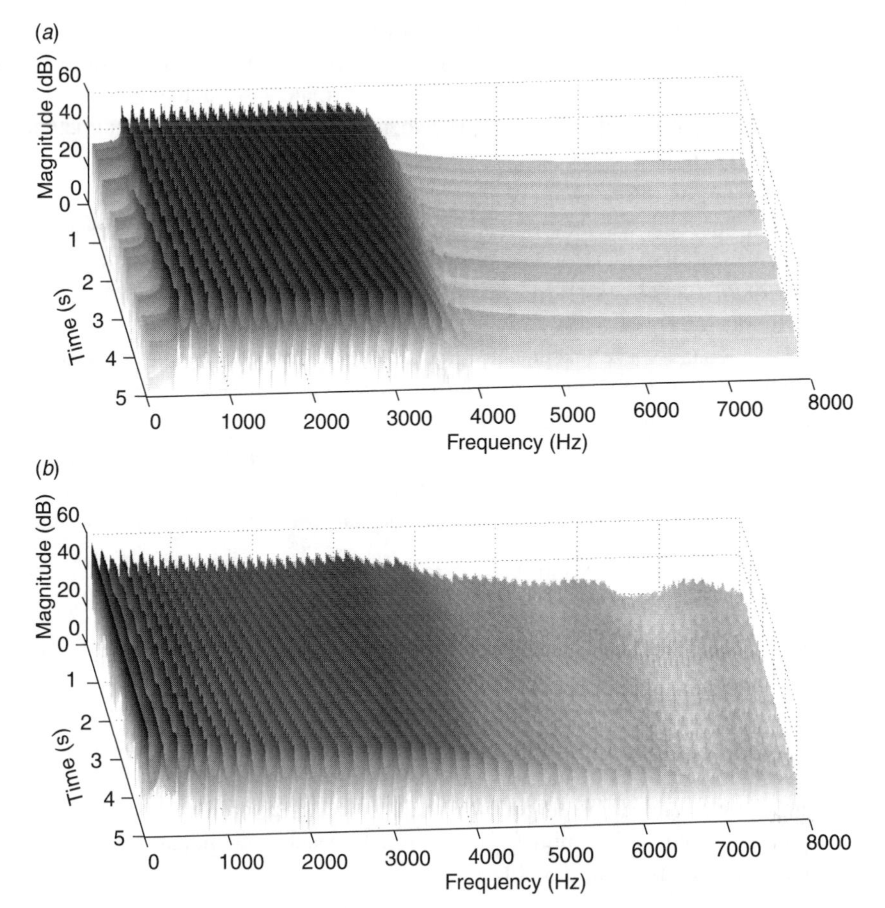

Figure 7.11 The effect of nonlinear characteristics (half-way rectification). (*a*) Harmonic (bandlimited) signal $e_{\mathrm{nb}}(n)$. (*b*) Harmonic (extended) signal $\hat{e}_{\mathrm{bb}}(n)$.

domain processing. Furthermore, the sine generators do not need an estimated value for the fundamental frequency with every sample or frame but, for example, whenever such an estimate is very reliable.[4]

7.5.2 Vocal Tract Transfer Function Estimation

Besides the generation of the excitation signal $\hat{e}_{\mathrm{bb}}(n)$ the wideband spectral envelope needs to be estimated. Several classes of estimation techniques have been suggested.

- One of the simplest approaches is linear mapping. In this case the feature vector containing the parameters of the bandlimited envelope is multiplied with a

[4]If the pitch estimation is performed by searching the short-term autocorrelation of a signal for a maximum in a predefined range [13] the reliability can be checked by comparing the maximum value with the short-term power of the signal (the autocorrelation value at lag zero). If this ratio is close to one, the signal shows a clear periodicity and, thus, the estimation can be classified as reliable.

matrix in order to estimate the wideband feature vector. The advantage of the linear mapping approach is its simplicity in terms of computational complexity and memory requirements. However, its performance is—due to the unrealistic assumption of a linear relation between bandlimited and wideband parameters—only very limited.

- Originating in the field of speech coding codebook approaches have been proposed. In codebook based methods the spectral envelope in terms of low order all-pole models of the current (bandlimited) frame is first mapped on one of N_{cb} codebook entries according to a predefined distance measure [23]. The codebook is usually trained with a large corpus of pairs of bandlimited and wideband speech sequences. Each entry of the codebook for bandlimited envelopes has its wideband counterpart. This counterpart is utilized as an estimate for the wideband spectral envelope of the current frame.
- The third class of estimation techniques is based on the application of neural networks. In this case the assumption of a linear relationship between the bandlimited parameter set and its wideband counterpart is dropped and also a nonlinear relation can be modeled.

Furthermore, combinations of these estimation techniques can be realized. Details about the above introduced approaches are described within the following sections. Common to all approaches is that the parameters of each method are trained using a large database of pairs of bandlimited and wideband speech sequences. As known from speech and pattern recognition the best results are achieved if the training database is as close as possible to the final application. Thus, boundary conditions such as recording devices and involved coding schemes should be chosen carefully. Better results are achieved if the training is performed speaker dependent. However, for most applications this is not possible.

Linear Mapping If the parameter set containing the bandlimited envelope information is described by a vector

$$\mathbf{x}(n) = [x_0(n),\ x_1(n), \ldots, x_{N_x}(n)]^{\mathrm{T}} \tag{7.40}$$

and its wideband counterpart by a vector

$$\mathbf{y}(n) = [y_0(n),\ y_1(n), \ldots, y_{N_y}(n)]^{\mathrm{T}}, \tag{7.41}$$

then a linear estimation scheme can be realized by a simple linear operation

$$\hat{\mathbf{y}}(n) = \mathbf{W}(\mathbf{x}(n) - \mathbf{m}_x) + \mathbf{m}_y. \tag{7.42}$$

In general, the entries of the vector $\mathbf{x}(n)$ and $\mathbf{y}(n)$ could be predictor coefficients, cepstral coefficients, line spectral frequencies, or any other set of features that describe the spectral envelope. However, since we will use a quadratic cost function cepstral coefficients are a good choice.[5] The multiplication of the bandlimited feature vector $\mathbf{x}(n)$ with the $N_y \times N_x$ matrix $\mathbf{W}$ can be interpreted as a set of N_y FIR filter operations. Each row of $\mathbf{W}$ corresponds to an impulse response which is convolved with the signal vector $\mathbf{x}(n)$ resulting in one element of the wideband feature vector $y_i(n)$. As common in linear estimation theory the mean values of the feature vectors $\mathbf{m}_x$ and $\mathbf{m}_y$ are estimated within a preprocessing stage. For obtaining the matrix $\mathbf{W}$ a cost function has to be specified. A very simple approach would be the minimization of the sum of the squared errors over a large database

$$F(\mathbf{W}) = \sum_{n=0}^{N-1} \left\| \mathbf{y}(n) - \hat{\mathbf{y}}(n) \right\|^2 \longrightarrow \min . \tag{7.43}$$

In case of cepstral coefficients this results in the distance measure described in Section 7.4.3 [see (7.30) and (7.31)]. If we define the entire data base consisting of N zero-mean feature vectors by two matrices

$$\mathbf{X} = [\mathbf{x}(0) - \mathbf{m}_x, \mathbf{x}(1) - \mathbf{m}_x, \ldots, \mathbf{x}(N-1) - \mathbf{m}_x], \tag{7.44}$$

$$\mathbf{Y} = [\mathbf{y}(0) - \mathbf{m}_y, \mathbf{y}(1) - \mathbf{m}_y, \ldots, \mathbf{y}(N-1) - \mathbf{m}_y], \tag{7.45}$$

the optimal solution [28] is given by

$$\mathbf{W}_{\text{opt}} = \mathbf{Y}\mathbf{X}^{\text{T}}(\mathbf{X}\mathbf{X}^{\text{T}})^{-1}. \tag{7.46}$$

Since the sum of the squared differences of cepstral coefficients is a well-accepted distance measure in speech processing often cepstral coefficients are utilized as feature vectors. Even if the assumption of the existence of a single matrix $\mathbf{W}$ which transforms all kinds of bandlimited spectral envelopes into their broadband counterparts is quite unrealistic, this simple approach results in astonishing good results. However, the basic single matrix scheme can be enhanced by using several matrices, where each matrix was optimized for a certain type of feature class. In a two matrices scenario one matrix $\mathbf{W}_{\text{v}}$ can be optimized for voiced sounds and the other matrix $\mathbf{W}_{\text{u}}$ for non-voiced sounds. In this case it is first checked to which class the current feature vector $\mathbf{x}(n)$ belongs. In a second stage the corresponding matrix is applied to generate the estimated wideband feature vector[6]

$$\hat{\mathbf{y}}(n) = \begin{cases} \mathbf{W}_{\text{v}}(\mathbf{x}(n) - \mathbf{m}_{x,\text{v}}) + \mathbf{m}_{y,\text{v}}, & \text{if the classification indicates a voiced frame,} \\ \mathbf{W}_{\text{u}}(\mathbf{x}(n) - \mathbf{m}_{x,\text{u}}) + \mathbf{m}_{y,\text{u}}, & \text{else.} \end{cases} \tag{7.47}$$

[5]Quadratic cost functions lead to rather simple optimization problems. However, the human auditory systems weights errors more in a logarithmic than in a quadratic sense. For that reason the cepstral distance (a quadratic function applied to nonlinearly modified LPC coefficients) is a good choice, since this distance represents the accumulated logarithmic difference between two spectral envelopes (see Section 7.4.3).

[6]Besides two different matrices $\mathbf{W}_{\text{v}}$ and $\mathbf{W}_{\text{u}}$ also different mean vectors $\mathbf{m}_{x,\text{v}}$ and $\mathbf{m}_{x,\text{u}}$, respectively $\mathbf{m}_{y,\text{v}}$ and $\mathbf{m}_{y,\text{u}}$, are applied for voiced and unvoiced frames.

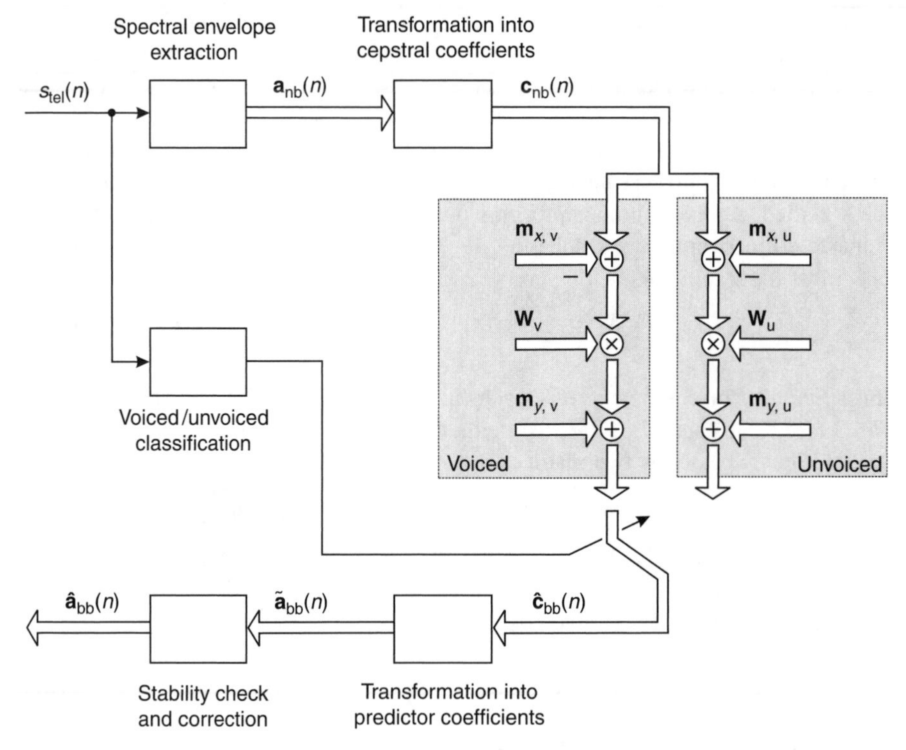

Figure 7.12 Structure of the linear mapping approach for estimating the broadband spectral envelope (using different parameter sets for voiced and unvoiced utterances).

The classification of the type of sound can be performed by analyses such as the zero-crossing rate [7] or the gradient index [20, 32]. Figure 7.12 shows the structure of the linear mapping approach using cepstral coefficients.

Since the first analysis stage results in predictor coefficients a feature vector transformation according to (7.21) is applied, resulting in a vector $\mathbf{c}_{nb}(n)$ containing the narrowband linear predictive cepstral coefficients. After narrowband mean value subtraction, application of the mapping matrices, and broadband mean value addition the resulting broadband cepstral vectors are transformed back into predictor coefficients. This can be achieved also in a recursive manner according to

$$\tilde{a}_{bb,i}(n) = \hat{c}_{bb,i}(n) - \frac{1}{i}\sum_{k=1}^{i-1} k\hat{c}_{bb,k}(n)\tilde{a}_{bb,i-k}(n),$$

$$\text{for } i \in \{1, \ldots, N_{bb,pre}\}.$$

$N_{bb,pre}$ denotes the length of the broadband predictor coefficients vector. Since stability of the resulting IIR filter cannot be guaranteed any more, a stability check needs to be performed. If poles outside the unit circle are detected, one can use, for example, the last valid coefficient vector or use one of the broadband mean vectors $\mathbf{m}_{y,v}$ or $\mathbf{m}_{y,u}$ instead. Additionally, one can compute the location of each pole and mirror the ones outside the unit circle into the unit circle. This approach, however, is quite costly since the computation of the pole locations is more expensive than just the detection of instability. A similar problem arises if a voiced segment is mistakenly classified as an unvoiced one (and vice versa). One can try to reduce the impact of these cases by performing some kind of reliability check (in addition to a stability check). If the resulting envelope produces extremely large results in the high or low frequency range, the estimated broadband envelope might be replaced by the narrowband one—leading to no extension for this segment.

Linear mapping can be applied as a postprocessing stage of codebook approaches. In this case the nonlinear mapping between bandlimited and wideband feature vectors is modeled as a piecewise linear mapping. We will describe this approach in more detail in Section 7.5.2.

Neural Network Approaches The type of neural network that is most often applied for the estimation of the vocal tract transfer function is the multilayer perceptron (MLP) in feed forward operation with three layers. As in the case of linear mapping, neural networks are often excited with cepstral coefficients, since the sum of squared differences between two sets of cepstral coefficients is a common distance measure in speech processing (see Section 7.4.3). Furthermore, such a cost function is a simple least square approach and standard training algorithms such as the well-known back-propagation algorithm can be applied to obtain the network parameters.

Before the cepstral coefficients are utilized to excite a neural network a normalization is applied. Usually the input features are normalized to the range $[-1, 1]$. To achieve this—at least approximately—the mean value of each cepstral coefficients is subtracted and the resulting value is divided by three times its standard derivation. If we assume a Gaussian distribution more than 99.7 percent of the values are within the desired interval. In Figure 7.13 the structure of a vocal tract transfer function estimation with a feed forward multilayer perceptron is depicted.

Since the feature vectors of the desired output (the cepstral coefficients of the broadband envelope) that have been used for training the neural network are normalized too, an inverse normalization has to be applied to the network output (as depicted in Fig. 7.13). As in the linear mapping approach it is not guaranteed that the resulting predictor coefficients belong to a stable all-pole filter. For this reason a stability check has to be applied, too.

In its very basic version the estimation of the vocal tract transfer function does not take any memory into account. This means that any correlation among succeeding frames is not exploited yet. This can be achieved if not only the normalized feature vector of the current frame is fed into the network but also the feature vectors of a few preceding and a few subsequent frames. For the latter case the network output can be computed only with a certain delay. That makes the usage of noncausal features inappropriate for several applications. However, if more than the current feature vector is fed into the network, the network parameters do not only model a direct mapping

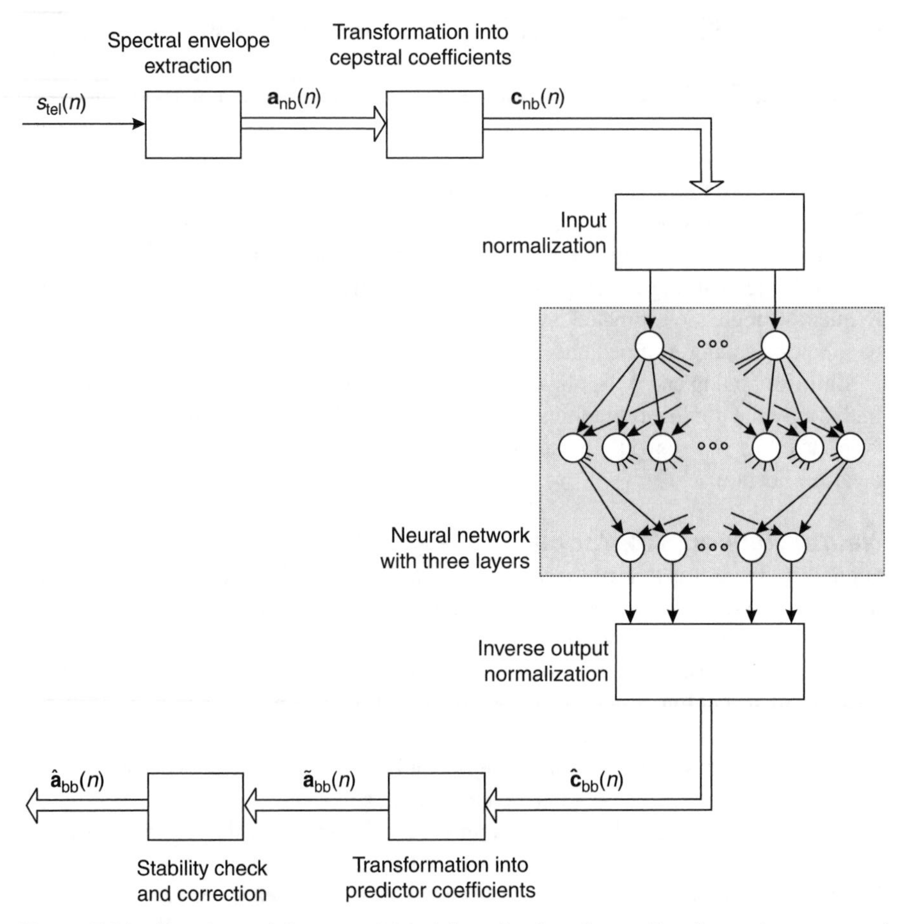

Figure 7.13 Structure of the vocal tract transfer function estimation using a neural network.

from the bandlimited to the broadband envelope. Also the temporal transition model between adjacent frames is learned.

The training of the neural networks is accomplished similar to the computation of the mapping matrices of the last section by providing pairs of bandlimited and broadband data. The training itself then has to be performed very carefully to avoid overtraining. Overtraining denotes the optimization on the training set only without further generalization. This can be observed by using a validation data set to control if the network still is generalizing its task or beginning to learn the training data set by heart. Figure 7.14 shows such a typical behavior.

The optimum iteration to stop the training is marked in Figure 7.14. This optimum is characterized by the minimum overall distance between the actual and the desired output of the artificial neural network produced using the validation data set as input.

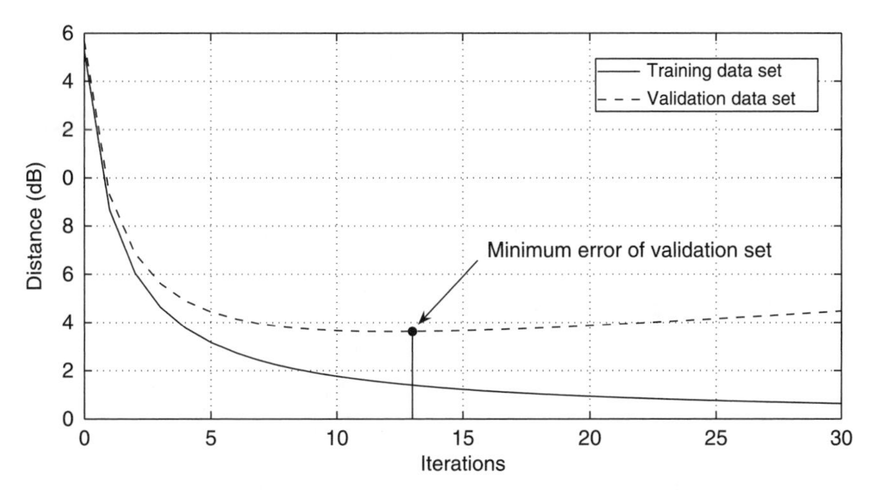

Figure 7.14 Typical progression of the distance between actual output and desired output of the artificial neural network during the training phase.

When it comes to the operation mode of the artificial neural network, after the training phase has been completed, there are two major characteristics of such a network that have to be mentioned. The first one is the low computational complexity needed by such an artificial neural network if not too many layers or nodes are used respectively. Such networks are able to learn complex tasks using comparatively few layers and neurons. This is an advantage over codebook approaches as described in the next section since the computational effort used to evaluate a distance measure for each codebook entry is omitted using artificial neural networks. On the other hand artificial neural networks do not offer possibilities to interfere manually the way codebook approaches do. For example by observing the distance between the narrowband spectral envelope of the input signal and the narrowband codebook entry producing minimum distance one is able to predict if the bandwidth extension runs out of the rudder completely and therefore switch off such a system.

Codebook Approaches A third approach for estimating the vocal tract transfer function is based on the application of codebooks [4]. A codebook contains a representative set of either only broadband or both bandlimited and broadband vocal tract transfer functions. Typical codebook sizes range from $N_{\text{cb}} = 32$ up to $N_{\text{cb}} = 1024$. The spectral envelope of the current frame is computed, for example, in terms of $N_{\text{pre,nb}} = 10$ predictor coefficients, and compared to all entries of the codebook. In case of codebook pairs the narrowband entry that is closest according to a distance measure to the current envelope is determined and its broadband counterpart is selected as the estimated broadband spectral envelope. If only one codebook is utilized the search is performed directly on the broadband entries. In this case the distance measure should weight the nonexcited frequencies much smaller than the excited

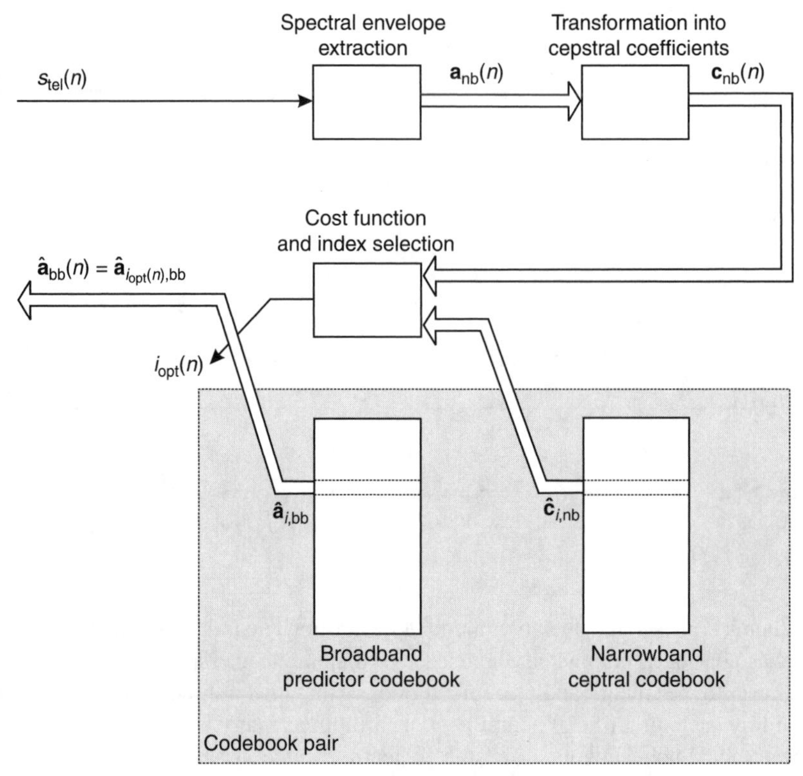

Figure 7.15 Basic structure of the vocal tract transfer function estimation based on codebook approaches.

frequency range from, for example, 200 Hz to 3600 Hz. The basic structure of a codebook approach utilizing a bandlimited and a broadband codebook is depicted in Figure 7.15. In the depicted structure a mixed codebook approach, consisting of a narrowband cepstral codebook and a broadband predictor codebook, is used.

As in the neural network approach the predictor coefficients can be transformed into another feature space, such as line spectral frequencies or cepstral coefficients (as depicted in Fig. 7.15), which might be more suitable for applying a cost function. However, also for predictor coefficients well-suited cost functions exist. The likelihood ratio distance measure, that is defined as

$$d_{\mathrm{lhr}}(n,\, i) = \frac{1}{2\pi} \int\limits_{-\pi}^{\pi} \left(\frac{|\hat{A}_{i,\mathrm{nb}}(e^{j\Omega})|^2}{|A_{\mathrm{nb}}(e^{j\Omega},\, n)|^2} - 1 \right) d\Omega, \tag{7.49}$$

is sometimes applied for this application. The quantities $A_{\mathrm{nb}}(e^{j\Omega}, n)$ and $\hat{A}_{i,\mathrm{nb}}(e^{j\Omega})$ denote the narrowband spectral envelopes of the current frame and of the ith codebook entry, respectively. The integration of the difference between the ratio of the two spectral envelopes and the optimal ratio

$$\left.\frac{|\hat{A}_{i,\mathrm{nb}}(e^{j\Omega})|^2}{|A_{\mathrm{nb}}(e^{j\Omega}, n)|^2}\right|_{\mathrm{opt}} = 1 \tag{7.50}$$

can be computed very efficient by using only the predictor coefficients and the autocorrelations matrix of the current frame

$$d_{\mathrm{lhr}}(n, i) = \frac{\hat{\mathbf{a}}_{i,\mathrm{nb}}^{\mathrm{T}} \mathbf{R}_{ss}(n) \hat{\mathbf{a}}_{i,\mathrm{nb}}}{\mathbf{a}_{\mathrm{nb}}^{\mathrm{T}}(n) \mathbf{R}_{ss}(n) \mathbf{a}_{\mathrm{nb}}(n)} - 1. \tag{7.51}$$

Since only the index $i = i_{\mathrm{opt}}(n)$ corresponding to the minimum of all distances (and not the minimum distance itself) is needed, it is sufficient to evaluate

$$\begin{aligned} i_{\mathrm{opt}}(n) &= \operatorname{argmin}\{\tilde{d}_{\mathrm{lhr}}(n, i)\} \\ &= \operatorname{argmin}\{\hat{\mathbf{a}}_{i,\mathrm{nb}}^{\mathrm{T}} \mathbf{R}_{ss}(n) \hat{\mathbf{a}}_{i,\mathrm{nb}}\}. \end{aligned} \tag{7.52}$$

Note that (7.52) can be computed very efficiently since the autocorrelation matrix $\mathbf{R}_{ss}(n)$ has Toeplitz structure. Beside the cost function according to (7.49), which weights the difference between the squared transfer function in a linear manner, a variety of others can be applied [12]. Most of them apply a spectral weighting function within the integration over the normalized frequency Ω. Furthermore, the difference between the spectral envelopes is often weighted in a logarithmic manner (instead of a linear or quadratic). The logarithmic approach takes the human loudness perception in a better way into account. In the approaches discussed in this chapter, we have again used a cepstral distance measure according to (7.30).

For obtaining the codebook entries iterative procedures such as the method of Linde, Buzo, and Gray (LBG algorithm [27]) can be applied. The LBG-algorithm is an efficient and intuitive algorithm for vector quantizer design based on a long training sequence of data. Various modifications exit (see [25, 30] e.g.). In our approach the LBG algorithm is used for the generation of a codebook containing the spectral envelopes that are most representative in the sense of a cepstral distortion measure for a given set of training data. For the generation of this codebook the following iterative procedure is applied to the training data.

1. *Initializing*:

   Compute the centroid for the whole training data. The centroid is defined as the vector with minimum distance in the sense of a distortion measure to the complete training data.

2. *Splitting:*

   Each centroid is split into two near vectors by the application of a perturbance.

3. *Quantization:*

   The whole training data is assigned to the centroids by the application of a certain distance measure and afterwards the centroids are calculated again. Step 3 is executed again and again until the result does not show any significant changes. Is the desired codebook size reached $\Rightarrow$ abort. Otherwise continue with step 2.

Figure 7.16 shows the functional principle of the LBG algorithm by applying the algorithm to a training data consisting of two clustering points [see Fig. 7.16*a*]. Only one iteration is depicted. Starting with step 1 the centroid over the entire training data is calculated which is depicted in Figure 7.16*b*. In step 2 this centroid is split into two near initial centroids by the application of a perturbance as can be seen in Figure 7.16*c*. Afterwards in step 3 the training data is quantized to the new centroids and after quantization the new centroids are calculated. This procedure is repeated until a predefined overall distortion is reached or a maximum amount of iterations or no significant changes occur. This final state of the first iteration is depicted in Figure 7.16*d*. Using the LBG algorithm this way, it is only possible to obtain codebooks with an amount of entries equal to a power of two. This could be circumvented by either splitting into more than just two new initial centroids or by starting with an initial guess of the desired n centroids or codebook entries, respectively.

Within these methods the summed distances between the optimum codebook entry and the feature vector of the current frame are minimized for a large training data base with Nfeat feature sets[7]

$$D = \sum_{n=0}^{N_{\text{feat}-1}} \min_i \{d(n, i)\} \longrightarrow \min. \tag{7.53}$$

Since the codebook entries should be determined by minimizing (7.53) on one hand but are also required to find the minimum distance for each data set on the other hand, iterative solutions, such as the LBG algorithm, suggest themselves.

During the generation of the codebook it can be assured that all codebook entries are valid feature vectors. As in the examples before, the term valid means that the corresponding all-pole filter consists only of poles within the unit circle in the z-domain. For this reason computational expensive stability checks—as required in neural network and linear mapping approaches—can be omitted. Furthermore, the transformation from, for example, cepstral coefficients into predictor coefficients, can be performed at the end of the training process. For that reason also the feature transformation that was necessary in linear mapping and neural network approaches can be omitted.

Another advantage of codebook based estimation schemes is the inherent discretization of the spectral envelopes. Due to this discretization (broadband spectral

[7]The term $d(n, i)$ denotes an arbitrary (nonnegative) cost function between the codebook entry with codebook index i and the current input feature vector with frame index n.

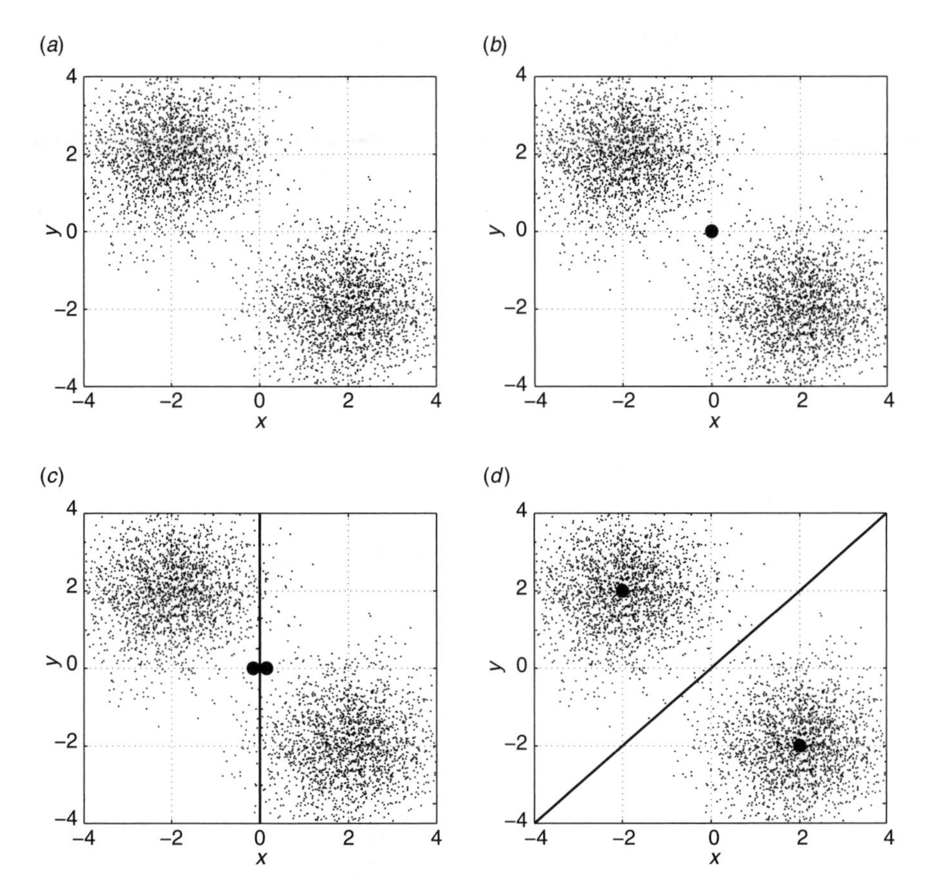

Figure 7.16 Example for the vector quantization of a training data with two clustering points following the LBG-algorithm. (*a*) Training data with vectors consisting of two features (*x* and *y*). (*b*) Initial centroid for training data. (*c*) Initial splitting of previous centroid during first iteration and initial cell border (step 2). (*d*) Final centroids at end of first iteration and final cell border (end of step 3).

envelope is one of N_{cb} prototypes) postprocessing schemes that take also the temporal properties of human speech generation into account can be applied. Examples for such postprocessing approaches are as follows.

- The application of a hidden Markov model [21] for modeling the temporal transition between the broadband prototype envelopes.
- The consideration of the codebook entry that has been selected during the codebook search of the previous frame for obtaining the current codebook entry [23]. In this case transitions between two rather different broadband envelopes are punished even if each of the succeeding codebook entries are optimum according to a distance measure without memory.

Combined Approaches Prosecuting consequently the idea presented at the end of the section on neural network approaches of classifying the current frame of speech signal first and then providing specific matrices leads to the awareness that linear mapping can be applied as a postprocessing stage of codebook approaches. In this case the nonlinear mapping between bandlimited and wideband feature vectors is modeled as a piecewise linear mapping. Figure 7.17 shows such a design.

First the narrowband codebook is searched for the narrowband entry or its index $i_{\mathrm{opt}}(n)$, respectively, producing minimum distance to the actual input feature vector considering a specific distance measure

$$i_{\mathrm{opt}}(n) = \underset{i}{\operatorname{argmin}}\{d(n, i)\}. \tag{7.54}$$

Here the cepstral distance measure as defined in (7.30)

$$d(n, i) = d_{\mathrm{ceps}}(n, i) = \|\mathbf{c}_{\mathrm{nb}}(n) - \hat{\mathbf{c}}_{i,\mathrm{nb}}\|^2 \tag{7.55}$$

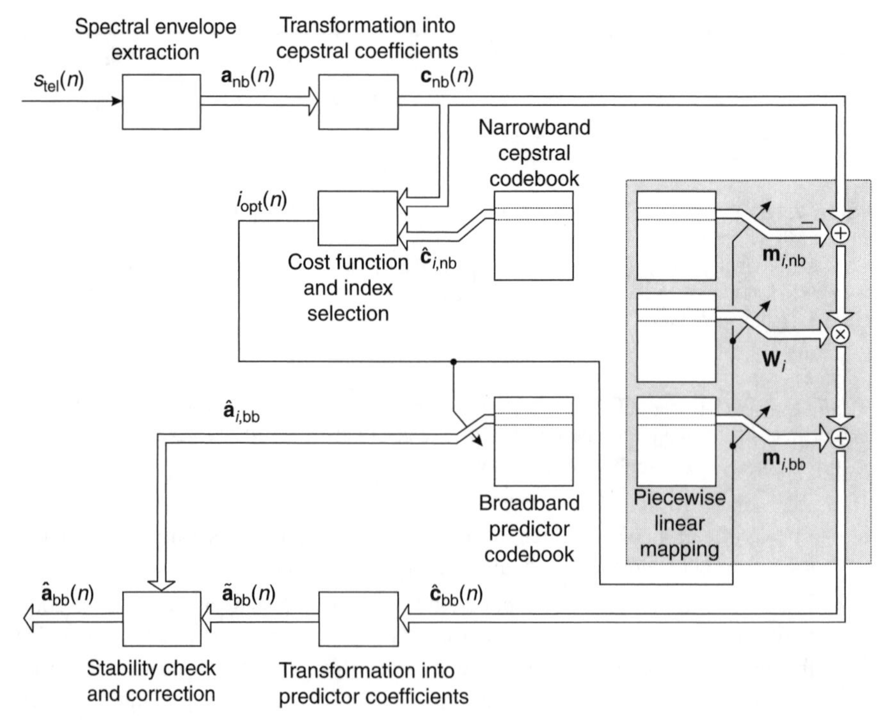

Figure 7.17 Combined approach using codebook preclassification and linear mapping.

has been used. An initial estimated broadband spectral envelope is generated by multiplying the matrix $\mathbf{W}_{i_{\text{opt}}(n)}$ with the cepstral representation of the mean value compensated input narrowband spectral envelope $\mathbf{c}_{\text{nb}}(n) - \mathbf{m}_{i_{\text{opt}}(n),\text{nb}}$. Afterwards the broadband mean value $\mathbf{m}_{i_{\text{opt}}(n),\text{bb}}$ is added, resulting in

$$\hat{\mathbf{c}}_{\text{bb}}(n) = \mathbf{W}_{i_{\text{opt}}(n)} \left(\mathbf{c}_{\text{nb}}(n) - \mathbf{m}_{i_{\text{opt}}(n),\text{nb}}\right) + \mathbf{m}_{i_{\text{opt}}(n),\text{bb}}. \tag{7.56}$$

After transformation of the vector containing the cepstral coefficients into a corresponding all-pole filter coefficients vector $\tilde{\mathbf{a}}_{\text{bb}}(n)$ stability needs to be checked. Depending on the result either the result of the linear mapping operation or the corresponding entry of the broadband predictor codebook entry $\hat{\mathbf{a}}_{i_{\text{opt}}(n),\text{bb}}$ is output as $\hat{\mathbf{a}}_{\text{bb}}(n)$

$$\hat{\mathbf{a}}_{\text{bb}}(n) = \begin{cases} \hat{\mathbf{a}}_{i_{\text{opt}}(n),\text{bb}}, & \text{if instability was detected,} \\ \tilde{\mathbf{a}}_{\text{bb}}(n), & \text{else.} \end{cases} \tag{7.57}$$

Note that stability of the broadband predictor codebook entries can be ensured during the training stage of the codebook.

The training of this combined approach can be split into two separate training stages. The training of the codebook is independent of the succeeding linear mapping vectors and matrices and can be conducted as described in the previous section. Then the entire training data is grouped into $N_{i,\text{cb}}$ sets containing all feature vectors classified to the specific codebook entries $\hat{\mathbf{c}}_{i,\text{nb}}$. Now for each subset of the entire training material a single mapping matrix and two mean vectors (narrowband and wideband) are trained according to the method described in Section 7.5.2. In Figure 7.18 the function principle of the combined approach using a preclassification by a codebook and afterwards doing an individual linear mapping corresponding to each codebook entry is illustrated as an example. The little dots at the bottom of each diagram represent data points in the two-dimensional input feature space (for this example we limit ourselves to a two-dimensional space). The big dot represents the centroid which is the Euclidean mean over all input vectors. The surface in Figure 7.18*a* represents the mapping of the input vectors to one feature of the output vectors—this would be the desired function for the extrapolation task. In pure codebook approaches we map the output feature of all vectors that fall into one cell [see part (*b*) of Figure 7.18]. If we now combine a codebook classification with linear mapping as illustrated in part (*c*) a plane is placed according to the data points within each cell with minimum overall distance to the original surface resulting in less error when processing an input vector which is close to a cell border. When comparing the approximations of parts (*b*) and (*c*) with the true mapping as depicted in part (*a*) the improvement due to the linear postprocessing stage is clearly visible.

Beside a postprocessing with linear mapping also individually trained neural networks can be utilized. Again, an improved performance can be achieved. Even if the computational complexity does not increase that much the memory requirements of combined approaches are significantly larger compared to single approaches.

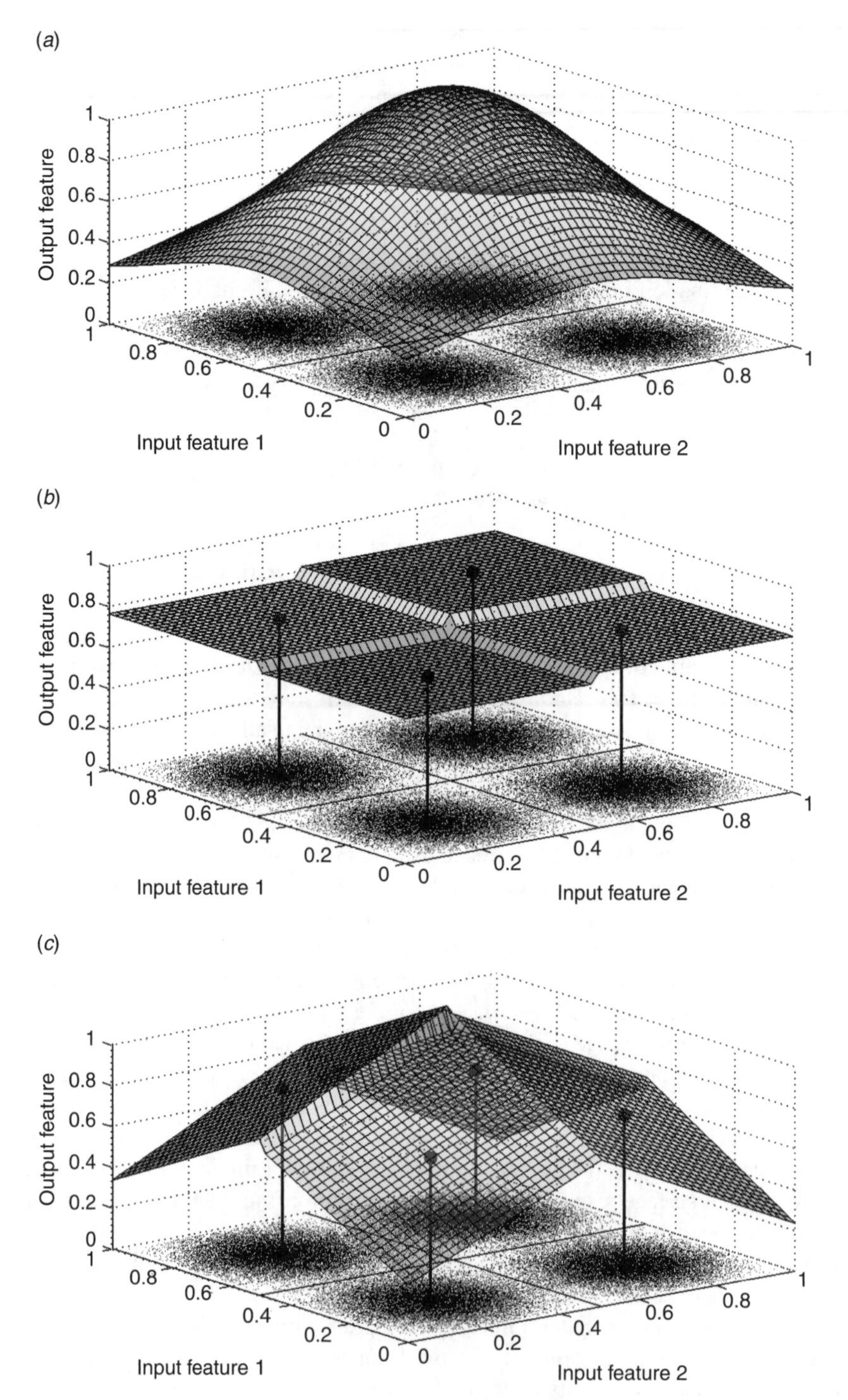

Figure 7.18 Illustration of the function of an approach using codebook classification and succeeding linear mapping. (a) True mapping function, (b) Codebook only approach, and (c) Combined approach using codebooks and linear mapping.

Concluding Remarks We have mentioned only the very basic approaches in this section and we do not claim for completeness. Several important approaches, such as Gaussian mixture models (GMMs) or mapping matrixes that operate directly on DFT-based input vectors, have not been treated. For these approaches we refer to books and Ph.D theses such as [15] or [20, 24].

7.6 EVALUATION OF BANDWIDTH EXTENSION ALGORITHMS

In many situations and particularly in the context of developing and improving bandwidth extension algorithms, performance measures are needed for comparison between different versions of a particular algorithm or different algorithms. Bandwidth extension systems are often evaluated utilizing objective measures, such as distance measures [12, 14] between the original broadband and the extended signal. These measures can be quantified in straightforward objective terms. We will describe a few of them in the next section. However, due to the complexity of current algorithms these methods deliver only a general sense of the speech quality.

A better, but also more expensive, way to evaluate the speech quality are subjective listening tests. These tests are the most reliable tool available for the evaluation of bandwidth extension algorithms. The challenge of these tests is to design them in such a way that the quality of the extension system can be measured in a reliable and repeatable manner. A variety of listening test have been published, each of them optimized for a special purpose. We will focus here only on comparison mean opinion scores (CMOS). For this kind of subjective test standards have been published by the International Telecommunication Union (ITU) [18].[8]

7.6.1 Objective Distance Measures

Some well known objective distance measures are L_2-norm based logarithmic spectral distortion measures [12], such as

$$d_{\mathrm{Eu,log}}(n) = \frac{1}{2\pi} \int_{-\pi}^{\pi} \left| 10 \log_{10} \frac{|S_{\mathrm{ext}}(e^{j\Omega},\, n)|^2}{|S_{\mathrm{bb}}(e^{j\Omega},\, n)|^2} \right|^2 d\Omega. \tag{7.58}$$

Here $S_{\mathrm{ext}}(e^{j\Omega}, n)$ denotes the short-term spectrum of the bandwidth extended signal and $S_{\mathrm{bb}}(e^{j\Omega}, n)$ is the original broadband spectrum. In order to achieve a reliable measurement, the distances $d_{\mathrm{Eu,log}}(n)$ between the original and the estimated

[8]Note, that the reason why we focus here only on the speech quality and not on the speech intelligibility using, for example, diagnostic rhyme tests [2, 37] is because of the nature of bandwidth extension algorithms. Even if a bandwidth extension system increases the bandwidth and makes the speech sound better it does not add any new information to the signal.

broadband envelope are computed for all frames of large database and averaged finally

$$D_{\text{Eu,log}} = \frac{1}{N} \sum_{n=0}^{N-1} d_{\text{Eu,log}}(n). \tag{7.59}$$

The distance measure presented above is applied usually only on the spectral envelopes of the original and the estimated signal, meaning that we can set

$$S_{\text{ext}}(e^{j\Omega}, n) = \hat{A}_{\text{bb}}(e^{j\Omega}, n), \tag{7.60}$$

$$S_{\text{bb}}(e^{j\Omega}, n) = A_{\text{bb}}(e^{j\Omega}, n). \tag{7.61}$$

For this reason they are an adequate measure for evaluating the vocal tract transfer function estimation. If the entire bandwidth extension should be evaluated distance measures which take the spectral fine structure as well as the characteristics of the human auditory system into account need to be applied.[9] For deriving such a distance measure we first define the difference between the squared absolute values of the estimated and the original broadband spectra of the current frame as

$$\Delta(e^{j\Omega}, n) = 20 \log_{10} |S_{\text{ext}}(e^{j\Omega}, n)| - 20 \log_{10} |S_{\text{bb}}(e^{j\Omega}, n)|. \tag{7.62}$$

One basic characteristic of human perception is that with increasing frequency the perception resolution decreases. We can take this fact into account by adding an exponential decay of a weighting factor for increasing frequency. Another basic characteristic is, that if the magnitude of the estimated spectrum is above the magnitude of the original one, there will occur bothersome artifacts. In the other case the estimated spectrum has less magnitude than the original one. This does not lead to artifacts that are as bothersome as the ones that occur when the magnitude of the estimated signal is above the original one. This characteristic implies the use of a non-symmetric distortion measure which we simply call spectral distortion measure (SDM)

$$d_{\text{SDM}}(n) = \frac{1}{2\pi} \int_{-\pi}^{\pi} \xi(e^{j\Omega}, n)\, d\Omega. \tag{7.63}$$

[9]If the logarithmic Euclidean distance is applied directly on the short-term spectra no reliable estimate is possible any more. For example, a zero at one frequency supporting point in one of the short-term spectra and a small but nonzero value in the other would result in a large distance even if no or nearly no difference would be audible.

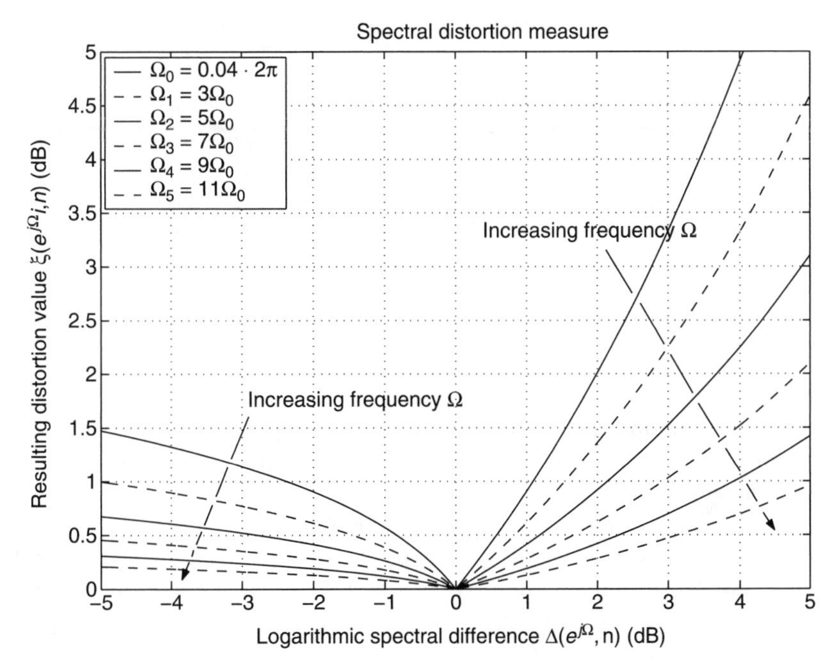

Figure 7.19 Branches of the spectral distortion measure for $\alpha = 0.1$ and $\beta \approx 0.0005$.

Where $\xi(e^{j\Omega}, n)$ is defined as

$$\xi(e^{j\Omega}, n) = \begin{cases} \Delta(e^{j\Omega}, n) \cdot e^{\alpha\Delta(e^{j\Omega},n)-\beta\Omega}, & \text{if } \Delta(e^{j\Omega}, n) \leq 0, \\ \ln[-\Delta(e^{j\Omega}, n) + 1] \cdot e^{-\beta\Omega}, & \text{else.} \end{cases} \tag{7.64}$$

Figure 7.19 shows the behavior of this distortion measure for different normalized frequencies Ω. It is clearly visible that positive errors (indicating that the enhanced signal is louder than the original one) leading to a larger distortion than negative ones. Furthermore, errors at lower frequencies are leading also to a larger distortion than errors at high frequencies. For computing the spectral distortion measure the integral in (7.63) is replaced by a sum over a significantly large number of FFT bins (e.g. $N_{\text{FFT}} = 256$). Subjective tests as described in the next section show that this measure shows a better correlation with human quality evaluations than standard distance measures. However, a variety of other distance measures that take not only the spectral envelope into account might be applied, too.

7.6.2 Subjective Distance Measures

In order to evaluate the subjective quality of the extended signals subjective tests such as mean-opinion-score (MOS) tests should be executed. As indicated at the beginning

Table 7.1 Conditions of a 7-level CMOS test

Score	Statement
−3	A is much worse than B
−2	A is worse than B
−1	A is slightly worse than B
0	A and B are about the same
1	A is slightly better than B
2	A is better than B
3	A is much better than B

of this section we will focus here on the evaluation of the progress during the design of a bandwidth extension algorithm. Thus, we will compare not only the enhanced and the original (bandlimited) signals but also signals that have been enhanced by two different algorithmic versions.

If untrained listeners perform the test it is most reliable to perform comparison ratings (CMOS tests). Usually about 10 to 30 people of different age and gender participate in a CMOS test. In a seven-level CMOS test the subjects are asked to compare the quality of two signals (pairs of bandlimited and extended signals) by choosing one of the statements listed in Table 7.1.

This is done for both versions of the bandwidth extension system. Furthermore, all participants were asked whether they prefer version A or version B of the bandwidth extension. For the latter question no "equal-quality" statement should be offered to the subjects since in this case the statistical significance would be reduced drastically if the subjective differences between the two versions are only small.

As an example for such a subjective evaluation we will present a comparison between two bandwidth extension schemes which differ only in the method utilized for estimating the vocal tract transfer function. For the first version a neural network as described in Section 7.5.2 has been implemented, the second method is based on a codebook approach as presented in Section 7.5.2. Two questions were of interest.

- Do the extension schemes enhance the speech quality?
- Which of both methods produces the better results?

Before we present the results of the evaluation some general comments are made. An interesting fact is that both the codebook as well as the neural network are not representing the higher frequencies appropriately where the behavior of the neural network is even worse. At least the power of the higher frequencies produced by the neural network and the codebook is mostly less than the power of the original signal so that bothersome artifacts do not attract attention to a certain degree. Further results are presented in [14].

The participants rated the signals of the network approach with an average mark of 0.53 (between equal and slightly better than the bandlimited signals) and the signals resulting from the codebook scheme with 1.51 (between slightly better and

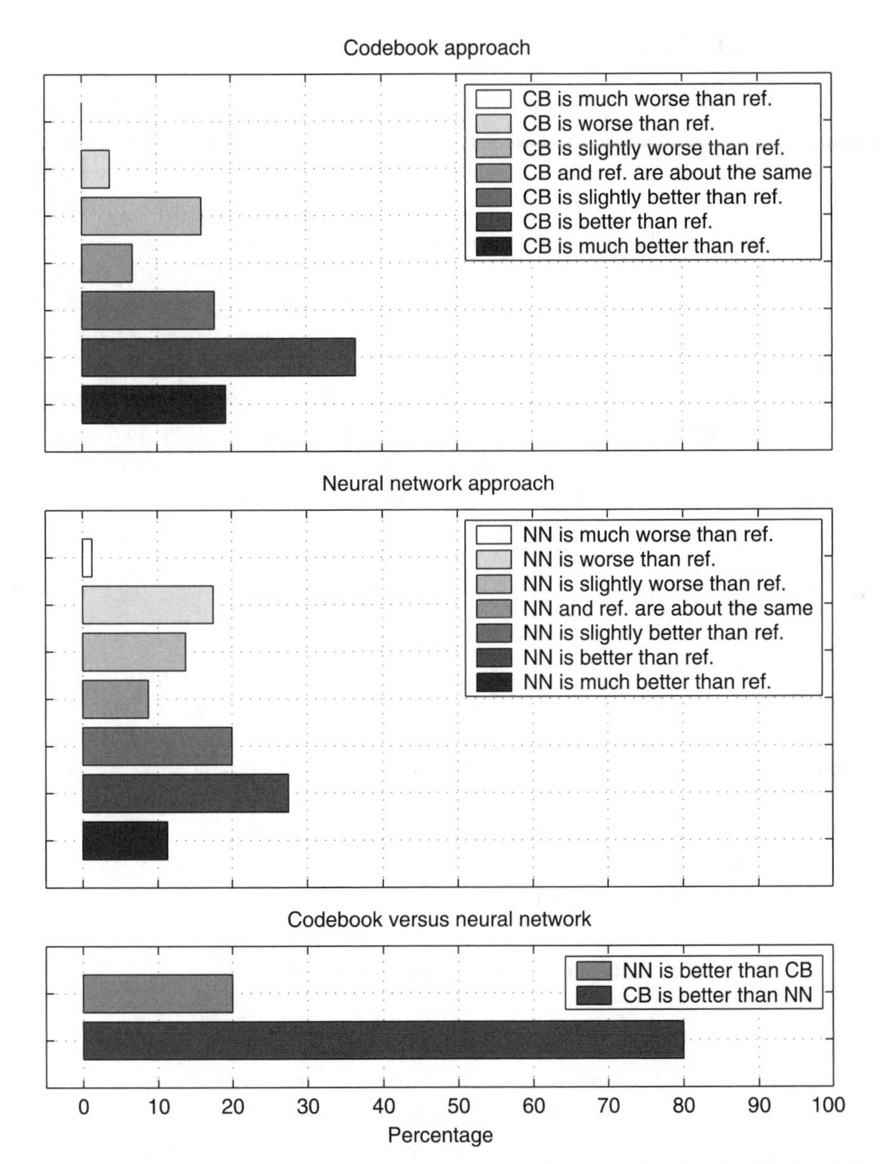

Figure 7.20 Results of the CMOS test (NN abbreviates neural network, CB stands for codebook, and ref. is denoting the bandlimited signal).

better than the bandlimited signals). So the first question can be answered positively for both approaches. When choosing which approach produces better results around 80 percent voted for the codebook based scheme. Figure 7.20 shows the results of the CMOS test. Note, that these results do not indicate that codebook approaches are generally better than neural networks. The results depend strongly on the specific implementation.

7.7 CONCLUSION

In this contribution a basic introduction into bandwidth extension algorithms for the enhancement of telephony speech was presented. Since nearly all bandwidth extension approaches can be split into one part that is generating the required excitation signal and another part that estimates the wideband vocal tract transfer function, we have described both tasks in detail in the main body of this chapter.

Even if the resulting quality is not as good as wideband coded speech, significant improvement (compared to pure transmission over the network) is possible. However, wideband coding and bandwidth extension are not competitors, since the latter one is also able to extend even wideband coded speech (e.g. from 7 kHz bandwidth in case of a transmission according to the ITU standard G.722 17 to 12 or 16 kHz). Due to the advantage of bandwidth extension schemes to enhance the speech quality of the incoming signal by bandwidth extension without modifying the network, research as well as product development in this topic will continue with increasing expense.

7.8 PROBLEMS

In this section we give some exercises that will help the reader to consolidate the techniques learned in this chapter. Attention is drawn to problems that come up when really designing a system for bandwidth extension. In the following we present the problems arising using a model based approach for bandwidth extension. The model based approach is characterized by the two steps of extending the excitation signal and the extension of the spectral envelope.

7.1 Consider a harmonic signal

$$x(n) = \cos(3\omega_0 n) + \cos(4\omega_0 n) \text{ with } \omega_0 < \frac{\pi}{12}$$

in the time domain.

a) Compute the extended excitation signal $y(n)$ after the application of a quadratic characteristic to the narrowband excitation signal $x(n)$.

b) At what frequencies do now components occur? Do they have equal amplitudes as it was the case with the input signal?

c) Compute the extended excitation signal $y(n)$ after the application of a cubic characteristic to the narrowband excitation signal $x(n)$.

d) At what frequencies do now components occur?

e) Assume a harmonic excitation signal sampled at 8 kHz with a fundamental frequency of 150 Hz which has been limited by a lowpass filter with a cutoff frequency of 3400 Hz. What problems occur if a quadratic characteristic is applied to extend the harmonics? What can be done against these effects?

7.2 You have extracted the following LPC feature vectors (without the first components) from a speech corpus consisting of broadband speech (b_i) and telephone limited narrowband speech (n_i)

$$\mathbf{n}_1 = [0.99, 0.1]^T, \qquad \mathbf{b}_1 = [0.99, \ -0.25, 0.75]^T,$$

$$\mathbf{n}_2 = [1.01, 0.4]^T, \qquad \mathbf{b}_2 = [1.02, 0.25, \ -0.75]^T,$$

$$\mathbf{n}_3 = [0.98, 0.6]^T, \qquad \mathbf{b}_3 = [1.01, \ -0.75, 0.25]^T,$$

$$\mathbf{n}_4 = [1.02, 0.9]^T, \qquad \mathbf{b}_4 = [0.98, 0.75, \ -0.25]^T.$$

a) Compute the linear mapping matrix **W** that tries to find a linear dependency $\hat{\mathbf{b}} = \mathbf{W}(\mathbf{n} - \mathbf{m}_{\mathbf{n}_i}) + \mathbf{m}_{\mathbf{b}_i}$, with $\mathbf{m}_{\mathbf{n}_i}$ and $\mathbf{m}_{\mathbf{b}_i}$ being the mean vectors of the narrowband and broadband feature vectors respectively.

b) Compute a codebook with two entries using the LBG-algorithm. As a distance measure use the cepstral distance (even though this would not be a preferable choice). Compute the narrowband and the broadband codebook separately. As an additive splitting perturbance use $\mathbf{p}_1 = [0.1, 0.1]^T$ and $\mathbf{p}_2 = [-0.1, \ -0.1]^T$ for the narrowband codebook and $\mathbf{p}_{1,\mathrm{b}} = [-0.1, \ 0.1, \ -0.1]^T$ and $\mathbf{p}_{2,\mathrm{b}} = [0.1, \ -0.1, \ 0.1]^T$ for the broadband codebook.

c) What is the difference between the two mappings when the codebook has either been trained on the narrowband data or on the broadband data?

d) Compute the narrowband codebook once again using the additive splitting perturbances $\mathbf{p}_3 = [0, \ 3]^T$ and $\mathbf{p}_4 = [0, \ -3]^T$. What problem occurs using this set of perturbances for the codebook training? What can be done to circumvent this problem?

e) Now use the linear mapping matrix and the mean vectors

$$\mathbf{W} = \begin{bmatrix} -0.2201 & -0.0089 \\ 35.5212 & -0.0579 \\ -22.3938 & -0.2896 \end{bmatrix}$$

$$\mathbf{m}_{\mathbf{n}_i} = [1, 0.5]^T$$

$$\mathbf{m}_{\mathbf{b}_i} = [1, 0, 0]^T$$

and the codebook mapping computed on the broadband features with the corresponding entries

$$\mathbf{c}_{1,\mathrm{n}} = [1, 0.35]^T, \quad \mathbf{c}_{1,\mathrm{b}} = [1, \ -0.5, 0.5]^T \ldots$$

$$\mathbf{c}_{2,\mathrm{n}} = [1, 0.65]^T, \quad \mathbf{c}_{2,\mathrm{b}} = [1, 0.5, \ -0.5]^T.$$

Compare the two approaches by evaluating the distance using the cepstral distance measure and the vector $\mathbf{n}_1$ as input.

REFERENCES

1. The New Bell Telephone, *Sci. Am.* 37(1), 1, (1877).
2. ANSI S3.2-1989, *Method for Measuring the Intelligibility of Speech over Communication Systems*, American National Standard, (1989).
3. I. N. Bronstein and K. A. Semendjajew, *Handbook of Mathematics*, Springer, Berlin, Germany, (2004).
4. H. Carl and U. Heute, *Bandwidth Enhancement of Narrow-Band Speech Signals*, Proc. EUSIPCO '94, vol. 2, pp. 1178–1181, (1994).
5. H. Carl, *Untersuchung verschiedener Methoden der Sprachcodierung und eine Anwendung zur Bandbreitenvergrößerung von Schmalband-Sprachsignalen*, Ph.D thesis, Ruhr-Universität Bochum, (1994) (in German).
6. M. G. Croll, *Sound Quality Improvement of Broadcast Telephone Calls*, BBC Research Report RD1972/26, British Broadcasting Corporation, (1972).
7. J. Deller, J. Hansen, and J. Proakis, *Discrete-Time Processing of Speech Signals*, Piscataway, NJ: IEEE Press, (2000).
8. M. Dietrich, *Performance and Implementation of a Robust ADPCM Algorithm for Wideband Speech Coding with 64 kbit/s*, Proc. Int. Zürich Seminar Digital Communications, (1984).
9. J. Durbin, *The Fitting of Time Series Models, Rev. Int. Stat. Inst.*, 28, pp. 233–244, (1960).
10. ETSI, *Digital Cellular Telecommunications System (Phase 2+); Enhanced Full Rate (EFR) Speech Transcoding*, ETSI EN 300 726 V8.0.1, November (2000).
11. B. Gold and N. Morgan, *Speech and Audio Signal Processing*, Wiley, New York, NY, USA, (2000).
12. R. M. Gray, et. al., *Distortion Measures for Speech Processing*, IEEE Trans. Acoust., Speech, Signal Processing, ASSP-28, No. 4, pp. 367–376, (1980).
13. W. Hess, *Pitch Determination of Speech Signals*, Springer, Berlin, Germany, (1983).
14. B. Iser and G. Schmidt, *Neural Networks Versus Codebooks in an Application for Bandwidth Extension of Speech Signals*, Proc. EUROSPEECH '03, vol. 1, pp. 565–568, (2003).
15. B. Iser, W. Minker, and G. Schmidt, *Bandwidth Extension of Speech Signals*, Springer, Berlin, Germany, (2007).
16. ITU, *General Performance Objectives Applicable to all Modern International Circuits and National Extension Circuits*, ITU-T recommendation G.151, (1988).
17. ITU, *7kHz Audio Coding within 64 kbit/s*, ITU-T recommendation G.722, (1988).
18. ITU, *Methods for Subjective Determination of Transmission Quality*, ITU-T recommendation P. 800, August (1996).
19. F. Itakura, *Line Spectral Representation of Linear Prediction Coefficients of Speech Signals,* J. Acoust. Soc. Am., 57, No. 1, (1975).
20. P. Jax, *Enhancement of Bandlimited Speech Signals: Algorithms and Theoretical Bounds*, Ph.D thesis, RWTH Aachen, (2002).
21. P. Jax and P. Vary, On Artificial Bandwidth Extension of Telephone Speech, *Signal Processing*, 83, No. 8, pp. 1707–1719, (2003).
22. K. D. Kammeyer, *Nachrichtenübertragung*, B. G. Teubner, Stuttgart, (1992) (in German).

23. U. Kornagel, *Spectral Widening of Telephone Speech Using an Extended Classification Approach*, Proc. EUSIPCO '02, 2, pp. 339–342, (2002).
24. U. Kornagel, *Synthetische Spektralerweiterung von Telefonsprache*, Ph.D thesis, Forschritt-Berichte VDI, 10, No. 736, Düsseldorf, Germany, (2004) (in German).
25. W. P. LeBlanc, B. Bhattacharya, S. A. Mahmoud, and V. Cuperman, *Efficient Search and Design Procedures for Robust Multi-Stage VQ of LPC Parameters for 4 kb/s Speech Coding*, IEEE Transactions on Speech and Audio Processing, 1, No. 4, pp. 373–385, (1993).
26. N. Levinson, *The Wiener RMS Error Criterion in Filter Design and Prediction,* J. Math. Phys., 25, pp. 261–268, (1947).
27. Y. Linde, A. Buzo, and R. M. Gray, *An Algorithm for Vector Quantizer Design*, IEEE Trans. Comm., COM-28, No. 1, pp. 84–95, (1980).
28. D. G. Luenberger, *Opitmization by Vector Space Methods*, Wiley, New York, NY, USA, (1969).
29. A. V. Oppenheim and R. W. Schafer, *Discrete-Time Signal Processing*, Prentice Hall, Englewood Cliffs, NJ, USA, (1989).
30. K. K. Paliwal and B. S. Atal, *Efficient Vector Quantization of LPC Parameters at 24 Bits/Frame*, IEEE Transactions on Speech and Audio Processing, 1, No. 1, 314, (1993).
31. P. J. Patrick, *Enhancement of Bandlimited Speech Signals*, Ph.D thesis, Loughborough University of Techology, (1983).
32. J. Paulus, *Variable Rate Wideband Speech Coding Using Perceptually Motivated Thresholds*, Proc. of IEEE Workshop on Speech Coding, Annapolis, MD, USA, pp. 35–36, (1995).
33. L. Rabiner and B. H. Juang, *Fundamentals of Speech Recognition*, Prentice Hall, Englewood Cliffs, NJ, (1993).
34. H. W. Schüsler, *Digitale Signalverarbeitung 1*, 4th edition, Springer, Berlin, Germany, (1994) (in German).
35. F. K. Soong and B.-H. Juang, *Optimal Quantization of LSP Parameters*, IEEE Transactions on Speech and Audio Processing, 1, No. 1, pp. 1524, (1993).
36. P. Vary and R. Martin, *Digital Speech Transmission: Enhancement, Coding and Error Concealment*, Wiley, New York, NY, USA, (2006).
37. W. Voiers, *Evaluating Processed Speech Using the Diagnostic Rhyme Test*, Speech Technology, pp. 30–39, (1983).

INDEX

Note: f denotes figures, t denotes tables

Adaptive Signal Processing: Next Generation Solutions. Edited by Tülay Adalı and Simon Haykin
Copyright © 2010 John Wiley & Sons, Inc.